Microbiology

laboratory exercises

Microbiology
laboratory exercises

second edition

Margaret Barnett
Grossmont College

illustrated by
Jane D. Venghaus

Boston, Massachusetts Burr Ridge, Illinios Dubuque, Iowa
Madison, Wisconsin New York, New York San Francisco, California St. Louis, Missouri

WCB/McGraw-Hill

A Division of The ***McGraw·Hill*** *Companies*

Project Team

Editor *Elizabeth Sievers*
Developmental Editor *Terrance Stanton*
Production Editor *Cathy Ford Smith*
Marketing Manager *Patrick E. Reidy*
Designer *K. Wayne Harms*
Art Editor *Jodi K. Banowetz*
Photo Editor *Nicole Widmyer*
Advertising Coordinator *Leslie Dague*

President and Chief Executive Officer *Beverly Kolz*
Vice President, Director of Editorial *Kevin Kane*
Vice President, Sales and Market Expansion *Virginia S. Moffat*
Vice President, Director of Production *Colleen A. Yonda*
Director of Marketing *Craig S. Marty*
National Sales Manager *Douglas J. DiNardo*
Executive Editor *Michael D. Lange*
Advertising Manager *Janelle Keeffer*
Production Editorial Manager *Renée Menne*
Publishing Services Manager *Karen J. Slaght*
Royalty/Permissions Manager *Connie Allendorf*

Copyedited by Sarah Aldridge

Cover Photo courtesy of North Wind Pictures Archives

ISBN-13: 978-0-697-16011-9
ISBN-10: 0-697-16011-4

Some of the laboratory experiments included in this text may be hazardous if materials are handled improperly or if procedures are conducted incorrectly. Safety precautions are necessary when you are working with chemicals, glass test tubes, hot water baths, sharp instruments, and the like, or for any procedures that generally require caution. Your school may have set regulations regarding safety procedures that your instructor will explain to you. Should you have any problems with materials or procedures, please ask your instructor for help.

Printed in China

12 13 14 15 CTP/CTP 14 13 12 11

To my parents
Mr. and Mrs. E. Faveluke
for their outstanding advice
which I pass along to you, the student,
"Go study!"

This book is also dedicated to the
memory of Editor-in-Chief
Edward G. Jaffe, who gave
me a chance to help others learn.

Contents

Microbiology Laboratory Exercises, 2nd edition is an up-to-date, student-tested, introductory laboratory manual for medical and general microbiology. It presents principles and techniques central to the study of microbiology. *Microbiology Laboratory Exercises* emphasizes performance and study methods, scientific investigation, and safety.

This manual is designed for learning. To help students develop self-confidence in the laboratory, exercises follow a logical progression. They gradually increase in difficulty from relatively simple observations and fundamental techniques to complex procedures demanding accurate collection and evaluation of scientific data.

By performing and evaluating the experiments in this manual, **students discover how microorganisms influence their lives.** While studying *Microbiology Laboratory Exercises* and completing an introductory microbiology course, students not only learn the fundamentals of a fascinating scientific discipline, they also develop confidence in their own ability to master a difficult subject.

An **Instructor's Manual** is available.

When mesophilic microbes dream...

New in This Edition

Innovations and additions in the second edition of *Microbiology Laboratory Exercises* include new illustrations and color photography, along with enhanced and updated tables and figures. The manual presents an added emphasis on safety and even more challenging opportunities for students to practice using scientific method.

Exciting new technologies that students tap when they discover the microbial world through the second edition of *Microbiology Laboratory Exercises* include **enzyme immunoassay, genetic engineering with bacterial transformation, monoclonal antibody testing, *Campylobacter* identification, and expanded virus experimentation.** Additional methodology includes testing bacterial motility and gelatin liquefaction capability.

An innovative feature that is available for the first time with this edition is **a computer simulation of microbial identification,** Wm. C. Brown Publishers' intriguing CD-ROM, *Identibacter interactus,* by Allan Konopka, et al., ISBN 0–697–29387–4. This thought-provoking compact disk augments *Microbiology Laboratory Exercises* Part 7, Identifying Bacterial Unknowns, by encouraging students to consider more than 50 tests—from stains to biochemical reactions—and simulate performing these procedures on over 60 unknown species. What a tremendous opportunity for additional practice in this essential field of laboratory investigation!

A Word to the Student

Microbiology is the study of organisms that are too small to be seen with the naked eye. Bacteria, protozoans, fungi, viruses, and selected multicellular invertebrates, such as parasitic worms, are generally examined during an introductory course in microbiology.

The microbial world influences every part of our existence. Through their ability to recycle the elements, microorganisms are central to the continued existence of life on earth. In addition, some microbes are indispensable for our health while others are uncontrolled killers.

Microbiology Laboratory Exercises, 2nd edition introduces you to the microbiology laboratory. There are exercises in this manual that allow you to observe living microorganisms. Factors that influence and control microbial growth are introduced, and you analyze selected metabolic reactions of the microbiota.

During your introduction to the field of medical microbiology, you observe and perform experiments with some of the leading spreaders of misery and disease that have plagued humanity throughout history. You may also be asked to identify a bacterial unknown, explore genetic engineering, and perform a selection of tests from the rapidly expanding field of immunology.

Employ *Microbiology Laboratory Exercises, 2nd edition* as a guide. By utilizing the detailed instructions and thorough posttests provided in this manual, you progress at your own pace. ***Microbiology Laboratory Exercises* introduces you to safe and fascinating methods of investigation.**

Laboratory Rules for Microbiology

In the microbiology laboratory, you may work with dangerous agents of death, disease, and decay. Use the following laboratory rules to protect yourself and your community from contamination.

These rules should become habits. Until they do, help each other to remember and employ them. If you do not follow instructions and utilize the laboratory rules for safe conduct, **your instructor** may ask you to leave the class.

1. **Wear a laboratory coat** or other long-sleeved covering over your street clothes.
2. **Tie long hair back and up.** Restrain fluffy or flyaway hair with a scarf, cap, headband, or other covering.
3. **Never eat or drink in the laboratory.** Keep hands, pencils, and other items away from your mouth.
4. **Wear shoes.**
5. **Do not place book bags, purses, or other belongings on the floor.**
6. **Keep unnecessary materials off the top of your workbench.**
7. **Read the assigned exercise**(s) before the laboratory session.
8. **Disinfect your bench top** before and after each laboratory session.
9. **Treat every microorganism used in the laboratory as if it were pathogenic.**
10. **All materials that are contaminated with microbial cultures should be placed in appropriate containers to be sterilized.**
11. **Do not take living cultures out of the laboratory.**
12. **Report all accidents and spills to your instructor.** Cover spills with paper towels and disinfectant for at least 15 minutes.
13. **Thoroughly wash your hands** with hot soapy water or disinfectant solution before leaving the laboratory.

Format of Exercises

Microbiology Laboratory Exercises, 2nd edition is designed for learning. The manual provides information and direction; you perform the work and reap the educational benefits. Most exercises include the following sections arranged to help you progress successfully in your study of microbiology:

1. A **necessary skills list** that directs you to previous exercises for background information.
2. **Materials and cultures** lists itemizing supplies you need for each experiment.
3. **Learning objectives** that tell you what you should know and be able to do after completing the exercise.
4. An **introduction** containing current information concerning the topic of the exercise.
5. **Procedures** to guide you through the experiments.
6. A **posttest (with answers)** for self-evaluation to indicate which portions of the exercise you have mastered and which you need to review.
7. A **laboratory report** where you illustrate your observations, record relevant data, and evaluate experimental results, demonstrating your command of the material.

In addition the text is supplemented by figures, tables, and appendices that illustrate and organize information. Use every part of *Microbiology Laboratory Exercises, 2nd edition* to help you achieve your educational goals.

Good luck.

Acknowledgments

This laboratory manual could not have been completed without the help of many knowledgeable and supportive people. I am especially indebted to the artist, Jane Venghaus, for bringing the material to life. I gratefully acknowledge the significant contributions of those who have reviewed this manual: Lisa Shimeld, *Crafton Hills College;* Beverly A. Roe, *Erie Community College;* John F. Ammons, *Mississippi Delta Community College;* Karen Kealy, *The University of Oklahoma;* Maha Nagarajan, Ph.D., *Wilberforce University;* Robert J. Ruzicka, *Cloud County Community College;* Bernice C. Stewart, Ph.D., *Prince Georges Community College;* the professional microbiologists, and my many students. Thank you for your perceptive suggestions and corrections.

I appreciate the help of my colleagues at Grossmont College, including microbiologists David Wertlieb and Ellen Lipkin, and biology consultants Arla Cox and Thomas Nicoll.

My special thanks go to the editors: Carol Mills, Liz Sievers, Terry Stanton, and the late Ed Jaffe of Wm. C. Brown Publishers. They have enabled me to put into print the content and format I have searched for in a microbiology laboratory manual. To all the other diligent, behind-the-scenes people, Cathy Smith, Jodi Banowetz, Gwen Woodard, Bea Sussman and many others, thank you.

The Compound Microscope

Necessary Skills

Knowledge of the metric system of linear measurement

Materials

Assigned compound light microscope
Prepared slides (1 each per group):
The letter *e* (or newsprint, scissors, slides, and coverslips)
Stained human blood smear
Bacteria, 3 morphological types
1 Dropper bottle of immersion oil per group
Lens paper

Primary Objective

Utilize all the powers of magnification of your compound light microscope.

Other Objectives

1. Summarize the major contributions of Anton van Leeuwenhoek and Ernst Abbé to microscopy.
2. Distinguish between a simple microscope and a compound microscope.
3. Given the magnifications of an ocular and an objective lens, calculate the total magnification of a compound microscope.
4. Explain how the use of immersion oil allows the microscopist to obtain higher resolution.
5. Describe the relationship between resolving power, numerical aperture, and the wavelength of light.
6. Describe the change in working distance as magnification increases.
7. Evaluate the advantage to the microscopist of parfocality.
8. Locate and name the parts of your own microscope; explain the function of each part.
9. List rules for the proper care of your microscope.

INTRODUCTION

One of the fundamental skills that you, as a microbiology student, must master is the ability to use a compound light microscope with accuracy and ease. In this exercise you learn to focus each **objective lens** (the lens nearest the specimen) on your microscope.

Development of the Microscope

The **microscope,** a device for magnifying objects that are too small to be seen with the naked eye, was invented in the seventeenth century. The first instruments were **simple microscopes,** one-lens magnifiers.

Anton van Leeuwenhoek, who published the first drawings of bacteria in 1676, ground his own lenses and constructed simple microscopes resembling the one shown in figure 1.1.

Compound microscopes, those employing two or more sets of lenses, were introduced later. The microscopist obtains much higher magnification by using a compound microscope. To calculate the **total magnification** of a compound microscope, **multiply** the magnification of the **objective lens** times the magnification of the **ocular lens** (eyepiece).

objective magnification × ocular magnification
= total magnification

But the microscopist needs to see a sharp image, not just a large one. Increasing the size of a blurred image does not reveal any further detail; this enlargement is known as **empty magnification.**

Ernst Abbé, working in the late nineteenth century, showed that **resolving** (making distinct) fine detail depends on the amount of light that is gathered by a lens. (More light gathering leads to improved resolution.) He noted that light is lost going through air before entering a lens, figure 1.2.

Abbé prevented this loss by placing a drop of **immersion oil,** a fluid with nearly the same **refractive index** (ability to bend light) as **glass,** on the specimen. He then lowered the objective lens into the oil. Now more light could be gathered by a lens, and resolution was greatly increased.

The **resolving power** (ability to distinguish close-together objects) of a microscope limits its useful magnification. Remember that magnification helps the microscopist only if it reveals more detail.

Numerical Aperture

Abbé developed the concept of **numerical aperture (NA),** a mathematical way of describing the light-gathering ability of a lens system.

Figure 1.1 A van Leeuwenhoek microscope. With a homemade, hand-held microscope, van Leeuwenhoek discovered protozoa, algae, yeast, and bacteria, as well as human red blood cells and spermatozoa.

The numerical aperture (NA) of an objective lens depends on two factors, the size of the cone of light the lens can receive and the medium between the lens and the specimen.

Generally, the medium surrounding a lens is air—but, as already noted, the medium is sometimes immersion oil. Only specially designed lenses can be used with immersion oil.

Check your own microscope; the numerical aperture is generally stamped on each objective lens. What is the numerical aperture of your microscope lens that is designed to be used with immersion oil? _______________

What are the numerical apertures of your other lenses?

____

Resolving Power

As the following formula shows, the resolving power (RP) of a lens depends upon its numerical aperture (NA) and upon the wavelength of light (λ) that the microscopist employs:

$$RP = \frac{\lambda}{2 \times NA}$$

An average wavelength of visible light is 550 nm, and the low-power objective lens on many microscopes has a numerical aperture of 0.25. Thus the smallest object that, theoretically, can be seen clearly with this light coming through the low-power objective lens is:

$$RP = \frac{550 \text{ nm}}{2 \times 0.25} = 1{,}100 \text{ nm} = \mathbf{1.1\ \mu m}$$

Calculate the theoretical resolving power of your microscope's oil-immersion lens, using an average wavelength of visible light.

Instead of the average wavelength of visible light, you may employ blue light that has a wavelength of 475 nm. How does this change in illumination alter the resolving power of your oil-immersion lens? Does the change improve resolution? _______________

Explain your answer.

Figure 1.2 (a) Rays that must pass through **air** are bent (refracted), and many do not enter the lens. (b) **Immersion oil** prevents the loss of light rays.

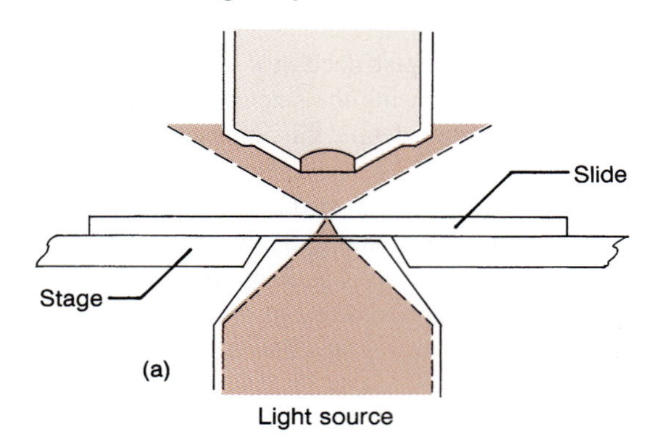

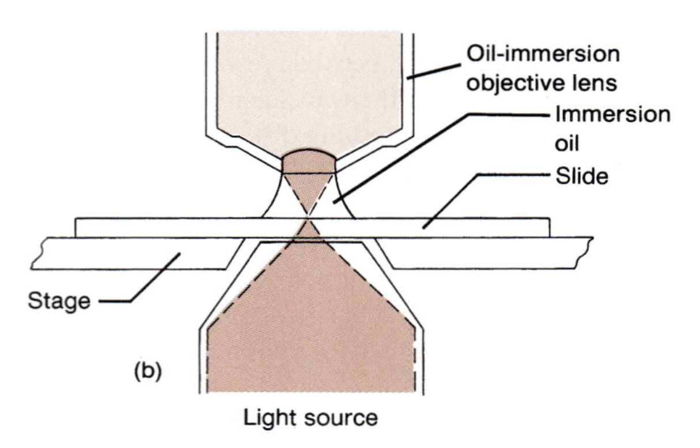

Other Features of Microscopy

Another aspect of microscopy is **working distance,** the distance separating a specimen and your objective lens. You decrease the working distance each time you turn to a higher magnification lens. To save time remember this relationship when focusing your microscope.

One more measurement that decreases as you increase magnification is the size of your **microscopic field.** This is the area of the microscope slide that you see through a lens.

To assist you in changing magnification, most microscopes are **parfocal** or nearly so. That is, the lenses are adjusted so that the specimen remains almost in focus after the microscopist rotates the nosepiece to utilize a different objective lens.

Thanks to parfocality, only minor adjustments with the fine-focus knob are necessary after you rotate to a new lens. But remember to increase illumination with the diaphragm lever when rotating a higher-power lens into the observation position.

One question most new microscopists ask is, "How do I know which lens to use?" A general answer is, whenever possible, use your naked eyes to locate the specimen on the slide and center the specimen over the stage aperture. Then, starting with low power (100×) or scanning power (40×), increase magnification to see more details.

While some large specimens are best observed using low magnifications, other very tiny objects require 1,000× magnification using the oil-immersion lens. Experience is the best teacher here.

Parts of the Microscope

You need to know the names of the parts of your microscope and how these parts function. A generalized microscope is illustrated in figure 1.3. Memorize the names of the parts, then learn their functions by referring to table 1.1. Become familiar with your own microscope and be able to name its parts.

Figure 1.3 The parts of the microscope

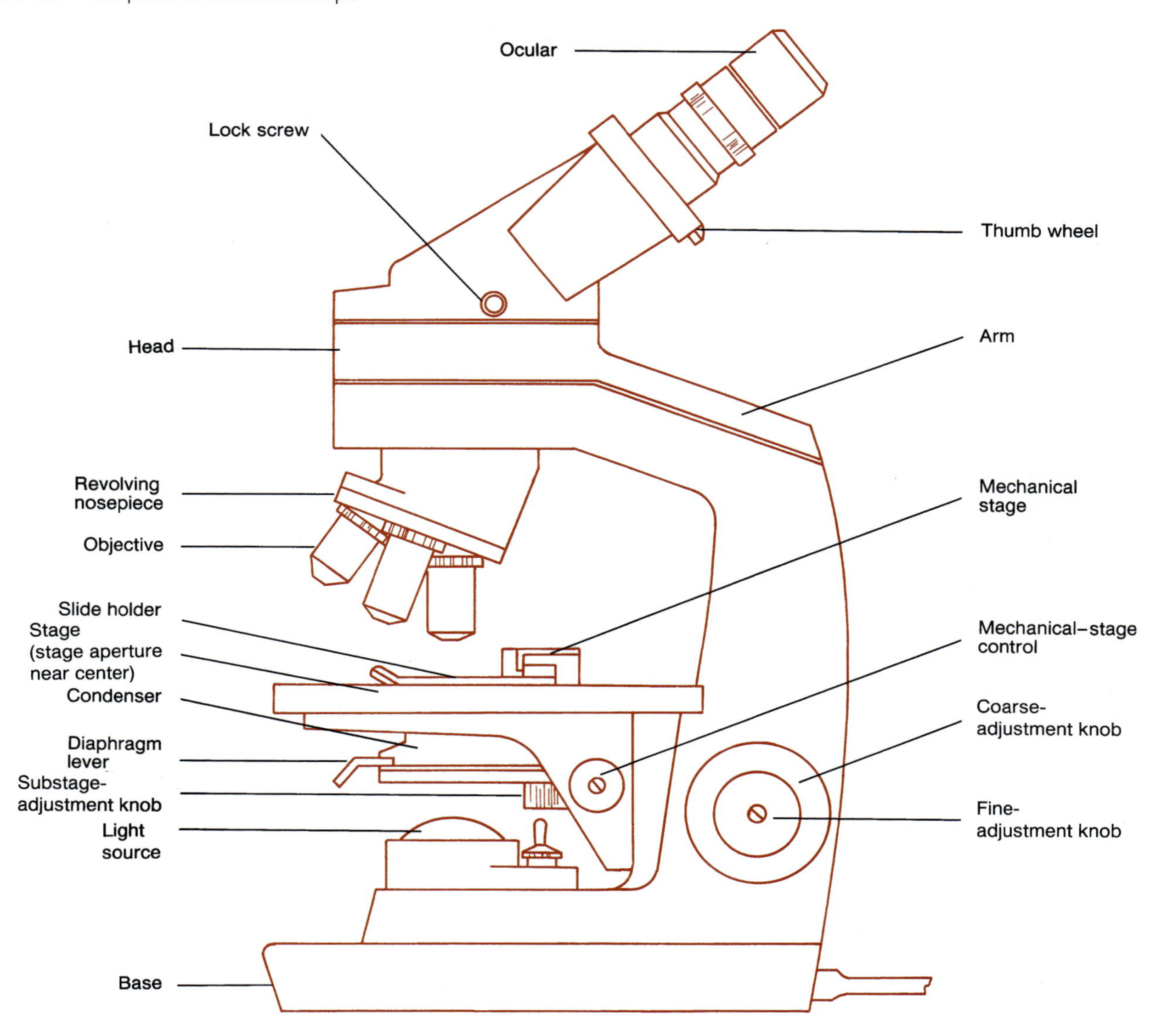

Table 1.1 The Parts of the Microscope and Their Functions

Part	Function	Part	Function
Ocular (eyepiece)	Magnifies image, usually 10× or 15×	Mechanical-stage controls	Move slide about on stage
Thumb wheel	Adjusts distance between oculars to match the microscopist's interpupillary distance	Stage	Holds slide
Head	Holds oculars	Stage aperture	Allows light to reach specimen
Lock screw	Secures head after rotation	Coarse-adjustment knob	Rapidly brings specimen into focus
Arm	Holds head and stage	Fine-adjustment knob	Slowly brings specimen into best focus
Revolving nosepiece	Rotates objective lenses into viewing position	Diaphragm lever	Controls amount of light entering stage aperture
Objectives	Magnify image, usually low (10×), high(43×), and oil-immersion (97×)	Condenser	Focuses light on specimen and fills lens with light
Slide holder	Fixed and movable parts secure slide on stage	Substage- (condenser-) adjustment knob	Raises and lowers condenser
Mechanical stage	Includes slide holder and is generally graduated to help in relocation of specimens	Base	Supports microscope
		Light	Illuminates specimen

Care of the Microscope

Proper care of your microscope is essential. Use the checklist in table 1.2 as a general guide to microscope handling. **Your instructor** may supplement this list.

Table 1.2 Care of the Microscope

1. Use both hands to carry the microscope. Keep the instrument upright; oculars sometimes fall out of an inverted microscope.
2. Use lens paper to clean all exposed glass in the optical and illumination systems of your microscope before and after each use. Remove all immersion oil. Wipe any fingerprints or mascara off the lenses.
3. Never remove parts of the microscope without the prior approval of **your instructor.**
4. Never use the coarse-adjustment knob to decrease the space between the objective lens and your specimen while looking through the instrument. You might accidentally grind the lens and the slide together. Instead, watch the procedure from the side of the microscope. By watching as you turn the coarse-adjustment knob, you can prevent abrasive contact between the lens and the slide.
5. When you are finished with your microscope, maximize the distance between the lens and the stage, remove the slide, clean all surfaces, clean the lenses with lens paper (be sure to remove all oil), put the lowest power in the observation position, center the mechanical stage, and wrap the electrical cord around the base. Cover the microscope and return it to the designated space.

PROCEDURES EXERCISE 1

Procedure 1

Examining the Letter *e* with the Low-Power Objective Lens

Note: Always clean all glass surfaces of your microscope with lens tissue before beginning your observations.

a **Your instructor** may demonstrate how to use the illumination system of your microscope.

b Illuminate the stage aperture of your microscope, then use the diaphragm lever to open the diaphragm all the way. This allows the maximum amount of light to enter your stage aperture.

c If you cut your own letter *e,* place it on a microscope slide, then cover the letter with a coverslip. Firmly fit the microscope slide with the letter *e* into the slide holder on the mechanical stage.

d As shown in the illustration, center the letter *e* over the stage aperture.

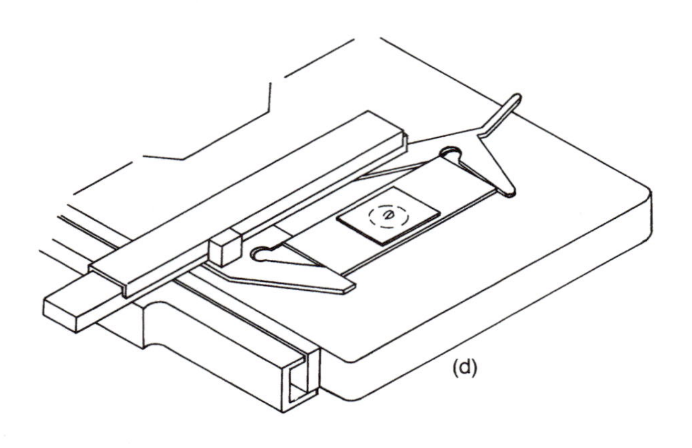

(d)

e As illustrated, turn the low- (10×) or scanning- (4×) power lens into the observation position.

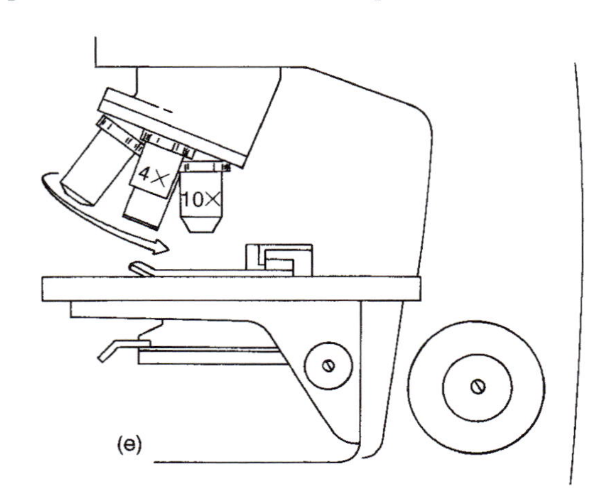

(e)

Figure 1.4 Formed elements of human blood

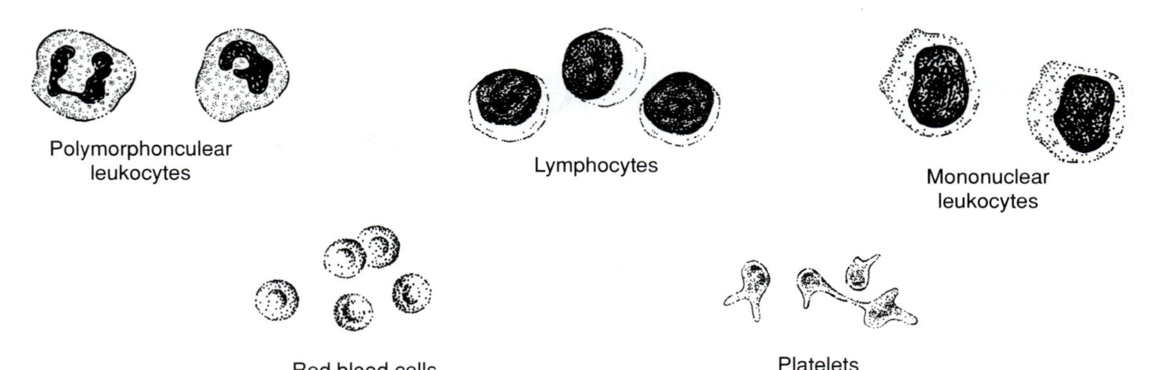

f While watching from the side, use the coarse-adjustment knob to decrease the distance between the specimen and the lens.

g Look through the oculars. Hold your head close to the eyepieces and *very still*. Do you see 1 circle of light? If not, adjust the distance between the eyepieces to match your interpupillary distance (the distance between the pupils of your eyes).

h As you look through the ocular lenses, focus with the coarse-adjustment knob and then the fine-adjustment knob. Alter the illumination with the diaphragm lever until you have obtained optimum resolution.

i In your laboratory report, draw the letter *e* both as it appears to your naked eyes and as it seems to be positioned when seen through the lens. Label the drawings, noting any magnification.

j While watching from the side of the microscope, use the mechanical-stage control knobs to move the slide to your right. Recenter the slide. Now *look through the lenses* and move the stage to the right. In your laboratory report, record where the *e* appears to go.

k Recenter the slide. While watching from the side, move the stage away from you. Recenter again and observe through the oculars while moving the stage away. Record the observed result in your laboratory report.

l Remove the slide from the slide holder and return it to the class set.

Procedure 2

Observing Stained Human Blood with the High-Power Objective Lens

a Obtain a slide with a stained smear of human blood on it, and center the blood smear over the stage aperture.

b Rotate the nosepiece of your microscope and look through the **high-power lens,** using your fine-adjustment knob to focus on the blood cells.

c Alter illumination with the diaphragm lever until you have maximum resolution.

d Is your microscope parfocal or nearly so? If it is, which way did you rotate the fine-adjustment knob to bring the specimen into sharp focus? You make this adjustment on your assigned microscope **every time** you switch from low to high power throughout this course, so note it now in your laboratory report.

e In the space provided in the laboratory report, draw representative human blood cells as they appear through the high-power lens. Refer to figure 1.4 for help in identifying common human blood cells.

f Remove the slide from the slide holder and return it to the class set.

Procedure 3

Observing 3 Morphological Types of Bacteria with the Oil-Immersion Lens

a Obtain a slide with 3 types of bacteria on it.

b Rotate the **low-power lens** into the observation position of your microscope.

c Hold the slide up to the room light. Observe 3 thin, colored films. These are stained smears of 3 morphological types of bacteria.

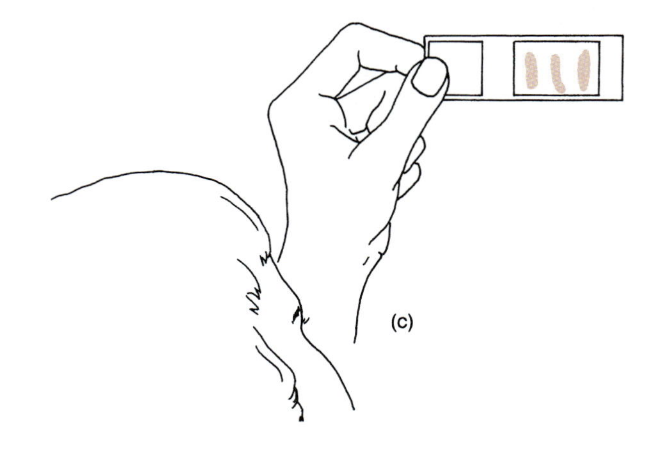

d Center the darkest smear over the stage aperture of your microscope.

e **While watching from the side,** decrease the distance between slide and lens. Adjust the diaphragm to allow a low light intensity.

f Look through the ocular lenses. Turn the coarse-adjustment knob **as slowly as possible,** increasing the working distance. Stop when you see *color.* The bacteria have been stained, so you are able to see a smear of color, even with low magnification.

g Improve the illumination and try to focus with the fine-adjustment knob. But remember that bacteria are very tiny, so you will probably observe only extremely small lines or dots.

Your instructor may demonstrate methods of focusing with the oil-immersion lens. The following method is one of several that are usually quick and successful.

h As in the illustration, rotate the lenses so that they are halfway between low- (10×) or scanning- (4×) power and the oil-immersion lens.

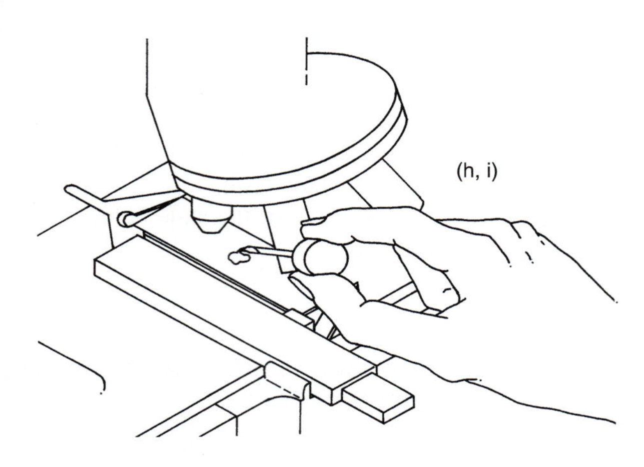

i Allow 1 drop of **immersion oil** to fall onto the smear. Do not touch the dropper to the slide. It might transfer dust and dirt back into the oil.

j Rotate the **oil-immersion lens** into the observation position and directly into the immersion oil.

k While watching from the side to prevent abrasive contact, bring the slide and lens together until they touch each other very gently.

l Look through the eyepieces and adjust the illumination. As slowly as possible, rotate the coarse-adjustment knob to increase the working distance. Adjust the illumination again.

m When *color* flashes by, stop. Go back slowly using the fine-adjustment knob to bring the bacteria into sharp focus. Improve resolution by adjusting illumination. Note in the laboratory report whether you increased illumination.

n By moving your mechanical stage and adjusting the fine focus, view each of the 3 types of bacteria. If you do not locate the bacteria, take out the slide. Look at it against the classroom light again. Locate the darkest area of the darkest smear; this may be a crystal of dye, but it should be easy to find under the lens. Center this heavily pigmented area over the stage aperture and under your low-power lens. Try to focus again. If you still encounter difficulty, ask **your instructor** for assistance.

o In your laboratory report, draw several cells of each of 3 types of bacteria. Note the total magnification that you employ. How does the three-dimensional twisting of spiral bacteria affect your ability to focus on these cells?

p Repeat the focusing procedure until you are satisfied with your ability to focus quickly and easily using the oil-immersion lens.

q Remove the oil from your slide and use lens paper to remove oil from your microscope lens. Return the slide to the class set. Store your microscope as directed by **your instructor.**

POSTTEST EXERCISE 1

Part 1

Matching (For answers see Appendix 1.)

Each answer may be used once, more than once, or not at all.

a. Anton van Leeuwenhoek   b. Ernst Abbé

1. _____ Which early microscopist published the first drawings of bacteria in 1676?
2. _____ Which microscopist showed the mathematical relationship between resolution and the ability of a lens to gather light?

Part 2

True or False (Circle one, then correct every false statement.)

1. T F The objective lens is the one nearest the observer's eye.
2. T F Immersion oil has nearly the same refractive index as glass.
3. T F A compound microscope contains two or more sets of lenses.
4. T F To calculate the total magnification of your microscope, add the magnifications of the objective and ocular lenses.
5. T F Increasing the size of a blurred image generally reveals further details.
6. T F Resolving power is the ability to rotate a new objective lens into the observation position and have the field remain in focus.
7. T F The numerical aperture of an objective lens depends on the size of the cone of light it can receive and also upon the medium in which the lens is suspended.

8. T F If a lens is designed to be used with immersion oil, it is able to gather more light when the oil is utilized than when air separates the lens from the specimen.
9. T F Most modern microscopes are parfocal. This means the microscopist can add infinitely more lens systems for greater magnification and resolution.
10. T F With the equation $RP = \lambda/(2 \times NA)$, as the RP becomes smaller, resolution becomes better and finer detail can be seen.
11. T F Use both hands to carry the microscope; keep the instrument upright.
12. T F Lens paper can be used to remove oil from the glass components of the microscope.
13. T F To save time, look through the microscope and rapidly bring the objective lens and the specimen together by rotating the coarse-adjustment knob.
14. T F Always be sure to oil your microscope lenses before returning the instrument to its designated space.
15. T F Objects appear upside down and backwards through a microscope.

Part 3

Completion and Short Answer

1. The _______________ is a device invented in the seventeenth century for magnifying objects that are too small to be seen with the naked eye.
2. The concept of the _______________ _______________ gives microscopists a mathematical way of describing the light-gathering ability of a lens system.
3. The space separating a specimen and the objective lens is called _______________ distance.
4. The area that you see through a lens is the microscopic _______________ .
5. Describe empty magnification.

6. Give the formula for total magnification.

7. By using the formula $RP = \lambda/(2 \times NA)$, evaluate which type of filter most improves resolution—a blue filter with short wavelength or a red filter with its longer wavelength.
8. List the names of at least 15 parts of your microscope, then briefly describe the function of each item on your list.

The Compound Microscope

	PART	*FUNCTION*
a.		
b.		
c.		
d.		
e.		
f.		
g.		
h.		
i.		
j.		
k.		
l.		
m.		
n.		
o.		
p.		
q.		
r.		

MICROSCOPY

Across

1 Window to invisible worlds
5 Fungus
6 Limb; branch
7 A microscope with well adjusted lenses is . . .
10 Hole; opening
13 Data; scientists' buildings blocks
14 A disk that cuts off light rays
17 Dipping
19 Crude; not refined
22 Resolving power (abbreviation)
23 Lens that is near the object
26 Refined; delicate
27 By numbers
28 Opposable wheel turner
31 Early microbe watcher
33 Lens that is near the eye
35 Persistent
38 Shelf on a microscope
40 View; behold
41 Expose; show
42 Lens; refractive index like immersion oil
43 Not wet or oily

Down

2 Composed of two or more
3 28.35 grams
4 Before others
5 Enlarge
8 Numerical aperture (abbreviation)
9 Glass that bends light rays
11 Uncomplicated
12 Amplification
15 Transparent gas
16 Viewing area
18 Millimeter (abbreviation)
19 Handle with . . .
20 Beam
21 Discover
24 Lamp
25 Microbe
29 Out of focus
30 Making individual parts visible
32 Worthless; not full
34 Price
36 Simple
37 A dab that is spread on a slide
39 Early illuminator

Laboratory Report **1**

Name ______________________________

Section ______________

The Compound Microscope

Procedure 1

Examining the Letter *e* with the Low-Power Objective Lens

(*i*) Draw the letter *e* as it appears to your naked eyes and as it seems to be positioned when seen through the microscope lenses. Note the total magnification used.

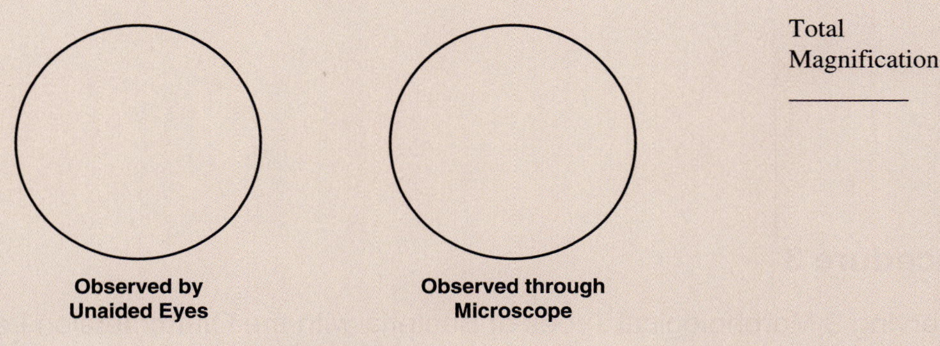

The Letter *e*

(*j*) When a microscope slide is actually being moved to the right by the mechanical-stage controls, where does it appear to move when observed through the lenses?

(*k*) When you move the microscope slide away from yourself, where does it appear to move if you watch the motion through the lenses?

Procedure 2

Observing Stained Human Blood with the High-Power Objective Lens

(*d*) Is your microscope parfocal or nearly so? ______________

Which way did you rotate the fine-adjustment knob to bring the specimen into sharp focus with the high-power objective lens after using the low-power lens? ______________

(*e*) Draw representative human blood cells as they appear through the high-power lens.

Total Magnification _________

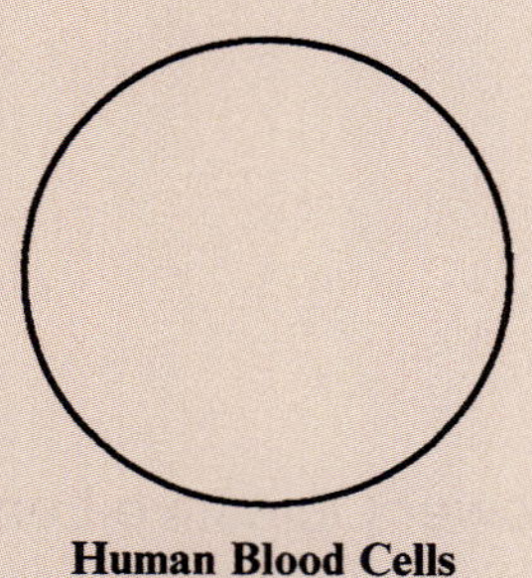

Human Blood Cells
High-Power Lens

Procedure 3

Observing 3 Morphological Types of Bacteria with the Oil-Immersion Lens

(*m*) When going from the low-power objective lens to the oil-immersion lens, did you need to increase the amount of illumination that came through the stage aperture so that you could see the bacteria better? ______________

(*o*) Closely observe and carefully draw some typical cells of each of the 3 morphological types of bacteria. Work to improve your ability to discern and record details.

Total Magnification _________

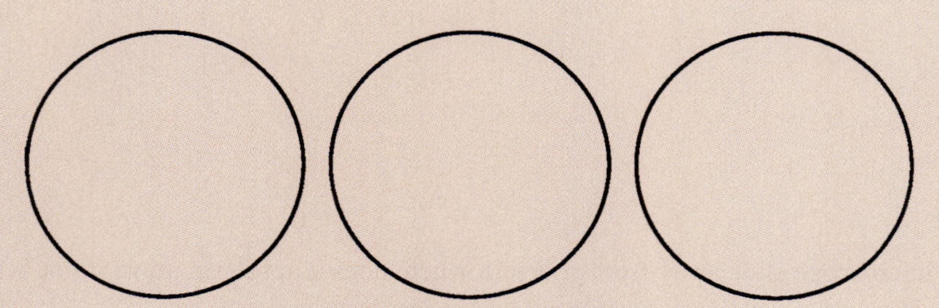

Three Morphological Types of Bacteria

Exercise 2

Microscopic Measurement

Necessary Skills

Microscopy, Knowledge of the metric system of linear measurement

Materials

Assigned compound light microscope
Prepared slides (1 each per group):
The letter *e* (or newsprint, scissors, slides, and coverslips)
Stained human blood smear
Bacteria, 3 morphological types (rods, cocci, spirals)
Several ocular micrometer disks per group
1 Stage micrometer per group
1 Dropper bottle of immersion oil per group
1 Ruler marked in millimeters per group
Lens paper
Calculator (optional)

Primary Objective

Accurately measure appropriate specimens with each objective lens of the compound light microscope.

Other Objectives

1. Define: ocular micrometer disk, stage micrometer.
2. Utilize a stage micrometer to calibrate a micrometer disk for use with low-power, high-power, and oil-immersion lenses.
3. Utilize mathematical formulas to evaluate the diameter of human blood cells.
4. Find the average dimensions of selected rods and cocci.

INTRODUCTION

In this exercise you learn to measure microscopic specimens. Since you cannot, generally, hold a ruler next to these tiny objects under your lenses, how do you accurately judge their dimensions?

Microscopists usually answer this question by placing a **micrometer disk** in the eyepiece. As shown in figure 2.1, the micrometer disk has lines etched onto it. (Your microscope may already have a scale marked in the eyepiece.) A **stage micrometer,** a slide with a scale that is generally divided into 0.01 mm segments, is then placed in the slide holder on the stage. Using the stage micrometer, the microscopist measures the distance between lines on the ocular micrometer disk.

In other words you, the microscopist, calibrate the scale of the micrometer disk by measuring its divisions with the stage micrometer.

Once you have determined the size of the divisions on the micrometer disk, you remove the stage micrometer and replace it with the specimen slide as shown in figure 2.2. Using the disk scale, you measure the specimen.

For example, if each disk division is 0.015 mm wide, and an **erythrocyte,** red blood cell, crosses only one-half of a division, how wide is the blood cell in millimeters? ______________ * in micrometers? ______________ **

But what happens when you rotate the next objective lens into place? As you perform the procedures in this exercise, note that when you increase magnification, the lines on the stage micrometer appear farther apart. Because of this apparent increase, you need to calibrate the scale on the micrometer disk separately for each objective lens. The steps for calibration are listed in the procedures section of this exercise.

*0.015 mm per division, times 0.5 divisions, equals 0.0075 mm.

**0.0075 mm, times 1,000 micrometers per mm, equals 7.5 μm.

Figure 2.1 Use a stage micrometer to calibrate an ocular micrometer disk.

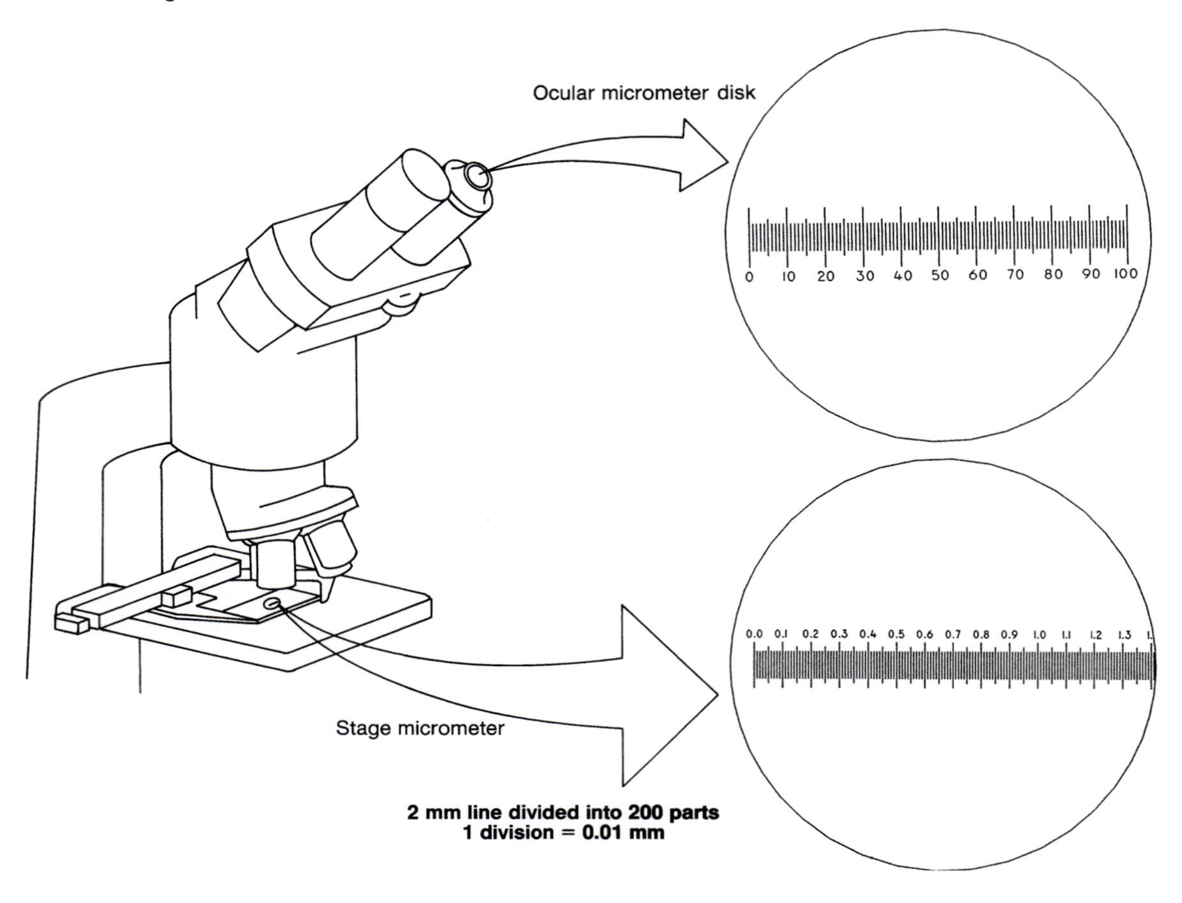

Figure 2.2 Human erythrocytes as seen at two magnifications through an eyepiece with an ocular micrometer disk

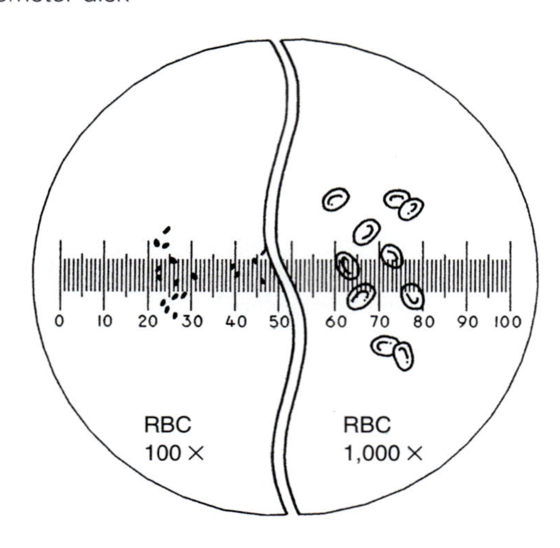

PROCEDURES EXERCISE 2

Procedure 1

Measuring with the Low-Power Lens

To measure specimens through a microscope, you first see how far apart the lines are in the ocular disk. Then you employ the ocular lines to measure your specimen. Examine figures 2.1 and 2.2 carefully while performing the following steps.

a (If your eyepiece has a scale etched into it, skip this step.) Obtain a micrometer disk and place it in the eyepiece of your microscope. **Your instructor** may demonstrate this step. Look through your eyepiece and rotate it until the scale is horizontal.

b Maximize distance from stage to objective lens with the coarse-adjustment knob. Place the **low-power lens** in the observation position.

c Obtain a stage micrometer; **handle it very carefully.** Examine the stage micrometer against the classroom light and notice the tiny scale etched onto it. Center this scale over the stage aperture and illuminate the slide with reduced light.

d With the coarse-adjustment knob, minimize the distance from the stage micrometer to the objective lens.

e Focus on the stage micrometer scale and center it in your field.

Your instructor may modify the following instructions for calibration of lenses.

f Using your gentlest touch, arrange the stage micrometer rule directly above the ocular micrometer scale, as shown in the illustration.

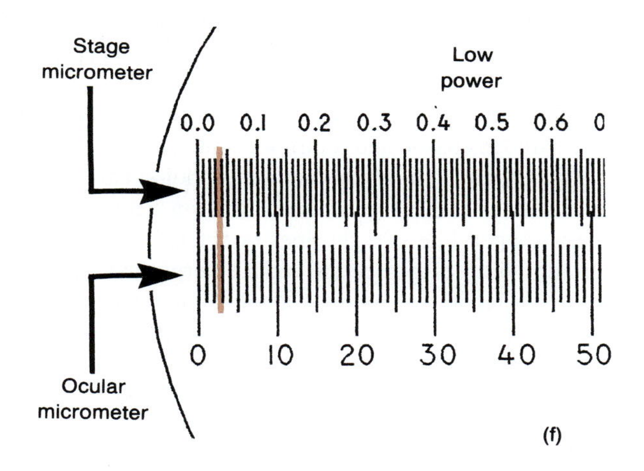

(f)

g Align the two scales, so the first line of the stage scale is centered and directly above the first line of the ocular disk scale if possible.

Remember that each of the narrowest divisions on the stage micrometer scale equals 0.01 mm. You are going to determine how many of these stage divisions equal 1 ocular division; stated another way, how many stage divisions **per** ocular division. Mathematically, this is written:

$$\frac{\text{stage divisions}}{\text{ocular division}}$$

h In your laboratory report, note which lines coincide exactly. How many spaces (divisions) are there *between* these matching lines? Refer to the illustration for help.

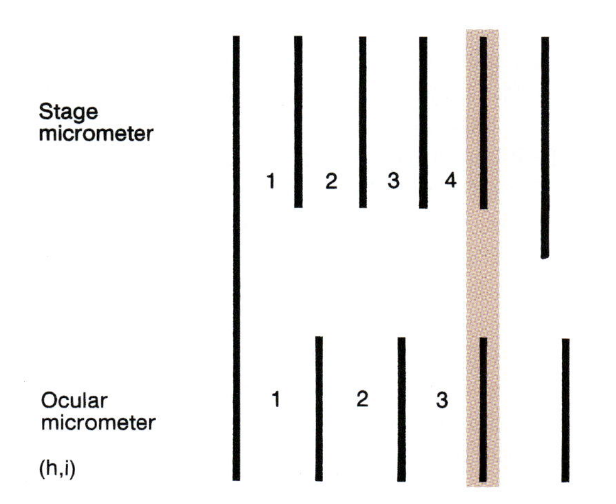

(h,i)

i Divide the number of the narrowest stage divisions by the number of ocular divisions. In the illustration there are 4 stage divisions per 3 ocular divisions, yielding 1.33 stage divisions per ocular division. That is:

$$\frac{4 \text{ stage divisions}}{3 \text{ ocular divisions}} = \frac{1.33 \text{ stage divisions}}{1 \text{ ocular division}}$$

j Each of the shortest stage divisions equals 0.01 mm, so in the example, each ocular division equals 0.01 mm × 1.33. That is 0.0133 mm.

Mathematically, this is written:

$$\frac{\text{stage divisions}}{\text{ocular divisions}} \times \frac{0.01 \text{ mm}}{\text{stage division}} = \textbf{millimeters per ocular division}$$

k In your laboratory report, note the length in millimeters of 1 ocular disk division as seen through **your** microscope.

l Now remove the stage micrometer and return it to the class set. Center the letter *e* under your low-power lens.

m Focus on the *e*. How many ocular divisions does it cross? In the illustration the *e* crosses 100 ocular disk divisions.

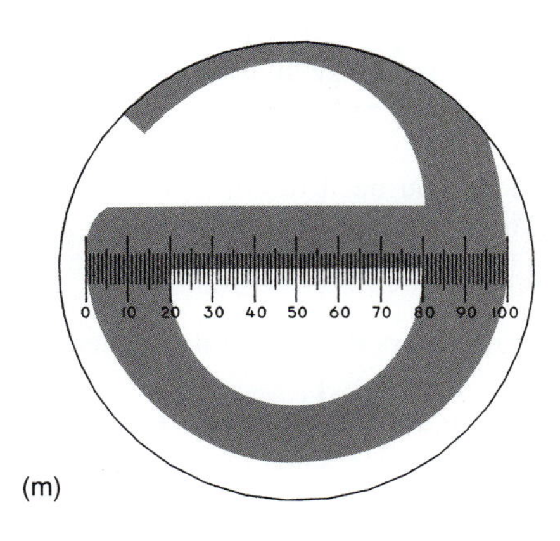

(m)

n To calculate the length of the *e*, multiply the size of each ocular disk division (0.0133 mm in the example) times the number of divisions across the *e*. In the example the *e* is 1.33 mm across.

o To check your work, remove the *e* from your microscope and measure it with a millimeter ruler. Do the measurements roughly agree? If not, recalibrate the ocular disk and measure the letter again. When the measurements agree, return your letter *e* slide to the class set.

You have successfully calibrated an ocular micrometer using your low-power objective. Calibration will be easier with the remaining objectives.

Procedure 2

Measuring with the High-Power Lens

a Center the finest lines of a stage micrometer scale over the stage aperture and focus through the ocular disk using low power. Rotate the **high-power objective** into the viewing position and refocus on the smallest divisions of the stage micrometer, aligning them as in the illustration.

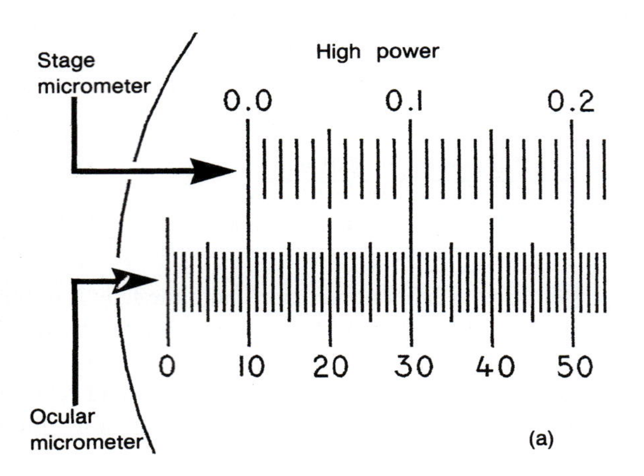

(a)

b Again, you can figure:

$$\frac{\text{stage divisions}}{\text{ocular divisions}} \times \frac{0.01 \text{ mm}}{\text{stage division}} = \textbf{millimeters per ocular division}$$

But for this lens you need to convert millimeters to micrometers. There are 1,000 micrometers in a millimeter. Therefore,

$$\text{millimeters} \times 1{,}000 = \text{micrometers}$$

You need to alter the formula from Procedure 1 to calculate micrometers per ocular division:

$$\frac{\text{stage divisions}}{\text{ocular divisions}} \times \frac{0.01 \text{ mm}}{\text{stage division}} \times \frac{1{,}000\ \mu\text{m}}{\text{mm}} = \textbf{micrometers per ocular division}$$

This can be simplified to:

$$\frac{\text{stage divisions}}{\text{ocular divisions}} \times \frac{10\ \mu\text{m}}{\text{stage division}} = \frac{\textbf{micrometers}}{\textbf{ocular division}}$$

Use this formula and enter the results in your laboratory report.

c Remove the stage micrometer and return it to the class set. Focus on cells in a prepared slide of stained human blood.

d Measure the diameters of 5 erythrocytes; enter the measurements, calculated in micrometers, into your laboratory report. Calculate the average diameter of human red blood cells. Return the slide of stained blood to the class set.

Procedure 3

Measuring with the Oil-Immersion Lens

a Repeat the calibration steps in the previous procedure but lower the **oil-immersion lens** directly into a drop of oil on the stage micrometer.

b After removing the stage micrometer, carefully clean it with lens tissue and return the slide to the class set.

c Obtain a slide showing 3 morphological types of bacteria. Focus the oil-immersion lens on bacterial rods (the cells shaped like straight lines or bricks). Measure the length and width of 5 separate rods, entering their dimensions in your laboratory report. Calculate their average size.

d Locate the cocci (spherical bacteria) and measure the diameters of 5 cells. What is their average diameter? Enter your calculations in the laboratory report.

e When you have completed your observations, clean the slide and return it to the class set. Thoroughly clean your microscope, removing all oil with lens tissue, and return the instrument to its assigned location.

POSTTEST EXERCISE 2

The posttest is intended to guide you through some of the important points in the introduction and procedures. Use it to reinforce your knowledge. However, the true test of your comprehension of this exercise is your ability to make microscopic measurements accurately.

True or False (Circle one, then correct every false statement.)

1. T F The microscopist measures ocular micrometer disk divisions with the stage micrometer.
2. T F A specimen is normally placed on top of the stage micrometer and measured with the tiny ruler that is etched onto the stage micrometer.
3. T F The stage micrometer scale is generally ruled to 0.0001 mm.
4. T F When utilizing a stage micrometer scale marked in 0.01 mm, the following formula holds true: (stage divisions/ocular divisions) × (0.01 mm/stage division) = millimeters per ocular division.
5. T F When utilizing a stage micrometer scale marked in 0.01 mm, the following formula also holds true: (stage divisions/ocular divisions) × (10 μm/stage division) = micrometers/ocular division.

Laboratory Report **2**

Name ______________________________

Section ____________

Microscopic Measurement

Procedure 1

Measuring with the Low-Power Lens

(*h*) Using the low-power lens, observe the stage micrometer through the ocular disk and draw part of the 2 scales here, showing which lines coincide.

Stage Scale **Ocular Scale** **Stage Micrometer Scale as Seen through the Ocular Disk and Low-Power Lens**

How many stage micrometer divisions are between the matching lines? ______
How many ocular disk divisions are between the matching lines? ______

(*i*) $\dfrac{\_\_\_\_\_\ \text{stage divisions}}{\_\_\_\_\_\ \text{ocular divisions}} = \dfrac{\_\_\_\_\_\ \text{stage divisions}}{1\ \text{ocular division}}$

(*k*) Each of the smallest stage divisions equals 0.01 mm. What is the length in millimeters of 1 ocular disk division as observed with the low-power lens of your microscope? ______ (Remember the equation from step [*j*].)
(*m*) How many ocular divisions does the *e* cross? ______
(*n*) Calculate the length of the *e*, multiplying the size of each ocular disk division times the number of divisions across the *e*. _____________ × _____________ = _____________
(*o*) Measure the *e* with a millimeter ruler. How long is it? ______ Does your answer in (*n*) roughly equal your answer here? ______ If not, find the source of any error before going on to the next procedure.

Procedure 2

Measuring with the High-Power Lens

(*a*) Using the high-power lens, observe the stage micrometer through the ocular disk and draw part of the 2 scales here, showing which lines coincide.

Stage Scale

Ocular Scale

Stage Micrometer Scale as Seen through the Ocular Disk and High-Power Lens

(*b*) (_______ stage divisions/ _______ ocular divisions) × (10 μm/stage division) = _______ μm/ocular division

(*d*) Fill in the chart.

Diameter of Human Red Blood Cells

Human Erythrocyte	Number of Ocular Divisions Crossed	Length of Cell in Micrometers
1.		
2.		
3.		
4.		
5.		

What is the average diameter of the human red blood cells that you measured? _________

Did you remember to include your units of measurement—mm or μm?

Procedure 3

Measuring with the Oil-Immersion Lens

(*a*) Using the oil-immersion lens, observe the stage micrometer through the ocular disk and draw part of the 2 scales here, showing which lines coincide.

Stage Scale

Ocular Scale

Stage Micrometer Scale as Seen through the Ocular Disk and Oil-Immersion Lens

(_______ stage divisions/ _______ ocular divisions) × (10 μm/stage division) = _______ μm/ocular division

(*c*) Fill in the chart.

Measurements of Selected Rods

Bacterial Rods	Length of Rod		Width of Rod	
	Ocular Divisions	μm	Ocular Divisions	μm
1.	_______	_______	_______	_______
2.	_______	_______	_______	_______
3.	_______	_______	_______	_______
4.	_______	_______	_______	_______
5.	_______	_______	_______	_______

What is the average length of the rods? _________

What is the average width of the rods? _________

Remember to include your units of measurement.

(*d*) Fill in the chart.

Diameters of Selected Cocci

Bacterial Cocci	Number of Ocular Divisions Crossed	Length of Cell in Micrometers
1.	______	______
2.	______	______
3.	______	______
4.	______	______
5.	______	______

What is the average diameter of the cocci? ______________

Exercise 3

Aseptic Transfer Techniques

Necessary Skills

Knowledge of the metric system of linear measurement

Materials
First Laboratory Session

Culture (48-hour, 30° C, Tryptic Soy Agar [TSA] slant, 1 per group):
- *Serratia marcescens*

Supplies (per student):
- Nonsterile 1 ml pipet
- Nonsterile 10 ml pipet
- Sterile 1 ml pipet
- Nonsterile 100 ml or larger beaker of water
- Forceps
- Capped test tube containing approximately 5 ml of sterile water
- 4 Capped test tubes of sterile Tryptic Soy Broth (TSB)
- 2 Empty, nonsterile test tubes with caps

Additional supplies (per group):
- Pipet fillers (recommended: Pi-Pump®, 1 each, blue and green)
- Dropper bottle of alcohol for flaming
- Container holding at least 1 sterile filter paper disk per student

Second Laboratory Session

Supplies (per class):
- Pipettors
- Nonsterile 1 ml pipets
- Nonsterile 10 ml pipets
- Nonsterile test tubes with caps
- 100 ml or larger beakers of nonsterile water
- Forceps
- Dropper bottles of alcohol for flaming
- Petri dish with nonsterile filter paper disks

Primary Objective

Understand and practice aseptic (sterile) technique.

Other Objectives

1. Define: pure culture, aseptic, sterile, contamination, mixed culture.
2. List three instances in which a microbiologist must utilize a pure culture.
3. Identify aseptic technique procedures that are performed before and after work with microorganisms.
4. Demonstrate aseptic technique precautions that should be taken while bacterial cultures are open.
5. Cite instances when a microbiologist must transfer bacteria from one culture to another.
6. Describe the instruments microbiologists most often use to transfer microorganisms.
7. Practice aseptic technique using inoculating loops and needles.
8. Obtain sterile pipets without contaminating the pipets that remain in the storage can.
9. While utilizing sterile technique, measure liquids and transfer them with a pipet and pipettor.
10. Safely alcohol-flame forceps and use the sterilized tips to transfer materials.
11. Explain how to protect yourself from laboratory-acquired infection.

INTRODUCTION

In this exercise you learn how to transfer bacteria from one container to another while maintaining the purity of the culture. A **pure culture** contains only one kind of bacterium; ideally, it consists of the descendants of one bacterial cell. As you demonstrate in Exercise 11, Ubiquity and Culture Characteristics, many microbes are present on your workbench, hands, and transfer instruments, as well as in the air. How, then, is it possible to avoid **contamination,** the introduction of unwanted organisms?

Table 3.1 Rules for Aseptic (Sterile) Technique

Before work:
- Remove all unnecessary items from your workbench.
- Put on your laboratory coat. Fasten it closed.
- Be sure your hair is under control.
- Never lay culture tubes on your workbench.

Before and after work:
- Wash your hands.
- Disinfect your area.
- Sterilize instruments.

While cultures are open:
- Do not talk.
- Work near a flame where rising air currents protect cultures from microbes carried on dust.
- Slant open test tubes toward the flame and away from your mouth and nose.
- Never lay caps or plugs on the bench top.
- Tilt an open Petri dish lid to form a barrier between the culture and your mouth and nose.
- Work quickly.

In general, while practicing sterile technique:
- Avoid producing **bacterial aerosols,** microbes floating in the air. (Aerosols generally result from the bursting of microbe-laden bubbles.)
- Keep cultures closed whenever possible.
- Keep all items away from your mouth.
- Remove your laboratory coat before leaving the area.

Aseptic Techniques

You learn to practice **aseptic techniques,** procedures that (1) protect the culture and (2) protect you and the environment. These procedures are also called **sterile techniques.**

Pure cultures are essential if a microbiologist is to identify bacteria with biochemical tests, perform antibiotic sensitivity testing, or maintain stock cultures. But in samples taken from nature, most microbes are found in **mixed cultures,** where two or more species live together. In later exercises of this laboratory manual, you learn to isolate bacteria; for the present, practice techniques that preserve purity.

Table 3.1 lists general rules for sterile technique; **your instructor** may wish to supplement the table.

When you transfer bacteria, remember they are so small that if you see a **colony** (ideally, the descendants of one cell growing together on a solid surface) you are probably viewing several million microbial cells. The tiniest visible inoculum generally contains enough cells to initiate abundant growth. You demonstrate this fact in the procedures section of this exercise.

But why does a microbiologist transfer bacteria from one culture to another? In maintaining stock cultures, the microbes are transferred because they need fresh nutrients and must escape their own toxic wastes. It is also necessary to transfer a culture when large amounts of an organism are required or when biochemical responses to various substances are investigated.

Figure 3.1 Instruments for microbial transfer

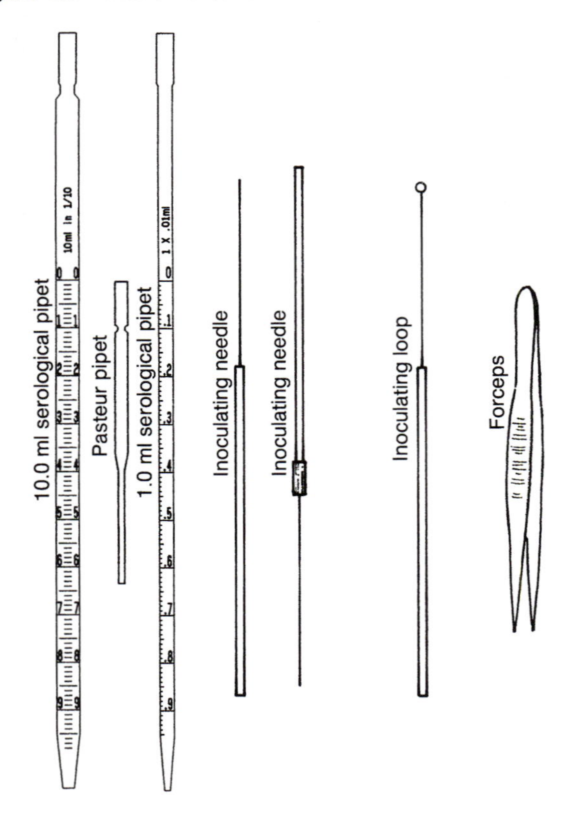

In this exercise you learn to manipulate the most common transfer instruments. These include inoculating needles and loops, pipets, and forceps such as those shown in figure 3.1.

Inoculating Loops and Needles

Microbiologists generally use inert metal wire, either Nichrome or platinum, to form inoculating loops and needles. To sterilize the loop or needle, you grasp its handle and hold the wire in a flame until it is red-hot. **Flaming** incinerates any organisms on the wire.

To prevent aerosols from forming, wires that form loops and needles are flamed from the handle connection to the tip. Boiling droplets of microorganisms spray into the air when a loop or needle full of bacteria is thrust into a flame. Flaming from the handle connection to the tip heats the fluid more gently, charring the bacteria before incinerating them.

After cooling, the needle or loop can be used to lift bacteria off surfaces or out of broths and transfer them to other media. The instrument is then resterilized.

Figure 3.2 Pipet-filling devices

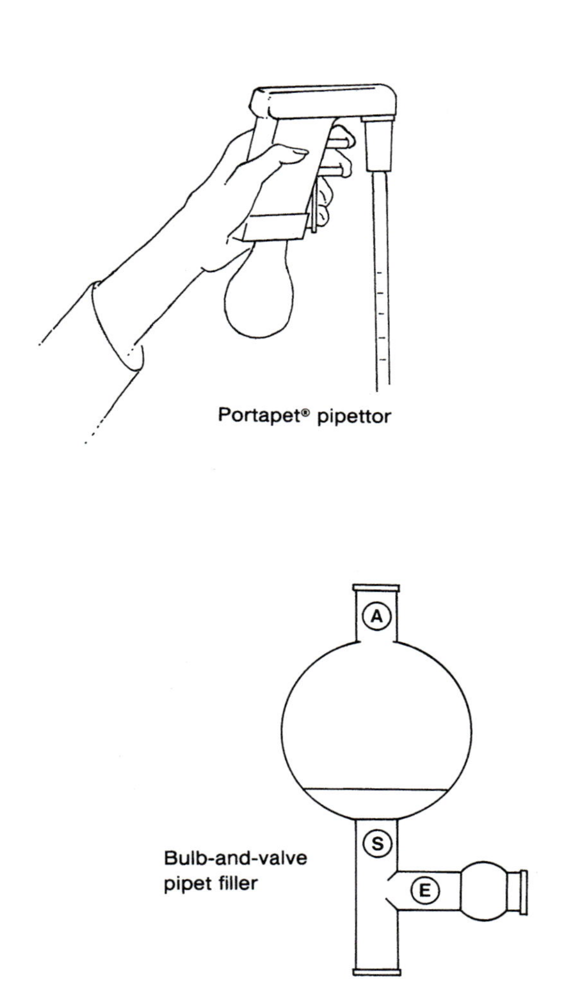

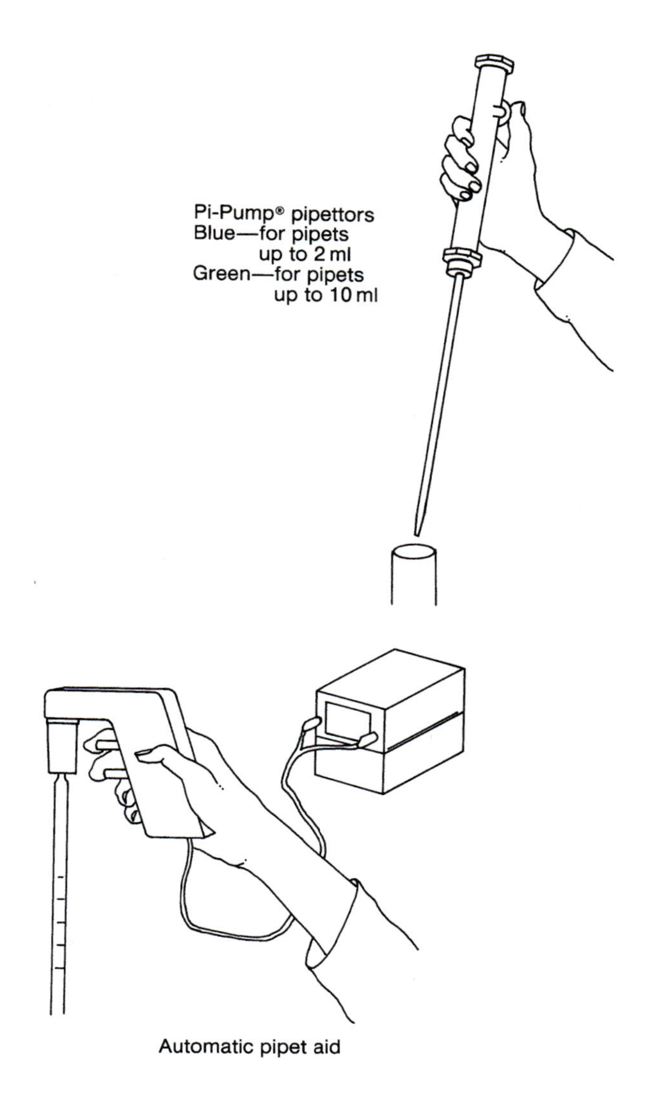

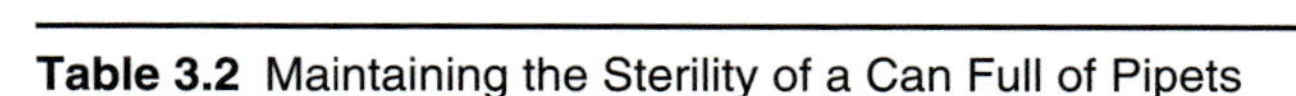

Table 3.2 Maintaining the Sterility of a Can Full of Pipets

1. Keep the can closed when not in use.
2. Do not reach over an open can of pipets.
3. Touch only 1 pipet at a time.
4. To withdraw a pipet, lift it out; do not drag the tip over the blunt ends of the other pipets.
5. After obtaining a sterile pipet, use it immediately. **Never** return a pipet to the can, even if you think the pipet is probably sterile.

Pipets

In this exercise you employ various pipets and pipettors to transfer liquids. After practicing with nonsterile pipets, you learn to manipulate those that are sterile.

Prepackaged, sterile, disposable pipets are available commercially. Many laboratories employ reusable, sterile, glass pipets, storing them in cans; see table 3.2.

Contaminated, reusable, glass pipets are often discarded into disinfectant in a pipet jar or large beaker. They are placed into the container gently to avoid breakage and tip down to prevent the formation of aerosols. Disposable pipets are also discarded in a sanitary manner to prevent the spread of microorganisms.

In a microbiology laboratory, it is generally unsafe to mouth pipet since organisms may accidentally be ingested. Manipulation techniques for several common pipettors are explained in this exercise. A selection of pipet fillers is illustrated in figure 3.2.

Alcohol-Flaming

You also learn a common but dangerous sterilization technique called **alcohol-flaming;** see figure 3.3. This method of incinerating microbes is utilized to sterilize the tips of forceps. If you flame the forceps tips until they are red-hot, heat transfers to the handle and burns your fingers. To

Figure 3.3 Alcohol-flaming

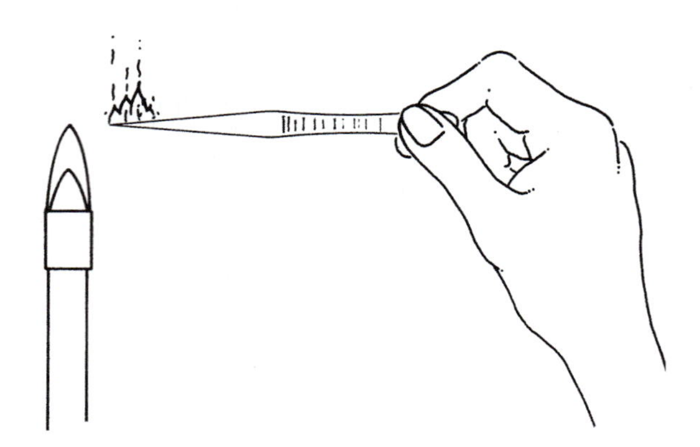

avoid this hazard, the microbiologist applies alcohol to the tips, passes them *through and out of the flame,* and allows the alcohol to burn off. This procedure is generally repeated three times.

The technique is dangerous because flaming alcohol can drip onto your workbench, igniting any flammable material it contacts. The flaming liquid may stream onto your hand if the forceps are held incorrectly. Be sure to study the directions in the procedures section before attempting alcohol-flaming.

Remember that in a microbiology laboratory every bacterial culture is treated as though it contains **pathogens** (disease-producing microorganisms). By practicing sterile technique and following the general rules for safe conduct in the laboratory, you protect yourself from laboratory-acquired infection.

PROCEDURES EXERCISE 3

Procedure 1

Nonsterile Pipetting with Pipettors

Instructions for using Pi-Pump® pipettors are given in this procedure. For use of bulb-and-valve pipet fillers, instructions are given on the package insert. Mouth pipetting is not recommended.

a Obtain 2 empty test tubes, 1 nonsterile 1 ml pipet, 1 nonsterile 10 ml pipet, and a beaker of water. Carefully examine the numbers inscribed on nonsterile 10 ml and 1 ml pipets. In your laboratory report, explain the meaning of the phrases on the pipets. Hint: notice that the smallest markings on the 1 ml pipet are 1/100ths and the smallest markings on the 10 ml pipet are 1/10ths.

Note that allowing fluid to fall from 0 to 1 in the 10 ml pipet releases exactly the same amount of liquid as can be held within the entire 1 ml pipet.

b One commonly used pipetting device is the Pi-Pump®. This brand of pipettor is color coded. The blue Pi-Pump® is used with pipets that hold up to 2 ml, the green Pi-Pump® with pipets holding up to 10 ml. Obtain a green Pi-Pump®.

c While holding a 10 ml pipet near its blunt end, attach the green Pi-Pump® to your 10 ml pipet, as shown in the illustration. Press the pipet into place gently but firmly.

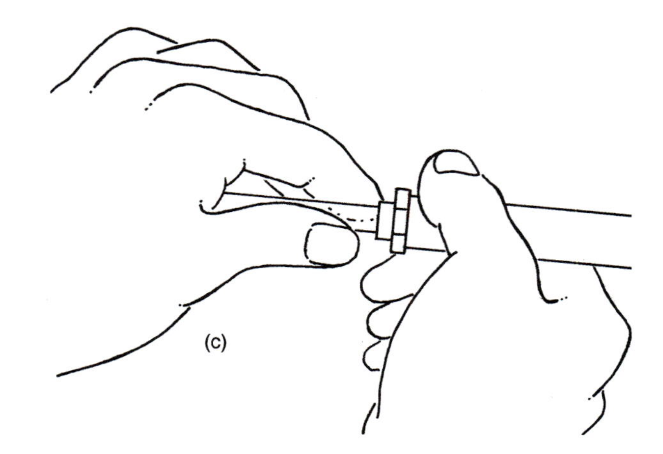

d Place the tip of the pipet into water. Rapidly draw fluid into the pipet by pulling the plunger up. Fill the pipet with more precision by rotating the knurled knob with your thumb. The pipet holds 10 ml when the meniscus of the fluid is aligned with the 0 line of the vertical pipet, as in the illustration.

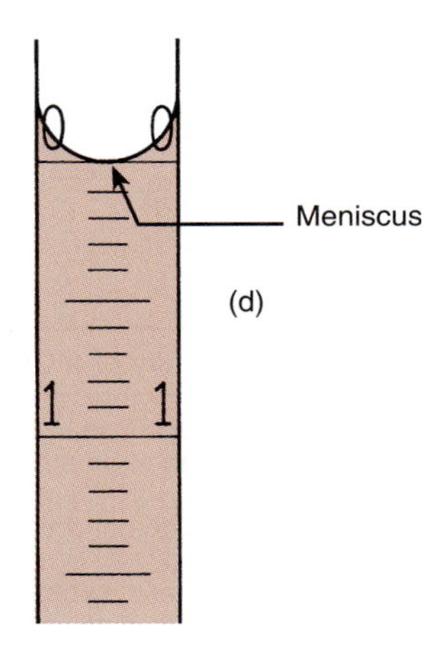

e Expel 1 ml (the meniscus moves from 0 to 1 on the pipet) into a clean test tube by briefly depressing the plunger. Expel the remaining 9 ml back into your beaker of water. Return the green pipet filler to the class set.

f Next, utilize the blue pipettor to draw 1 ml of fluid into the 1 ml pipet. Obtain a blue Pi-Pump® and attach the pipettor to the pipet. Turn the knurled wheel one-half rotation counterclockwise. This intake gives you a little extra push of air to empty the pipet later.

Figure 3.4 Processes utilized in Procedure 2, first laboratory session

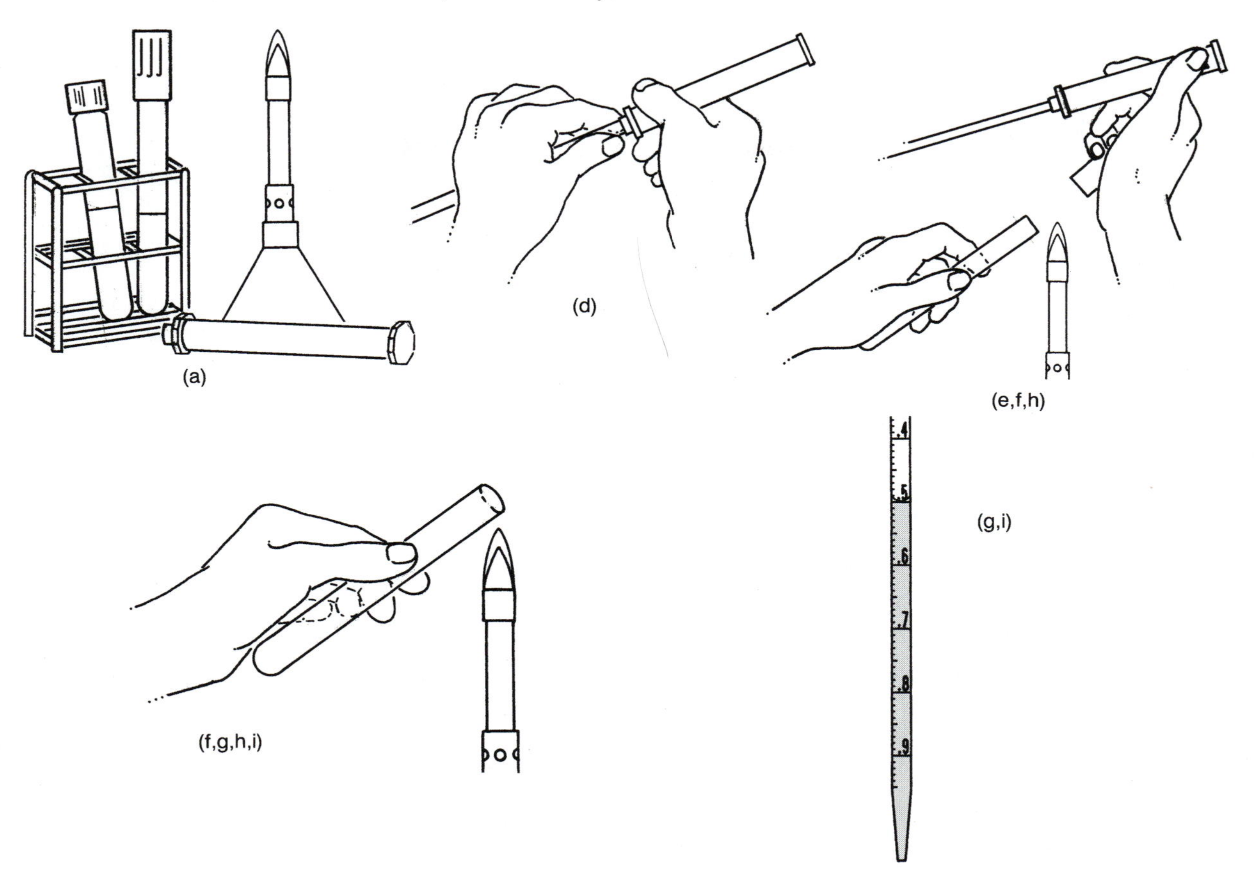

g Dip the pipet into water and fill the device to the 0 line. Deliver 1 ml (from 0 to empty) into a clean test tube.

h In your laboratory report, comment on the use of the excess air to blow out the last drop of fluid from your pipet. Practice these techniques before attempting the next procedure with sterile pipets.

i **Your instructor** may discuss proper recycling of the nonsterile pipets and test tubes. After you are thoroughly familiar with the markings on the pipets, follow **your instructor's** directions for pipet and test tube disposal. Return the pipettor and beaker of water to the class set.

Procedure 2

Sterile Pipetting with Pipettors (figure 3.4)

First Laboratory Session

This procedure presents one method of sterile pipetting. **Your instructor** may demonstrate other techniques. You may employ individually packaged, disposable pipets instead of sharing a can of sterile glass pipets. Assemble all the materials you need for this procedure **before** exposing a sterile pipet to the air in the laboratory.

a Obtain a tube of sterile Tryptic Soy Broth and a tube of sterile water. Place them in your test tube rack. As in figure 3.4, set the rack near your flame. Have a blue Pi-Pump® (or other pipettor) ready for a sterile 1 ml pipet.

b Label the broth "0.5 ml sterile H_2O in TSB." Add your name and the date. **Your instructor** may suggest further labeling.

c Loosen the lids on the 2 tubes.

At this point, **review** the general rules for sterile technique and handling sterile pipets, which are listed in table 3.1.

d Using aseptic technique, obtain a sterile 1 ml pipet. Quickly, touching only the upper portion of the pipet, attach it to the pipettor.

e Pick up the tube of sterile H_2O with your nondominant hand. Hold the pipet and pipettor in your dominant hand.

f Remove the lid of the H_2O tube with the small finger of the pipetting hand and hold the lid as illustrated in figure 3.4. Briefly flame the lip of the H_2O tube.

The following steps suggest that you hold the tube in one hand, the pipettor in the other. You may find it easier to replace your tube of water in the rack and use both hands for operating the pipettor.

Figure 3.5 Processes utilized in Procedure 3, first laboratory session

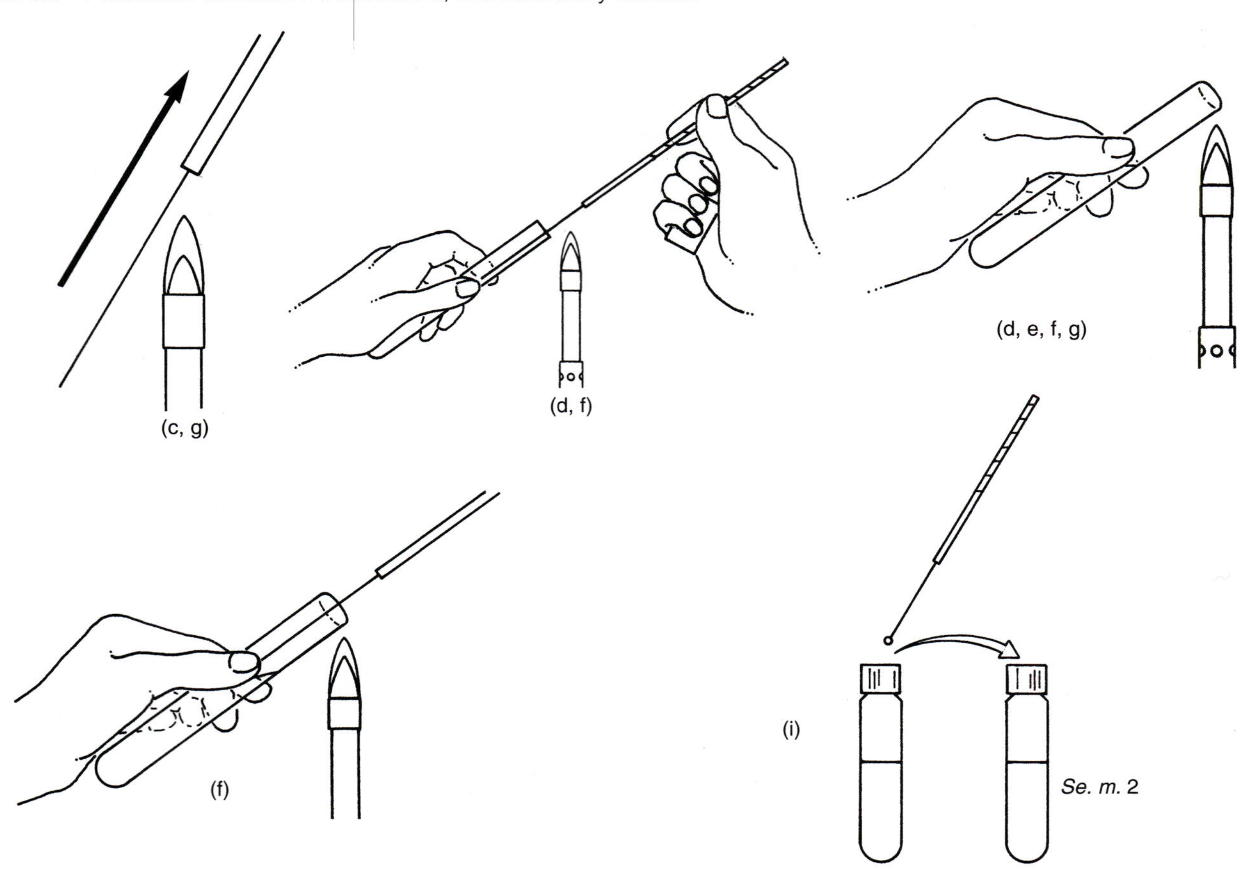

g After rotating the knurled wheel about one-half turn, draw 0.5 ml of sterile H_2O into the pipet. Briefly flame the lip of the tube; close it; place it back in your rack.

h Pick up the tube of Tryptic Soy Broth with your free hand. With the little finger of your pipetting hand, remove the lid of the Tryptic Soy Broth tube. Holding the lid in your curled finger, briefly flame the lip of the tube.

i By rotating the toothed wheel, deliver 0.5 ml of sterile H_2O into the test tube. Quickly flame the lip of the tube; close it; place it back in your rack.

j Follow **your instructor's** directions for disposal of the used pipet and tube of water. Return the pipettor. Close your tube of Tryptic Soy Broth firmly. (Loosen screw caps one-quarter turn for gas exchange.)

In the future, when you pipet bacteria, a fluid culture may inadvertently come in contact with the pipettor, contaminating it. Follow **your instructor's** directions for disinfection of the instrument. Would your culture still be pure?

k Incubate your tube of Tryptic Soy Broth at room temperature until the next laboratory session.

Second Laboratory Session

a Inspect your tube of broth labeled "0.5 ml sterile H_2O in TSB." Note in your laboratory report if the broth shows any sign of bacterial growth such as sediment or turbidity (cloudiness).

b If there is growth in the Tryptic Soy Broth, review the list of aseptic transfer technique rules in the introduction of this exercise. Repeat aseptic transfer technique with the nonsterile equipment provided until you are confident of your abilities. Sterile transfer is an important skill that you must master.

c If there is no growth in the tube of sterile Tryptic Soy Broth, congratulate yourself on your successful application of sterile transfer technique.

d Follow **your instructor's** directions for safe disposal of these tubes.

Procedure 3

Sterile Transferring to Broths with Inoculating Needles and Loops (figure 3.5)

First Laboratory Session

a Obtain an agar slant culture of *Serratia marcescens* and 2 tubes of sterile Tryptic Soy Broth. Place the tubes in your test tube rack near the Bunsen burner flame. Label the tubes "*Se. m.* 1" and "*Se. m.* 2." Add your name and the date to each label; loosen lids. **Your instructor** may suggest further labeling.

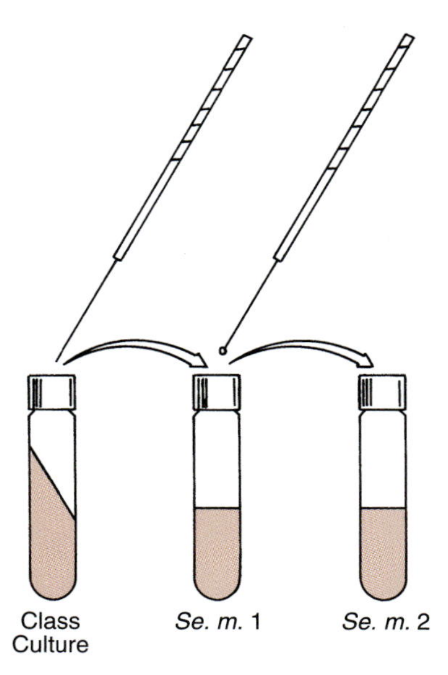

b Review the rules for aseptic technique and apply them during the following transfers. If the handle of your inoculating needle is all metal, wrap tape around the upper portion of the handle to insulate it.

c As in figure 3.5, hold the inoculating needle with your dominant hand. Flame the wire of the inoculating needle from its handle connection to its tip. Remove the tip from the flame. While keeping the inoculating needle in the air and near the flame, cool the needle for about 15 seconds.

d Pick up the culture tube in your free hand. Using the small finger of your dominant hand, remove the lid from the culture. Briefly flame the lip of the tube; touch the needle to the agar slant to be sure it is cool. **Without gouging the agar,** use your inoculating needle to remove a barely visible bit of bacterial growth from the surface of the slant.

e Briefly flame the lip of the tube; close it; return it to the rack.

f With your free hand, quickly pick up the sterile broth labeled "*Se. m.* 1." Remove its lid with the small finger of your dominant hand; briefly flame the lip of the tube. Dip the tip of your inoculating needle into the broth. (Remember to keep the broth tilted toward the flame, away from your nose.) You do not need to stir the needle.

g Remove the needle from the broth. Flame the lip of the tube; replace its lid; place the tube in the test tube rack. Flame the wire of the inoculating needle from the handle connection to the tip of the wire. Tighten the lid of the tube.

h **Your instructor** may demonstrate a safe procedure for mixing the bacteria in your tube of Tryptic Soy Broth. Use caution so that the fluid does not spill out of the test tube.

i After loosening lids, use sterile technique to transfer 1 loopful of bacterial suspension into the tube of sterile broth labeled "*Se. m.* 2." Remember to flame the loop before and after use, to keep the tube closed as much as possible, to flame the lip of the tube before and after use, and to keep tubes slanted away from your nose and toward the flame. Keep your work near the flame and your head away from it.

Figure 3.6 A technique for handling two test tubes

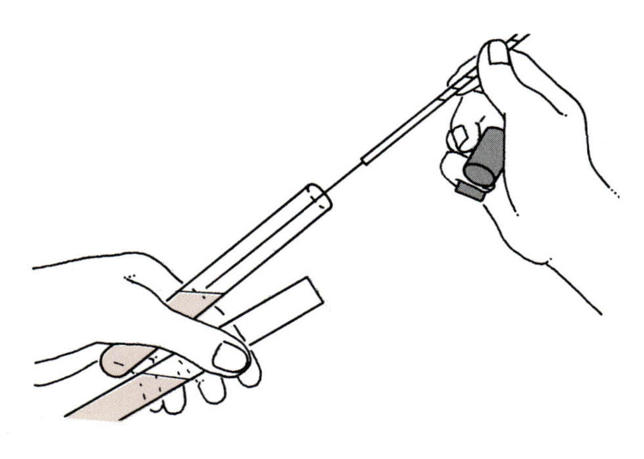

j After practice, and with **your instructor's** approval, you may wish to hold 2 tubes in one hand and 2 caps in the other, as you perform sterile transfers (see figure 3.6).

k Incubate your Tryptic Soy Broth cultures of *Serratia marcescens* at room temperature for no more than 48 hours. (Sedimentation and/or cell lysis [bursting] may occur with longer incubation.) Remember to loosen screw caps one-quarter turn.

Second Laboratory Session

a Examine your 2 cultures of *Serratia marcescens.* Recall the inoculation procedure and compare the amount of growth in the 2 tubes.

b In your laboratory report, note whether there is considerably more growth in 1 broth than in the other.

c Follow **your instructor's** directions for disposal of these contaminated tubes.

Procedure 4

Alcohol-Flaming Forceps for Sterile Transfer (figure 3.7)

> **Warning:** If improperly performed, alcohol-flaming can result in fire damage to you and/or your belongings. Read and understand the directions thoroughly before performing this procedure.

First Laboratory Session

In this procedure you burn alcohol off the tips of forceps 3 times, and then you use the forceps to transfer a sterile paper disk into a tube of Tryptic Soy Broth. If the broth remains sterile after incubation, you have succeeded in demonstrating aseptic technique.

Figure 3.7 Processes utilized in Procedure 4, first laboratory session

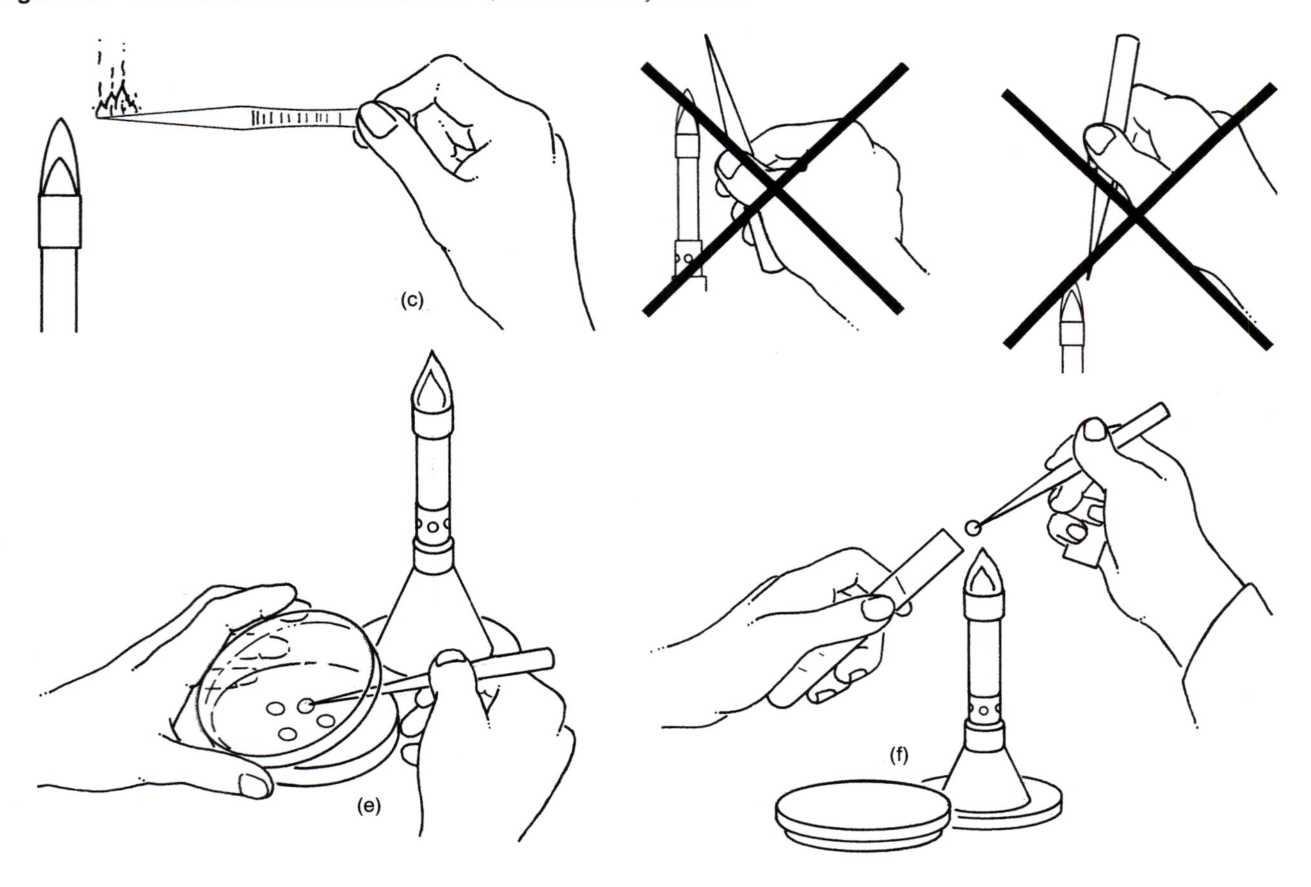

a Obtain forceps, alcohol, a container of sterile paper disks, and a tube of sterile Tryptic Soy Broth.

b Label the tube "TSB with sterile disk." Add your name and the date to the label. **Your instructor** may suggest further labeling. Place the tube in your rack and near the flame. **Clear your bench top.** Loosen the lid of the tube.

c As in figure 3.7, hold the forceps ***horizontally.*** Apply a small amount of alcohol to the tips; pass the tips through and **out of** the flame. Allow the alcohol to burn off the forceps.

d Keeping the forceps horizontal, repeat the alcohol-flaming procedure twice more.

e Continue to work quickly and near the flame. Touching only 1 disk, use the alcohol-flamed tips to pick up a sterile disk. Immediately cover the container of disks.

f With your free hand, pick up the tube of Tryptic Soy Broth. Remove its lid with the little finger of your dominant hand. Briefly flame the lip of the tube. With the forceps, drop the paper disk into the tube. Reflame the lip; close the tube.

If the forceps had touched a contaminated surface, you would now alcohol-flame the tips 3 times.

g Incubate the Tryptic Soy Broth at room temperature until the next laboratory session. Return the forceps, alcohol, and container of sterile paper disks to the class set.

h In your laboratory report, explain what would happen if you tilted your burning forceps either up or down during alcohol-flaming.

Second Laboratory Session

a After examining the tube of Tryptic Soy Broth labeled "TSB with sterile disk," comment in the laboratory report on your success using the aseptic technique of alcohol-flaming.

b Follow **your instructor's** directions for safe disposal of the tube of Tryptic Soy Broth. Practice alcohol-flaming if you need to improve your technique.

POSTTEST EXERCISE 3

Part 1

Matching

Each response may be used once, more than once, or not at all. Some statements may require more than one response.

a. Pure culture
b. Aseptic
c. Sterile
d. Contamination
e. Mixed culture

1. _____ Devoid of life
2. _____ A combination of various microbial species living together
3. _____ One kind of bacteria, ideally the descendants of one cell, living together
4. _____ The introduction of unwanted organisms
5. _____ Free from disease-producing microorganisms

Part 2

True or False (Circle one, then correct every false statement.)

1. T F Aseptic techniques are performed only for their traditional value. In a modern microbiology laboratory, microbes are electronically removed from the air, workbench, inoculation instruments, and hands.
2. T F Sterile techniques are generally utilized with broth cultures; aseptic techniques are reserved for solid media work.
3. T F To help identify a bacterial species, a microbiologist usually performs biochemical tests on the mixed culture.
4. T F Microbiology laboratories often maintain stock cultures. These stocks are pure cultures.
5. T F Antibiotic sensitivity testing is performed with pure cultures.
6. T F In nature most microorganisms are found in mixed cultures.
7. T F An aerosol is a visible clump of bacteria suspended in proteinaceous matter such as sputum.
8. T F To preserve purity, handle cultures as little as possible. Open them all at the beginning of the laboratory period; close them once, before leaving the area.
9. T F When transferring bacteria from a solid surface, be sure to gouge the agar, digging out the entire colony where it has burrowed into the medium.
10. T F Inoculating needles and loops are generally constructed of Nichrome or platinum wire.
11. T F Flaming sterilizes by incinerating organisms.
12. T F In a microbiology laboratory, it is generally safe to mouth pipet since most microorganisms are not killed by exposure to exhaled air.
13. T F Alcohol-flaming can be employed to sterilize the tips of forceps.
14. T F During alcohol-flaming, the tips of the forceps are **not** held in the burner flame until they are red-hot.
15. T F Alcohol-flaming should always be performed over a stack of papers so that the dripping alcohol does not burn the workbench.
16. T F Factors that limit the amount of growth in a broth culture include size of inoculum, availability of nutrients, temperature and duration of incubation, presence of oxygen, concentration of toxins, and so on.

Part 3

Completion and Short Answer

1. Aseptic techniques protect the _____________ and also protect _____________ and the _____________ .
2. List three steps of sterile technique that are performed before and after working with microorganisms.
   a.
   b.
   c.
3. List six rules of sterile technique that are especially important to remember while cultures are open.
   a.
   b.
   c.
   d.
   e.
   f.
4. List six rules for proper use of a can full of sterile pipets.
   a.
   b.
   c.
   d.
   e.
   f.
5. In maintaining stock cultures, the bacteria are transferred because they need _______________ , and they must escape _______________ .

Laboratory Report **3**

Name ______________________________

Section ______________

Aseptic Transfer Techniques

Procedure 1

Nonsterile Pipetting with Pipettors

(*a*) The phrase "1 ml in 1/100" appears on most 1 ml pipets. What does it mean?

The phrase "10 in 1/10 ml" appears on most 10 ml pipets. What does it mean?

(*c*) When attaching a pipet pump to a pipet, why is it advisable to hold the pipet near the blunt end rather than at the pointed tip?

(*h*) Was there enough excess air in your pipettor to blow the last drop of fluid out of your pipet? _______
Do the 2 test tubes appear to contain approximately the same amount of water? _______
How does the amount of pressure necessary to fill a 1 ml pipet compare with the amount needed to fill a 10 ml pipet?

Procedure 2

Sterile Pipetting with Pipettors

First Laboratory Session

(*c*) Why should you routinely loosen lids on tubes before exposing a sterile pipet to the laboratory air?

(*f*) Why not lay caps from sterile tubes on the bench top with the **open end up** while you work?

Why not lay the caps from sterile tubes on the bench top with the **open end down** while you work?

Second Laboratory Session

(*a*) Are there any signs of bacterial growth in your tube of broth labeled "0.5 ml sterile H_2O in TSB"? _______________

Procedure 3

Sterile Transferring to Broths with Inoculating Needles and Loops

First Laboratory Session

(*a*) Why do you place tubes to be used in sterile transfers near the Bunsen burner flame?

(*k*) You have transferred a needle tip of bacteria into the tube labeled "*Se. m.* 1" and then a loopful of liquid from "*Se. m.* 1" into the tube labeled "*Se. m.* 2." Do you expect to see growth in tube 1? _______
Do you expect to see growth in tube 2? _______
Do you expect to see as much growth in tube 2 as in tube 1? _______
Why is it necessary to cool the inoculating instrument thoroughly before obtaining bacteria from a solid surface?

Second Laboratory Session

(*b*) Compare the amount of growth in the tubes labeled "*Se. m.* 1" and "*Se. m.* 2." Is there considerably more growth in 1 tube than in the other? _______________
What factors other than the size of the initial inoculum might limit the amount of bacterial growth in a tube of Tryptic Soy Broth? Hint: What factors might limit the size of a human population living on a small island?

Procedure 4

Alcohol-Flaming Forceps for Sterile Transfer

First Laboratory Session

(*h*) Explain what happens if you tilt your forceps either up or down during alcohol-flaming.

Second Laboratory Session

(*a*) Was there growth in the tube labeled "TSB with sterile disk"? _______
What sources of contamination can you think of that might cause growth in the tube?

Exercise 4

Preparing a Wet Mount and Observing Microorganisms

Necessary Skills

Microscopy, Microscopic measurement

Materials

Cultures (1 per class):
- *Rhodospirillum rubrum* (1- to 2-week, Tryptic Soy Broth [TSB])
- Mixed culture of algae
- Mixed culture of protozoa
- Hay infusion (hay or grass aerated at least 1 week in pond or fish-tank water)

Supplies:
- Labeled Pasteur pipets with bulbs (at least 1 for each culture)
- Basins of disinfectant (for used depression slides, other microscopic slides, coverslips, and pipets)
- A few small vials of glucose (dextrose) per class
- Spatulas for glucose (optional)
- Toothpicks
- Petroleum jelly
- At least 4 microscope slides per group
- At least 10 coverslips per group
- 1 Depression slide per group
- Protoslo® (Methylcellulose, 1.5% aq.)
- Lens paper
- Wax marking pencil
- Calibrated ocular micrometer disks (or ocular micrometer disks and stage micrometer)
- Immersion oil

Stained microscope slides of the following bacteria:
- *Treponema pallidum*
- *Clostridium*
- *Corynebacterium*
- *Neisseria*
- *Staphylococcus*
- *Micrococcus*

Primary Objective

Learn to prepare wet mounts and hanging-drops, and recognize some of the most common microscopic life-forms including typical bacteria.

Other Objectives

1. Be able to focus the microscope on living procaryotic and eucaryotic single-celled and multicellular organisms.
2. State two altered characteristics of stained microorganisms.
3. Observe and differentiate Brownian movement and true motility.
4. Distinguish between procaryotic and eucaryotic cells.
5. Observe, identify, and draw the common shapes and arrangements of procaryotic cells.
6. Describe positive and negative chemotaxis, phototaxis, and aerotaxis.
7. Define: photosynthetic, saprophytic, nitrogen fixation, hay infusion, biodetritus.
8. Evaluate and compare the advantages of hanging-drop and wet mount preparations.
9. Observe and describe the distinguishing characteristics of algae, protozoa, and multicellular invertebrates.

INTRODUCTION

In this exercise you prepare wet mounts and hanging-drops, and you learn to recognize some of the **free-living** microorganisms that surround us. Here you observe tiny, **photosynthetic** (able to make cellular molecules using light energy and CO_2) and **chemosynthetic** (able to construct cellular molecules using preformed organic molecules for carbon and energy) microbes. **Saprophytic** (living on dead, decaying organic material) microorganisms, not **parasitic** (living at the expense of other creatures) ones, are the main concern of this exercise.

Since many of these organisms consist of a single cell that has approximately the same refractive index as water, they are nearly transparent and are difficult to distinguish from the background.

The microscopist chooses between two general types of preparations, **stained** or **unstained.** Stains are usually applied to dead cells. Living, unstained microorganisms may be observed in either a **wet mount** or a **hanging-drop** preparation.

Observing Microorganisms

The compound light microscope is generally used with stained smears. A stained cell is easily observed, but the heat fixation associated with most staining procedures shrinks, and often kills, microbes.

For the best estimates of size and true motility (a function for which cells must be alive), a microbiologist examines living microorganisms.

To prepare a **wet mount,** place a drop of liquid containing **specimens** (organisms you observe), onto a glass microscope slide. Lower a glass or plastic coverslip onto the drop.

The **hanging-drop,** as the name implies, is a bit more complex. In it the specimen hangs suspended from a coverslip into the indentation of a glass depression slide.

Microbial Motion

In this exercise you observe **Brownian movement,** a random bouncing that results from microorganisms colliding with molecules that are in constant motion in the liquid environment. You learn to distinguish this endless jiggling from **true motility,** which is directed movement.

Not all microorganisms are motile, but those that "swim" appear to proceed in one direction for a considerable distance. Occasionally, they may also spin or roll. Motile organisms display positive or negative **chemotaxis, phototaxis,** or **aerotaxis** (the ability to move toward or away from chemicals, light, or oxygen, respectively).

Challenges

Two limitations challenge you as you examine unstained bacteria. First, focusing in oil on a wet mount preparation moves the coverslip and the fluid containing your specimen. The organisms you want to observe flow away. For this reason, you are limited to a maximum of 400× magnification.

Your second challenge is that most of the microorganisms are transparent. They seem to disappear under full illumination. To overcome this obstacle, use the iris diaphragm lever of your microscope to reduce light until your "brightfield" is dull gray. Microbes appear shaded in contrast to this background. You need to readjust the light intensity repeatedly to observe each specimen accurately.

In this exercise you observe many microorganisms. It is impossible to identify all of the microbes in one laboratory session or in one lifetime. For now, concentrate on learning to prepare and observe both a wet mount and a hanging-drop. Increase your powers of observation by drawing accurate, detailed sketches of the microorganisms in your samples.

Basic characteristics of both procaryotic and eucaryotic microorganisms follow. These brief descriptions provide a few hints to guide your observations; see figure 4.1.

Procaryotic Cells

Procaryotic cells, bacteria, are tiny and lack a membrane-bound nucleus. Throughout the world, the most widely accepted general reference work on procaryotic classification is ***Bergey's Manual of Systematic Bacteriology.***

Cyanobacteria

Members of the *Bergey's Manual* section entitled Cyanobacteria have simple nutritional requirements. Using CO_2 as a carbon source, they grow profusely in extreme environments such as acid hot springs. They are also abundant in soil, as well as in fresh and salt water. Many **fix** (incorporate into cellular material) nitrogen from air. Of all the known organisms on earth, only cyanobacteria and a few other bacteria fix gaseous nitrogen. Many cyanobacteria release gaseous oxygen.

The cells of most cyanobacterial genera are either oval or **filamentous** (threadlike). They are often encased in a sheath or gelatinous mass. **Cyanobacteria** are generally green but lack **chloroplasts** (membrane-bound organelles containing photosynthetic pigments). Many cyanobacteria display **gliding motility.** Watch for it.

Common, Heterotrophic Bacteria

Along with representative cyanobacteria, you also examine **heterotrophic** (requiring organic molecules for a carbon source) bacteria. Most heterotrophic bacteria are barely visible under 400× magnification. Use the oil-immersion lens.

Observing the shape and arrangement of cells helps microbiologists identify bacteria. As illustrated in figure 4.2, **cocci** (spherical or oval bacteria) may divide on one, two, or three planes. The divisions produce **singles, diplococci** (pairs), **chains, tetrads** of four, or **packets** of eight cells. If the divisions are at random angles, **grapelike clusters** result. In this exercise you examine stained smears of ***Staphylococcus, Micrococcus,*** and ***Neisseria*** for evidence of these arrangements.

Rods (line or bar-shaped bacteria) are also arranged in singles, pairs, and chains. Some rods, especially the **diphtheroids** (those morphologically resembling ***Corynebacterium diphtheriae,*** which causes diphtheria) form **palisades,** a "picket fence," parallel arrangement of cells. **X's Y's,** and **pleomorphic** (many-shaped) cells are also typical of diphtheroids; see figure 4.2. Watch for these arrangements of rods when you observe ***Corynebacterium.***

Figure 4.1 Some common microscopic organisms with suggested magnifications

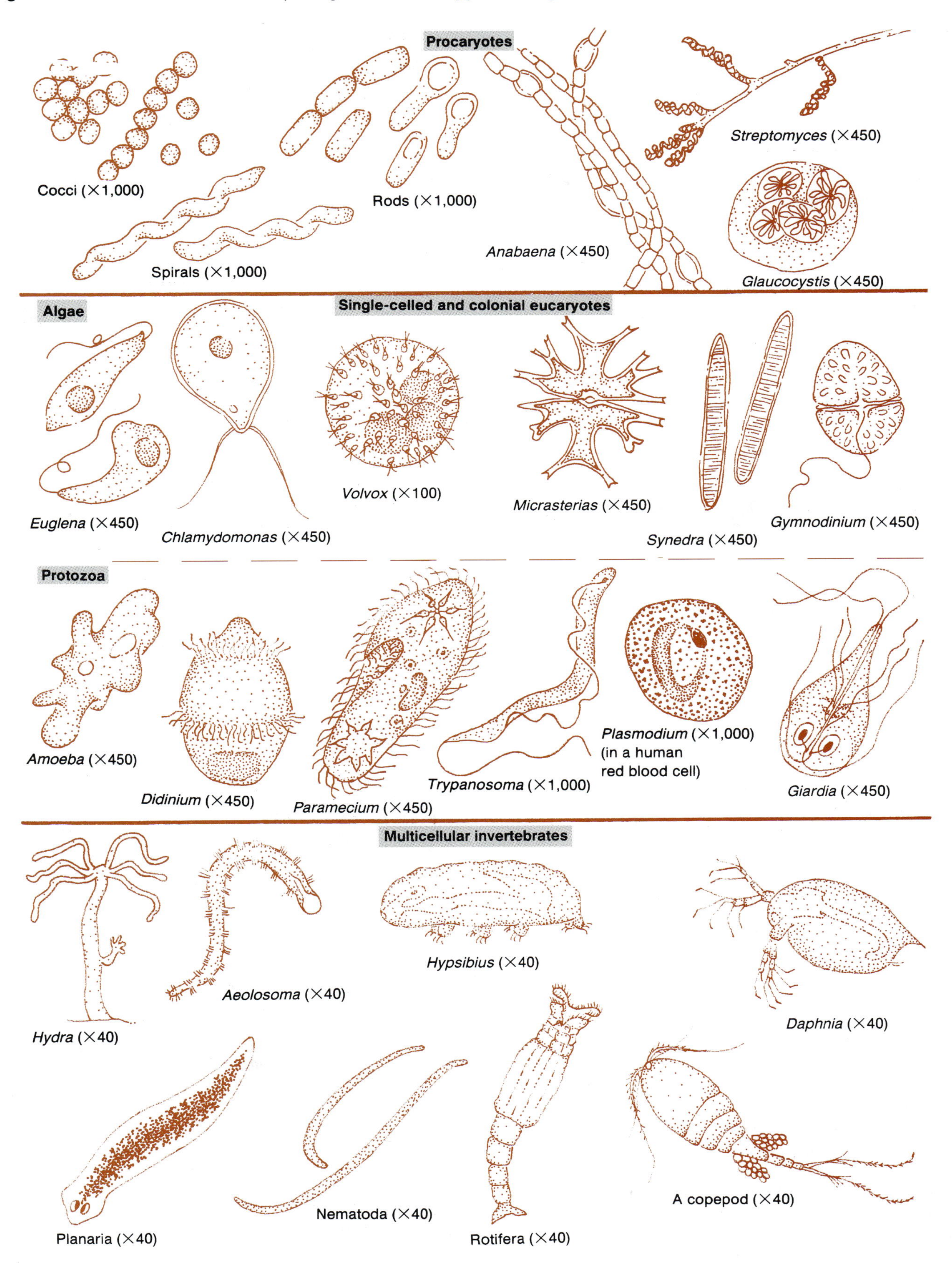

Figure 4.2 Common arrangements of bacterial rods and cocci

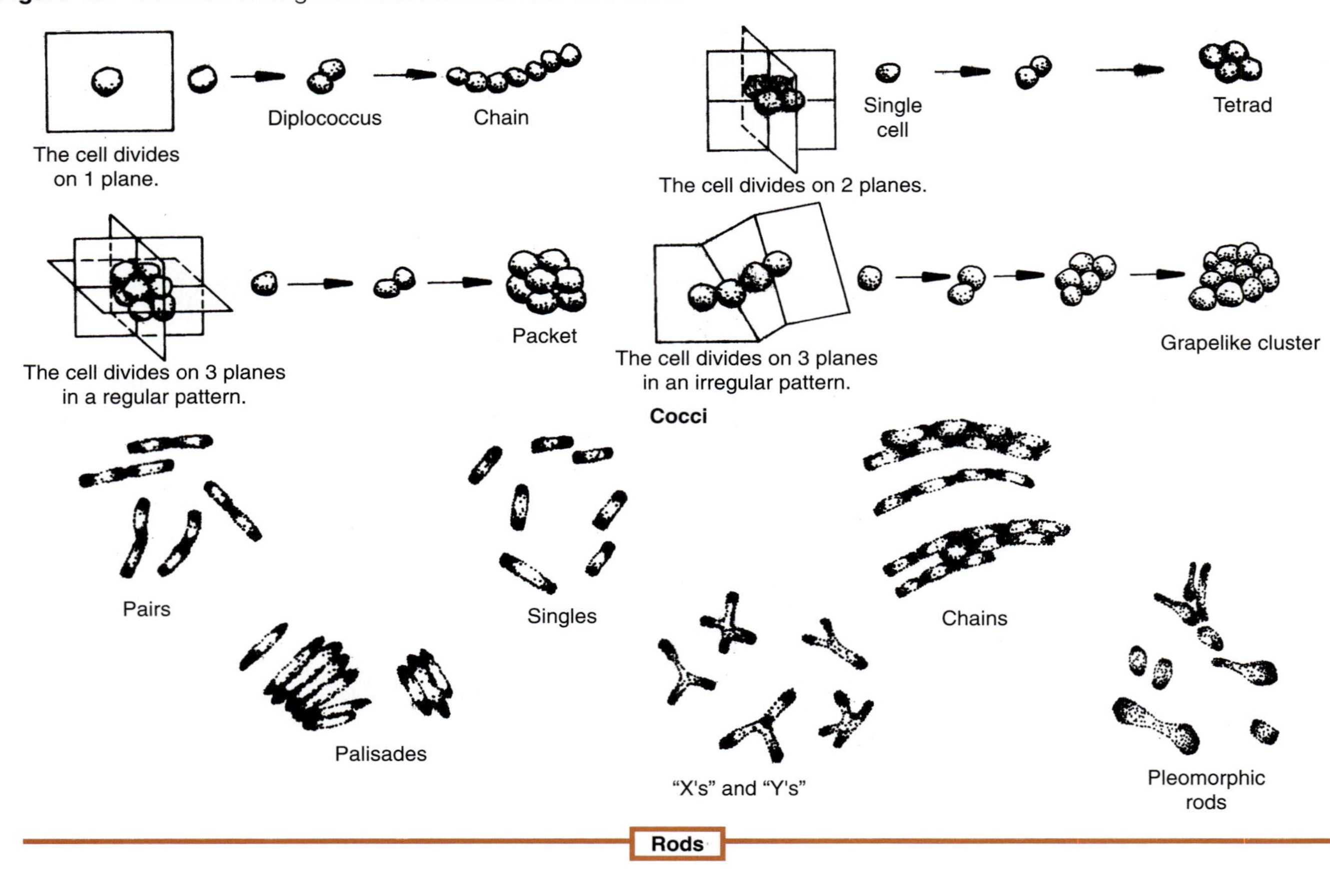

Figure 4.3 Endospore-formers

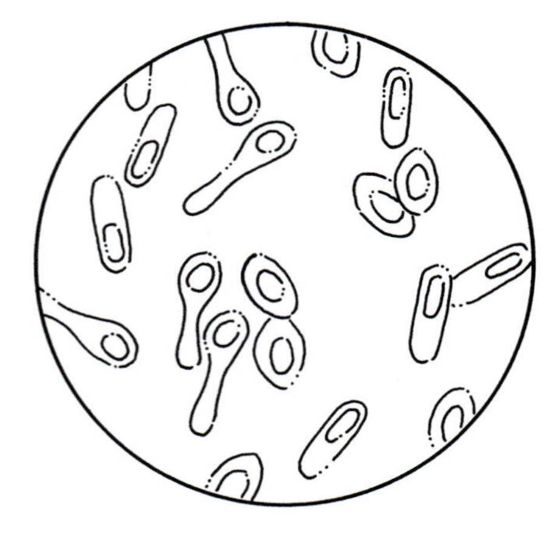

Figure 4.4 Spirilla and spirochetes

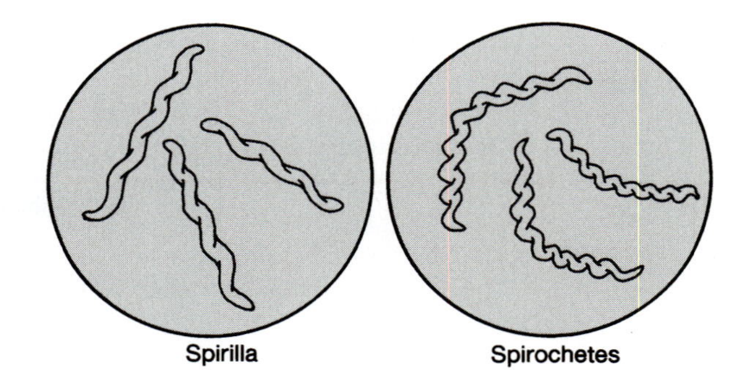

Some bacteria form **endospores,** the most highly resistant life-forms known to exist; see figure 4.3. Endospores resist staining. Through the oil-immersion lens, they are visible as **refractile** (shiny) circles or ovals. In this exercise you observe endospores among cells of ***Clostridium.***

Some bacterial genera have **flagella.** These threadlike organelles propel a cell through liquid. Bacterial flagella are too thin to be seen through a brightfield, light microscope without special stains.

Two forms of spiral-shaped bacteria are common—flexible **spirochetes** and rigid **spirilla;** see figure 4.4. In living material this difference is readily apparent. You observe living spirilla **(*Rhodospirillum rubrum*)** in this exercise. You also examine prepared slides of a spirochete, ***Treponema pallidum,*** the causative agent of syphilis.

A clinical microbiologist observes bacteria daily, generally deciding if they are rods, cocci, spirilla, or spirochetes. In this exercise you observe only a small sampling of the many life-forms that are explored in later exercises of this laboratory manual. Learning to recognize common forms now simplifies your task later in this course.

Eucaryotic Cells

Eucaryotic cells are almost always larger than procaryotic cells and usually contain a microscopically visible, membrane-bound nucleus.

Protozoa are unicellular, generally motile, eucaryotes. They are normally unpigmented. However, the protozoan you are examining may have just devoured a meal of algae, making color determination difficult. To identify protozoans, compare their shapes to the illustrations in figure 4.1 and see Exercise 6, Fungi, Protozoa, and Microscopic Animalia.

Algae, also eucaryotes, range in size from single cells to giant sea kelp. Single-celled algae are often motile. Algae contain green, brown, red, and/or golden pigments within chloroplasts. See Exercise 5, Algae.

Multicellular invertebrates, animals such as worms, rotifers, and copepods, contain many tiny, eucaryotic cells. To identify these creatures, watch for well-defined, multicellular organs such as an eyespot or a digestive tube. Observe external structures, which may include segmentation and, sometimes, a head. See Exercise 6, Fungi, Protozoa, and Microscopic Animalia.

Learn to distinguish **fibers, crystals,** and **biodetritus** (fragments of disintegrating, decomposing, once-living material) from living cells. Recognize that most life-forms have specific, characteristic shapes, some of which are shown in figure 4.1. **Bubbles** present a challenge since they often resemble **amoeboid** cells (see figure 4.1) in general shape. You can recognize bubbles by their heavy black outlines and lack of any internal structures.

Life establishes order. The energy of life dispels **entropy,** which is chaos. Thus, the object you are trying to identify is probably not alive if it is jaggedly irregular. The majority of living, microscopic organisms you are likely to encounter resemble the drawings in figure 4.1 or in other figures in this laboratory manual.

The purpose of this exercise is to familiarize you with the use of the wet mount and hanging-drop techniques. For the best use of laboratory time, concentrate your efforts on easily identifiable microbes. Leave any difficult determinations for later in the course.

PROCEDURES EXERCISE 4

Procedure 1

Preparing and Observing Wet Mounts of Algae and Protozoa

a Take a clean microscope slide. As in the illustration, label the center "A" and "P" for algae and protozoa.

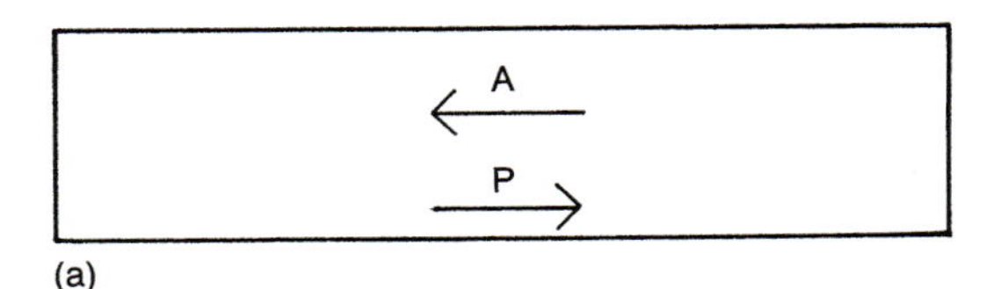

(a)

b Use a Pasteur pipet with a bulb to deliver 1 or 2 drops of **algal culture** to the left side of the slide. The greatest variety of algae is usually in the bottom or top layers of liquid, not in the middle.

c Hold a coverslip at a 45° angle to your slide. Lower the coverglass gently onto the drop. This angling helps prevent bubbles.

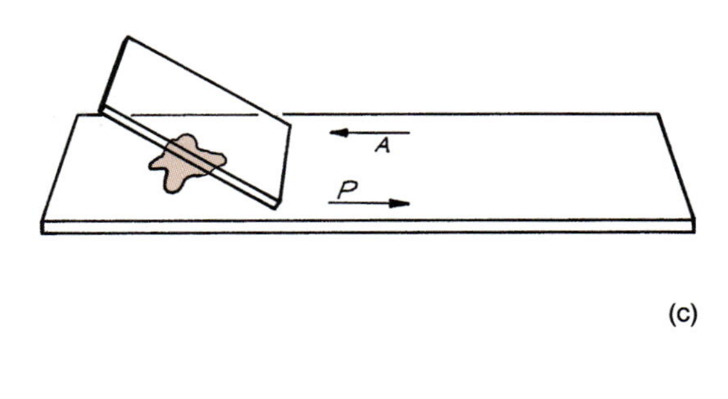

(c)

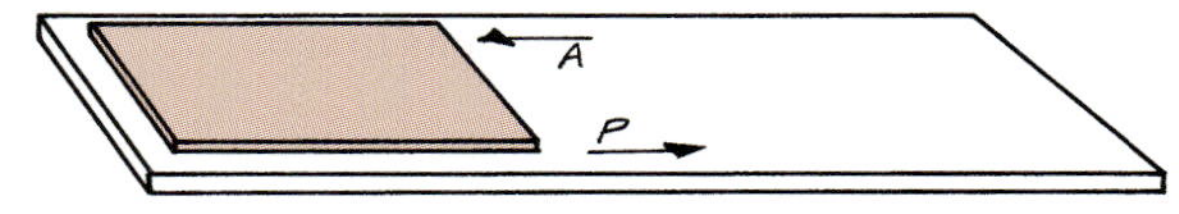

d Deliver 1 or 2 drops of **protozoan culture** to the right side of your slide. Generally, the best protozoan samples are taken from the bottom of the container. Angle a coverslip onto the drop.

e Set up your assigned microscope. Clean its glass surfaces with lens paper.

f Begin searching the waters with your scanning or low-power objective after using your iris diaphragm lever to reduce the light.

g Some microscopic life-forms are best examined under low power, and they may appear in either water sample, so before increasing magnification, search each specimen carefully for microscopic invertebrates such as worms, *Daphnia* (water fleas), copepods, and other "large" microorganisms. Enter drawings and descriptions of your observations in the laboratory report. Include as much detail as possible in each drawing. Note the magnification and the source of the sample for each entry.

h Increase magnification and adjust light intensity. Keep the background of the microscopic field dull gray. Watch for single-celled organisms that resemble the pictures in figure 4.1.

i Compare the size and shape of typical bacterial cells to the size and complexity of the algal and protozoan cells. Note that there are probably many algal cells in the protozoan water, and vice versa. You are sure to find bacteria in both samples. Observe and draw protozoans and bacteria feeding on biodetritus.

j Do you see evidence of Brownian movement? Describe your observations in your laboratory report.

k If a slide begins to dry, place a drop of water from the same culture next to the coverslip. The fresh sample runs under the coverslip by capillary action.

l Do you see motile organisms moving to the edge of the coverslip? Organisms that require oxygen are often found in high numbers displaying positive aerotaxis around the rim of a wet mount preparation. **Anaerobes,** organisms that are poisoned by oxygen, may display negative aerotaxis by clustering under the center of the coverslip. Describe your observations in the laboratory report.

m In your algal sample, look for green, golden, or brown algae with chloroplasts. Also try to observe the very small and/or thin, green, procaryotic cyanobacteria. Note any gliding motility. Enter detailed drawings and descriptions of your observations in the laboratory report.

n With a spatula or toothpick, drop a few crystals of glucose into the liquid at one edge of the coverslip on the mixed protozoa sample. For the next few seconds, observe positive and negative chemotaxis among the organisms as they react to the sugar. Note your observations in the laboratory report. After completing this procedure, wipe your microscope lenses thoroughly with lens tissue.

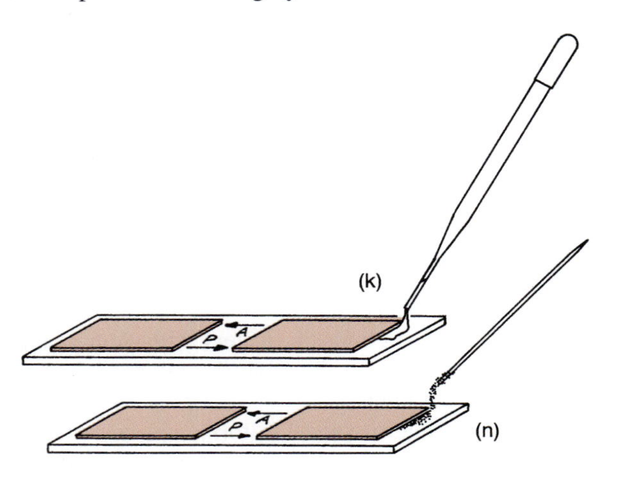

Procedure 2

Preparing and Observing a Hanging-Drop of a Hay Infusion

a To prepare your hanging-drop, obtain a depression slide, a coverslip, a wax marking pencil, and a toothpick with a small amount of petroleum jelly on the tip. The object of making a hanging-drop preparation is to suspend a fluid sample within an airtight seal of petroleum jelly, so that the sample does not dry before you have time to examine it thoroughly.

b With your wax marking pencil, draw a short line on 1 flat side of the coverslip. Later, finding this line helps you focus the microscope on the sample.

Your instructor may demonstrate the following method or another acceptable procedure for preparing a hanging-drop.

c Use a toothpick to spread a very thin film of petroleum jelly on the outside rim of the palm of your nondominant hand.

Read (*d*) and study the illustration before continuing.

d With the clean forefinger and thumb of your dominant hand, grasp the center of a coverslip. Gently scrape jelly from your palm onto each of the 4 edges of the coverslip. The jelly forms a small "wall," a ridge of sealant on each edge. The sealant is on the same side of the coverslip as the short line you drew. If petroleum jelly gets onto the bottom or center of your coverslip, discard the slip and try again. Now, perform step (*d*).

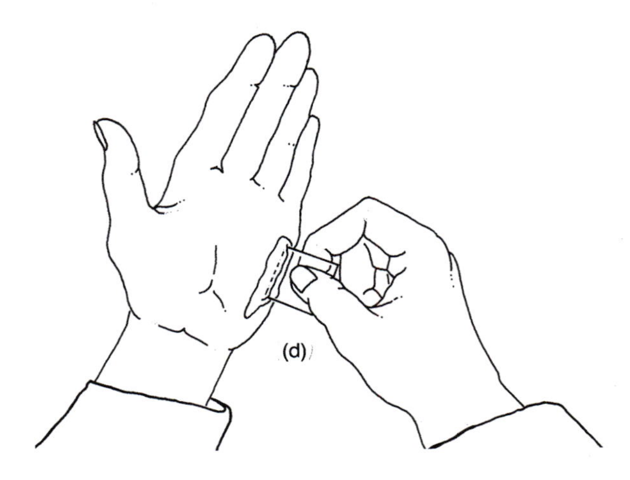

e With a Pasteur pipet, obtain a few drops of fluid from the bottom of the **hay infusion.** Deliver them to the center of the "walled" coverslip. The drops should be surrounded by glass, then jelly. To avoid bubbles, keep the sample away from the jelly.

f Lower a depression slide over the sample on the coverslip, so that the drop goes into the depression and the petroleum jelly forms a complete seal between the slide and the coverslip.

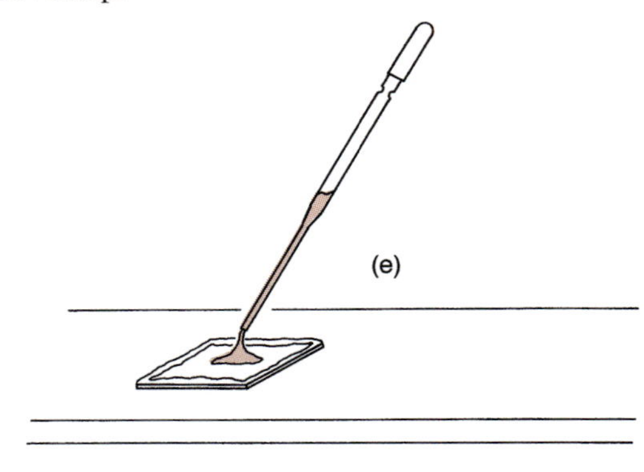

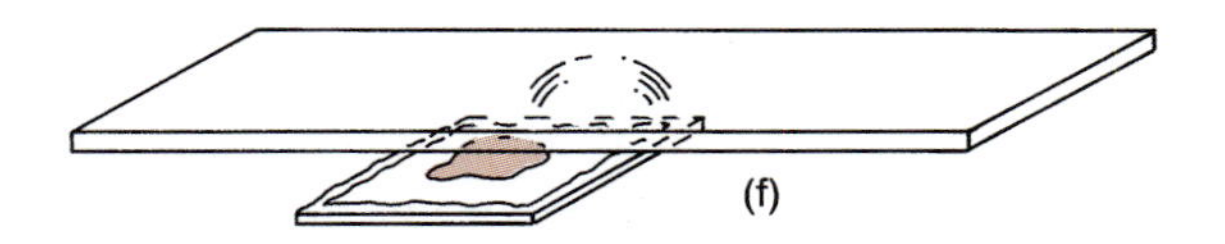

g Gently turn the depression slide over.

h Examine your hanging-drop with reduced illumination using the lowest power of your microscope. Focus first on the short line you drew. The line is in the same plane as the microorganisms. Watch for multicellular invertebrates, as well as large algal and protozoan cells. Enter your observations in the laboratory report.

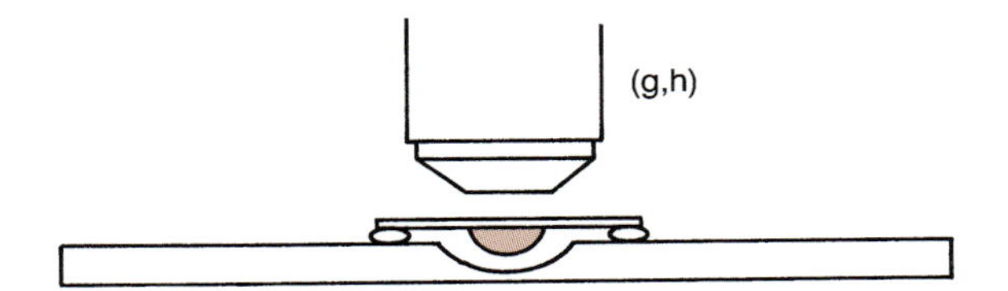

i Make detailed observations with each of the higher powers, readjusting the illumination and recording your findings in the laboratory report.

j After completing this procedure, discard your depression slide and its coverslip into disinfectant.

k Clean your microscope. Wipe the lenses thoroughly with lens tissue.

Procedure 3

Preparing and Observing Wet Mounts of *Rhodospirillum rubrum*

a Prepare a wet mount of *Rhodospirillum rubrum.* Collect liquid from the bottom of the container. If any of the fluid touches your hands, wash immediately.

b Thoroughly examine the sample. In your laboratory report, describe the true motility exhibited by *Rhodospirillum rubrum.*

c After successfully completing steps (*a*) and (*b*), discard the slide and coverslip as directed by **your instructor.** Wipe your microscope lenses thoroughly with lens tissue.

Procedure 4

Examining Stained Bacteria

a With your low-power and high-power lenses, locate a stained area on a prepared microscope slide of *Treponema pallidum.* Use your oil-immersion lens to examine the smear of the spirochete.

When alive, these very thin bacteria are best observed by darkfield microscopy. The cells are approximately 0.13–0.15 μm wide × 10–13 μm long. Thus the width of unstained treponemes is below the resolution capabilities of a compound, light, brightfield microscope. With brightfield microscopy, *Treponema pallidum* cells are invisible unless stained by special techniques that thicken the cells.

b With your calibrated ocular micrometer disk, measure the width and length of several stained *Treponema pallidum* cells. In your laboratory report, note how your figures compare with the average width and length of living *Treponema pallidum* cells.

c Note how tightly the spirochetes are coiled. Do the cells appear to be flexed or rigid? Record your observations in your laboratory report.

d Observe *Clostridium* and *Corynebacterium* cells using the oil-immersion lens of your microscope. Compare the shapes and arrangements of these rods. In which culture do you see endospores? Which culture demonstrates palisades, X's, Y's, and pleomorphism? Draw and describe your observations in the laboratory report.

e Observe the circular cocci of *Staphylococcus* and *Micrococcus.* Now observe the kidney-bean shape of *Neisseria.* Do you see the flattened adjacent sides of the diplococci? Are capsules present? Record your observations in the laboratory report.

f Which genus displays tetrads; which genus is characterized by large, grapelike clusters of cells? After completing this procedure, wipe your microscope lenses thoroughly with lens tissue. Clean your microscope and store it as directed by **your instructor.**

POSTTEST EXERCISE 4

Part 1

Matching

Each answer may be used once, more than once, or not at all. Some questions may require more than one answer.

a. Algae
b. Protozoa
c. Bacteria (including cyanobacteria)
d. Multicellular invertebrates

1. _____ Single-celled organisms are found in which group(s)?
2. _____ List the procaryotic cells.
3. _____ Which group(s) contain(s) endospores?
4. _____ If an organism lacks a nuclear membrane but is green and produces oxygen, it is probably a member of which group?
5. _____ Select organisms with a eucaryotic cell or cells.

Part 2

True or False (Circle one, then correct every false statement.)

1. T F A photosynthetic organism is able to make cellular energy out of light energy.
2. T F A saprophytic organism usually parasitizes living tissue, causing disease and death.
3. T F A chemosynthetic organism is able to construct cellular molecules out of carbon dioxide and oxygen.
4. T F Bacteria moving toward oxygen are displaying negative phototaxis.

5. T F To focus on unstained, transparent cells, reduce illumination of the field far below the levels used with stained cells.
6. T F Many cyanobacteria fix gaseous nitrogen.
7. T F Heterotrophic organisms require organic molecules for a carbon source.
8. T F Pairs of cocci are called packets.
9. T F Some rods are called diphtheroids because they resemble *Corynebacterium diphtheriae.*
10. T F Endospores are resting or survival structures.
11. T F Unstained *Treponema pallidum* is too thin to be observed with a compound, light, brightfield microscope.
12. T F Spirochetes are rigid, inflexible, spiral-shaped bacteria; spirilla are flexible, spiral-shaped bacteria.
13. T F Worms are multicellular invertebrates.
14. T F Rotifers are protozoa.
15. T F The energy of life produces order and dispels entropy.
16. T F *Daphnia* is an alga.

Part 3

Completion and Short Answer

1. Explain three advantages of examining unstained specimens as opposed to stained preparations.

   a.

   b.

   c.

2. Distinguish between Brownian movement and true motility.

3. What is biodetritus?

Laboratory Report **4**

Name ______________________________

Section ______________

Preparing a Wet Mount and Observing Microorganisms

Procedure 1

Preparing and Observing Wet Mounts of Algae and Protozoa

(*g*) Draw and describe some of the relatively large organisms you observed in the algal and protozoan samples using your scanning and/or low-power lenses. Note the magnification and source of the sample for each entry. Attempt preliminary identification of the organisms.

Total Magnification ________

Source ________

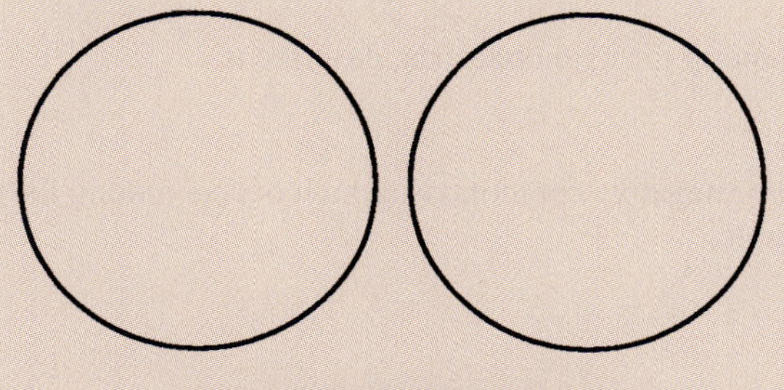

Algal and Protozoan Samples

Total Magnification ________

Source ________

(*i*) Find a field where bacteria and protozoa are feeding on biodetritus. Draw the field. Note magnification used and the source of your sample. Describe the comparative size and complexity of the organisms.

Total Magnification ________

Source ________

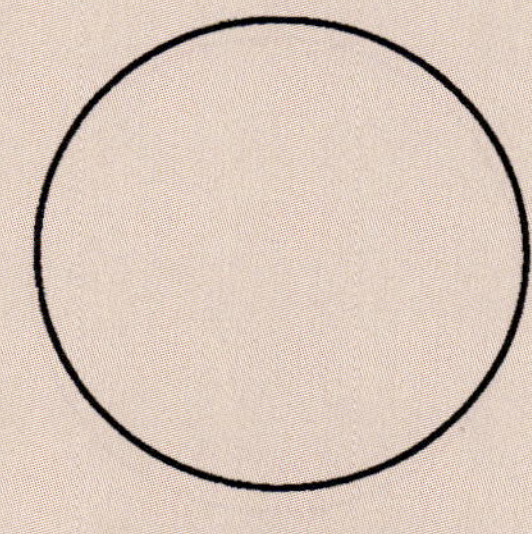

Feeding Behavior

(*j*) Describe Brownian movement.

(*l*) What evidence do you see of aerotaxis?

(*m*) Draw, describe, and try to identify various types of algae. Remember to note the magnification used and the source of your sample.

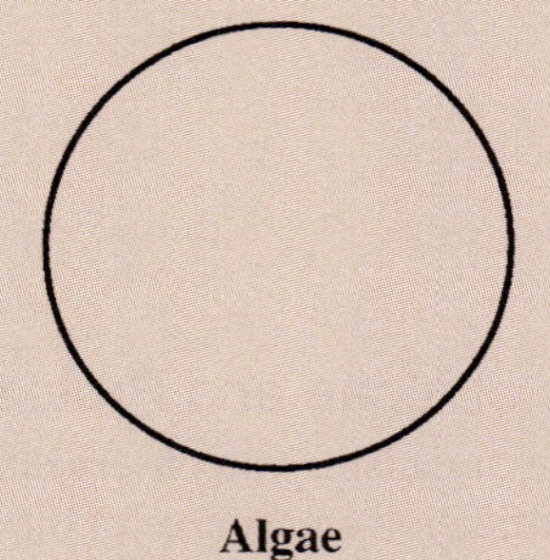

Total Magnification ________

Source ________

Algae

(*m*) If you observed gliding motility among the cyanobacteria, describe it.

(*n*) Describe evidence of positive and/or negative chemotaxis, which occurs among the organisms when you place glucose on the edge of the mixed protozoa sample.

Procedure 2

Preparing and Observing a Hanging-Drop of a Hay Infusion

(*h,i*) Examine your hanging-drop of the hay infusion. Draw and label some of the organisms you observe with each power of your microscope.

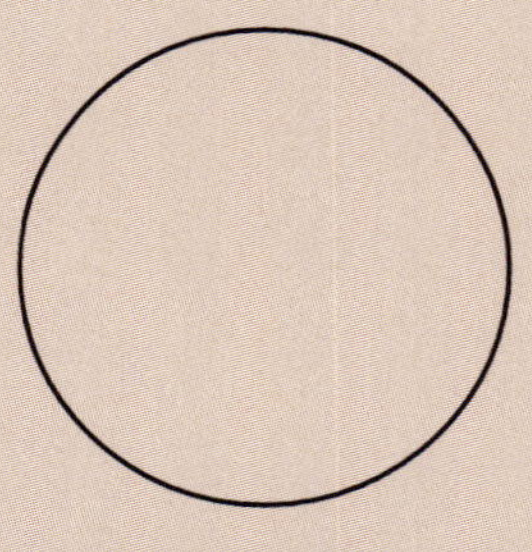

Low Power

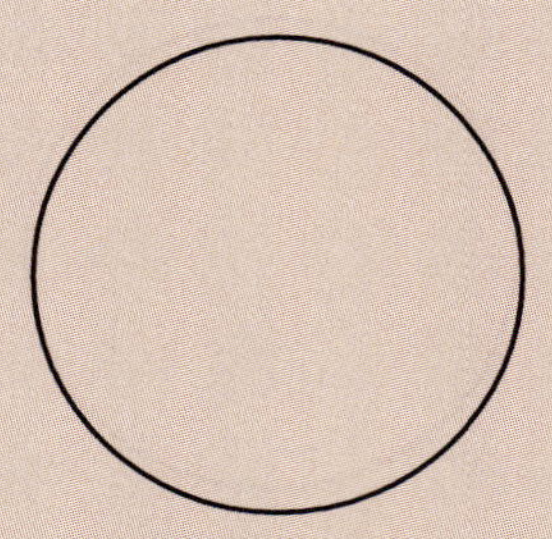

High Power

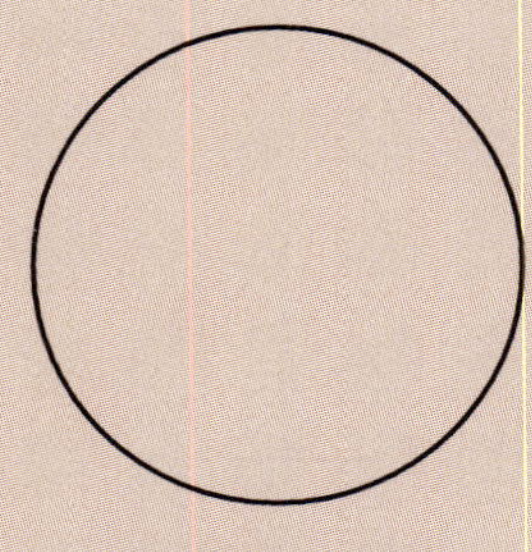

Oil Immersion

Hay Infusion

Procedure 3

Preparing and Observing Wet Mounts of *Rhodospirillum rubrum*

(*b*) Draw *Rhodospirillum rubrum* and describe its motility. Note magnification employed.

Total
Magnification ________

Description of motility:

Rhodospirillum rubrum

Procedure 4

Examining Stained Bacteria

(*a,b*) Draw a typical oil-immersion lens field of *Treponema pallidum.* Measure the width and length of 3 treponemes, then calculate the average width and length of stained treponemes.

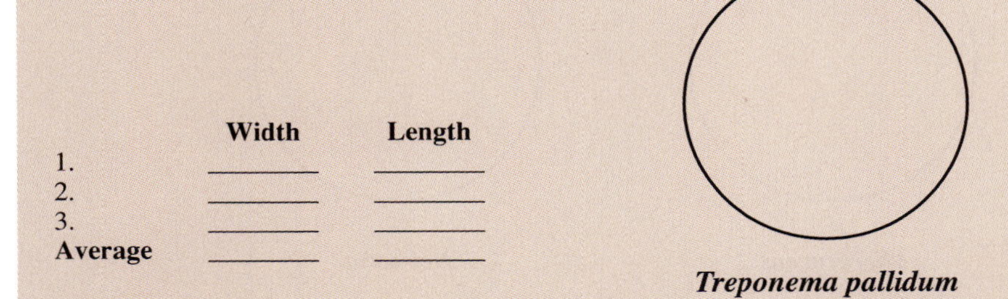

	Width	Length
1.	________	________
2.	________	________
3.	________	________
Average	________	________

Treponema pallidum

(*b*) Unstained *Treponema pallidum* cells usually are 0.13–0.15 μm wide × 10–13 μm long. How do your stained cells compare with these averages?

(*c*) Unstained, living *Treponema pallidum* cells are tightly coiled and move with graceful, flexing motility. Describe the evidence of flexibility that remains in the stained preparation.

(*d*) Draw typical oil-immersion lens fields of *Clostridium* and *Corynebacterium.* Label various shapes and arrangements of these cells. Label endospores, palisades, X's, Y's, club-shaped cells, and coccoid forms if you observe them.

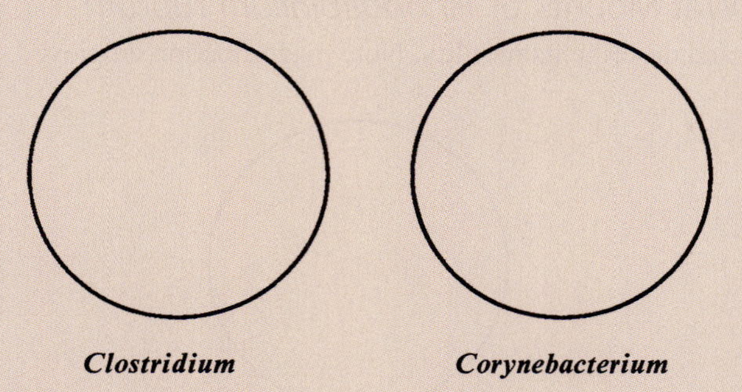

Which genus has the most pleomorphic cells? _______________
(*e,f*) Carefully note the kidney-bean shape of *Neisseria.* Do you see the flattened adjacent sides of the diplococci? _________

A clear, oval, unstained space surrounding pairs of bacteria is evidence of capsule formation. Are you able to see capsules surrounding any of your diplococci? _______________
Draw typical oil-immersion lens fields of *Staphylococcus, Micrococcus,* and *Neisseria.* Label diplococci, tetrads, and grapelike clusters if you observe them.

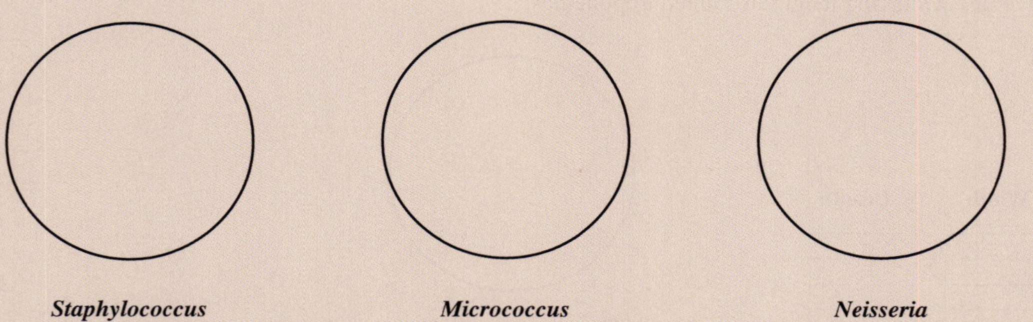

Exercise 5

Algae

Necessary Skills

Microscopy, Preparing a wet mount

Materials

Cultures (per class):
- *Euglena*
- Mixed algae or pond water

Supplies:
- 1 Labeled dropper with bulb per class culture
- 1 Pair forceps for mixed algae sample
- At least 2 slides per group
- At least 3 coverslips per group
- Methylcellulose, 1.5% aq. (Protoslo®) in dropper bottle
- Toothpicks
- Reference books on algal identification

Primary Objective

Begin to understand the ecological and economic significance of algae.

Other Objectives

1. List the features of plants that algae lack.
2. Define: algae, producers, intertidal, niche, neritic, phytoplankton.
3. Discuss the habitats of algae.
4. Analyze the connection between cultural eutrophication and algal blooms.
5. Identify products that are derived from algae.
6. Begin identification of algal genera.

INTRODUCTION

The term **algae** refers to eucaryotic single-celled and multicellular organisms that live in water and are **photosynthetic.** Photosynthesis is a process that employs light energy as a power source for combining carbon dioxide and water to help form organic molecules.

In this exercise you examine and begin to identify genera of **marine** (of the ocean or sea) and/or freshwater algae. To help you, figure 5.1 contains illustrations of many common algae.

Algal Structures

Even a cursory examination of figure 5.1 shows that algae are extremely diverse. They share few characteristics beyond the **lack** of structures that would enable them to survive prolonged dehydration and support an erect structure. These are features of land plants.

Even the giant kelps, largest of all algae and some of the longest organisms on earth, have only a rudimentary vascular system. Buoyant in water, they lack supporting tissue. The simple algal body is called a **thallus;** algae are **thallophytes** and the scientists who study them are **phycologists** (algologists).

Algal Ecology

Nearly 70% of the earth's surface is covered by water. In this aqueous environment, algae are the major **producers** (organisms that use photosynthesis to produce organic molecules). These "grasses of the oceans" are indispensable in a tremendous food web, supporting other marine life and producing oxygen.

Algae are found in aqueous habitats, from the perpetual ice of glaciers to 70° C hot springs. They are found in fresh water, sea water, and in environments as salty as the Great Salt Lake in Utah.

Marine algal habitats are illustrated in figure 5.2. Some algae inhabit the **intertidal** zones where they are alternately exposed to desiccation by low tides and then flooded by high tides. Others find a **niche** (functional relationship with the environment) in the **neritic** zone, the shallow, near-shore area below the intertidal zone. Many form **phytoplankton,** the free-floating photosynthetic organisms found wherever light penetrates above deep water.

Figure 5.1 The algae. (a) Representative flagellate algae

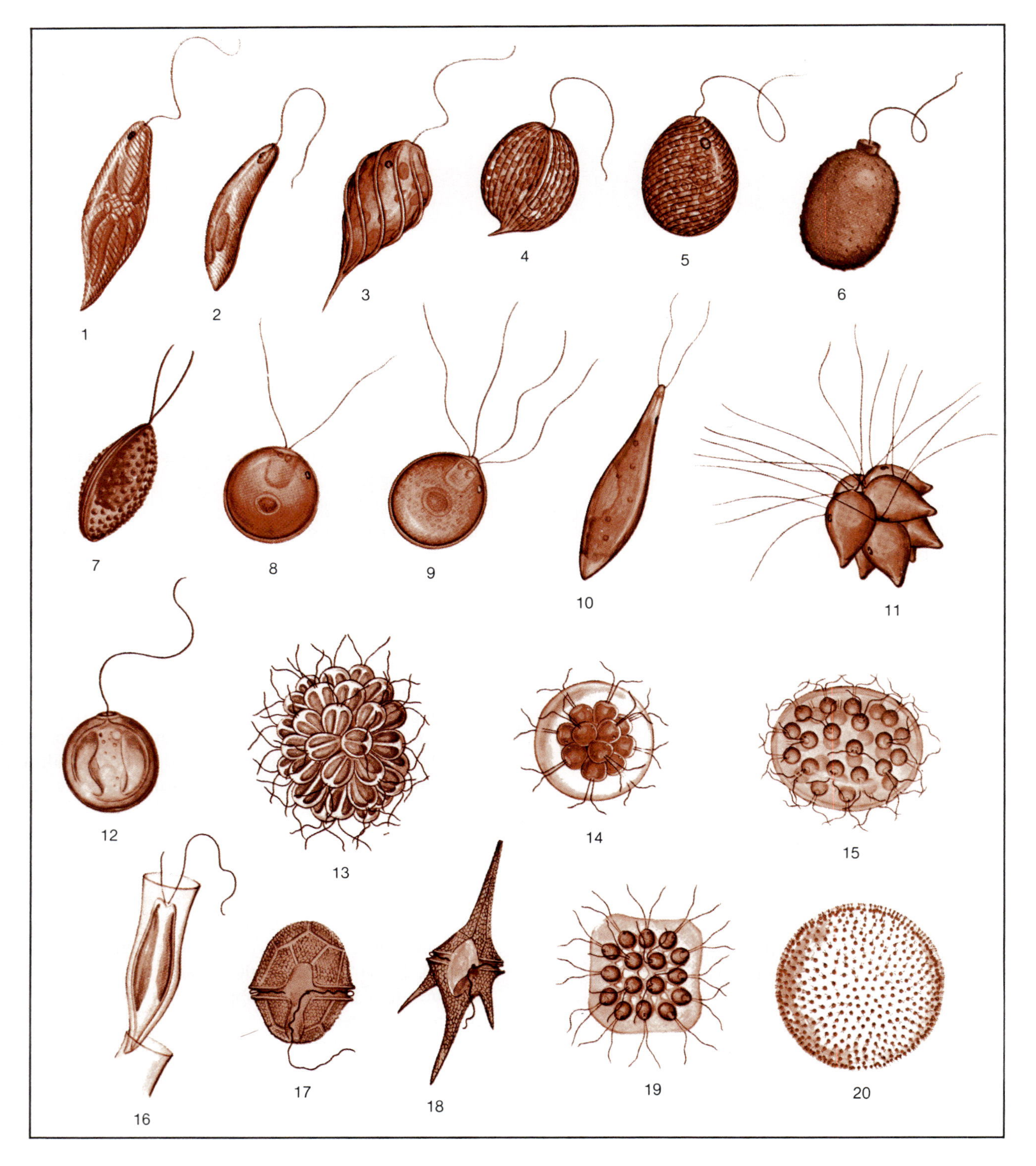

Courtesy of the U.S. Environmental Protection Agency, Office of Research & Development, Cincinnati, Ohio 45268.

1. *Euglena* (700×)
2. *Euglena* (700×)
3. *Phacus* (1,000×)
4. *Phacus* (350×)
5. *Lepocinclis* (350×)
6. *Trachelomonas* (1,000×)
7. *Phacotus* (1,500×)
8. *Chlamydomonas* (1,000×)
9. *Carteria* (1,500×)
10. *Chlorogonium* (1,000×)
11. *Pyrobotrys* (1,000×)
12. *Chrysococcus* (3,000×)
13. *Synura* (350×)
14. *Pandorina* (350×)
15. *Eudorina* (175×)
16. *Dinobyron* (1,000×)
17. *Peridinium* (350×)
18. *Ceratium* (175×)
19. *Gonium* (350×)
20. *Volvox* (100×)

Figure 5.1 Continued (b) Representative filamentous algae

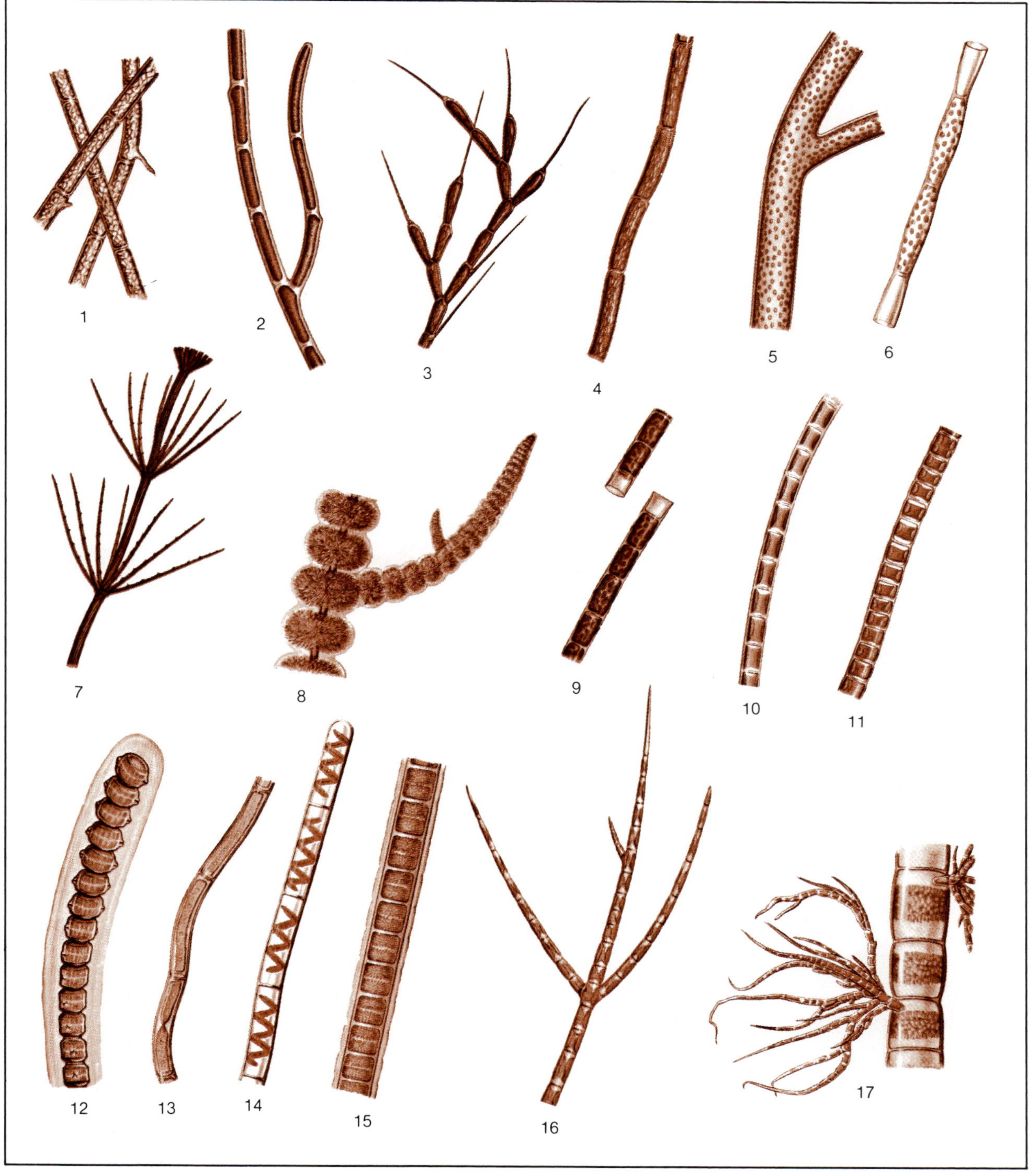

Courtesy of the U.S. Environmental Protection Agency, Office of Research & Development, Cincinnati, Ohio 45268.

1. *Rhizoclonium* (175×)
2. *Cladophora* (100×)
3. *Bulbochaete* (100×)
4. *Oedogonium* (350×)
5. *Vaucheria* (100×)
6. *Tribonema* (300×)
7. *Chara* (3)
8. *Batrachospermum* (2)
9. *Microspora* (175×)
10. *Ulothrix* (175×)
11. *Ulothrix* (175×)
12. *Desmidium* (175×)
13. *Mougeotia* (175×)
14. *Spirogyra* (175×)
15. *Zygnema* (175×)
16. *Stigeoclonium* (300×)
17. *Draparnaldia* (100×)

Figure 5.1 Continued (c) Representative nonflagellated and nonfilamentous algae

Courtesy of the U.S. Environmental Protection Agency, Office of Research & Development, Cincinnati, Ohio 45268.

1. *Chlorococcum* (700×)
2. *Oocystis* (700×)
3. *Coelastrum* (350×)
4. *Chlorella* (350×)
5. *Sphaerocystis* (350×)
6. *Micractinium* (700×)
7. *Scendesmus* (700×)
8. *Actinastrum* (700×)
9. *Phytoconis* (700×)
10. *Ankistrodesmus* (700×)
11. *Pamella* (700×)
12. *Botryococcus* (700×)
13. *Tetraedron* (1,000×)
14. *Pediastrum* (100×)
15. *Tetraspora* (100×)
16. *Staurastrum* (700×)
17. *Staurastrum* (350×)
18. *Closterium* (175×)
19. *Euastrum* (350×)
20. *Micrasterias* (175×)

Figure 5.1 Continued (d) Representative diatoms

Courtesy of the U.S. Environmental Protection Agency, Office of Research & Development, Cincinnati, Ohio 45268.

1. *Diatoma* (1,000×)
2. *Gomphonema* (175×)
3. *Cymbella* (175×)
4. *Cymbella* (1,000×)
5. *Gomphonema* (2,000×)
6. *Cocconeis* (750×)
7. *Nitschia* (1,500×)
8. *Pinnularia* (175×)
9. *Cyclotella* (1,000×)
10. *Tabellaria* (175×)
11. *Tabellaria* (1,000×)
12. *Synedra* (350×)
13. *Synedra* (175×)
14. *Melosira* (750×)
15. *Surirella* (350×)
16. *Stauroneis* (350×)
17. *Fragillaria* (350×)
18. *Fragillaria* (750×)
19. *Asterionella* (175×)
20. *Asterionella* (750×)
21. *Navicula* (750×)
22. *Stephanodiscus* (750×)
23. *Meridion* (750×)

Figure 5.2 Diagrammatic section of marine environment

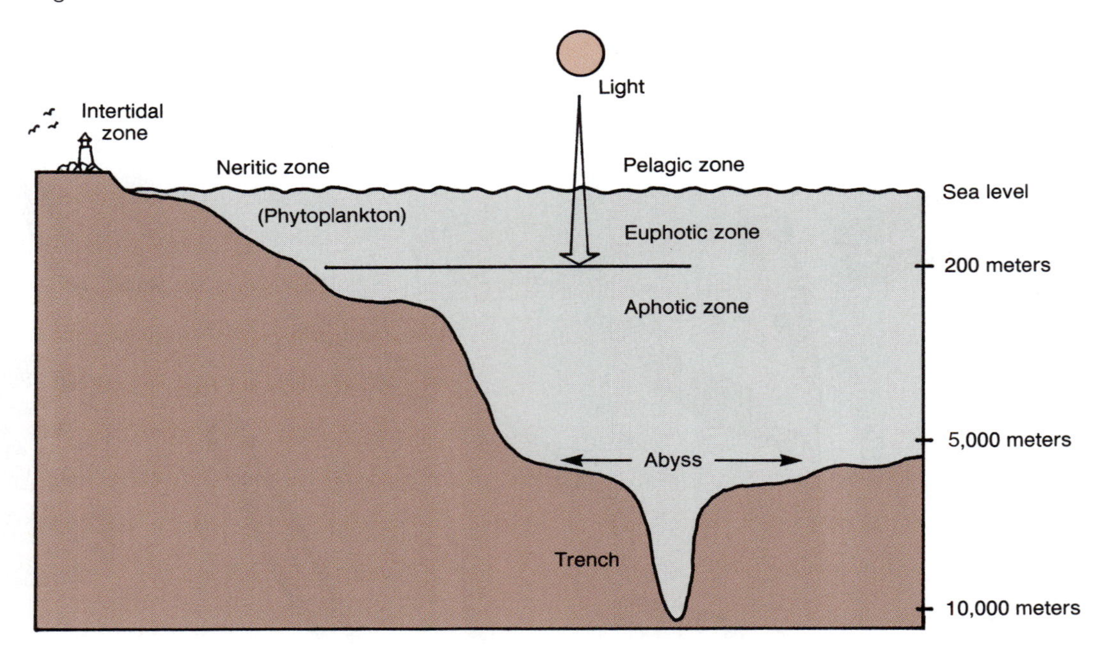

The growth of algae is usually limited by **light, nutrients,** and **temperature.** When increased amounts of these limiting factors are available, tremendous **blooms** (growths so dense they color the water) of algae sometimes occur.

The activities of human beings often alter the three limiting factors. When a coastline is used as a sewer for a large, crowded human population, silt is deposited, making the neritic zone more shallow. The fouled water is warmed by the sun and receives nutrients from human waste. This process is called **cultural eutrophication.** It greatly accelerates **natural eutrophication** (the gradual pollution of water by geological and biological processes of silt formation and encroaching vegetation).

In response to the long days of summer heat and light, **red tides,** algal blooms, can develop along coastlines where cultural eutrophication occurs. Generally, red tides are caused by rapid proliferation of the dinoflagellates ***Gymnodinium*** or ***Gonyaulax.***

Both of these genera produce water-soluble **toxins** (poisons) that are structurally similar to curare, the paralyzing substance used by some South American Indians to poison arrow tips. Red tides cause massive fish and shellfish kills. This affected marine life then poisons other animals that ingest it.

Algal Products

While some algae produce toxins, others form a significant part of the diet and economic wealth of people throughout the world. Algae are especially valued in Asia where many genera serve as foods, condiments, or medicines. In agriculture seaweed fertilizers compare favorably with manure. In industry fossil diatoms are employed as filters and as thickeners.

Agar (also called agar-agar) is a complex carbohydrate produced by red algae, primarily ***Gelidium*** and ***Gracilaria.*** Although agar is used to produce edible gels, it is indigestible and acts as a laxative.

The solidifying properties of agar were explained in the 1800s by Frau Fanny Eilshemius Hesse to her husband. He then introduced the substance to his colleague, the famous medical microbiology pioneer Robert Koch. Agar is now utilized to solidify media in microbiology laboratories all over the world.

PROCEDURES EXERCISE 5

Procedure 1

Observing Motility in *Euglena*

a Prepare a wet mount from the class culture of *Euglena.* Examine the algae with the low and medium powers of your microscope.

b If all the algae are moving very rapidly, use methylcellulose to slow them. To do this, add a fresh drop of *Euglena* culture to a drop of methylcellulose on a microscope slide. Mix the 2 liquids with a toothpick before adding a coverslip.

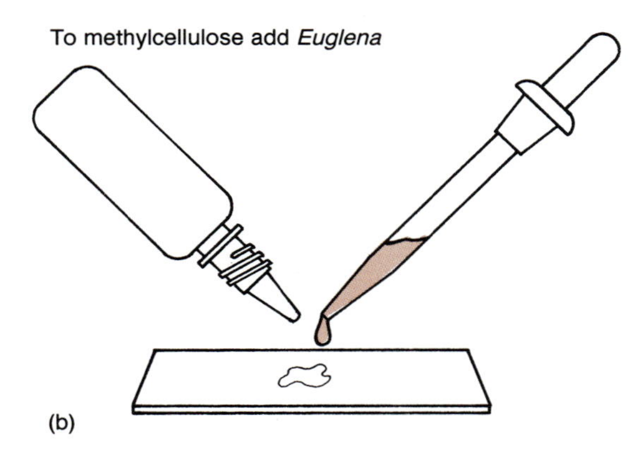

c Observe 3 types of motility: (i) forward motion, (ii) spinning, and (iii) "euglenoid motion," a lengthening and shortening of *Euglena's* body. Locate the pigmented **stigma** (eyespot) near the swollen structure at the base of the flagellum. It is present in many but not all euglenoids.

d Draw and describe your observations in your laboratory report. Note whether the flagellum leads or follows the algal cell. Remember that the flagellum is rotating in and out of your shallow depth of focus. Careful and continual adjustment of your fine-focus knob is required if you are to follow flagellar motion.

e Discard or recycle your slide and coverslip as directed by **your instructor.**

Procedure 2

Identifying Algae

a Prepare a wet mount of the mixed algae or pond water culture. Use forceps to obtain filamentous forms. Use figure 5.1 and supplementary texts to help you identify various genera of microscopic algae.

b In your laboratory report, draw and identify a number of the algae you observe.

c After completing your work, clean your microscope thoroughly. Store your microscope as directed by **your instructor.**

d Discard or recycle slides and coverslips as directed by **your instructor.**

POSTTEST EXERCISE 5

Part 1

True or False (Circle one, then correct every false statement.)

1. T F The term *algae* refers to single-celled and multicellular marine organisms only.
2. T F Most algae have structures that enable them to survive prolonged desiccation.
3. T F Organisms that use photosynthesis to form organic molecules are called producers.
4. T F Cultural eutrophication is the intelligent use of the oceans by concerned citizens.
5. T F An oceanic algal bloom of *Gymnodinium* or *Gonyaulax* may form a red tide.

Part 2

Completion and Short Answer

1. The ________________ zone of the ocean is washed by tides; the _______________ , shallow zone is near shore. Algae floating freely in the ocean make up the ________________ .
2. Algal ________________ are growths so dense that they color the water.
3. Agar, isolated from the ________________ , primarily *Gelidium* and *Gracilaria,* is a solidifying agent.

Laboratory Report **5**

Name ______________________________

Section ____________

Algae

Procedure 1

Observing Motility in *Euglena*

(*b*) If you utilized methylcellulose, describe its effect on the algae.

(*c*) Observe and describe 3 types of motility demonstrated by *Euglena.*

(*d*) From your observations, draw a large, typical specimen of *Euglena,* showing as much internal and external detail as possible. Label the stigma if it is present in your specimen. Does the flagellum lead or follow the algal cell? ____________

Total
Magnification ________

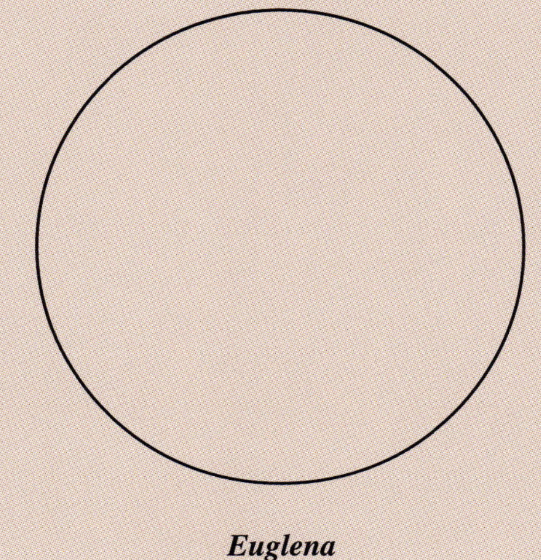

Euglena

Procedure 2

Identifying Algae

(*b*) Observe, draw, and identify as many different genera of algae as time permits. Next to each drawing, note the source of the algal cells (from which class culture were they obtained) and the magnification you employed.

Exercise **6**

Fungi, Protozoa, and Microscopic Animalia

Necessary Skills

Microscopy, Microscopic measurement, Aseptic transfer techniques, Preparing a wet mount and hanging-drop

Materials

Cultures (room temperature or 30° C, Sabouraud Dextrose Agar, 1 per set):
- *Saccharomyces cerevisiae* (48-hour slant)
- Hay infusion (hay or grass aerated 1 week in pond or fish-tank water)
- Pond water samples with floating vegetation
- Mixed protozoa culture

Supplies:
- 1 Labeled Pasteur pipet and bulb per sample
- Several dropper bottles of methylcellulose, 1.5% aq. (Protoslo®)
- Toothpicks
- Several pairs forceps
- Several scalpels
- Container of disinfectant for used depression slides and their coverslips
- Petroleum jelly
- Lens paper
- Ocular micrometer disks
- Stage micrometers, if necessary
- Lactophenol cotton blue stain
- At least 5 microscope slides per group
- At least 7 coverslips per group
- At least 1 depression slide per group

Prepared, stained microscope slides:
- *Rhizopus*
- *Penicillium*
- *Entamoeba histolytica* cysts in stool
- *Giardia lamblia* in stool
- *Plasmodium* species in blood

Prepared specimens:
- *Ascaris lumbricoides* nematodes
- *Clonorchis sinensis* (or *Opisthorchis* species) flukes

Additional supplies:
- Wall charts and/or reference books on fungi, protozoans, and microscopic animals

Primary Objective

Recognize the main fungal, protozoan, and microscopic animalian groups.

Other Objectives

1. Give examples of the organisms studied by mycologists.
2. Describe the characteristics that fungi share.
3. Summarize the general characteristics of protozoa.
4. Define: phagocytosis, pinocytosis.
5. Describe the two life-forms in which many parasitic protozoa are able to exist. Draw and label two pathogenic genera showing the dual life-forms of each.
6. List the four traditional taxonomic groups of protozoa and indicate the mode of motility associated with each group.
7. Observe and draw free-living protozoans.
8. Observe, measure, draw, and label examples of *Entamoeba histolytica, Giardia lamblia,* and *Plasmodium.*
9. Describe the usual habitats of microscopic metazoans.
10. Outline a taxonomic guide to the microscopic metazoans.
11. Improve your ability to observe microscopic detail in living microscopic animals and to draw what you see.
12. Observe and describe preserved specimens of *Ascaris* and *Clonorchis.*

INTRODUCTION

This exercise introduces you to **mycology** (the study of **fungi**) by showing you yeasts and molds. You are introduced to **protozoology** by locating free-living protozoa in

Figure 6.1 Five classes of fungi

Class	Characteristic Sexual Reproduction Structure	Plant or Animal Pathogenic Genus	Notes
Oomycetes (water fungi)	Oospores	*Phytophthera* (potato blight)	Often not included in fungal classification schemes
Zygomycetes	Zygospores	*Rhizopus*	Naked sexual spores
Ascomycetes	Ascospores	*Aspergillus*	Sexual spores contained in an ascus (sac)
Basidiomycetes	Basidiospores	*Amanita* (deadly mushroom)	Sexual spores borne on a basidium (base)
Deuteromycetes (Fungi Imperfecti)	No sexual spores identified	*Candida* (candidiasis) *Epidermophyton* (ringworm)	Most often, these are classified into class Ascomycetes when the sexual stage is discovered

pond water, and **parasitology** by examining prepared microscope slides of parasitic protozoa in human blood and feces. Then you become familiar with the basics of **invertebrate zoology** by observing microscopic animals.

Fungi

A **mycologist** is a scientist who studies fungi. Mycologists examine **macroscopic** (visible to the naked eye) forms such as the mushrooms, bracket fungi, puffballs, stinkhorns, and truffles. They also investigate **microscopic** fungi including the **yeasts** (single-celled fungi) and the filamentous (threadlike) **molds.**

What characteristics are shared by the diverse life forms calls "fungi"? Their eucaryotic cells have walls that are usually composed of **chitin,** a polysaccharide. Since they are *not* **photosynthetic** (able to transform light energy into cellular energy), the fungi are either **saprophytic,** living on dead, decaying plant and animal matter, or they are **parasites,** invading living organisms.

Fungi are classified by their sexual spores; see figure 6.1. In nature these structures are usually not as visible as **hyphae,** threadlike, branching tubes, and **fruiting bodies,** structures that generally bear **asexual** spores.

Therefore, when determining the genus and species of a fungus, a mycologist routinely examines the asexual macroscopic and microscopic characteristics of the fungal **thallus.** Thallus is another word for body or colony. A thallus is also called a **mycelium** if it contains hyphae rather than yeast cells.

Figure 6.2 illustrates the microscopic appearance of some frequently encountered filamentous fungi as well as a few of the pathogens. It will be helpful for you to examine this illustration carefully before observing stained slides of molds.

Figure 6.2 The microscopic appearance of some common filamentous fungi—see legend for figure 6.2 on page 58.

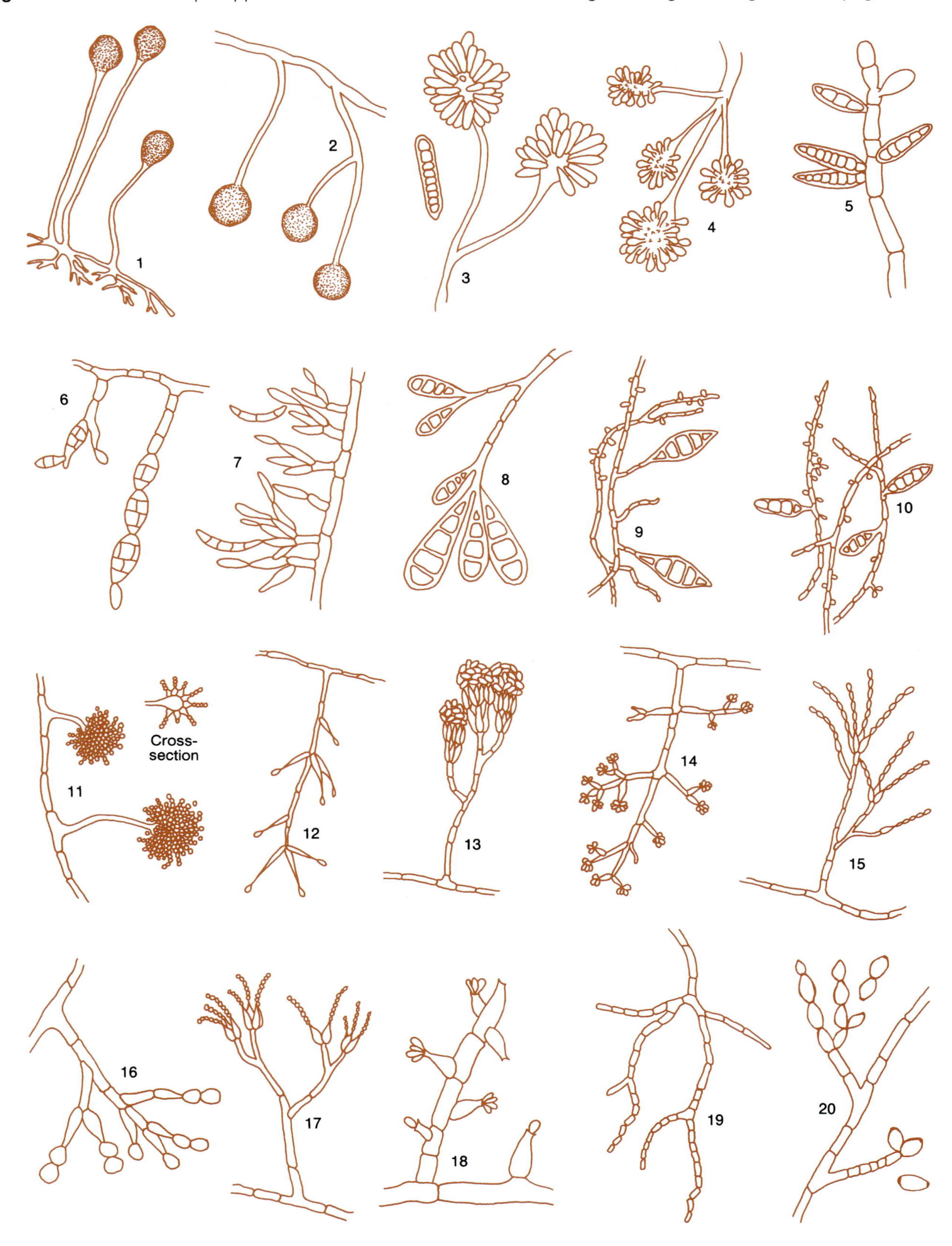

Figure 6.2 Continued Descriptions of a typical thallus surface and human pathogenic potential follow each genus.

1. ***Rhizopus*** Dense, gray (reverse colorless), rapid coenocytic (no walls between cells) growth. Black fruiting bodies are visible. Mucormycosis.
2. ***Mucor*** Dense, tan to gray, cottony, rapid coenocytic growth. Visible fruiting bodies. Mucormycosis.
3. ***Syncephalastrum*** Gray, cottony or woolly, rapid coenocytic growth.*
4. ***Cunninghamella*** White, rapid coenocytic growth.*
5. ***Helminthosporium*** Surface gray or grayish-pink. Reverse black.*
6. ***Alternaria*** Woolly, white, gray, brown, or dark green. Black or brown reverse. Allergies.
7. ***Fusarium*** White to rose or lavender, cottony. Keratitis, onychomycosis.
8. ***Epidermophyton*** After two weeks, heaped and folded yellowish to golden tan. Reverse dark orange. Dermatophyte.
9. ***Trichophyton*** White and fluffy, or powdery and tan. Reverse becomes yellow to brown within a few weeks and forms red rings. Dermatophyte.
10. ***Microsporum*** Velvety buff to brown. Reverse may form dark, red-brown rings. Dermatophyte.
11. ***Aspergillus*** Varies greatly. White, tan, yellow, rose, dark brown, or green. Fruiting bodies visible. Reverse often different color than surface. Toxin producer. Allergies, aspergillosis.
12. ***Verticillium*** Velvety, white. Reverse colorless or slightly yellow.*
13. ***Gliocladium*** Rough, dark green or white to pink. Reverse colorless.*
14. ***Trichoderma*** White, becoming green at maturity. Woolly.*
15. ***Paecilomyces*** Brown to pink. Unlike *Penicillium,* which it may resemble microscopically, the thallus is rarely green.*
16. ***Scopulariopsis*** White, wrinkled thallus becomes powdery, tan to brown within two weeks. Never green. Onychomycosis.
17. ***Penicillium*** Powdery; varies, but usually bluish green.*
18. ***Phialophora*** Dark gray or brown. Reverse dark brown. Chromomycosis.
19. ***Geotrichum*** Gray to white, thin, cobweb appearance. Geotrichosis.
20. ***Cladosporium*** Varies, but usually dark green to brown, folded, velvety. Reverse black.*

*Rarely or never pathogenic, generally seen as a laboratory contaminant.

The legend that accompanies figure 6.2 describes the macroscopic appearance of the thallus for each fungus pictured. Note, however, that for some genera, the thallus changes from day to day as it matures. Occasionally, the front and reverse sides of the thallus are different colors.

Protozoa

Certain characteristics are shared by all organisms traditionally called protozoans. Each protozoan is a single, eucaryotic cell. Protozoa are wall-less, mostly motile, colorless **chemoheterotrophs** (cells that derive both energy and carbon from organic molecules). In other words, they resemble animals but are unicellular.

Protozoa feed by both **phagocytosis** (ingestion of particles) and **pinocytosis** (ingestion of dissolved food, liquids). Most parasitic and many free-living protozoa have two life-forms. Under favorable conditions, the protozoan exists as a motile, feeding **trophozoite.** When exposed to more adverse conditions, the organism forms a resistant **cyst,** as shown in figure 6.3.

Figure 6.3 Life cycle of *Giardia,* a representative protozoan

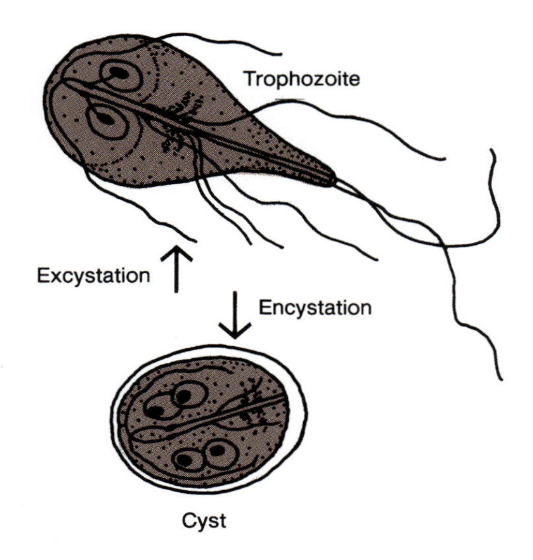

Figure 6.4 Amoeboid motion—motility by means of pseudopodia

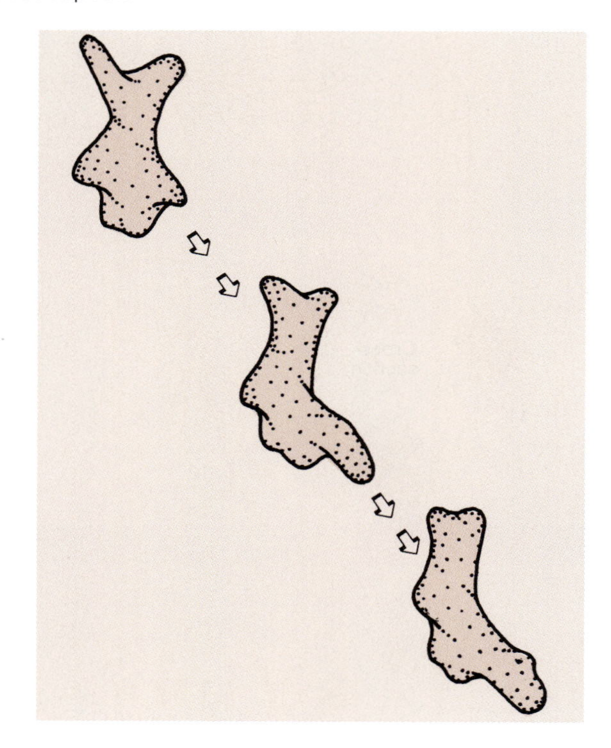

Traditionally, the protozoa are divided into four taxonomic groups primarily on the basis of motility. The groups include **Sarcodina** (amoeboid motion; see figure 6.4), **Mastigophora** (flagella), **Ciliata** (cilia), and **Sporozoa** (mature stages nonmotile); see table 6.1 and figure 6.5.

Figure 6.5 Protozoa

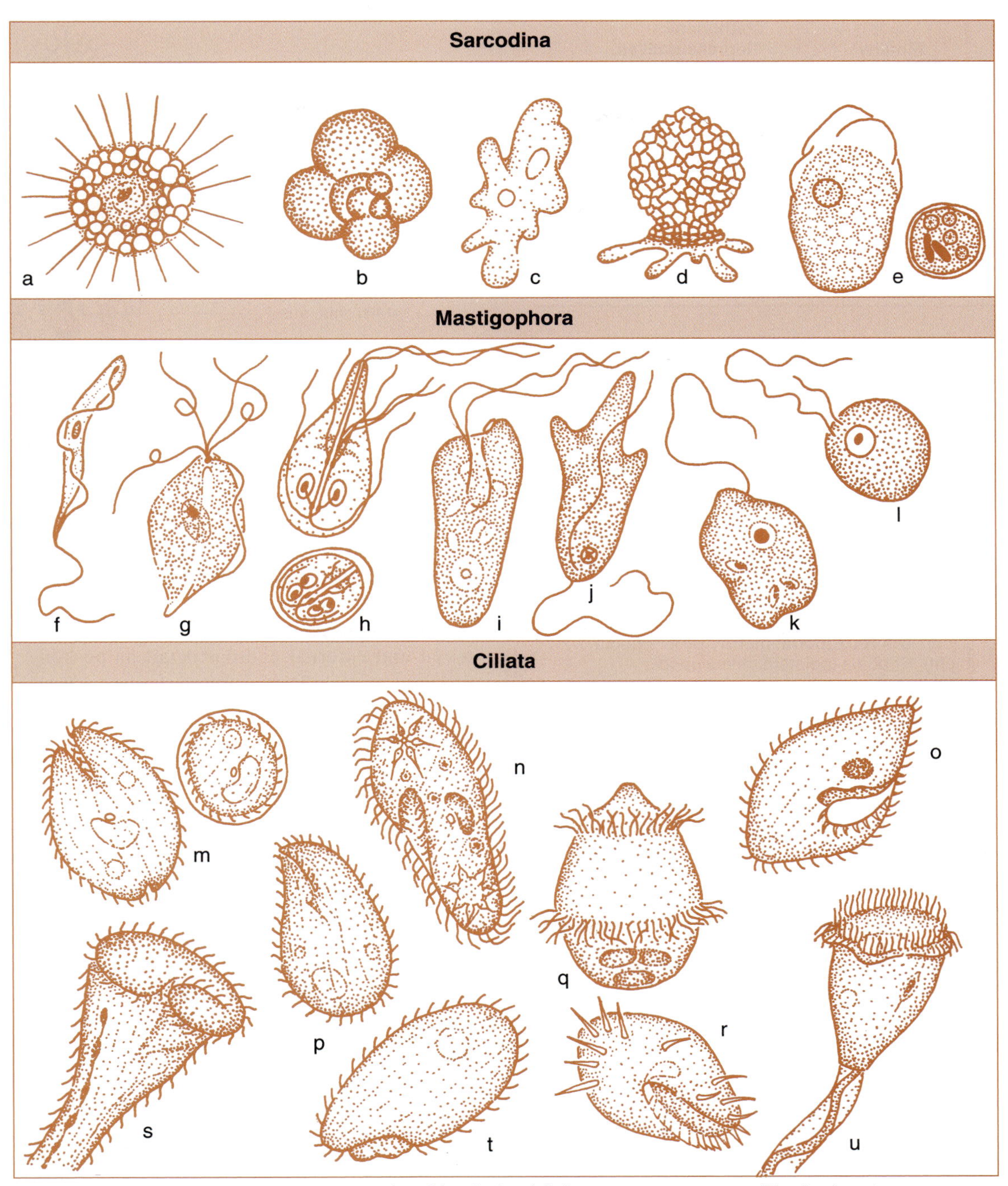

a. Radiolarian
b. Foraminiferan
c. *Amoeba proteus*
d. *Difflugia*
e. *Entamoeba histolytica**
f. *Trypanosoma brucei**
g. *Trichomonas vaginalis**
h. *Giardia lamblia**
i. *Chilomonas*
j. *Cercomonas*
k. *Oikomonas*
l. *Monas*
m. *Balantidium coli**
n. *Paramecium*
o. *Blepharisma*
p. *Chilodonella*
q. *Didinium*
r. *Euplotes*
s. *Stentor*
t. *Tetrahymena*
u. *Vorticella*

*This protozoan species causes disease in humans

Table 6.1 Traditional Protozoan Groups

Group	Motility	Pathogenic Representatives
Sarcodina	Amoeboid	*Entamoeba histolytica*
Mastigophora	Flagella	*Trypanosoma brucei* varieties *gambiense* and *rhodesiense* *Trichomonas vaginalis* *Giardia lamblia*
Ciliata	Cilia	*Balantidium coli*
Sporozoa	None in mature stages	*Plasmodium,* species that cause malaria include *falciparum, vivax, malariae,* and *ovale*

Table 6.2 Current Taxonomic System for Protozoa

I. Phylum **Sarcomastigophora:** flagella, pseudopodia, or both
  A. Subphylum **Mastigophora:** flagella in adult stage
    1. Class **Phytomastigophorea:** plantlike flagellates; examples, *Euglena, Volvox*
    2. Class **Zoomastigophorea:** flagellates lacking chromoplasts; examples, *Trichomonas, Trypanosoma*
  B. Subphylum **Opalinata:** cytostomeless body with rows of ciliumlike organelles; parasitic
  C. Subphylum **Sarcodina:** pseudopodia; flagella in developmental stages of some
    1. Superclass **Rhizopoda:** motility by lobopodia (blunt), filopodia (pointed), reticulopodia (filamentous), or cytoplasmic flow; examples, *Amoeba, Entamoeba*
    2. Superclass **Actinopoda:** mostly planktonic; pseudopodia in form of axopodia (permanent, needlelike, with an axial rod)

II. Phylum **Labyrinthomorpha:** mostly marine; small group

III. Phylum **Apicomplexa:** apical complex (a set of organelles at the anterior end) in some stage; cilia and flagella absent from adult stages; examples, *Plasmodium, Toxoplasma*

IV. Phylum **Myxozoa:** parasites of fishes and invertebrates

V. Phylum **Microspora:** parasites of invertebrates and lower vertebrates

VI. Phylum **Ascetospora:** parasites of invertebrates and a few vertebrates

VII. Phylum **Ciliophora:** cilia; 2 types of nuclei; heterotrophic; contractile vacuole; examples, *Paramecium, Balantidium*

Modified from Hickman, Jr., Roberts, Hickman, *Biology of Animals,* 6th ed., 1993, Mosby, Times Mirror Co.

Currently, protozoologists, using ever-growing evidence from electron microscopic studies, are able to combine the Sarcodina and Mastigophora, while dividing the Sporozoa into four new groups; see table 6.2. The new groupings include many flagellated algae as well as the slime molds.

In this exercise you observe free-living Sarcodina, Mastigophora, and Ciliata. These three traditional groups are presented in figure 6.5. Figure 6.6 shows you some of the structures found inside the common ciliate ***Paramecium.***

Since all the Sporozoa are parasitic, you do not see them moving about freely in pond water samples. For example, see the life cycle of the sporozoan ***Plasmodium*** in figure 6.7. This genus causes **malaria,** one of humanity's leading causes of death.

Figure 6.6 *Paramecium*

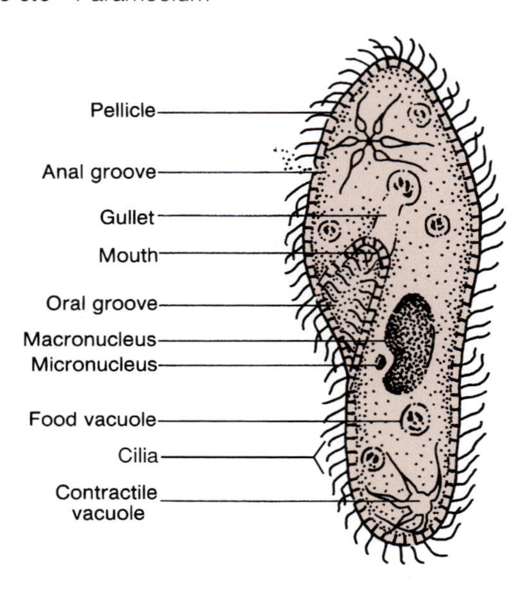

To observe parasitic protozoa, you examine slides of ***Entamoeba histolytica, Giardia lamblia,*** and ***Plasmodium*** species. Note which are in blood and which are in stool; measure the trophozoites and/or cysts. Observe as much internal and external cellular detail as possible. While improving your skills as a microscopist, you are learning to recognize some of the world's most widespread and devastating parasites.

Microscopic Animalia

Microscopic animals have **cells** arranged to form **tissues.** Most also have **organs,** parts that perform specific functions and are formed of combined tissues. Microscopic animals are members of a broad taxonomic division, the **Metazoa.** Metazoans are multicellular animals that develop from eggs and possess cell layers. You are a metazoan.

Microscopic metazoans generally inhabit the seas and other moist environments. Some parasitize our bodies. In this exercise you examine free-living representatives of the group in a hay infusion. Some of the most frequently encountered free-living microscopic metazoans are illustrated in figure 6.8.

You also examine prepared slides and preserved specimens of parasitic metazoans. Representative parasitic metazoans are diagramed in figure 6.9. Most of these parasites are microscopic. Certain macroscopic creatures are included in figure 6.9 because their microscopic eggs or larval forms are recovered from the human host and identified by a clinical microbiologist during diagnosis.

Zoologists, faced with a tremendous diversity of living animals, have devised many plans that impose order on metazoan multiformity. One such taxonomic guide to microscopic life-forms is presented in table 6.3.

As you complete the procedures section of this exercise, you discover delightfully entertaining, free-living,

Figure 6.7 The life cycle of *Plasmodium*

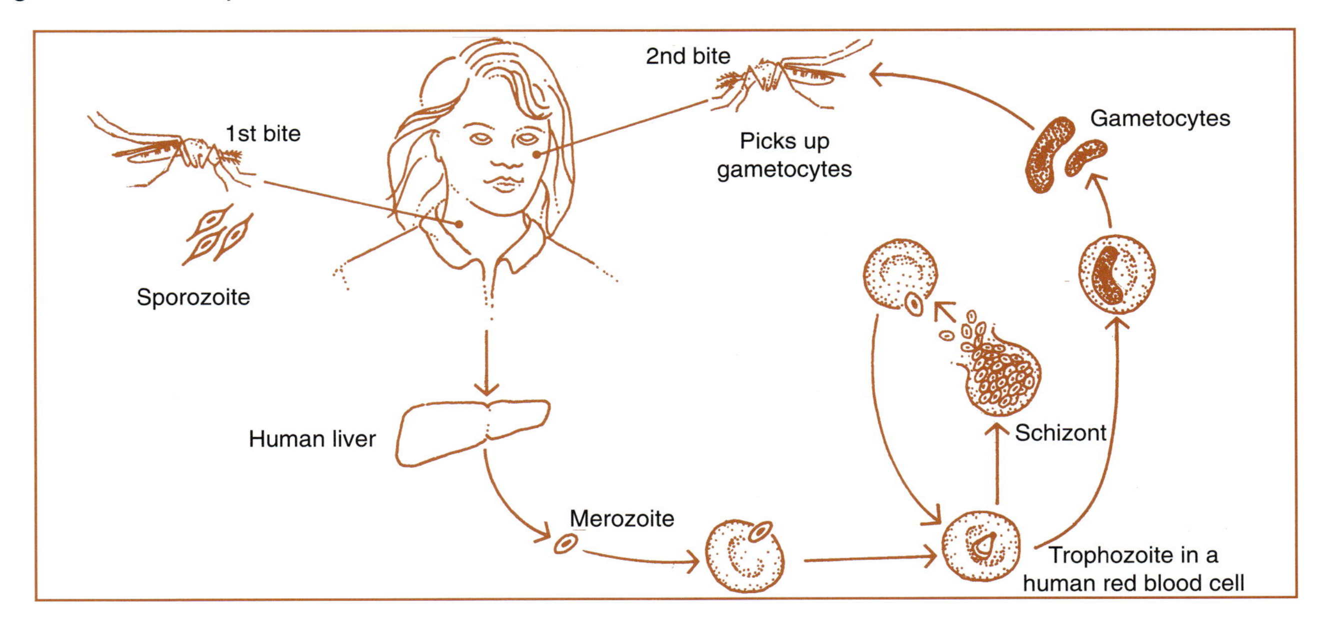

Table 6.3 Taxonomic Guide to Microscopic Metazoans in the Kingdom Animalia

Outline of Selected Microscopic Animals

I. Phylum **Porifera:** no organs, body with pores and canals (sponges)
II. Phylum **Cnidaria** (Coelenterata): polyps and/or medusae
III. Phylum **Platyhelminthes:** body flattened dorsoventrally[1] (flatworms)
   A. Class **Turbellaria:** free-living (planarians)
   B. Class **Trematoda:** "leaf-shaped" body (flukes)
   C. Class **Cestoda:** scolex, neck, and proglottids (tapeworms)
IV. Phylum **Rotifera:** cylindrical body with "toes," mastax, and ciliated corona
V. Phylum **Gastrotricha:** cylindrical body with "toes," scales, and rows of cilia
VI. Phylum **Nematoda:** tubular, thread-shaped body (roundworms)
VII. Phylum **Arthropoda:** jointed legs, body segmentation
   A. Class **Arachnida:** 4 pairs of legs, fused cephalothorax,[2] no mandibles[3]
   B. Class **Crustacea:** gills, mandibles, maxillae,[4] exoskeleton contains limy salts
   C. Class **Insecta:** antennae, mandibles, maxillae, head, thorax, abdomen, and 3 pairs of legs
VIII. Phylum **Tardigrada:** 4 pairs of unjointed legs, body unsegmented and cylindrical ("water bears")

[1]Front to back
[2]Head and "chest"
[3]Jaws
[4]Accessory appendages of the jaw

multicellular invertebrates (animals without backbones) in the hay infusion and pond water. On prepared microscope slides you observe a few of the microscopic animals that plague humans.

Take your time. Refine your microscopic technique until you are able to discern and record some of the amazing detail of these minute creatures.

PROCEDURES EXERCISE 6

Procedure 1

Examining Yeast

a Place a few drops of lactophenol cotton blue in the center of a clean microscope slide.

b Using sterile technique, mix a small amount of *Saccharomyces cerevisiae* culture from the tip of your inoculating needle into the dye on the slide. Remember to flame your needle after the transfer of yeast.

c Carefully angle a coverslip onto the suspended yeast and dye mixture.

d Set the slide aside for at least 5 minutes to give the lactophenol cotton blue time to stain the yeast cells.

e After examining a slide for Procedure 2, you should be able to return to the yeast preparation and use your high-power or oil-immersion lens to observe budding yeast cells.

f Record your observations in your laboratory report.

Procedure 2

Examining Prepared Fungal Slides

a Through your low- and high-power lenses, scrutinize a prepared slide of *Penicillium,* comparing it with your slide of *Rhizopus.* How do your observations compare with the details shown in figure 6.2?

b Enter your drawings and comments in the laboratory report.

Figure 6.8 Representative, free-living, microscopic metazoans

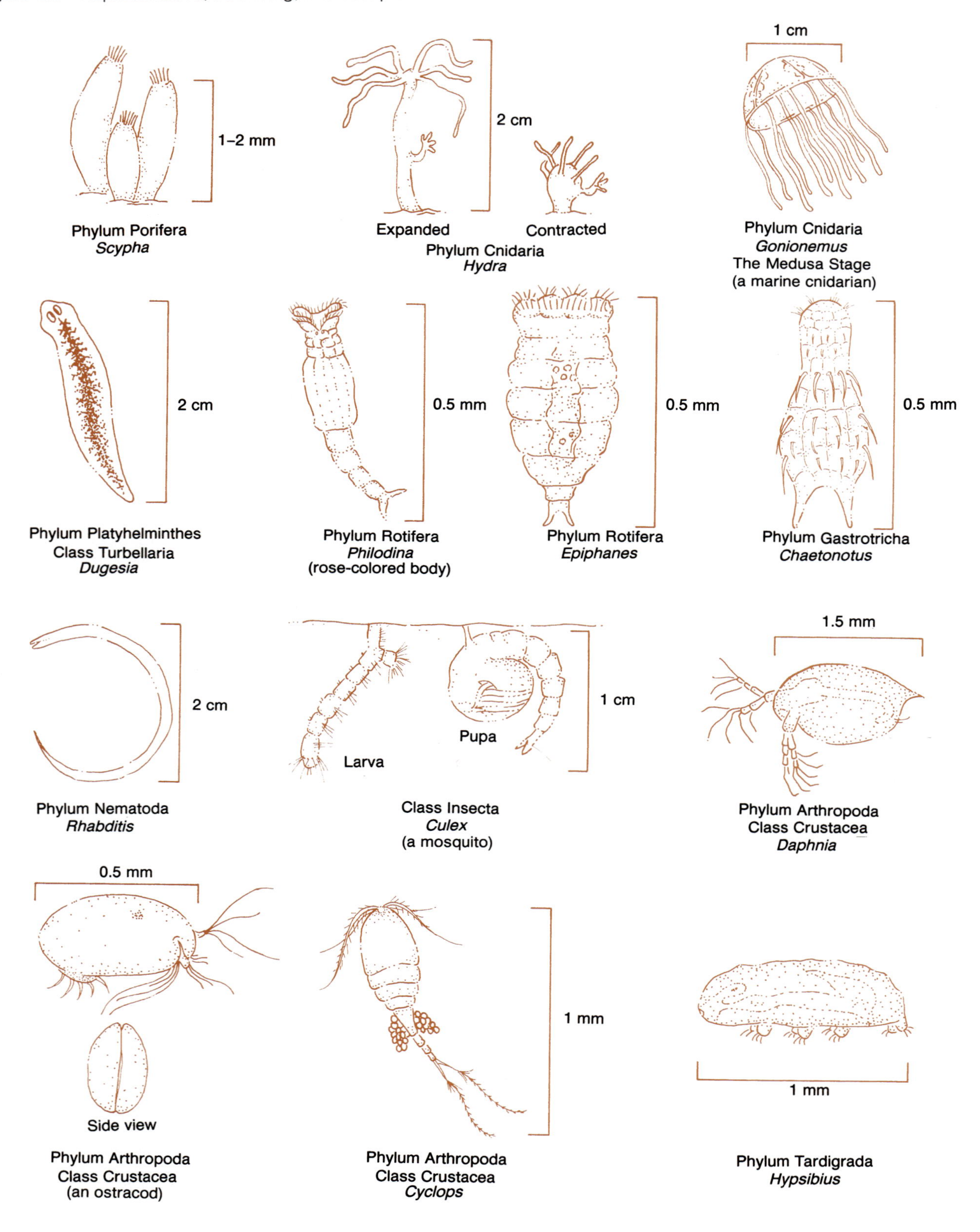

Figure 6.9 Representative parasitic metazoans

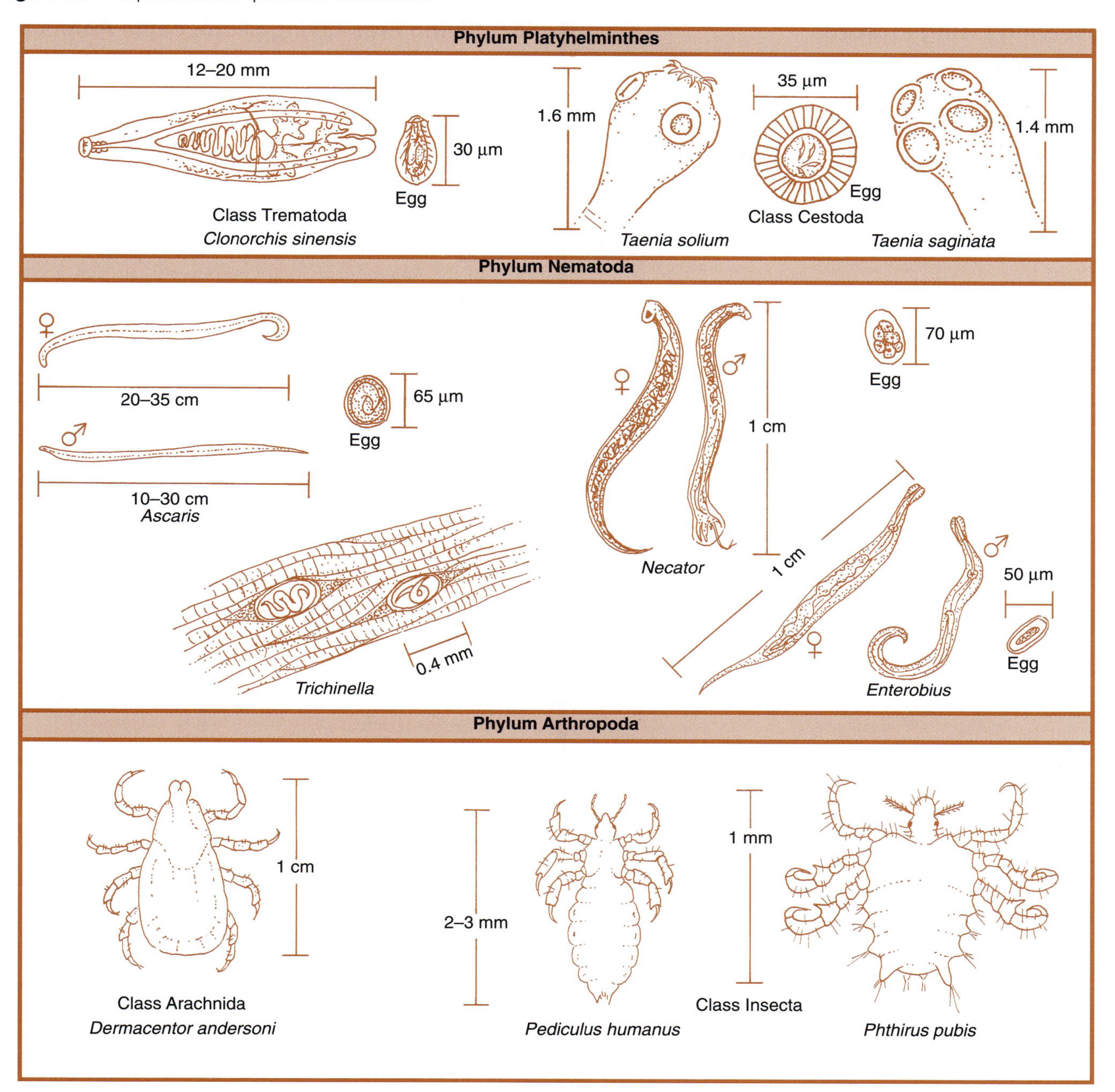

Procedure 3

Observing Living Protozoa

a Prepare wet mounts from the pond water samples. One method that is often successful is to remove a floating leaf from the pond water sample with a pair of forceps and a scalpel. Use these instruments to scrape a slime of bacteria and protozoa off the bottom of the leaf onto your slide. Add a coverslip to the wet mount.

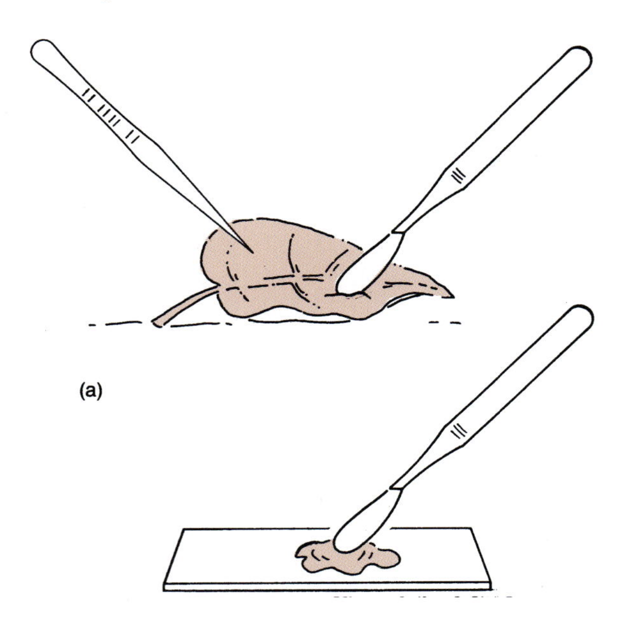

b If the organisms are moving too fast for detailed observation, use a toothpick to mix your next wet mount preparation into a drop of methylcellulose.

c In the laboratory report, enter detailed drawings of protozoa that you observe in the pond water samples.

d Sampling from the bottom of the jar, prepare a wet mount and examine the mixed protozoa culture. Try to find amoebas, flagellates, and ciliates. Share your findings with your classmates. Enter detailed drawings in your laboratory report. Draw only what you actually observe, but use the labeled figures in this exercise to help increase your powers of observation.

Procedure 4

Observing Prepared, Stained Slides of Protozoan Parasites

a Observe, measure, and draw in your laboratory report representative trophozoites and/or cysts of each of the following protozoans, then return each slide to the class set.

(i) *Entamoeba histolytica,* cysts in stool

(ii) *Giardia lamblia,* cysts in stool (trophozoites are found only in watery stools and may not be present)

(iii) *Plasmodium* species, trophozoites in blood

Procedure 5

Identifying Free-Living Microscopic Animals

a Prepare wet mounts from the hay infusion and pond water. Sample material floating on the top and sediment from the bottom.

b Use figure 6.8 and supplementary texts to help you identify various phyla of microscopic animals. Can you identify the genera of several of these microscopic invertebrates?

c In your laboratory report, draw and identify a variety of the invertebrates that you observe. Be sure to note clearly where you found each animal. Also, list the magnification represented by each drawing. Try to improve your eye for detail with each illustration you prepare.

d If you find animals of particular interest, prepare a hanging-drop suspension with liquid from the same source. With a bit of luck, you may be able to observe the close relatives of your original organism.

e Share your best finds with other members of the class.

Procedure 6

Examining Prepared Specimens of Microscopic Animals

a Examine preserved specimens of the human parasites *Ascaris lumbricoides* and *Clonorchis sinensis* or *Opisthorchis* species.

b In your laboratory report, describe and draw the nematode and the fluke.

POSTTEST EXERCISE 6

Part 1

True or False (Circle one, then correct every false statement.)

1. T F All fungi are macroscopic.
2. T F The cell walls of fungi are usually composed of chitin.
3. T F Fungi are not photosynthetic eucaryotes; they are generally saprophytic.
4. T F Ascomycetes are often called "water fungi."
5. T F Members of the class Zygomycetes form ascospores when members of opposite mating types contact each other.
6. T F Members of the deadly genus *Amanita* are mushrooms.
7. T F Fruiting bodies of some fungi bear asexual spores.
8. T F The body or colony of a fungus is called a hypha.
9. T F Phagocytosis is the ingestion of particulate material.

10. T F Protozoa in the trophozoite stage generally are capable of phagocytosis and pinocytosis.
11. T F During trying times, under adverse environmental conditions, cowardly protozoan cysts form trophozoites.
12. T F Amoeboid motion is characterized by rows of cilia joyously waving synchronously.
13. T F The outer protective covering of *Paramecium* is a pellicle.
14. T F The sexual stages of *Plasmodium* are in the *Anopheles* mosquito.
15. T F A metazoan develops from an egg and possesses cell layers.
16. T F Microscopic metazoans are all parasitic.
17. T F A tapeworm has no mouth, eyes, ears, or digestive tract.
18. T F *Clonorchis,* the human liver fluke, is a trematode, a parasitic, leaf-shaped animal.
19. T F *Ascaris* is a nematode that can infect humans.

Part 2

Completion and Short Answer

1. Macroscopic fungal forms include the ____________, ____________, ____________, ____________, and ____________.
2. The mycelium of a thallus contains ____________ cells that may be coenocytic.
3. Define the term mycologist.
4. List five characteristics shared by organisms that are traditionally called protozoans.
   a.
   b.
   c.
   d.
   e.
5. Fill in the name of the group and its mode of motility (if any) for each of the four traditional protozoan groups.

Traditional Protozoan Groups

	Name of Group	Motility
a.		
b.		
c.		
d.		

Laboratory Report **6**

Name ____________________

Section ____________

Fungi, Protozoa, and Microscopic Animalia

Procedure 1

Examining Yeast

(*f*) Draw cells of *Saccharomyces cerevisiae* as you observe them in your wet mount preparation of living material stained with lactophenol cotton blue. Label budding cells.

Total Magnification ________

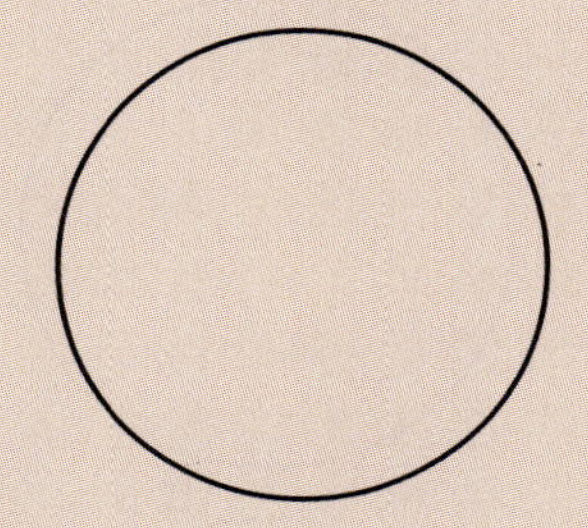

Saccharomyces cerevisiae

Procedure 2

Examining Prepared Fungal Slides

(*b*) Draw hyphae and fruiting bodies that you observe in *Penicillium* and *Rhizopus*.

Total Magnification ________

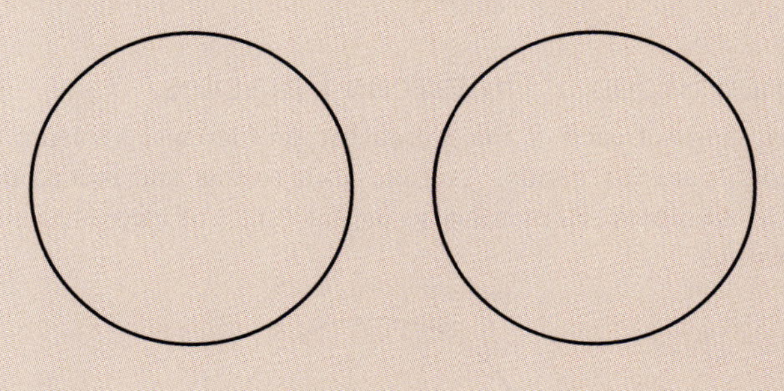

Total Magnification ________

Penicillium ***Rhizopus***

Procedure 3

Observing Living Protozoa

(*c,d*) Draw at least 3 unique examples of protozoans that you observe living in the pond water and/or mixed protozoa culture. Employ this laboratory manual and other reference works to help you label the drawings. Try to identify the genera that you observe. For each drawing record the following information:

(i) Source of the sample (pond water or mixed protozoa culture)
(ii) Total magnification employed
(iii) Type of motility observed (cilia, flagella, or pseudopodia), if any
(iv) Genus (if identified)

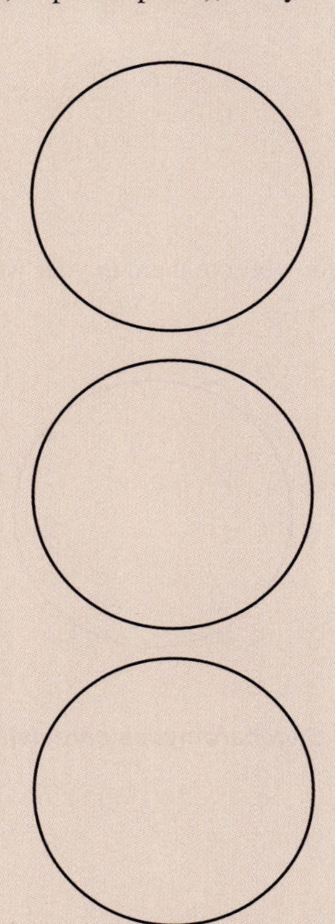

1. (i)
(ii)
(iii)
(iv)

2. (i)
(ii)
(iii)
(iv)

3. (i)
(ii)
(iii)
(iv)

Procedure 4

Observing Prepared, Stained Slides of Protozoan Parasites

(*a*) Observe and make detailed drawings of each of the 3 parasitic protozoans. Measure the diameter or length and width of at least 3 representative cells of each parasitic genus. Average your results and record the average size of the cells you observe. Note the magnification that you employ. Remember to include units of measurement.

(i) *Entamoeba histolytica,* cysts in stool

Diameters

____

____

____

Average diameter ________

Total
Magnification ________

Entamoeba histolytica

(ii) *Giardia lamblia,* cysts and/or trophozoites in stool

Length	Width
________	________
________	________
________	________

Average length and width ________ × ________

Total
Magnification ________

Giardia lamblia

(iii) *Plasmodium* species, merozoites inside red blood cells

Diameters

____

____

____

Average diameter ________

Total
Magnification ________

Plasmodium

Procedure 5

Identifying Free-Living Microscopic Animals

(*c*) Draw and identify a variety of the invertebrates that you observe. Be sure to state where you found each animal. Also, list the magnification represented by each drawing. Try to improve your eye for detail with each illustration you prepare.

Procedure 6

Examining Prepared Specimens of Microscopic Animals

(*b*) Describe and draw the nematode and the fluke. Measure each, and remember to include units of measurement.

Size ________

Ascaris lumbricoides

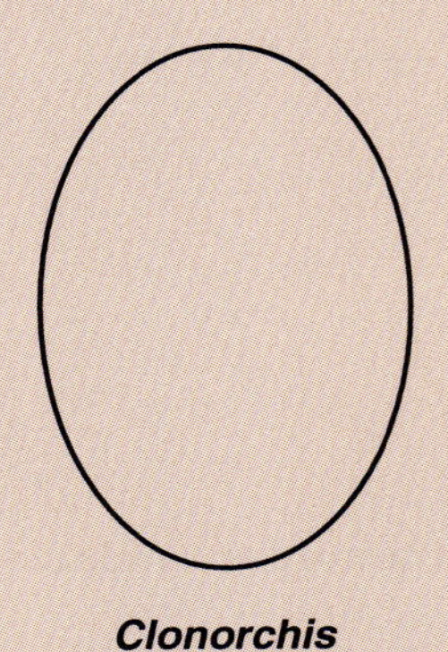

Size ________

Clonorchis sinensis **or** ***Opisthorchis***

Preparing a Smear and Simple Stain

Necessary Skills

Microscopy, Microscopic measurement, Aseptic transfer techniques

Materials

Cultures (1 per set):
- *Staphylococcus epidermidis* (24-hour, 35° C, Tryptic Soy Agar slant)
- *Escherichia coli* (24-hour, 35° C, Tryptic Soy Broth)

Staining equipment (per group):
- Stain rack and pan
- Bibulous paper (optional)
- Wash bottle
- Lens paper
- At least 2 microscope slides
- Transparent tape (optional)
- Inoculating loop and needle
- Bunsen burner and striker
- Wax marking pencil, indelible ink pen, or diamond tip pen
- Methylene blue chloride stain in dropper bottle
- Crystal violet stain in dropper bottle
- Forceps, slide holder, or clothespin
- Disposable gloves (optional)
- Access to *Bergey's Manual*
- Immersion oil

Primary Objective

Prepare and stain bacterial smears; evaluate cell morphology.

Other Objectives

1. Discuss the reasons microscopists stain bacteria.
2. List four characteristics of microbial morphology.
3. Define: cation, anion, chromophore.
4. Name several basic dyes and an acid dye.
5. Prepare smears of bacteria and heat fix them.
6. Stain, observe, and describe bacteria.

INTRODUCTION

Why does a microscopist stain smears of bacteria before observing the cells with a brightfield microscope? The answer becomes apparent when you try to examine unstained bacteria and note, with some difficulty, that they are almost transparent.

It is nearly impossible to distinguish any details in or on unstained microbiota. Stains are used to enhance contrast. Microbiologists generally employ **aniline dyes** (compounds derived from benzene, a coal tar derivative) to stain microbes. Cellular **morphology** (shape, arrangement, size, and structure) is carefully evaluated by the experienced microscopist because it yields information that helps identify microorganisms.

Bacterial Morphology

The **shapes** of stained bacteria can usually be identified as **rods, cocci** (spheres), or **spirals** (spirochetes or spirilla). These shapes are illustrated in figure 7.1. Generally, short rods are called **coccobacilli.** Bacteria that assume many shapes are often labeled **pleomorphic rods.**

The **arrangement** of bacteria, their physical placement in relation to other cells of their species, is also noted by the microbiologist. Common arrangements include **singles; diplococci** and **diplobacilli,** two microbes together; **chains** of either rods or cocci; **tetrads,** groups of four cocci; and grapelike **clusters** of cocci. The cells of a single species usually display several of these groupings. Note the variety displayed in your culture of ***Staphylococcus epidermidis.***

Figure 7.1 Bacteria can usually be identified as rods, cocci, or spirals.

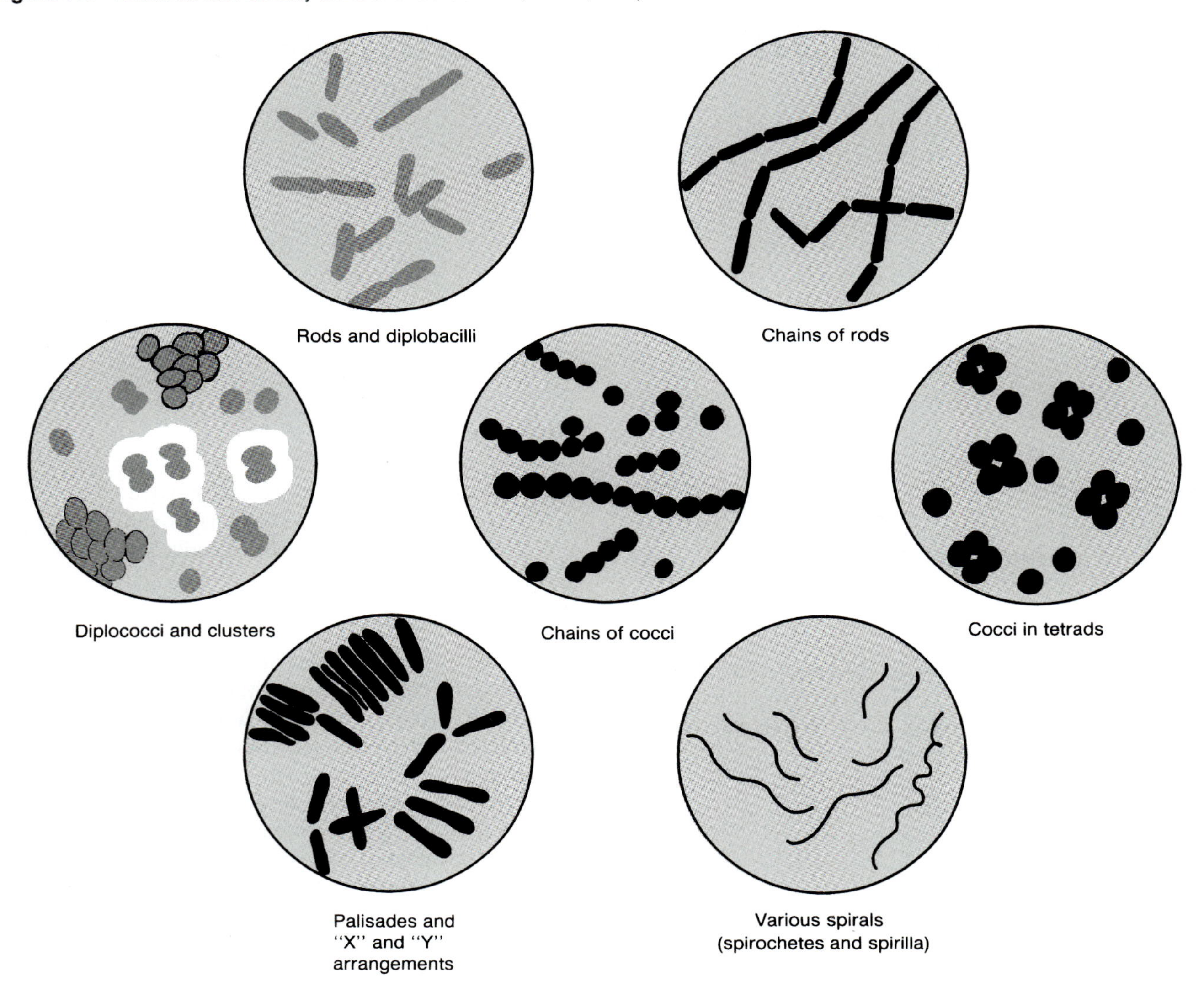

Some pleomorphic rods are grouped into **palisades,** which resemble fingers pressed tightly or a closely spaced picket fence. Other rods form "X" and "Y" arrangements.

Bacteria are very small but are found in a wide range of sizes. Cell **size** is an important aspect of a thorough morphological characterization. The use of **micrometers** (a microscopic ruler and scored ocular disk) to measure microbiota is described in Exercise 2, Microscopic Measurement.

The study of morphology includes an examination of **structures** that are revealed by staining. Microbiologists employ special techniques to stain bacterial endospores, granules, capsules, and flagella, as well as other structures.

Some bacteria do not stain evenly; after dye is applied, dark spots may appear. The cells look **beaded.** As illustrated in figure 7.2, a beaded rod closely resembles a short chain of cocci.

If the spots of stain are on the ends of the cell, the phenomenon is called **bipolar staining.** Coccobacilli exhibiting bipolar staining may appear to be diplococci. See if your laboratory strain of ***Escherichia coli*** displays bipolar staining with methylene blue.

In this exercise you receive labeled cultures. To be sure that you are correctly characterizing cell morphology, do a little research in your microbiology textbook or in *Bergey's Manual.*

The Chemistry of Staining

The chemistry of staining is based on the principle that unlike charges attract, while similar charges repel each other. A bacterial cell carries electrical charges. In an aqueous environment, with the pH at approximately 7, the net electrical charge produced by most bacteria is **negative.**

The negatively charged cells attract molecules carrying positive charges and repel those that are negatively charged. This is the chemical situation that occurs when bacterial cells are covered with common microbiological stains. How electrical charges relate to staining is illustrated in figure 7.3.

Figure 7.2 Beaded rods resemble short chains of cocci, and bipolar beading in coccobacilli can be mistaken for diplococci. (Photograph courtesy of DIFCO Laboratories, Detroit, MI 48232)

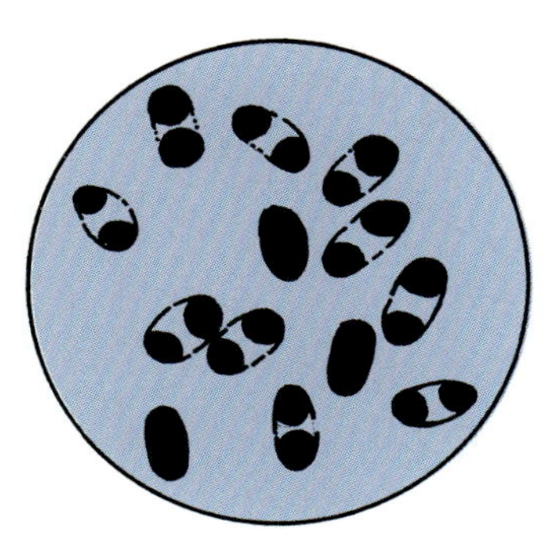

Bipolar staining in coccobacilli

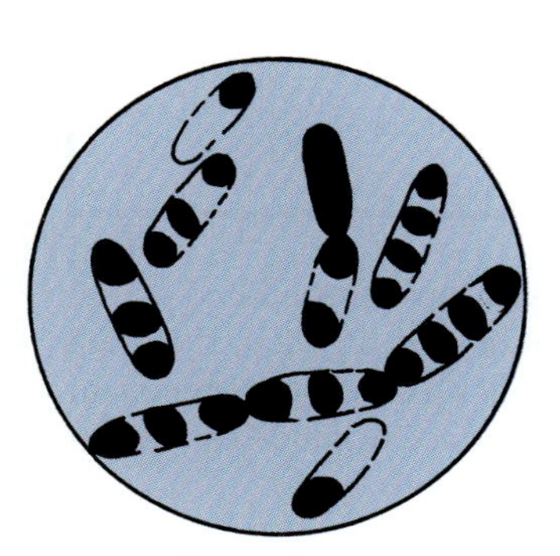

Beaded rods

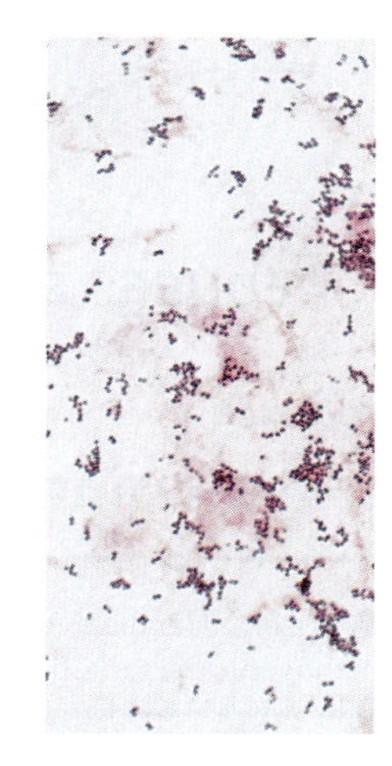

Singles, pairs, short chains, and clusters of the coccus, *Staphylococcus aureus*

Each dye is a salt containing two ions, one with a positive charge, the **cation,** and the other with a negative charge, the **anion.** Either one of the two ions can be the **chromophore,** the part of the molecule that is brightly colored.

The most frequently employed microbiological dyes are called **basic dyes;** their chromophores are positively charged. The chromophores are attracted to, and subsequently color, bacteria. A few of the basic dyes you view in this course are methylene blue, crystal violet, and safranin. You prepare a **simple stain** by coloring bacteria with a single, basic dye.

Acid dyes such as nigrosine are generally repelled by bacteria because of the negative charges on their chromophores. Acidic dyes color the background and leave cells colorless.

Preparing Stains

To prepare a simple stain, the microscopist employs a perfectly clean microscope slide. The slide must be so clean that a drop of water spreads out on it. Water **coalescing** (beading) on the slide indicates the presence of an oily film that clumps bacteria. Adequate evaluation of clumped cells is difficult or impossible.

It is helpful to use new slides since even a small amount of oil from a fingerprint leaves a film, and minute scratches on the surface of an old slide may be mistaken for rods. However, after disinfection, slides can be recycled by scrubbing them with cleanser, then drying and polishing them with a clean, lint-free cloth. An ammonia-water rinse also helps remove traces of oil.

The microbiologist spreads a thin film of bacteria on the slide and allows the **smear** to dry. The preparation is then heated gently to **fix** (attach) microorganisms firmly onto the slide.

Adequate **heat fixation** shrinks cells slightly, but it helps bacteria adhere to the slide through several rinses. Excessive heat warps cells; applying heat to the smear

Figure 7.3 Negatively charged bacterial cells attract positively charged chromophores in an aqueous suspension (in water) of methylene blue$^+$ and chloride$^-$.

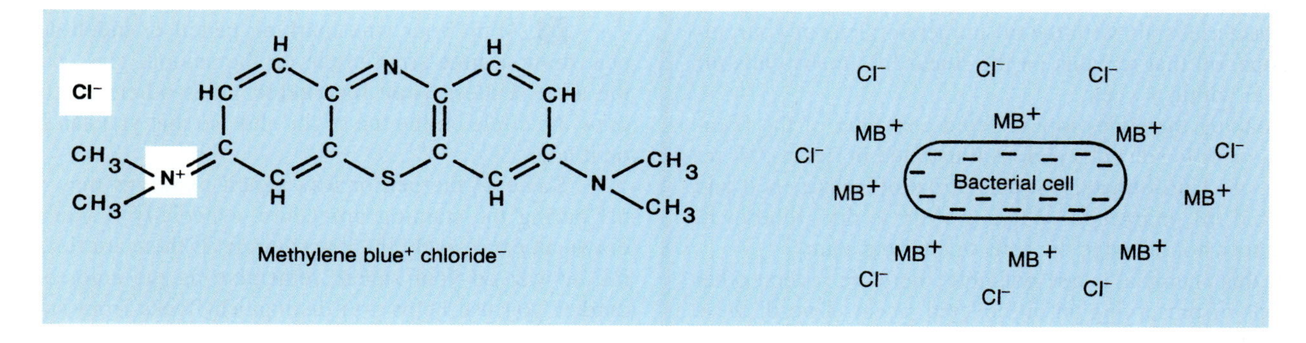

Figure 7.4 Three ways to preserve smears covered with immersion oil: (a) Soak the slide for a few minutes in a covered Coplin jar of xylol. (b) Cover the oily smear with a coverslip, and for future use dispense oil onto the top of the coverslip. (c) Separate upright slides in a slide box lined with absorbent tissue.

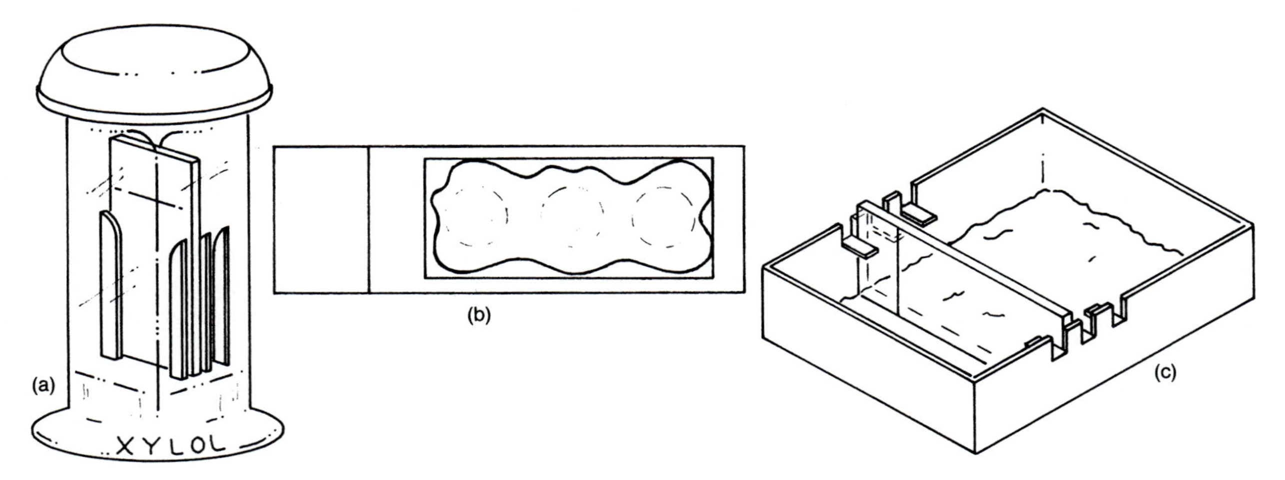

before it is completely dry also distorts cells. After heat fixation, the smear is stained.

A stained smear of bacteria should be somewhat difficult to see with the naked eye. A dark spot of color indicates that the bacteria are piled on top of each other and are much too close together to be adequately evaluated.

Preparation of serviceable smears and stains takes practice. Repeat the procedures in this exercise until every stained smear you prepare reveals well-isolated, minimally distorted, satisfactorily colored microorganisms.

Staining techniques fit into even the busiest schedule. You can start and stop the staining process at almost any point. For instance, if you place labeled smears in a locked drawer to dry, they will be ready for heat fixation and staining any time within the next few hours or years. Similarly, you can generally store stained or unstained smears for years in a dark, dry place. When you are ready to examine them, simply begin preparations wherever you stopped.

Storing Stained Microscope Slides

If the stained smears on microscope slides are clean, they may last for months in a safe, dark place. But after examining a bacterial stain with the oil-immersion lens of your microscope, oil that remains on the smear lifts many cells off the slide within a week.

Soaking the slides in a covered container of the flammable, organic solvent **xylol,** as illustrated in figure 7.4, removes oil. Unfortunately, it also strips off wax pencil and indelible ink markings. When a slide is dry, relabel it. Xylol does not dislodge the cells or alter the stain.

If this chemical is not available, there are several other methods of preserving an oil-covered smear. Two of these are also illustrated in figure 7.4.

Before beginning the procedures in this exercise, study figure 7.5, which is a flowchart of smear preparation.

PROCEDURES EXERCISE 7

Procedure 1

Preparing a Smear from a Broth Culture

Warning: Handle every culture as if it contained pathogenic bacteria.

a First, prepare the slide. If you have frosted-end slides, label the etched portion of the glass with a pencil. If frosted slides are unavailable, employ a wax marking pencil or wrap transparent cellophane tape once around one end of the slide. Write on the tape with a pencil.

(b)

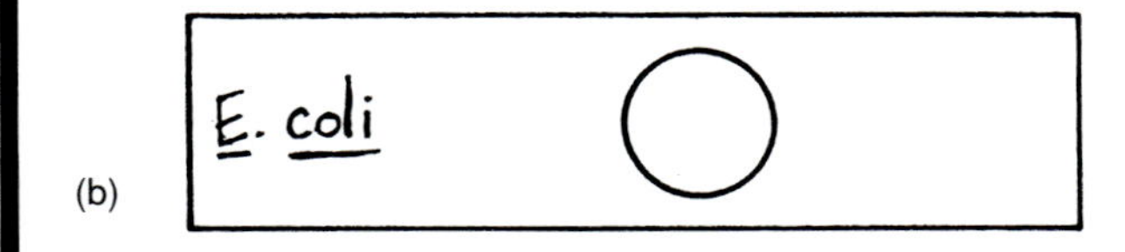

b With your wax marking pencil or indelible ink pen, draw a dime-sized circle on the smooth glass part of the slide. Unless **your instructor** gives other directions, draw the circle on the top of the slide so that you can focus on it later.

Some of the pen or wax pencil markings may wash off during the staining procedure, especially if you have drawn an excessively thick wax circle. If there are flecks of ink or wax on your smear, remembering the color of the marker you use helps you distinguish flecks from stained bacteria.

c Obtain a broth culture of *Escherichia coli*. Examine figure 7.5 before you continue with this procedure.

Figure 7.5 A flowchart of smear preparation

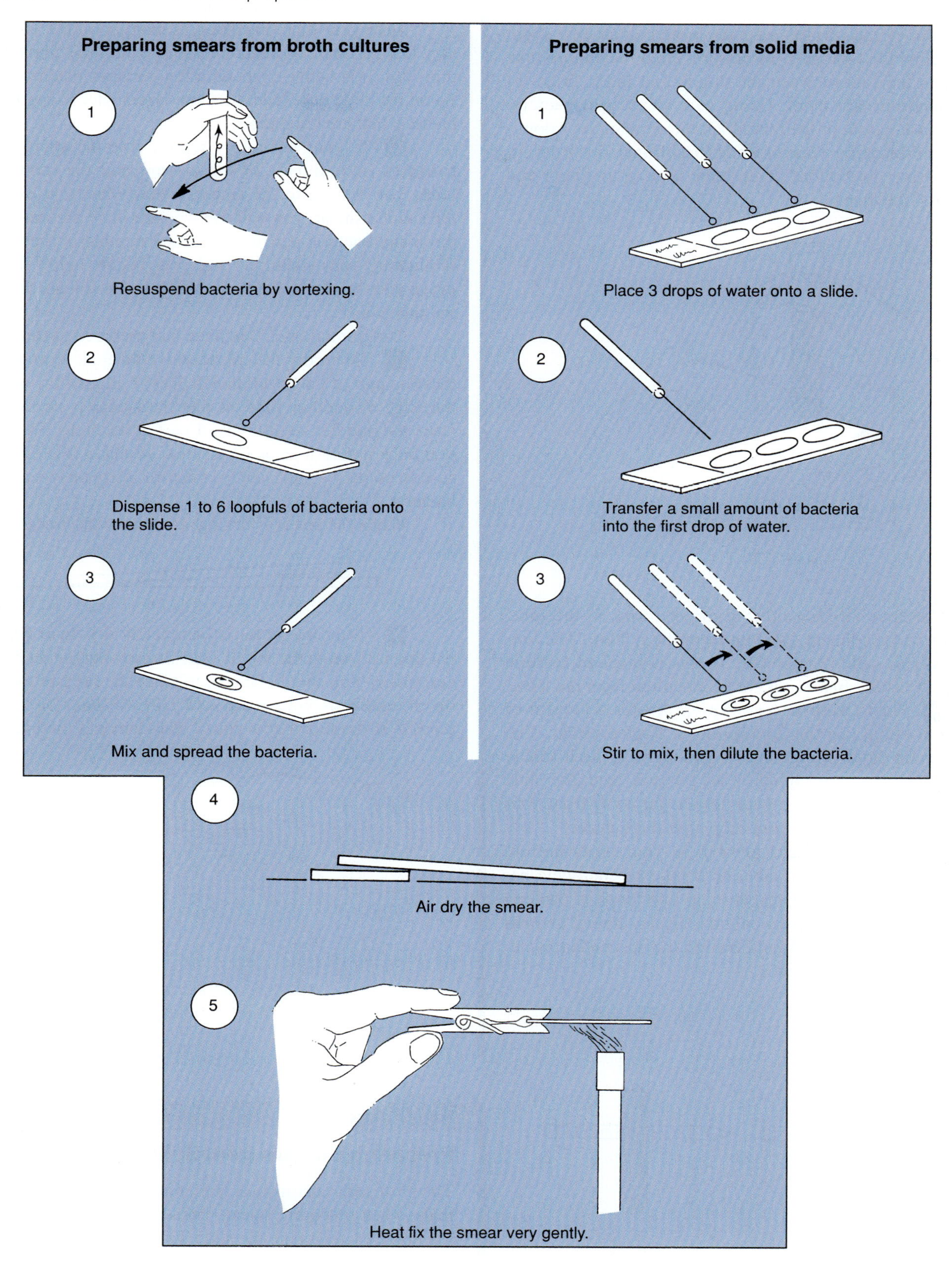

d Resuspend the bacteria in your broth by **vortexing** (swirling the liquid in the tube). To create a **vortex** (whirlpool effect), firmly grasp the broth culture tube about 1 cm below the bottom of its closure. Hold the culture tube tightly in the fold between the thumb and index finger of your nondominant hand. Using care to prevent bacteria from splashing out of the tube, flick the wrist of your dominant hand bringing your extended forefinger toward your body. Strike the bottom of the tube with your finger and come past the tube.

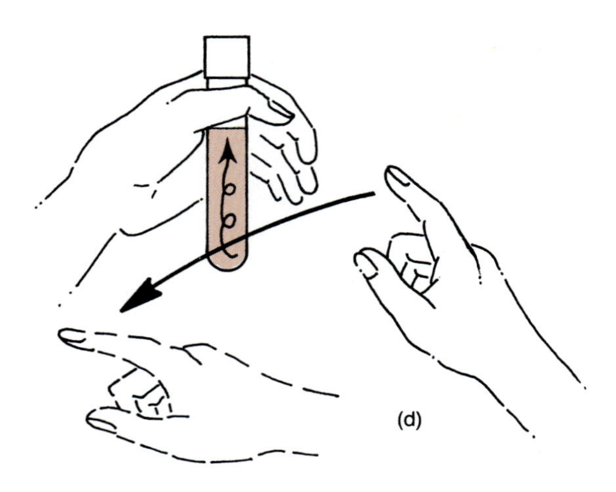
(d)

Continue striking until a vortex forms in the tube. You may see sediment swirl up from the bottom. The vortex separates cells. Remember, always **resuspend** bacteria in a broth culture **before** taking an inoculum from the tube.

e Using **aseptic technique,** transfer a loopful of the broth culture to the appropriately labeled circle. You may have to tap the liquid out of the loop to make it adhere to the slide. To avoid inhaling an aerosol, have the slide near your flame, and keep your nose as far away from your work as possible. **Flame** your loop after the transfer.

f If there is **light growth** in your broth culture (you can read the printed label on a yellow pencil held behind the dispersed cells), dispense up to 6 loops of bacteria into the same smear. When cultures display **heavy growth** (you cannot see a pencil held behind the suspended bacteria), 1 loop of liquid suffices.

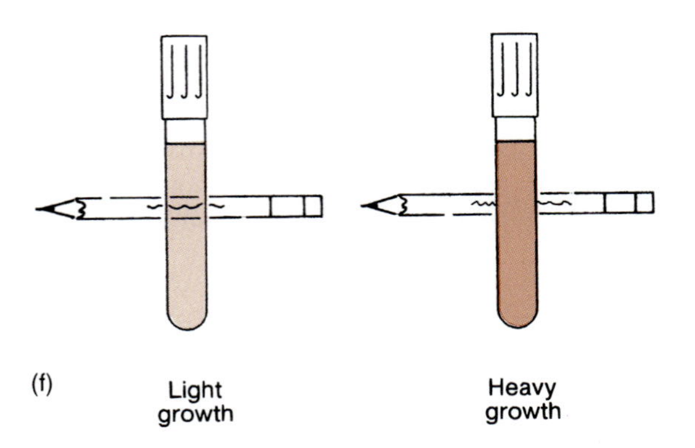

(f)

g If necessary, transfer more bacteria to the smear. **Flame** your loop after each transfer.

Take your time. Dispense as many loopfuls as necessary to your circle. Spread the liquid within the circle into a thin film. It is better to spend a few minutes placing bacteria on the slide now, than to waste hours later trying to find bacteria that are not there.

h In your laboratory report, note the approximate **tubidity** (cloudiness) of the culture and the number of loops you deliver to the smear. Later, when you examine the stained smear, you will see whether there are too few or too many bacteria per field, or if the number is acceptable. Evaluating your technique and your results helps you improve your smear preparation. You prepare smears throughout this course.

Congratulations! You have just prepared a smear.

i To air dry the smear, move the slide to an out-of-the-way part of your workbench. Rest 1 edge of the slide on the edge of another slide so that the smear is drying on a slant and puddles slightly into 1 part of its circle. This gives you thick and thin areas on a smear so that a few fields are sure to contain a sufficient number of adequately separated bacteria.

While the smear is drying, continue with Procedure 2.

(i)
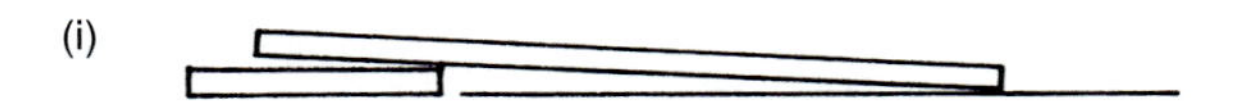

j After the smear is completely dry, **heat fix** it by passing it through the top of your Bunsen burner flame several times. You should be able to place the back of the slide on the palm of your hand and *feel the heat but not be burned.* If the slide burns you, it also damages the bacteria.

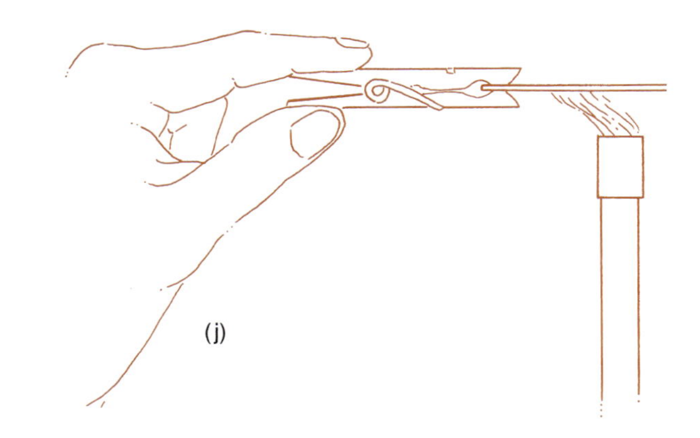
(j)

Procedure 2

Preparing Smears from Solid Media

The most common mistake beginning microbiology students make when preparing smears from solid media is to transfer too many bacteria onto the slide. In this procedure you avoid this problem by diluting the bacteria; see figure 7.5.

a Label a slide for *Staphylococcus epidermidis*. Draw 3 dime-sized circles on the slide.

b Place a drop of water on your **loop** and touch the wet loop onto the first circle on your slide. Similarly, dispense drops into the second and third circles.

c Quickly, before the drops dry, flame and thoroughly **cool** your inoculating **needle.**

d **Without gouging the agar,** transfer a barely visible amount of the bacteria into the first circle.

Flame your needle.

e Now you are going to dilute the bacteria. Use a sterile inoculating **loop** to mix the bacteria in the first drop of water, and transfer a loopful of suspended organisms into the second circle. Mix this suspension and transfer a loopful of these diluted cells into the last circle.

Mix the suspension, then **flame** your loop.

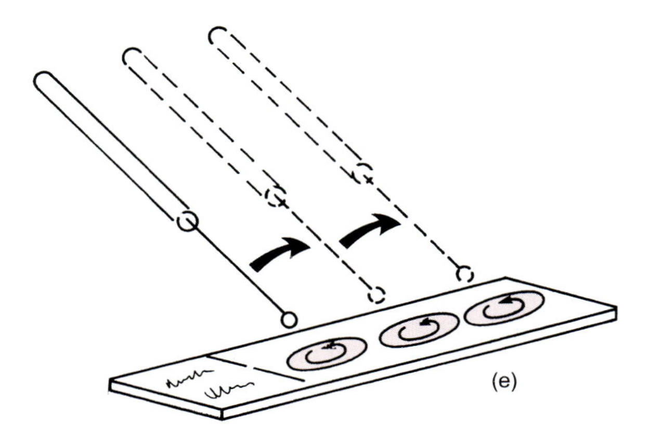

(e)

f Place the slide in an out-of-the-way place on your laboratory bench. Allow it to air dry at a slight slant by resting 1 end on another slide.

g After it is completely dry, heat fix it gently but thoroughly. When the slide cools, it is ready to stain.

In the future it will not be necessary to prepare 3 dilutions each time you stain solid growth. Making the dilutions simply emphasizes for you the necessity of transferring a very small amount of growth to the smear.

Procedure 3

Simple Staining

Your instructor may modify the following instructions. Before actually staining the smears, you must be sure that your work area is ready.

a Clear all unnecessary items off the laboratory bench. Place your **stain rack** over a **stain pan** as shown in the illustration. The pan may be placed in a **flat dish** to protect the bench top from spills and splashes. Be sure your **wash bottle** has plenty of water in it. Have your filled **stain bottle** and **forceps, slide holder,** or **clothespin** ready.

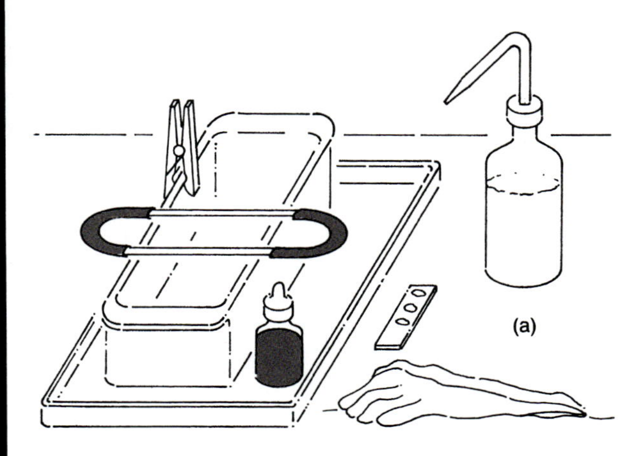

(a)

b If you wish to avoid dyed fingers, use a slide holder or wear thin, disposable **gloves.** The gloves must fit your hands well enough so that they do not become a fire hazard.

c Place the slide that you prepared in Procedure 1 on the stain rack. Cover the entire slide, except the labeled area, with methylene blue **or** crystal violet. Allow methylene blue to remain on the smear for exactly 1 minute. Crystal violet requires only 30 seconds.

Flooding the slide with excess stain, as you have, helps prevent the formation of crystals that may be mistaken for bacteria.

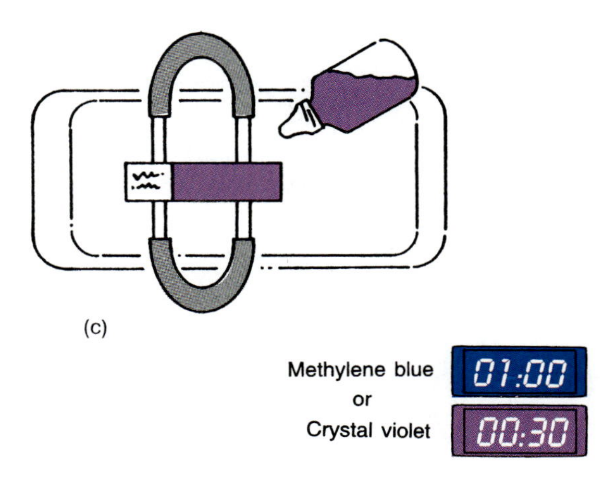

(c)

Methylene blue 01:00
or
Crystal violet 00:30

d After the time interval, use your forceps, slide holder, or clothespin to tilt the slide so that excess stain pours off into the stain pan.

e While holding the slide at an angle over the stain pan, gently rinse excess color off the slide with water from your wash bottle. Aim the water **around** the smear, not directly onto it. You do not want to wash off the smear.

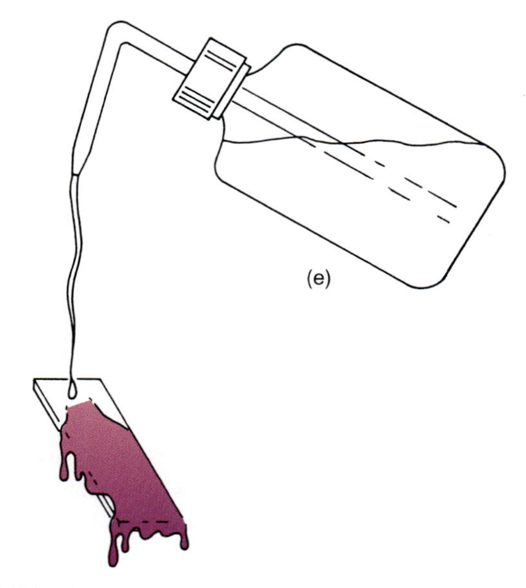

f Shake excess water off the slide, into the pan. Since water itself is a decolorizer, it should not be allowed to remain on the stained smear very long. In your laboratory report, record the number of seconds you stained the smear.

g Air dry your slide, or blot it very gently by placing it between paper towels or in a book of **bibulous** (absorbent) paper. Close the paper over the slide. Do not press down! Open the papers and carefully lift them off the slide. Avoid rubbing the slide on the paper; it can remove your smear.

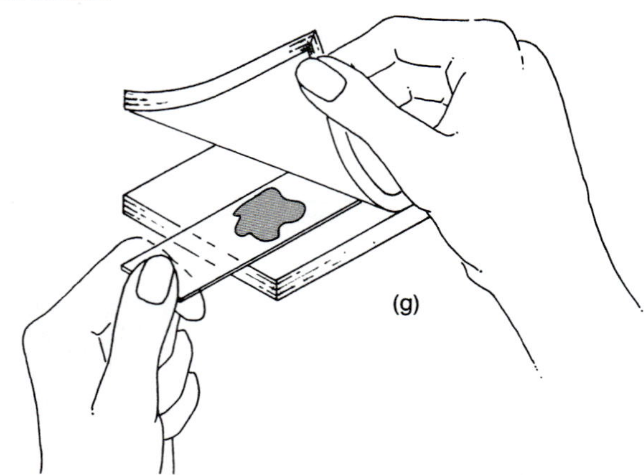

h While your slide is drying, stain the slide you prepared in Procedure 2.

i Examine the stained smears with the oil-immersion lens of your microscope.

j In your laboratory report, draw a typical field of bacteria from each smear. Comment on the density of cells in each smear. Describe the morphology of the bacteria in each culture.

k Describe the color intensity of the cells. Comment on how to adjust the amount of time you expose smears to methylene blue or crystal violet to best suit your needs.

l After completion of this procedure, clear your area. Dispose of your cultures safely as directed by **your instructor.** Leave your workbench cleaner than you found it. Clean and store your microscope as directed by **your instructor.**

POSTTEST EXERCISE 7

Part 1

Number the following steps in the correct order for preparation of a simple stain of bacteria. Assume that the bacteria have produced heavy growth in a broth culture. Each letter may be used once, more than once, or not at all.

a. _____ Dispense dye onto the smears, then wait and time the reaction.
b. _____ Deliver one loopful of bacteria to the slide; spread the drop.
c. _____ Flame the loop.
d. _____ Air dry the smear on a slight slant.
e. _____ Draw circles on a perfectly clean slide.
f. _____ Vortex the broth culture.
g. _____ Cool the loop.
h. _____ Examine stained bacteria with the oil-immersion lens of the microscope.
i. _____ Use heat fixation to attach the smears firmly to the slide.
j. _____ Dry the stained slide.
k. _____ Gently rinse dye off the slide with water.
l. _____ Dispense a drop of water off your loop into the circle.
m. _____ Hold the wet smear over the flame to dry.

List the selected letters in the correct order. ____________
____

Part 2

True or False (Circle one, then correct every false statement.)

1. T F Aniline dyes are derived from benzene, a coal tar derivative.
2. T F Coccobacilli are cocci; palisades are groups of four cocci.
3. T F In the neutral environment of most microbiological dyes, bacteria generally carry a net negative charge.
4. T F A cation has a positive charge; chromophores are always anions.
5. T F The chromophore of a basic dye carries a negative charge.
6. T F An aqueous solution is made with water.
7. T F Smears should be air dried before they are heat fixed.

8. T F A slide is clean enough for bacteriological smears if water coalesces on it.
9. T F Heat fix bacterial smears by holding the slide in the hottest part of the flame for 2 minutes.
10. T F Before vortexing bacterial cells in a broth culture, always remove the test tube closure.

Part 3

Short Answer

1. When evaluating the morphology of a bacterial species, what characteristics are considered?

2. Name at least seven common arrangements of bacteria.
   a.
   b.
   c.
   d.
   e.
   f.
   g.
   h.
   i.

3. Explain how a basic stain colors bacteria.

4. Define or describe the following microbiological terms.
   a. Vortex
   b. Light growth (in broth culture)
   c. Tetrads of cocci
   d. Bipolar staining
   e. Acid dyes
   f. Xylol

Laboratory Report **7**

Name ______________________________

Section ____________

Preparing a Smear and Simple Stain

Procedure 1

Preparing a Smear from a Broth Culture

(*h*) How many loopfuls of *Escherichia coli* broth did you employ? ________
How turbid was the broth?

Procedure 3

Simple Staining

(*f*) Record the amount of time each smear was exposed to the dye you selected.

	Stained Employed	Seconds of Exposure
Staphylococcus epidermidis		
Escherichia coli		
	Duration of Staining	

(*j*) Draw and color a representative field of bacteria from each smear.

Staphylococcus epidermidis ***Escherichia coli***

Comment on the density of the cells in each of your smears.

Describe the morphology of *Staphylococcus epidermidis*.

Describe the morphology of *Escherichia coli*.

(*k*) Evaluate the color intensity of your stained bacteria. Were all of your smears adequately tinted? _________ Is your technique adequate? _________ If not, how can you improve it?

Exercise 8

Gram Staining

Necessary Skills

Microscopy, Aseptic transfer techniques, Preparing a smear and simple stain

Materials

Cultures (24-hour, 35° C, 1 Tryptic Soy Broth and 1 Tryptic Soy Agar slant of each per set):
Staphylococcus epidermidis
Escherichia coli
Supplies:
Staining equipment including:
At least 4 microscope slides
Gram's stain set including:
Crystal violet
Gram's iodine
Ethyl alcohol (95%)
Gram's safranin
Other staining equipment and supplies
Prepared Gram-stained slides:
Staphylococcus aureus
Salmonella species

Primary Objective

Learn to prepare and interpret a Gram stain.

Other Objectives

1. Summarize the history and importance of the Gram stain.
2. Identify the primary stain, mordant, decolorizer, and counterstain used in the Gram staining procedure.
3. Describe gram-positive and gram-negative cell walls.
4. List pathological conditions and diseases caused by selected gram-positive bacteria.
5. List pathological conditions and diseases caused by selected gram-negative bacteria.
6. Define gram-variable and gram-nonreactive. Give examples of genera that display each characteristic.
7. Review seven items, other than cell wall structure, that alter the results of a Gram stain.
8. Practice preparing and interpreting Gram stains until the procedure is nearly effortless and your results are reliable.

INTRODUCTION

In this exercise you prepare and interpret Gram stains. The Gram stain is the most important staining procedure in microbiology.

In 1884 Hans Christian Gram was studying the **etiology** (causes and origins) of respiratory diseases. He discovered a staining procedure that differentiated ***Streptococcus pneumoniae*** from human lung tissue, making recognition of the pathogen much easier at autopsy.

His **Gram stain** technique is now performed thousands of times daily by microbiologists worldwide; see figure 8.1. The procedure divides most of the bacteriological assortment of rods, cocci, and spirals into two large, additional groupings. After a Gram stain, cells that looked identical become separable as purple **gram-positive** organisms and pink **gram-negative** ones. Except for gram-positive bacteria, yeasts, and a few molds, almost all cells are gram-negative.

Clinical microbiologists generally perform Gram stains on pure cultures of bacteria. Often this staining is the first step toward identification of a pathogen. **Gram-positive bacteria** can cause boils, wound infections, diphtheria, septic sore throat, scarlet fever, gas gangrene, some pneumonia (including that caused by Gram's original *Streptococcus pneumoniae*), and so on. The **gram-negative bacteria** are etiologic agents of dysentery, typhoid, bubonic plague, and many other diseases.

Figure 8.1 Typical Gram stain results (Photographs courtesy of DIFCO Laboratories, Detroit, MI 48232)

Gram stains:

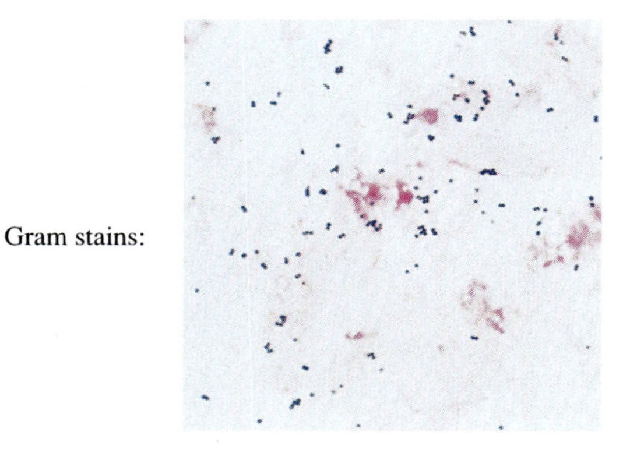

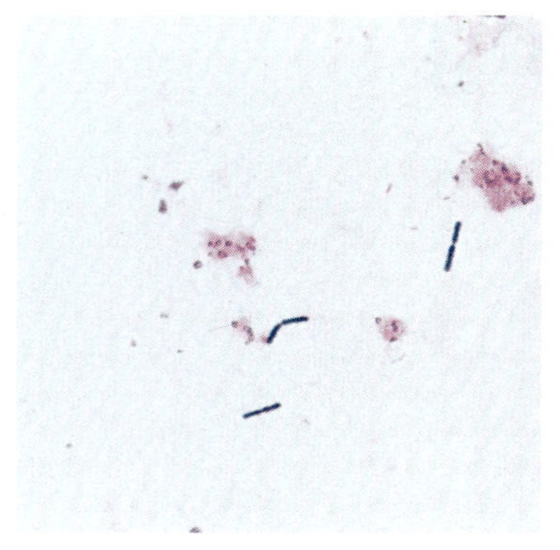

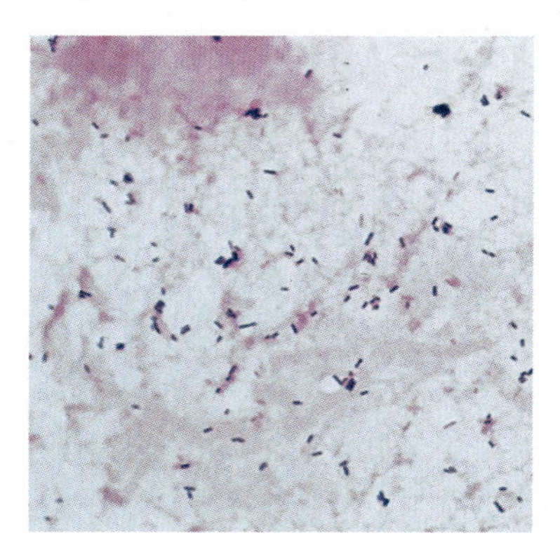

Source:	Chest wall abscess	Pudding cup	Ground turkey
Observation:	Gram-positive cocci in singles, pairs, and clusters	Gram-positive rods and small, pleomorphic, gram-negative rods	Gram-positive rods and gram-negative rods
Culture grew:	*Staphylococcus aureus*	*Bacillus cereus* and *Salmonella pullorum*	*Listeria monocytogenes* and *Salmonella pullorum*

Figure 8.2 The Gram stain procedure

Step	Reagent	Time	Color: Gram-positive	Color: Gram-negative
Primary stain Brief H_2O rinse	Crystal violet	1 min	Purple	Purple
Mordant Brief H_2O rinse	Gram's iodine	1 min	Purple	Purple
Decolorizer Brief H_2O rinse	95% ethanol	Brief	Purple	Colorless
Counterstain Brief H_2O rinse	Safranin	1 min	Purple	Pink

Gram Staining Technique

The Gram stain technique is a **differential stain,** a procedure that distinguishes between various microorganisms according to their ability to retain certain dyes. The Gram stain depends on the ability of some bacterial cells to resist decolorization longer than others; see figure 8.2.

In the Gram stain, first a **primary stain,** crystal violet, is applied to a dried, heat-fixed smear of bacteria. Then a **mordant** (a substance that intensifies the reaction between cells and stain), Gram's iodine, is poured onto the smear. The mordant reacts with crystal violet and the bacteria.

After a quick water rinse, a **decolorizer,** usually 95% ethyl alcohol, is briefly applied. A decolorizer removes color from cells. Gram-negative cells decolorize faster with alcohol than do gram-positives. Results are based on the rate of decolorization, not some absolute reaction. This critical step requires practice and judgment based on experience.

After another water rinse, you have created a differential stain with purple gram-positive bacteria and colorless gram-negatives. To see the gram-negative cells, it is helpful to add the pink **counterstain,** safranin.

Cell Wall Structure

The cell walls of gram-positive bacteria are structurally different from those of gram-negatives. A gram-positive cell wall has a thick layer of peptidoglycan associated with teichoic acids. In gram-negative cell walls, lipoprotein and lipopolysaccharide are found in conjunction with a thin peptidoglycan layer; see figure 8.3. The differences can be verified through chemical analysis, but using the Gram stain procedure is much quicker, easier, and less expensive.

Besides the gram-positives and the gram-negatives, there are other groupings of bacteria. The **gram-variables,** including ***Neisseria*** and ***Moraxella,*** are chemically gram-

Figure 8.3 Gram-positive and gram-negative cell wall structure

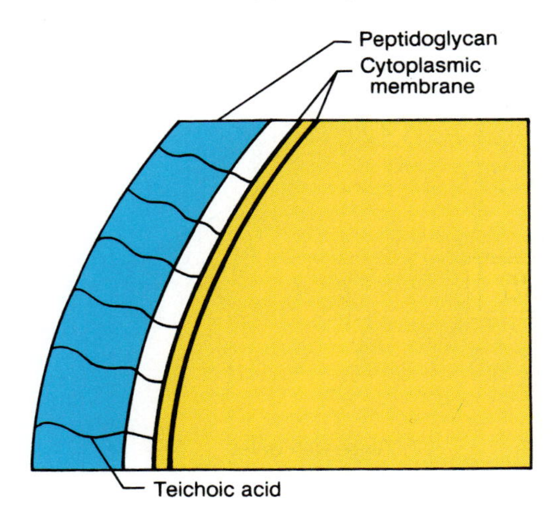

Gram-positive cell wall

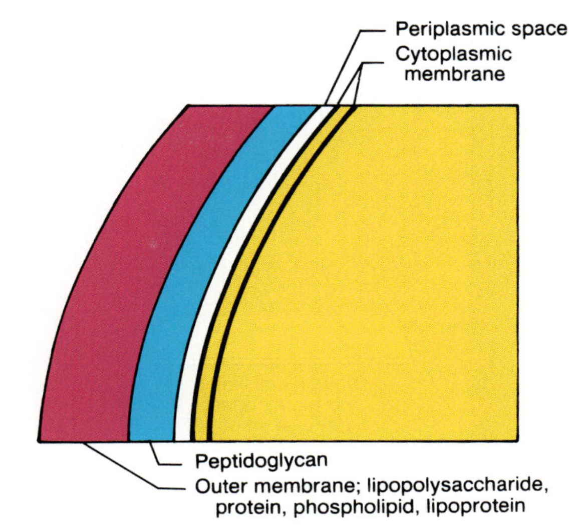

Gram-negative cell wall

Table 8.1 How to Limit Variables that Affect Gram Stain Results

1. **Use actively growing cells.** Old cells lose their ability to hold the stain; they appear gram-negative.
2. **Prepare thin smears and adjust decolorization time.** Where cells are crowded, they resist decolorization. Bacteria in thin smears decolorize much faster than bacteria in thick smears.
3. **Examine well-dispersed cells; avoid stain crystals.** The Gram reactions of clumped bacteria or cells adjacent to crystals of dye are unreliable.
4. **Avoid overheating the cells** during heat fixation. Excessive heat disrupts cell walls making gram-positive bacteria appear gram-negative.
5. **Use fresh staining reagents.** Old reagents give variable results and form crystals.
6. **Adjust timing to your reagents.** For example, many laboratories decolorize with fast-acting acetone-alcohol instead of 95% ethanol.
7. **Do not rinse too long.** Water is a decolorizer.
8. **Remember, gram-positivity is a characteristic that can be lost.**

negative, but they resist decolorization. The **gram-nonreactives** stain poorly or do not stain. ***Mycobacterium,*** which has waxy cell walls, is often gram-nonreactive. (Species of *Mycobacterium* that accept the Gram stain are gram-positive.) The Gram stain is inappropriate for the genus ***Mycoplasma,*** which lacks a cell wall.

Factors Affecting Gram Stain Results

The chemical nature of bacterial cell walls is the primary factor in determining the outcome of a Gram stain. Controlling other variables that may affect your results is explained in table 8.1.

You can see from table 8.1 that this procedure takes practice and careful attention to detail. You must become proficient at preparing and interpreting Gram stains. The technique is fundamental to microbiology. It is performed many times daily in clinical laboratories worldwide and is used repeatedly in the remainder of this laboratory manual. Refer to figure 8.4 and practice until your results are reliable.

PROCEDURES EXERCISE 8

Procedure 1

Preparing Smears for Gram Stains

a Use the techniques described in Exercise 7, Preparing a Smear and Simple Stain, to make thin smears from broth cultures of *Staphyloccocus epidermidis* and *Escherichia coli.*

b Prepare thin smears from slant cultures of *Staphylococcus epidermidis* and *Escherichia coli.*

c While the smears are drying, continue with Procedure 2, then heat fix the dried smears. Allow slides to cool before applying stain.

Procedure 2

Examining Prepared Gram-Stained Slides

a Obtain a prepared Gram-stained slide of *Staphylococcus aureus.* This bacterium is the most common etiologic agent of pimples, boils, and wound infections. Examine it with the oil-immersion lens of your microscope. Record your observations in your laboratory report.

b Obtain a prepared Gram-stained slide of *Salmonella* and examine it microscopically under oil. This genus is the causative agent of salmonellosis, a disease marked by painful, violent diarrhea and cramps. Record your observations in your laboratory report.

Figure 8.4 Gram stain procedure

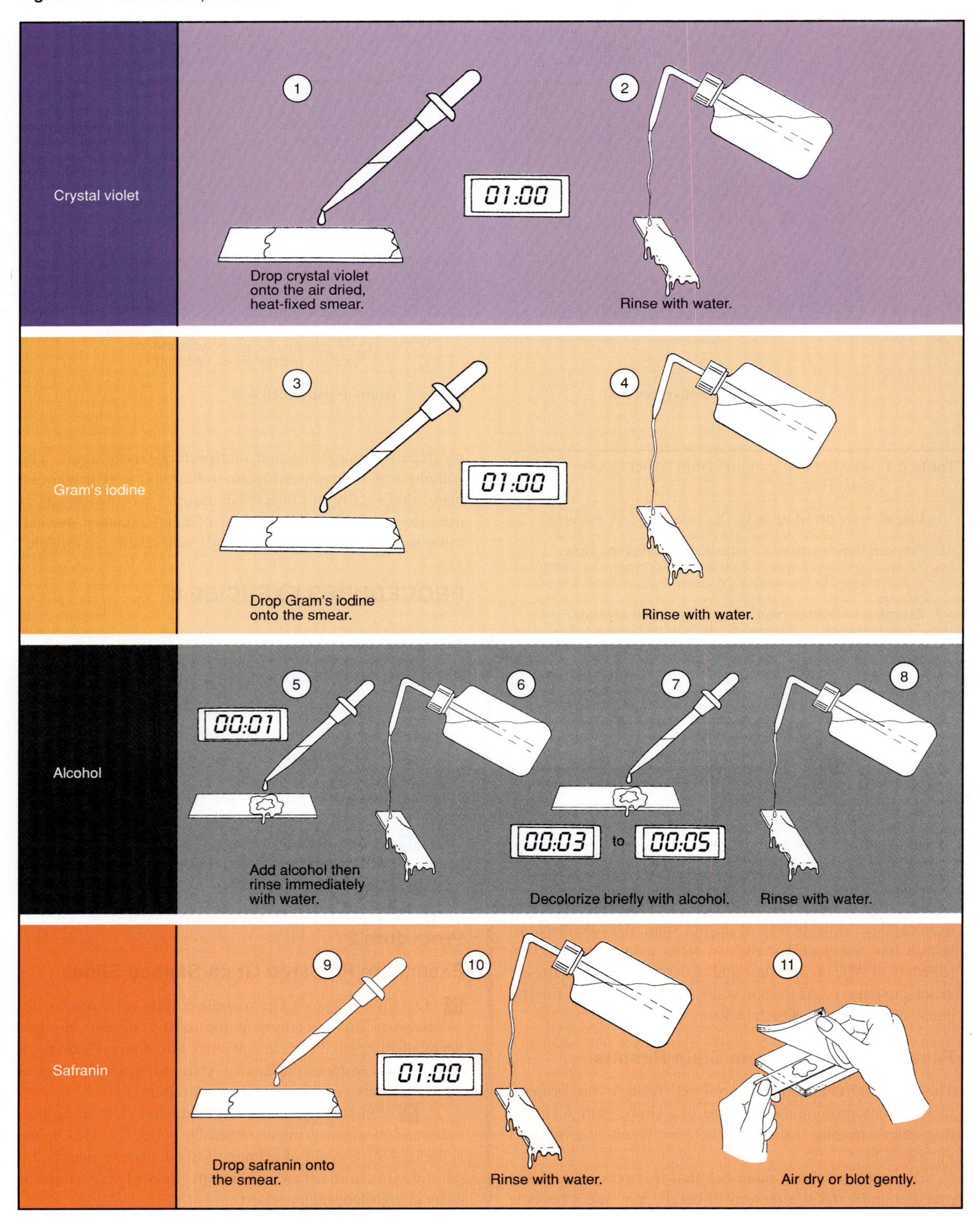

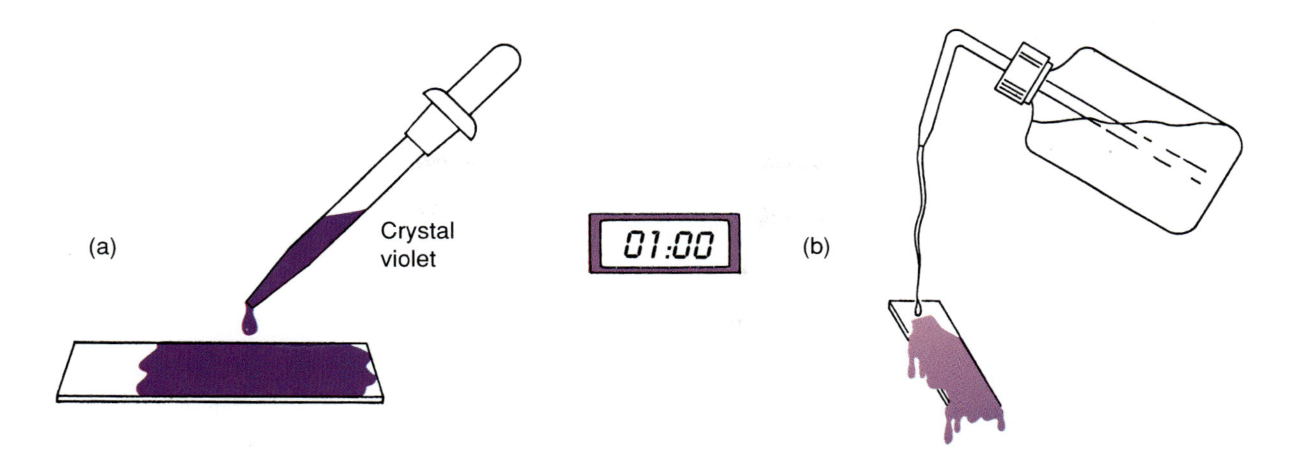

Procedure 3

The Gram Stain Procedure

Your instructor may suggest methods other than those listed here. Refer to figure 8.4 before preparing your Gram stains.

a Use only thoroughly dried and heat-fixed smears for the Gram stain. Cover a cooled slide with **crystal violet** and stain for 1 minute.

b Quickly and gently rinse off the dye with water. Shake excess water off the slide. (In some laboratories Gram's iodine is used instead of water to rinse off crystal violet.)

c Apply **Gram's iodine.** Leave it on the smear for 1 minute.

d Quickly and gently rinse off the mordant with water. Shake off the excess water.

Understand decolorization methodology before attempting it. Read and understand the next two paragraphs **before** decolorizing and/or follow **your instructor's** directions.

e Decolorize the smear for less than 1 second with 95% **alcohol** by applying alcohol, then immediately rinsing the slide gently with water. This removes superfluous color.

f Shake excess water off the slide, then reapply alcohol. This time, allow decolorizer to remain on the smears for approximately 3 to 5 seconds. Rock the slide. Leave the alcohol on the smear until precisely that moment when you no longer see purple stain lifting freely out of the smear. Immediately rinse your slide gently with water. (Now that you have read these directions, perform the decolorization.)

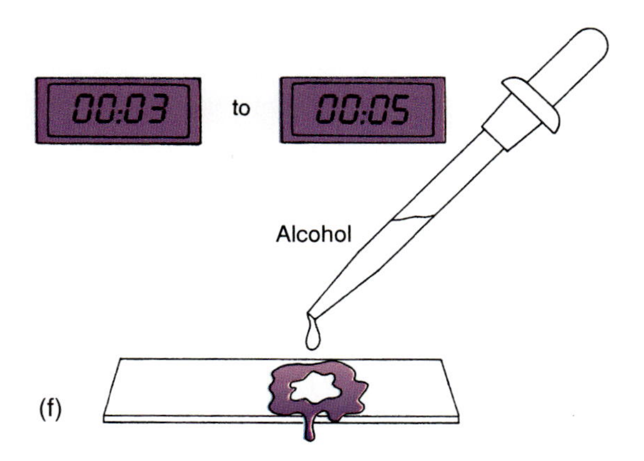

g Shake off any excess water and counterstain with **safranin** for 1 minute. Rinse the slide gently and briefly with water.

h Gently blot the slide dry or allow it to air dry.

i Hold the slide up to the light and examine the smears with your naked eye. Locate areas that look like faint residues of colored water drops on a windowpane. It is these spots that are most likely to contain your bacteria. Circle the spots on the back of the slide. After applying immersion oil for microscopic examination of the smear, these areas become invisible to the unaided eye.

j Locate the circled areas with your low-power lens. Examine the areas with the oil-immersion lens of your microscope. Record your observations in the laboratory report.

k Repeat the Gram stain technique and microscopic examination with your other slides. Record your observations in your laboratory report.

POSTTEST EXERCISE 8

Part 1

Matching

Each response may be used once, more than once, or not at all.

a. Pink b. Purple c. Colorless

In the Gram stain procedure, what is the color

1. _____ of gram-positive bacteria after the primary stain?
2. _____ of gram-negative bacteria after the primary stain?
3. _____ of gram-positive bacteria after decolorization?
4. _____ of gram-negative bacteria after the counterstain?
5. _____ of gram-negative bacteria after the mordant?

Part 2

True or False (Circle one, then correct every false statement.)

1. T F The primary stain in the Gram stain is crystal violet.
2. T F The decolorizer is applied before the mordant in the Gram stain procedure.
3. T F The decolorizer in the Gram stain is Gram's iodine.
4. T F Water rinses should not be left on a smear too long during the Gram stain procedure because water slowly decolorizes stained cells.
5. T F The cell walls of gram-positive bacteria have a thick layer of peptidoglycan associated with teichoic acids.
6. T F The cell wall of a gram-negative bacterium contains lipoprotein and lipopolysaccharide.
7. T F The peptidoglycan layer of a gram-negative cell wall is thicker than the corresponding layer of a gram-positive cell wall.
8. T F *Neisseria* and *Moraxella* are chemically gram-positive, but they stain gram-negative, so they are called gram-nonreactive.
9. T F *Mycobacterium* has waxy cell walls and is often gram-nonreactive.
10. T F Gram-positivity is a characteristic that is easily lost.

Part 3

Completion

1. The general term for a procedure that separates groups of bacteria by depending on the chemical affinities of various bacteria for different dyes is a ________________________ .
2. If you performed a Gram stain on *Streptococcus pneumoniae* in lung tissue, what color would the bacteria be? _______________ What color would the human tissue be? _______________
3. Gram-negative cells are decolorized by alcohol _______________ (faster, slower) than gram-positive cells.
4. List seven pathological conditions caused by gram-positive bacteria.
   a.
   b.
   c.
   d.
   e.
   f.
   g.
5. List three diseases caused by gram-negative bacteria.
   a.
   b.
   c.
6. Name two genera that are gram-variable.
   a.
   b.
7. ______________________________ is a genus for which the Gram stain is inappropriate.
8. The primary factor in determining the outcome of a Gram stain is the ______________________________ .
9. List four factors that can make gram-positive bacteria appear gram-negative in your stain preparation.
   a.
   b.
   c.
   d.
10. List two factors that may cause gram-negative bacteria to appear gram-positive in your Gram stain preparation.
    a.
    b.

Part 4

Prepare an outline that shows how to perform a Gram stain. Include the amount of time, the reagent, and the name for each step of the procedure.

Laboratory Report **8**

Name ______________________________

Section ____________

Gram Staining

Procedure 2

Examining Prepared Gram-Stained Slides

(*a*) Draw and color a typical oil-immersion lens field of *Staphylococcus aureus* as it appears on a prepared Gram-stained slide.

Gram reaction ______________________
Cell shape______________________
Arrangement ______________________
____

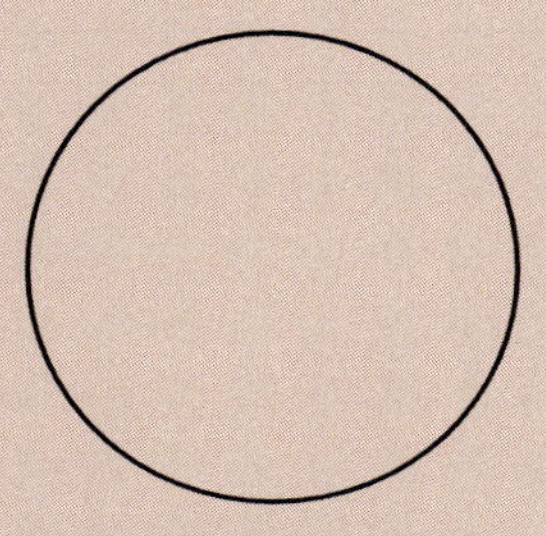

Staphylococcus aureus

(*b*) Draw and color a typical oil-immersion lens field of *Salmonella* as it appears on a prepared Gram-stained slide.

Gram reaction ______________________
Cell shape______________________
Arrangement ______________________
____

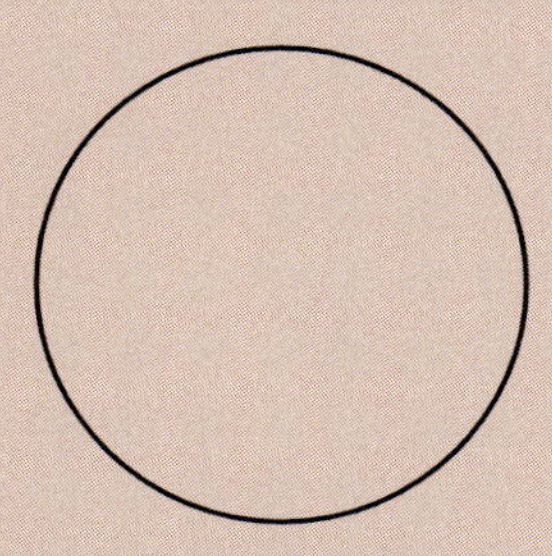

Salmonella species

Procedure 3

The Gram Stain Procedure

(*j,k*) Examine your Gram-stained broth culture and slant culture smears. Draw and color typical oil-immersion lens fields of uncrowded *Staphylococcus epidermidis* and *Escherichia coli*.

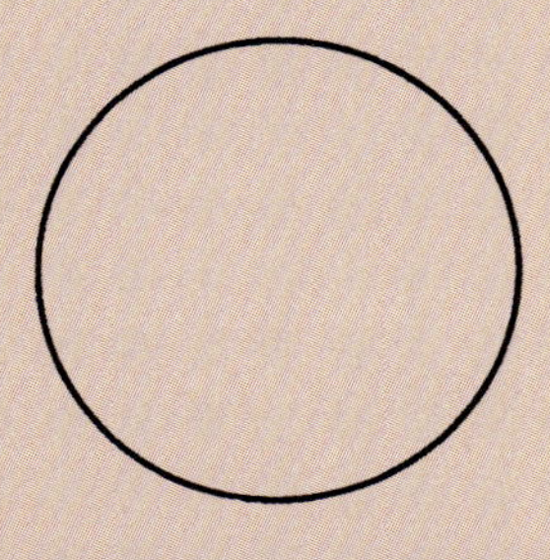

Staphylococcus epidermidis

Gram reaction ____________________
Cell shape ______________________
Arrangement ____________________
____

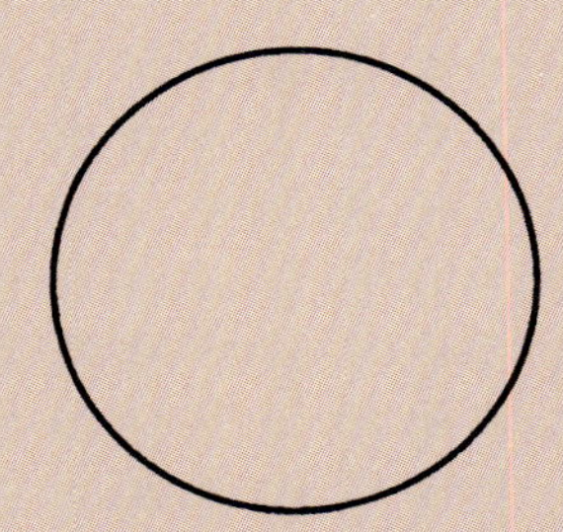

Escherichia coli

Gram reaction ____________________
Cell shape ______________________
Arrangement ____________________
____

Exercise 9

Acid-Fast Staining

Necessary Skills

Microscopy, Aseptic transfer techniques, Preparing a smear and simple stain

Materials

Cultures (1 Tryptic Soy Broth and 1 Tryptic Soy Agar slant of each per set):
- *Staphylococcus epidermidis* (24-hour, 35° C)
- *Mycobacterium smegmatis* (24- to 48-hour, 35° C)

Supplies (per group):
- At least 2 microscope slides
- Acid-fast stain reagent set including:
  - Carbolfuchsin
  - Acid-alcohol (3% HCl in 95% ethanol)
  - Methylene blue
- Ring stand or ring and support
- Paper towel
- Other staining equipment and supplies

Primary Objective

Prepare mixtures of acid-fast and non-acid-fast bacteria, stain them with the Ziehl-Neelsen procedure, and learn to interpret this acid-fast stain accurately.

Other Objectives

1. Name genera that include acid-fast species.
2. Evaluate differences between various methods of acid-fast staining.
3. List the reagents utilized in a Ziehl-Neelsen acid-fast stain.
4. Name three diseases caused by acid-fast organisms.
5. Explain the role mycolic acids theoretically play in the acid-fast staining reaction.
6. List clinical specimens that may contain pathogenic, acid-fast bacteria.

INTRODUCTION

In 1882 **Paul Ehrlich** developed the **acid-fast stain** while working with ***Mycobacterium tuberculosis*** bacilli, the bacterial rods that cause tuberculosis. The acid-fast procedure is a **differential stain,** one that distinguishes one group of bacteria from another. Acid-fast staining renders *Mycobacterium tuberculosis* easily distinguishable from most other bacteria.

Mycobacterium and many ***Nocardia*** species are called **acid-fast** because during an acid-fast staining procedure they retain the primary dye despite decolorization with the powerful solvent **acid-alcohol.** Nearly all other genera of bacteria are **non-acid-fast,** easily decolorized by the acid-alcohol.

Acid-Fast Staining Procedures

Ehrlich's method was improved by later microbiologists including **Ziehl and Neelsen.** Both the Ehrlich and the Ziehl-Neelsen acid-fast staining techniques require heat to drive a primary dye past lipid in the cell wall. Other acid-fast staining protocols avoid heating by employing wetting agents such as detergents. In the **Kinyoun** acid-fast procedure, a high concentration of phenol in the primary dye acts as both a mordant and a wetting agent, making heating unnecessary.

Acid-fast staining has also been modified to accommodate fluorescence microscopy. Auramine O is generally employed as the primary, fluorescent dye.

In this exercise you perform the **Ziehl-Neelsen acid-fast stain;** see figure 9.1. This procedure employs a red primary dye, **carbolfuchsin,** which contains basic fuchsin, phenol, ethanol, and water. After this solution is steamed into the cells for about 5 minutes, the smear is cooled, rinsed with water, and decolorized. The decolorizer is **acid-alcohol** (3% concentrated hydrochloric acid in 95% ethanol).

Figure 9.1 The Ziehl-Neelsen acid-fast stain

Acid-fast stain			
Procedure	Reagent	Cell color	
		Acid-fast	Non-acid-fast
Primary dye	Carbolfuchsin	Red	Red
Decolorizer	Acid-alcohol	Red	Colorless
Counterstain	Methylene blue	Red	Blue

After decolorizing, only **acid-fast** cells remain red. The hydrochloric acid and ethanol solution removes color from non-acid-fast cells. These colorless bacteria are then dyed blue with methylene blue.

Cell Wall Structure and Acid-Fast Staining

In the cell walls of mycobacteria and many species of *Nocardia,* are lipoidal **mycolic acids.** Normally, these lipids prevent dye chromophores from coloring the cell. Some mycobacteria, including ***Mycobacterium tuberculosis,*** contain so much mycolic acid that they are nearly impossible to stain with the Gram stain. Others, like ***Mycobacterium smegmatis*** (originally isolated from smegma) and ***Mycobacterium phlei*** (isolated from hay and grass), are protected by less mycolic acid. These mycobacteria are gram-positive. Mycolic acids are not found in the cell walls of most bacterial genera other than *Mycobacterium* and *Nocardia.*

One theory concerning the ability of the acid-fast stains to differentiate bacteria holds that mycolic acids are somewhat permeable to dye that is dissolved in alcohol and applied with heat and/or a wetting agent. After the bacteria are cooled and/or the wetting agent is rinsed off, mycolic acids in the cell walls coalesce, forming a barrier. The lipids prevent acid-alcohol from decolorizing protoplasm. Other bacteria, lacking a high lipid concentration in their cell walls, are easily decolorized.

Medical Significance of Acid-Fast Bacteria

The acid-fast stain aids in diagnosis of both **tuberculosis,** a leading killer in undeveloped countries, and **leprosy** (also called Hansen's disease), caused by ***Mycobacterium leprae.*** Another acid-fast pathogen is ***Nocardia asteroides,*** the **etiologic** (causative) agent of **nocardiosis,** a disease that generally resembles tuberculosis.

Because of the lipid in their cell walls, bacteria in the genera *Mycobacterium* and *Nocardia* are somewhat resistant to many common disinfectants. For this reason, clinical specimens that may contain tuberculosis, leprosy, or nocardiosis germs must be handled with care.

While the tubercle bacillus can infect every human organ, the specimen that most frequently contains *Mycobacterium tuberculosis* is sputum. The pathogen is also found in other body fluids (urine, synovial fluid, spinal fluid, etc.) and biopsy tissue. Sputum and other body fluids may also contain *Nocardia asteroides.* Scrapings from skin or nasal mucosa, or tissue from a biopsy of earlobe skin, are the usual specimens for *Mycobacterium leprae.*

PROCEDURES EXERCISE 9

Procedure 1

The Acid-Fast Stain (Broth Cultures)

Your instructor may modify these instructions. Many microbiologists employ hot plates or boiling water baths instead of open flames in this staining technique. Refer to figure 9.2 when you prepare a Ziehl-Neelsen acid-fast stain.

(a)

S. e.
BROTH MIX
M. S.

a Draw 3 dime-sized circles on a microscope slide. Label them "*S.e.,*" "Broth Mix," and "*M.s.*" Resuspend bacteria in the staphylococcal broth culture. Depending on broth culture turbidity, place from 1 to 6 loopfuls of *Staphylococcus epidermidis* in the first circle. **Sterilize** your loop after each transfer.

Figure 9.2 The Ziehl-Neelsen acid-fast stain procedure

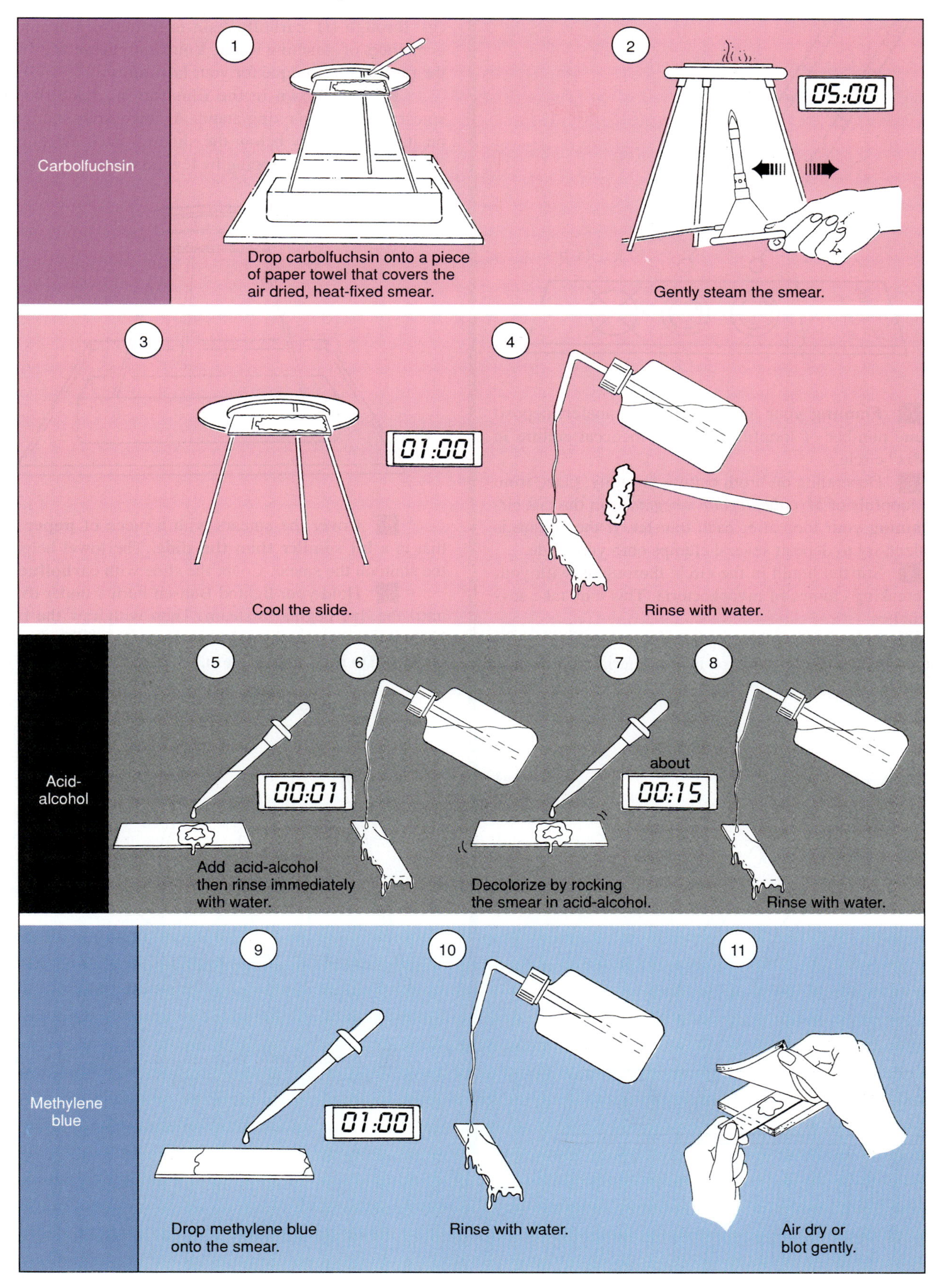

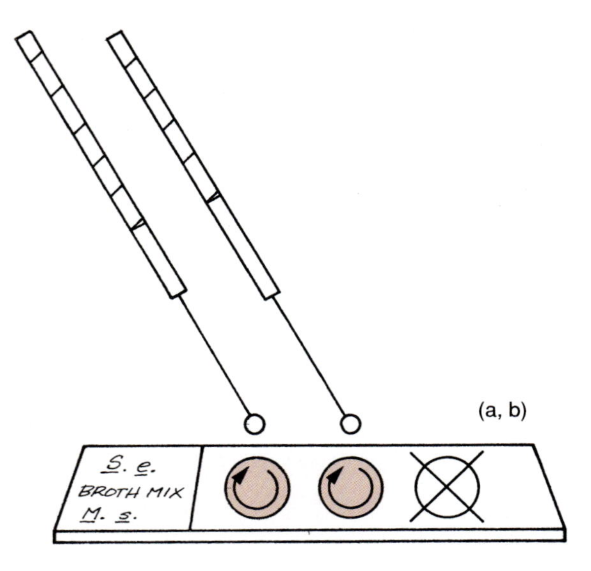

b **Flaming** your loop after each transfer, deposit the same number of loopfuls of staphylococcal culture in the "Mix" circle.

c Depending on broth culture turbidity, place from 1 to 6 loopfuls of *Mycobacterium smegmatis* in the last circle, **flaming** your loop after each transfer. If the culture is flocculent, try to deposit several clumps onto your slide.

d Stir the liquid in the circle thoroughly with your loop, breaking clumps of mycobacteria. This may take several minutes.

e **Sterilizing** your loop after each transfer, place the same number of loopfuls of mycobacterial culture into the "Broth Mix" circle. Break apart any clumps on the slide.

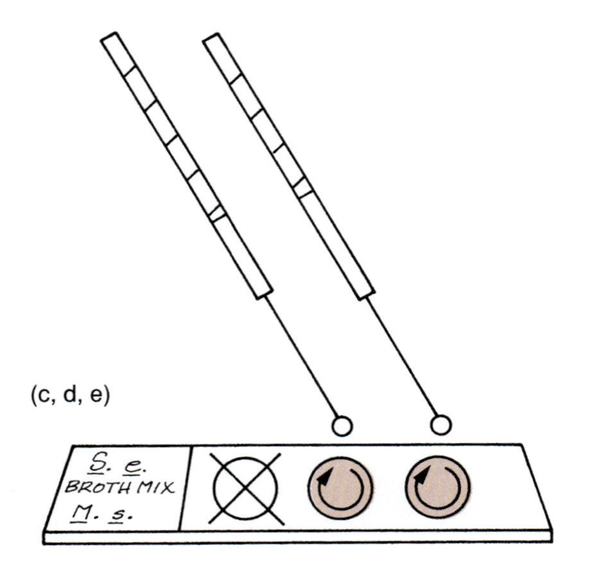

f Air dry, then heat fix the smears.

Instead of employing the method indicated in steps (*g*) through (*i*), carbolfuchsin is sometimes heated over containers of steaming water. **Your instructor** will indicate the preferred technique for your laboratory.

g As shown in the illustration, place the slide, smears up, on your ring stand. Arrange your staining pan on its staining tray below the slide to catch dripping dye. (Prevent excessive dripping by leveling your ring stand.)

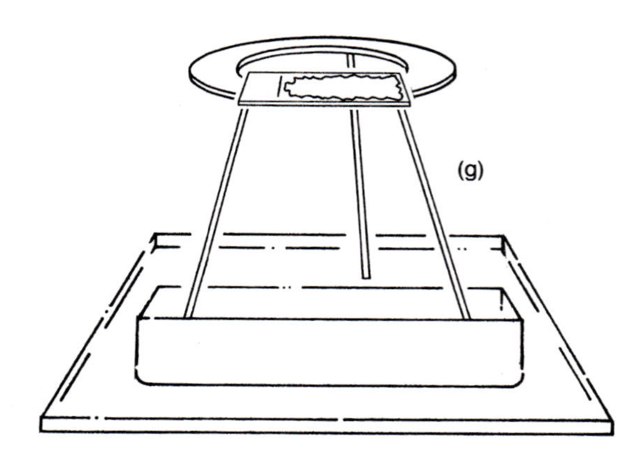

h Cover the smears with a piece of **paper towel** that is **a bit smaller than the slide.** The towel helps hold the stain on the smear. Cover the slide with **carbolfuchsin.**

i Hold your lighted Bunsen burner under the slide until the dye begins to steam. Then withdraw the burner. Keep the slide steaming by intermittently holding the flame under it. **Do not boil the stain.**

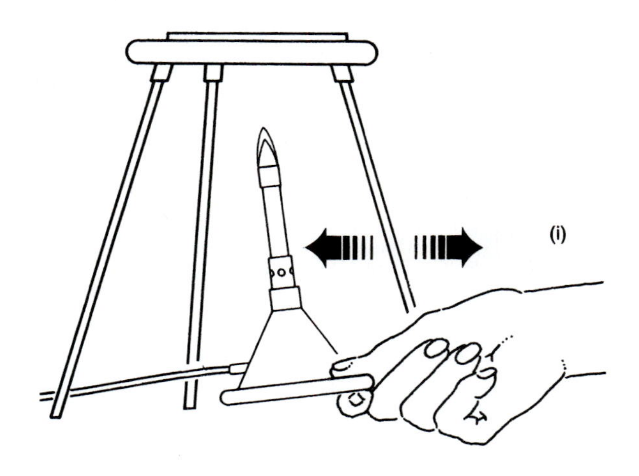

j Steam gently for 5 minutes adding carbolfuchsin as necessary to keep the smear wet.

k Cool the slide for about 1 minute. Rinse it briefly with water from your wash bottle, removing the paper towel; use forceps if necessary. Do not rinse the towel down a drain.

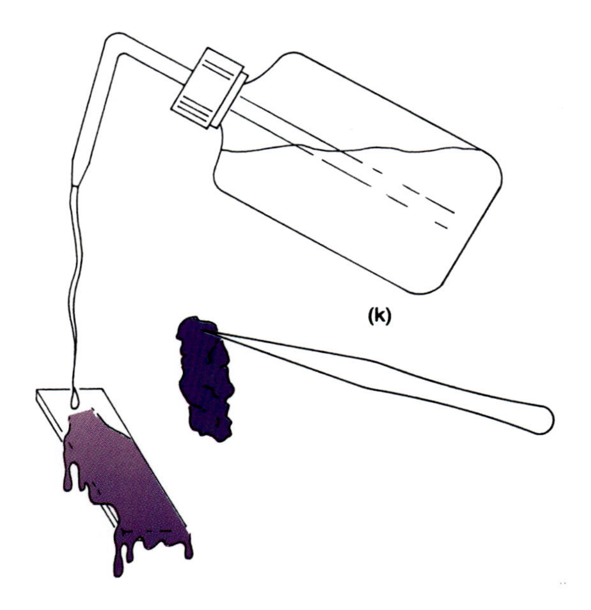

l Rinse quickly with **acid-alcohol,** then water, to remove excess stain.

m Cover the slide with acid-alcohol. Rock the slide for about 15 seconds until the red color is no longer rinsing freely from the cells.

n Rinse the slide briefly with water from your wash bottle.

o Cover the slide with **methylene blue** for 1 minute. Rinse with water from your wash bottle.

If necessary, remove any burnt-on dye from the back of your slide by wiping the back of the slide with a clean paper towel dampened with acid-alcohol. Do not allow the solvent to touch your stained smear.

p Blot or air dry the smears. Add immersion oil and examine the smear with the oil-immersion lens of your microscope.

q **Acid-fast rods appear Bing cherry red** (bright pale red or hot-pink to intense purple). They are often in clumps. **Non-acid-fast bacteria stain light blue** in this Ziehl-Neelsen staining procedure.

r Record your results in your laboratory report.

When cleaning your stain pan, do not allow bits of paper towel to clog the drain.

Procedure 2

The Acid-Fast Stain (Slant Cultures)

a Draw 3 dime-sized circles on a microscope slide. Label them "*S.e.*," "Slant Mix," and "*M.s.*"

b Place a drop of water in each circle.

c Use aseptic technique and your **slant** cultures. Always remember to **flame** and **cool** your needle before placing it into your culture tube. Transfer small, barely visible amounts of *Staphylococcus epidermidis* on the sterile, **cooled** inoculating needle into the appropriate circles on the microscope slide.

d Use the sterilized, cooled inoculating needle to transfer small amounts of *Mycobacterium smegmatis* into the appropriate circles.

e *Mycobacterium* rods, particularly those grown on a solid surface, tend to clump together in characteristic palisade arrangements. You must break apart the larger clumps with your needle so that individual rods become distinguishable. Use your inoculating needle for several minutes to stir and disrupt clumps of *Mycobacterium* on your slide.

f Air dry and heat fix the smears.

g Perform the Ziehl-Neelsen acid-fast staining procedure on the smears as outlined in steps (*g*) through (*o*) of Procedure 1.

h Blot or air dry your smears. Add immersion oil and examine the smears with the oil-immersion lens of your microscope. **Watch for bright red acid-fast bacilli and light blue non-acid-fast cocci.**

i Record your results in your laboratory report.

POSTTEST EXERCISE 9

Part 1

Matching

Each answer may be used once, more than once, or not at all.

a. Carbolfuchsin
b. Methylene blue
c. Acid-alcohol
d. Water
e. Ethyl alcohol
f. Auramine O

1. _____ Which is the decolorizer in the Ziehl-Neelsen acid-fast staining procedure?
2. _____ Which is the primary dye used in the Ziehl-Neelsen technique for compound, light microscopy?
3. _____ Which primary dye is used for fluorescence microscopy acid-fast staining?
4. _____ Which is used briefly to remove excess primary dye, decolorizer, and counterstain?
5. _____ Which is the counterstain in the Ziehl-Neelsen acid-fast staining procedure?

Part 2

True or False (Circle one, then correct every false statement.)

1. T F The Kinyoun acid-fast procedure does not utilize heat.
2. T F Carbolfuchsin contains basic fuchsin, phenol, ethanol, and methylene blue.
3. T F Acid-alcohol contains 3% concentrated NaOH (sodium hydroxide) in 95% ethanol.
4. T F Mycolic acids are lipids found in the cell membranes of acid-fast bacteria.
5. T F *Mycobacterium tuberculosis* rods contain enough mycolic acid so that they are nearly impossible to stain with the Gram stain.
6. T F Cells that contain mycolic acids are more difficult to decolorize with acid-alcohol than cells that do not contain these lipids.
7. T F Bacteria in the genera *Mycobacterium* and *Nocardia* are somewhat resistant to many common disinfectants.
8. T F The only clinical specimen that contains *Mycobacterium tuberculosis* is sputum.
9. T F Urine may contain *Mycobacterium tuberculosis.*
10. T F *Mycobacterium leprae* may be found in scrapings from skin or nasal mucosa, or a biopsy of earlobe skin.

Part 3

Completion

1. In 1882 _______________ developed the acid-fast stain while working with _______________ _______________ bacilli.
2. Three diseases caused by acid-fast bacteria are _____________________ , _____________________ , and _____________________ .
3. The usual specimen that contains *Mycobacterium tuberculosis* is _______________ .
4. *Nocardia asteroides* may be found in sputum or other _______________ _______________ .

Laboratory Report **9**

Name ______________________________

Section ______________

Acid-Fast Staining

Procedure 1

The Acid-Fast Stain (Broth Cultures)

(*r*) Draw and color a typical oil-immersion lens field from each smear.

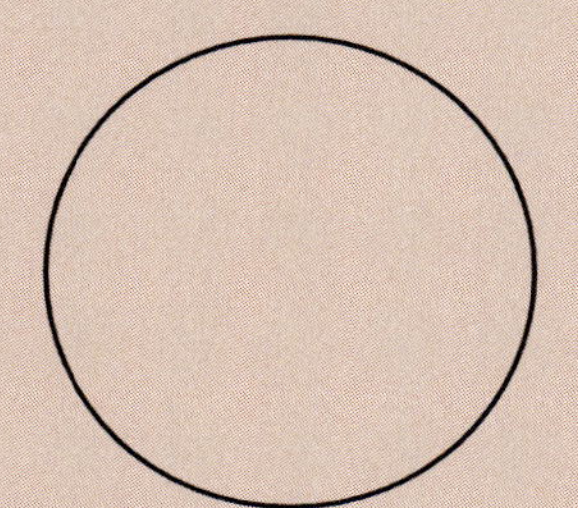

Staphylococcus epidermidis

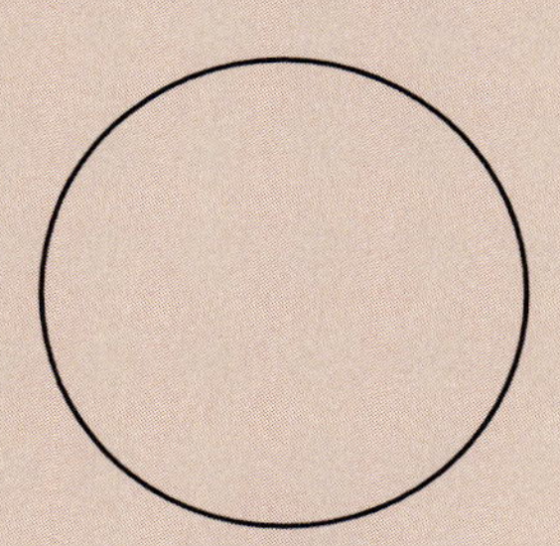

Broth Mix

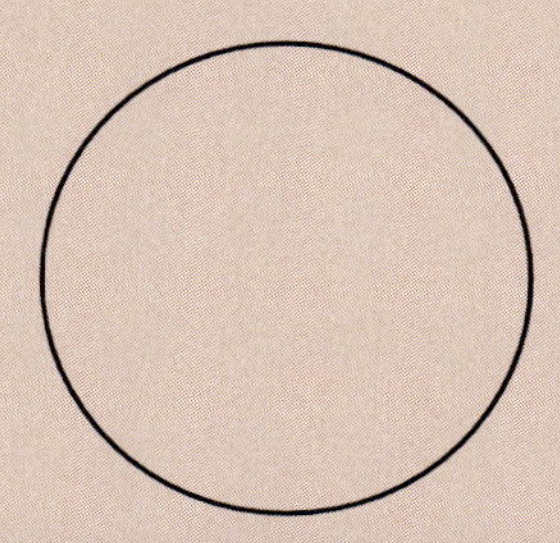

Mycobacterium smegmatis

Acid-Fast Stain of Bacteria from Broth Cultures

Are you satisfied with your results? ___________ If not, what can you do to improve your technique the next time you prepare an acid-fast stain from a broth culture?

Procedure 2

The Acid-Fast Stain (Slant Cultures)

(*i*) Draw and color a typical oil-immersion lens field from each smear.

Staphylococcus epidermidis

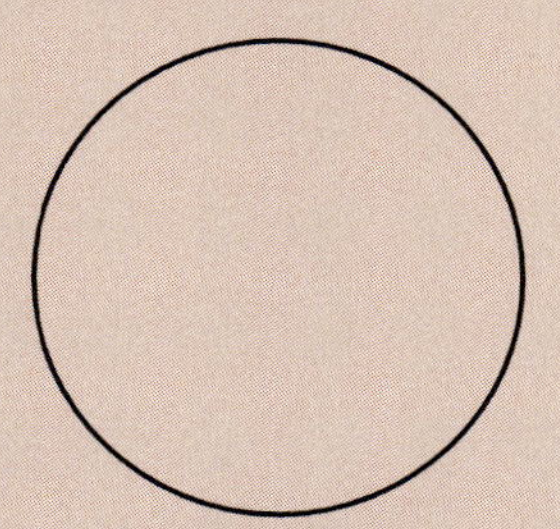

Slant Mix

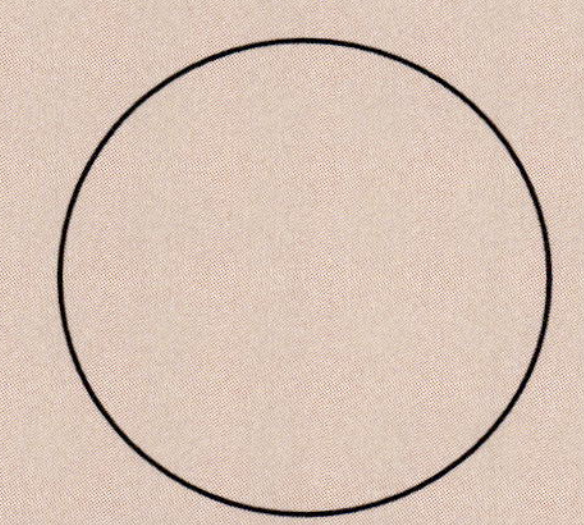

Mycobacterium smegmatis

Acid-Fast Stain of Bacteria from Solid Surfaces

Are you satisfied with your results? ___________ If not, what can you do to improve your technique the next time you prepare an acid-fast stain from a culture grown on a solid medium?

Exercise 10

Endospore Staining

Necessary Skills

Microscopy, Aseptic transfer techniques, Preparing a smear and simple stain

Materials

Cultures (1 per set):
- *Staphylococcus epidermidis* (48-hour, 35° C, Tryptic Soy Agar [TSA] slant)
- *Bacillus subtilis* (3 to 5 day, 35° C, Tryptic Soy Agar slant)

Supplies (per group):
- Microscope slide
- Schaeffer-Fulton endospore-staining reagents:
  - Malachite green, 5% aq.
  - Safranin, 0.5% aq.
- Paper towel
- Ring stand or support and ring
- Access to acid-alcohol
- Other staining equipment and supplies

Primary Objective

Prepare mixtures of endospore-forming rods and non-endospore-forming cocci, stain them with the Schaeffer-Fulton procedure, and learn to interpret this endospore stain accurately.

Other Objectives

1. List the bacterial genera that form endospores.
2. Define: cryptobiotic state, sporangium, saprophytes.
3. Illustrate various characteristics of endospore containment.
4. Name the genus and specific epithet of four endospore-forming bacteria that cause disease.
5. List four types of antibiotics produced by endospore-formers and indicate the mode of action for each.

Table 10.1 Endospore-Forming Rods and Cocci

Genus	Description
Bacillus	Aerobic or facultative, flagellated or nonmotile, gram-positive rods
Sporolactobacillus	Peritrichously flagellated, gram-positive rods; ferment lactose; lack cytochromes
Clostridium	Anaerobic or microaerotolerant, flagellated or nonmotile, gram-positive rods
Desulfotomaculum	Flagellated, anaerobic, sulfate- and sulfite-reducing, gram-positive rods; produce hydrogen sulfide
Sporosarcina	Aerobic, flagellated or nonmotile, tetrad or packet-forming, gram-positive cocci
Oscillospira	Anaerobic, flagellated, gram-negative rods or filaments divided into numerous disk-shaped cells

INTRODUCTION

In this exercise you investigate the staining properties of bacterial **endospores.** These hard, dry structures form inside certain genera of rods and cocci; see table 10.1. When the **vegetative** (feeding, dividing) **cell** dies, a free endospore remains, as shown in figure 10.1. The endospore preserves life.

Endospores

Endospores exist in a **cryptobiotic** state, one with no measurable metabolic activity.

These metabolically inert life-forms are resistant to the harmful effects of desiccation and ultraviolet radiation. They survive lengthy exposure to strong acids, bases, and disinfectants. Endospores resist staining with the aniline dyes commonly employed by microbiologists.

As shown in figure 10.2, endospores are located centrally, subterminally, or terminally. If the diameter of the endospore is larger than the diameter of the cell, the

Figure 10.1 An endospore

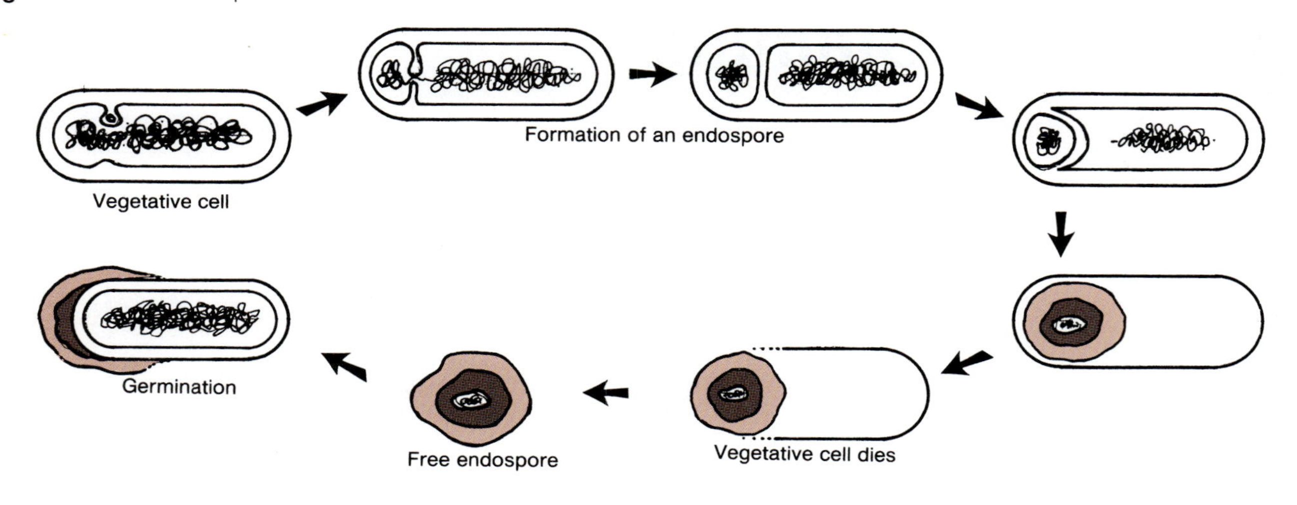

Figure 10.2 Location of an endospore inside a sporangium

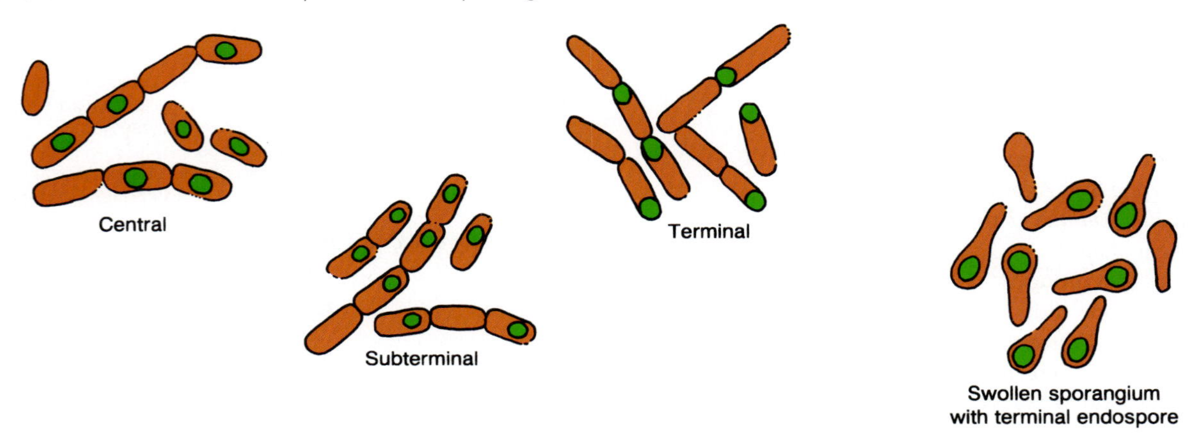

sporangium (vegetative cell that holds an endospore) is swollen. Because these characteristics of endospore containment are stable for each species, microbiologists employ these features to help identify the endospore-formers.

The Schaeffer-Fulton Endospore Stain

The **Schaeffer-Fulton** endospore-staining procedure employs heat to drive the primary dye, **malachite green,** into the endospore. Once dyed, it is difficult to decolorize an endospore. A water rinse strips malachite green from vegetative cells, leaving them colorless. The counterstain, **safranin,** is then employed to color vegetative cells pink. The resulting preparation shows green endospores, which may be either free or contained within pink bacterial cells.

Medical Importance of Endospore-Forming Bacteria

Many endospore-forming bacteria are harmless soil **saprophytes,** organisms that live on dead matter, but a few are important pathogens. ***Clostridium botulinum*** produces the most deadly, biological nerve toxin known. A soft drink bottle of pure **botulism** toxin could contain enough poison to kill every human on earth. ***Clostridium tetani*** toxin kills with the rigid paralysis of **tetanus** (lockjaw). ***Clostridium perfringens*** and other clostridia cause deadly **gas gangrene.**

Bacillus anthracis is the etiologic agent of **anthrax,** a farm animal disease that is transmissible to people. Other *Bacillus* and *Clostridium* species poison food causing **gastroenteritis,** inflammation of the stomach and intestines. Foods that are left at room temperatures after cooking are especially hazardous, since normal cooking temperatures often leave endospores unharmed and ready to germinate. Refrigeration retards germination.

The endospore-formers *Bacillus* and *Clostridium* are also important medically because they supply many **antibiotics,** metabolic products that kill or inhibit the growth of other microbes. Antibiotics produced by endospore-formers include the bacitracins, cyclic peptides that

Figure 10.3 The Schaeffer-Fulton endospore-staining procedure

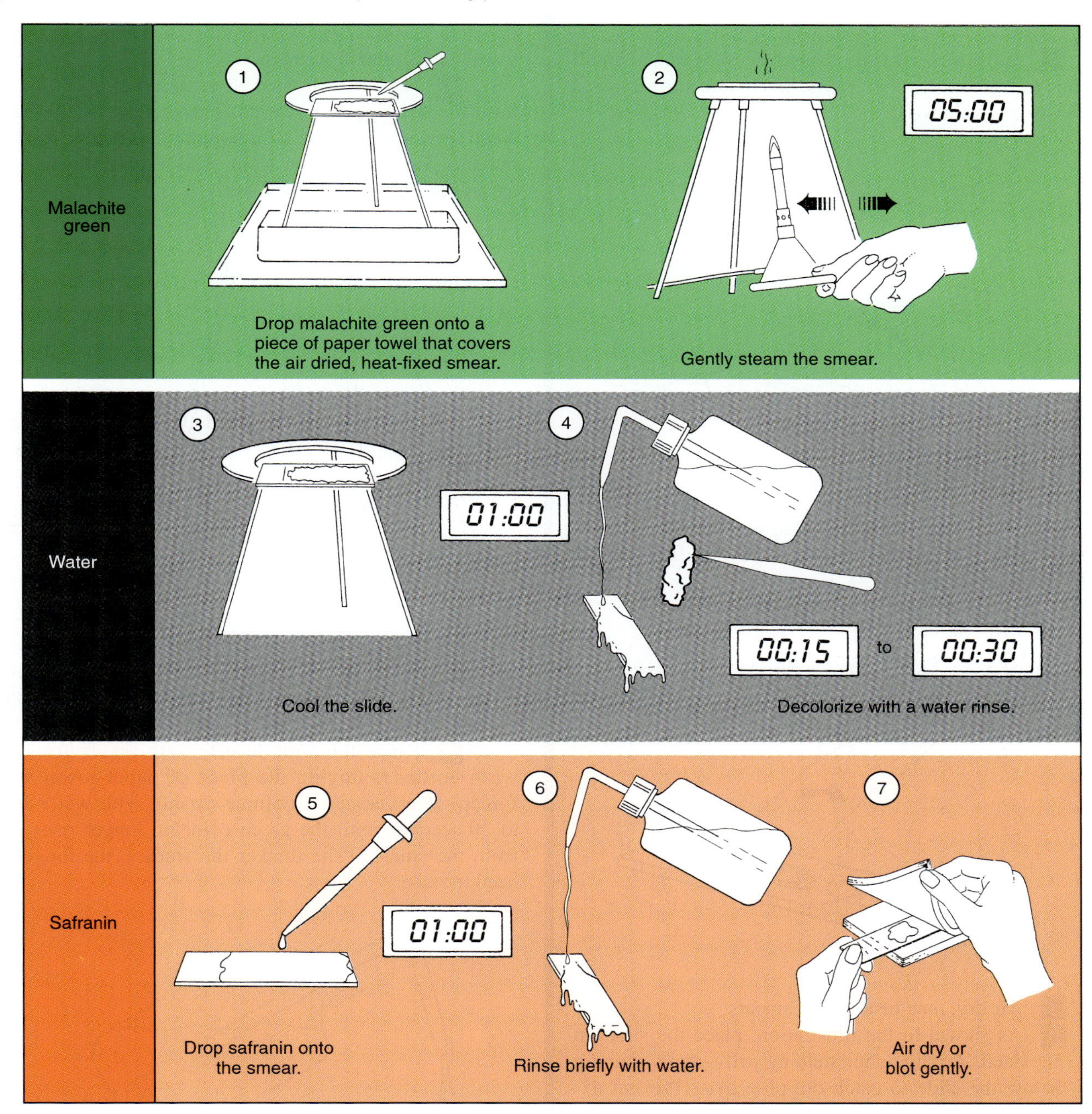

inhibit cell wall synthesis. Also produced are gramicidin, polymyxin, and tyrocidin-type peptide antibiotics that alter cell membranes.

In this exercise you perform the **Schaeffer-Fulton staining procedure,** using heat to drive malachite green into endospores. Refer to figure 10.3 when you prepare a Schaeffer-Fulton endospore stain.

Repeat the Schaeffer-Fulton procedure, making notes on any variations in timing, thickness of smear, concentration of staining reagent, and so on. Use these notes to help you produce repeatable, accurate results.

PROCEDURES EXERCISE 10

Procedure 1

The Schaeffer-Fulton Endospore Stain

Your instructor may suggest modifications of the Schaeffer-Fulton procedure that are more suitable for your laboratory than the methods described here. Many microbiologists employ hot plates or boiling water baths instead of open flames in this staining technique.

a Draw 3 dime-sized circles on a microscope slide. Label them "*S.e.,*" "Mix," and "*B.s.*" With your inoculating loop, transfer a drop of water into each circle.

b With a sterilized, cooled needle, mix a small amount of *Staphylococcus epidermidis* into the first and second drops as shown in the illustration.

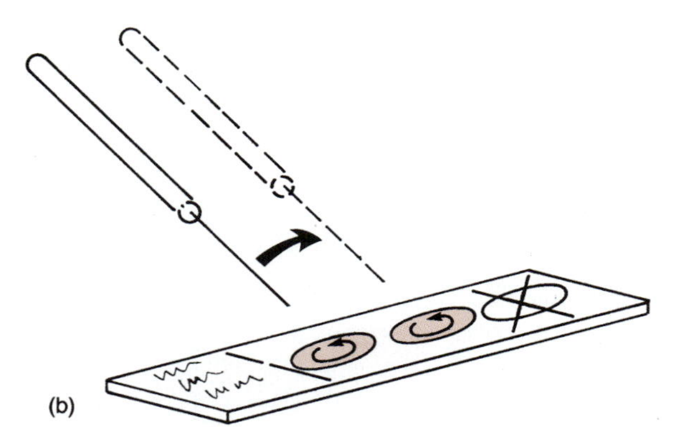

c With your flamed, cooled needle, mix a small amount of *Bacillus subtilis* into the third, then the second, circles. Flame your needle.

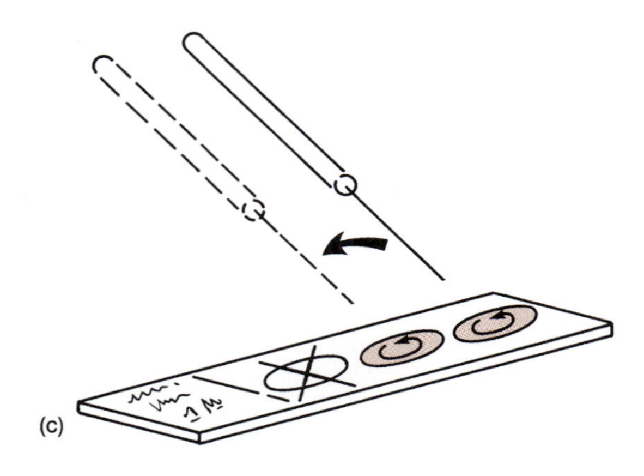

d Air dry, then heat fix the smears.

e As shown in the illustration, place the slide on your ring stand. Arrange your staining pan—on its staining tray—below the slide to catch dripping dye. (Prevent excessive dripping by leveling your ring stand.)

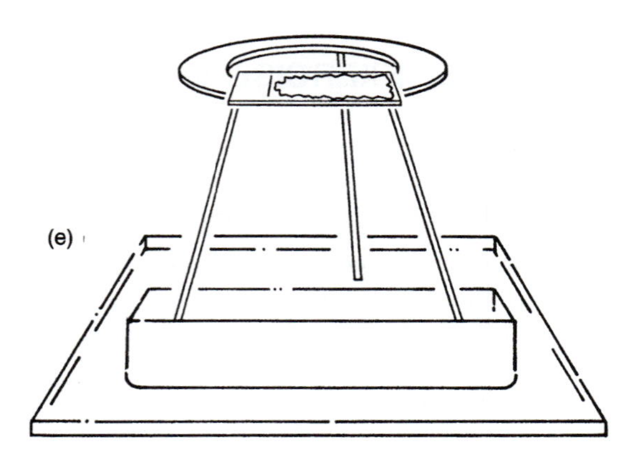

f Cover the smears with a piece of **paper towel** that is a bit **smaller than the slide.** The towel helps hold the stain on the smear. Cover the entire slide and piece of towel with **malachite green.**

g Hold your lighted Bunsen burner under the slide until the dye begins to steam. Then withdraw the burner. Keep the slide steaming by intermittently holding the flame under it. **Do not boil the stain.** Add more malachite green as needed to keep the smear wet.

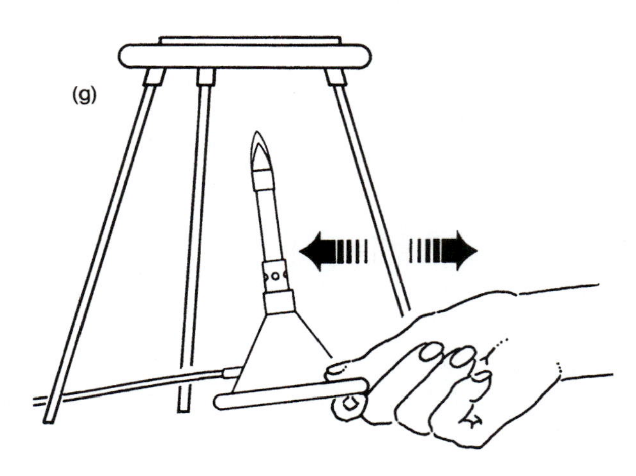

h Steam gently for 3 to 5 minutes.

i Cool the slide on the ring stand for about 1 minute. In your laboratory report, explain why you cool the slide.

j Rinse the slide briefly with **water** from your wash bottle, removing the piece of paper towel with a forceps if necessary. Continue rinsing with water for 15 to 30 seconds until the green color no longer rises freely from the smears. The thicker the smears, the longer you need to rinse.

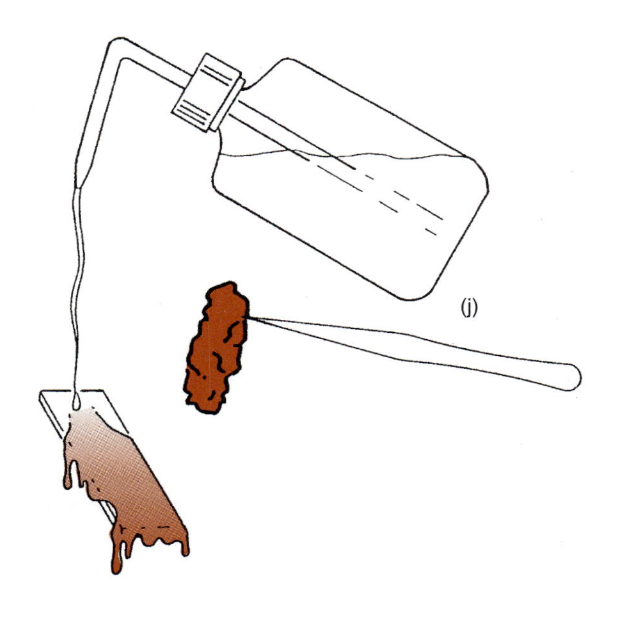

k Stain the slide with **safranin** for 1 minute.

l Briefly rinse with water from your wash bottle.

m If necessary, remove any burnt-on dye from the back of your slide by wiping the back of the slide with a clean paper towel dampened with acid-alcohol. Do not allow the solvent to touch your stained smear.

n Blot or air dry the smear. Add immersion oil, and examine the smear with the oil-immersion lens of your microscope.

o Record your results in your laboratory report. Carefully note the difference between the staining properties of cocci and the oval to spherical endospores that they superficially resemble.

When cleaning your stain pan, do not allow bits of paper towel to clog the drain.

POSTTEST EXERCISE 10

Part 1

Matching

Each answer may be used once, more than once, or not at all.

a. Safranin
b. Methylene blue
c. Malachite green
d. Water

1. _____ Which is the decolorizer in the Schaeffer-Fulton endospore-staining procedure?
2. _____ Which is the primary dye used in the Schaeffer-Fulton endospore-staining procedure?
3. _____ In the Schaeffer-Fulton endospore stain, which of these is the counterstain?
4. _____ Which of these is a common aniline dye used by microbiologists but not employed in the Schaeffer-Fulton endospore-staining procedure?

Part 2

True or False (Circle one, then correct every false statement.)

1. T F Bacterial endospores are hard, dry structures that burst forth as external buds on certain genera of flowering rods.
2. T F The Schaeffer-Fulton endospore-staining procedure employs ethyl alcohol and phenol to drive malachite green into endospores.
3. T F All endospores are terminally located in swollen sporangia.
4. T F The most deadly biological nerve toxin known is botulism toxin.
5. T F *Clostridium tetani* toxin kills with the rigid paralysis of tetanus.
6. T F *Bacillus anthracis* is the etiologic agent of anthrax.
7. T F Endospores in cooked food may germinate at room temperatures and lead to food poisoning.
8. T F After staining with the Schaeffer-Fulton endospore procedure, endospores appear pickle green while vegetative cells are periwinkle blue.

Part 3

Completion

1. Endospores preserve life in a/an __________________ state with no measurable metabolic activity.
2. Endospores are resistant to the lethal effects of
____,
____,
____,
____, and
____.
3. Endospore-forming bacteria produce the __________ antibiotics that inhibit cell wall synthesis.
4. The endospore-formers produce gramicidin, polymyxin, and tyrocidin-type peptides that inhibit microbial growth by ____________________
____
____.

Laboratory Report **10**

Name ____________________

Section ____________

Endospore Staining

Procedure 1

The Schaeffer-Fulton Endospore Stain

(*i*) After steaming the slide for 3 to 5 minutes, you cool the slide before flooding it with cold water. Why? (Hint: What happens to hot glass that is plunged into cold water?)

(*o*) Draw and color a typical oil-immersion lens field from each smear. Clearly show the placement of endospores that are contained within vegetative cells. Indicate whether the sporangia are swollen. Label free endospores, vegetative cells, and endospores within sporangia.

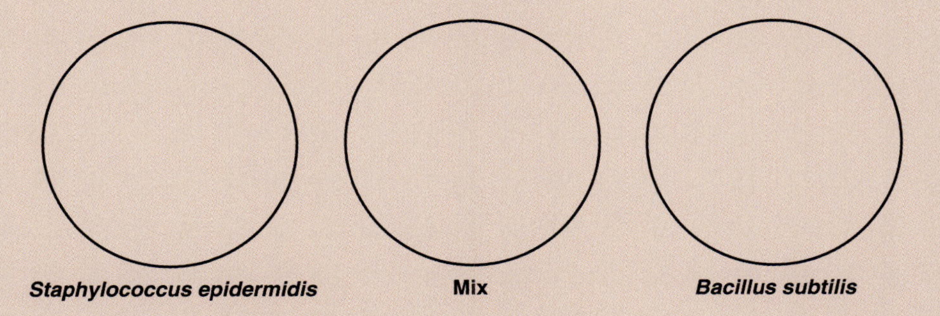

Compare and contrast the appearance of bacterial cocci, represented by *Staphylococcus epidermidis,* and the endospores of *Bacillus subtilis.*

Are you satisfied with the results of your endospore stain? ________ If not, how can you improve your results the next time you prepare an endospore stain?

Ubiquity and Culture Characteristics

Necessary Skills

Knowledge of the metric system of linear measurement

Materials

First Laboratory Session

Cultures (24- to 48-hour, 35° C, Tryptic Soy Broth, 1 per set):
Micrococcus luteus
Bacillus subtilis
Enterococcus faecalis
Supplies:
Procedure 1
3 Tryptic Soy Agar plates per student plus 2 per class to serve as uninoculated, negative controls
2 Moist, sterile cotton-tipped applicators in a capped tube per student
2 Dry, sterile cotton-tipped applicators in a capped tube per student
Sterile test tubes, a few per class
Sterile diluent in test tubes, a few per class
Procedure 2 (per group)
3 Tryptic Soy Agar slants
3 Tryptic Soy Broths

Second Laboratory Session

Supplies:
Metric rulers

Primary Objective

While exploring the ubiquity of microorganisms, learn to employ the scientific method of investigation and to utilize the descriptive vocabulary of microbiology.

Other Objectives

1. Define: ubiquitous, microbial colony, taxonomy.
2. List conditions of incubation that affect microbial growth.
3. Measure the abundance of microbial life in an environment by sampling, plating, and estimating the numbers of colonies that develop on a plate.
4. Estimate the diversity of microbes in an environment by sampling, plating, and recognizing various growth characteristics.
5. Explain why a microbiology student needs to know the descriptive vocabulary associated with microbial growth.
6. Prepare broth cultures and recognize bacterial growth patterns in liquid media.
7. Prepare slant stroke cultures and identify various bacterial growth characteristics on slants.
8. Describe eight characteristics of bacterial colonies.
9. Inspect isolated microbial colonies and use microbiological terminology to describe seven or eight characteristics of each colony.

INTRODUCTION

Microorganisms are said to be **ubiquitous;** that is, they seem to be present nearly everywhere. They are in the air we breathe, the food we eat, and on the lips we kiss. In this exercise you design experiments to discover the numbers and diversity of microbes around us.

Microbial Growth

Individual microbes are too small to be seen with the naked eye. But if these cells arrive in an environment that provides appropriate nutrients, chemicals, temperature, and other **conditions of incubation,** they divide. Given time, microbes on a solid, nutrient surface may multiply until they form a **colony** (macroscopic microbial growth resulting from the multiplication of a single organism or small cluster of similar cells). If the growth is in a nutrient liquid, the **broth** (liquid medium) may become cloudy.

Thus you can test to see if organisms are in a particular environment by providing them with favorable conditions of incubation and then evaluating their growth.

When a bacterial or fungal cell falls onto a Petri dish of solid growth medium and finds suitable conditions of incubation, it may form a colony. You can **estimate how many** microbes are in an environment by sampling the environment, plating the sample, and counting the colonies that

Table 11.1 Scientific Method of Investigation

Step 1. First, you must **recognize a problem** or situation. (Example: There are fruit flies in the laboratory.)
Step 2. Then, **gather information** concerning it. (Example: Consult reference books and perhaps an entomologist concerning the habits of fruit flies.)
Step 3. Once you are well informed on the topic, **formulate a hypothesis that can be tested.** (Example: Fruit flies are attracted by sweet-smelling disinfectant.)
Step 4. **Make a prediction** concerning the outcome of your investigation. (Example: Fruit flies will not infest the laboratory if bleach water is used as a disinfectant instead of the sweet-smelling product.)
Step 5. Design and carry out experiments and/or observation schedules that **test the hypothesis. Gather data.** (Example: Count the flies in one-half of the laboratory. Use only bleach water as a disinfectant on this side of the room for 1 month. Count the flies in the bleach water side of the room again at the end of the month.)
Step 6. Include **a control.** The control is a test that is almost exactly the same as your main experiment. Only one central factor being tested is altered. (Example: Count flies in the other half of the room. Continue use of the sweet-smelling disinfectant on this side. Count flies here at the end of the month. Compare the numbers, before and after the month, on both sides of the room.)
Step 7. **Arrive at a conclusion by evaluating your data.** (Example: After the month, both sides of the room have about the same number of fruit flies, and there are more flies than when the experiment started.)
Step 8. **Reevaluate** your experimental design. (Example: Some other attractor must be drawing flies. Repeat the experiment but alter the lure.)
Step 9. **Formulate another hypothesis and test it.** (Example: Fruit flies are attracted by tubes of fresh media stored in unsealed cupboards around the laboratory.)

develop. Since the appearance of each colony reflects the biochemistry of whichever species formed it, you can **estimate the diversity** of microbial life in the environment by counting how many different types of colonies grow.

Growth patterns of microorganisms reflect **conditions of incubation** such as **time, incubation temperature, humidity, pH, salinity,** availability of **nutrients,** and so on. But, if these conditions are well defined, the features displayed by growing bacteria and fungi often give the microbiologist a first hint as to the identity of an unknown culture.

Scientific Method of Investigation

In this exercise you apply the **scientific method of investigation** to test a **hypothesis** (trial idea, unproven theory, supposition) concerning the ubiquity of microbes. The scientific method is summarized in table 11.1. As demonstrated in the table, it can be used to help find solutions for problems.

The scientific method of investigation has given the twentieth century a vast amount of knowledge based on observed and tested facts. Like a careful detective solving a mystery, the researcher who utilizes the scientific method avoids confusing feelings and prejudices with observable or experimentally repeatable phenomena.

In this exercise you work together in groups with your colleagues to test whether bacteria are ubiquitous. You may feel that they are not.

Table 11.2 Typical Environmental Exposures for Tryptic Soy Agar

Air sample: Place a Petri plate open in the classroom for about 1 hour. Place a Petri plate open in a garden (in the shade) for about 30 minutes.
Soil sample: Dilute a few grains of rich, garden soil in about 1 ml of sterile diluent; moisten a cotton-tipped applicator with the dilute soil; swab the entire agar surface of a plate.
"Clean" surfaces (counters, plates, skin): Rub a moistened cotton-tipped applicator over a 2–10 cm diameter circular area (state the size of the area and do not vary the size in any one experiment); swab the entire agar surface of a Petri plate. The diameter you choose depends on the expected number of bacteria on the surface. For ease of counting, try to isolate fewer than 300 colonies per agar plate.
Objects: Lightweight objects such as hairs, feathers, fingernail parings, saliva droplets, paper, small insects, etc., generally adhere to the surface of an agar plate.
Notes:
1. Remember to **invert** your plates for incubation and for storage. In this position water from condensation does not drip onto developing colonies. Instead, it helps keep the agar moist.
2. Label the outer surface of the **bottom** (the side with the agar in it) of each agar plate.
3. Incubate agar plates containing body temperature samples at 35°–37° C. Samples from room temperature environments are best incubated at 30° C or ambient temperatures.

For example, you may feel that because you washed your face before coming to class and you are a "clean" person, your lips are not covered with microorganisms. To show that microbes are not ubiquitous, you kiss a Petri plate of Tryptic Soy Agar and incubate the plate at body temperature until the next laboratory period. What results will you observe? Try it and find out. See table 11.2 for further suggestions.

The Terminology of Growth Characteristics

To report the growth characteristics of bacteria accurately, you must learn the descriptive vocabulary of a microbiologist. Descriptions should be precise and convey specific meanings to other scientists.

The terms introduced in this exercise are frequently utilized in ***Bergey's Manual of Systematic Bacteriology.*** This tome, now in four volumes, is accepted worldwide as a source book for bacterial **taxonomy,** the science of classification.

Later in this course, you employ *Bergey's Manual* to help identify unknowns, so master the vocabulary now.

Broths

Bacteria develop many growth patterns in broths. Some of these growth characteristics are illustrated in figure 11.1. Bacteria suspended in a broth may cause **turbidity** (cloudiness), or they may be **flocculent** (floating in clumps). They may form a **ring** around the top rim of the medium, or they may float in a heavy **pellicle** on the surface. Cells often sink to the bottom of the tube, forming a **sediment.** Some genera even form **streamers** of growth that float down from the surface into the broth.

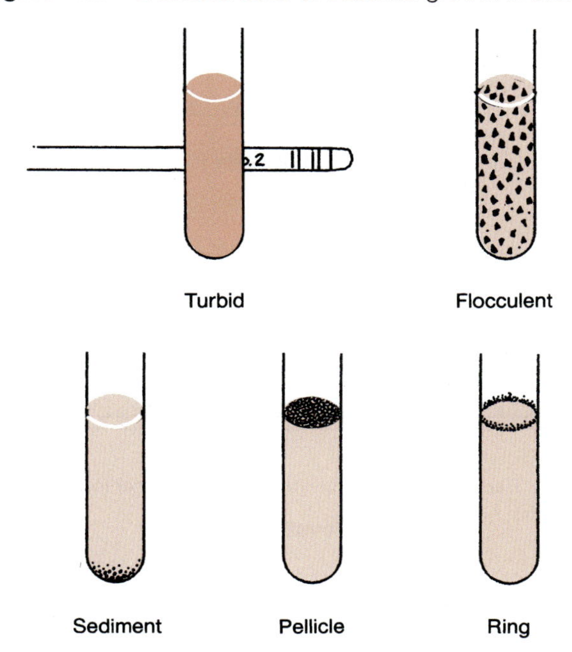

Figure 11.1 Characteristics of bacterial growth in broth

Combinations of these features are common. For instance, your broth culture may be turbid with heavy sediment and slight ring formation. Various growth patterns are characteristic of different bacterial species and are often used to help identify them.

Agar Slants

Growth characteristics of bacteria on agar slants are illustrated in figure 11.2. Each slant is inoculated with a single straight stroke of the loop.

After incubation, the bacteria may have **spread,** covering the entire surface with **heavy** or **effuse** (thin) growth. They may be **beaded,** in small colonies. If the growth resembles the branches of a tree, it is called **arborescent;** if it looks more like the roots, it is termed **rhizoidal.** If the edges of the stroke are smooth, microbiologists describe the growth as **filiform;** if spiny or saw-toothed like a serrated knife, the growth is called **echinulate.**

Various growth characteristics of bacteria on slants are demonstrated by each species, so they are used to help identify unknowns. They may be found in combinations. For example, your slant stroke may be spreading at the bottom, filiform in the middle, and beaded at the top.

The **pigmentation** and **surface attributes** of slant growth are also evaluated. These features are shared by colonial growth and are explained in table 11.3.

Isolated Colonies on Plates

In this exercise you become familiar with many growth characteristics of bacterial colonies; see figure 11.3. A complete description of the growth patterns of a bacterial colony includes the features listed in table 11.3.

Figure 11.2 Characteristics of bacterial growth on slants

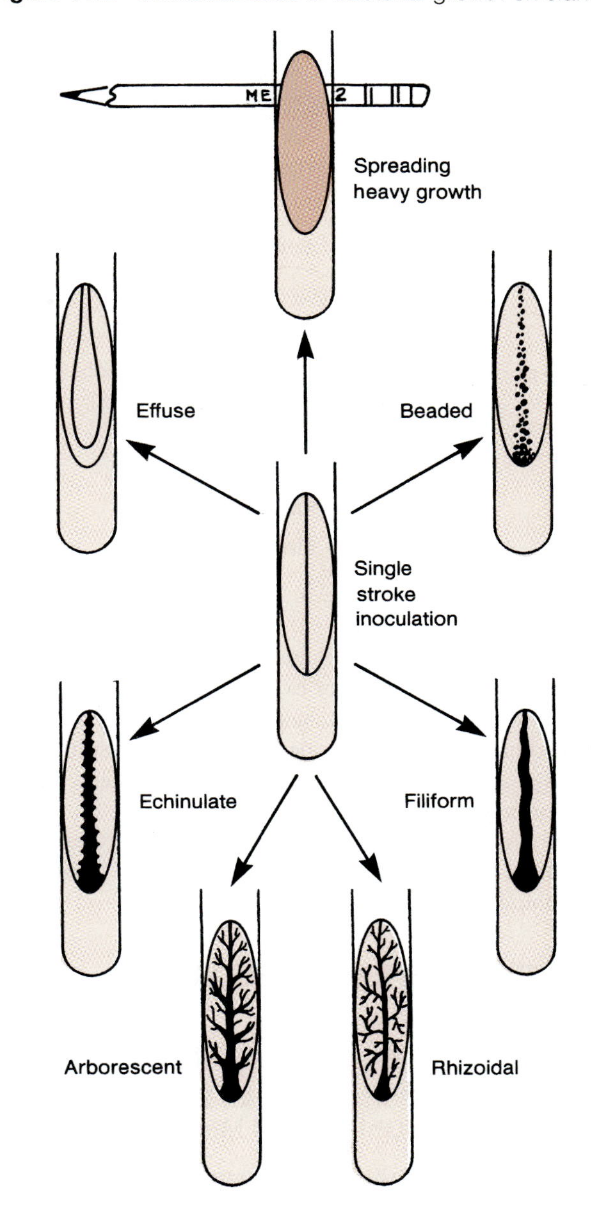

Table 11.3 Growth Characteristics of Bacterial Colonies

1. **Size:** colony diameter measured in mm or cm
2. **Whole colony shape:** circular, irregular, rhizoid, punctiform, etc.
3. **Edge or margin:** entire, lobate, erose, undulate, filamentous, etc.
4. **Elevation:** flat, raised, convex, pulvinate, umbonate, etc.
5. **Surface:** wrinkled, rough, concentric rings, dull, glistening, waxy, etc.
6. **Pigmentation:**
   a. **Color:** red, yellow, cream, white, none, etc.
   b. **Water-solubility:** water-soluble, tints both the agar and the colony; or nonwater-soluble, only the colony is colored
7. **Opacity:** transparent, translucent, opaque
8. **Odor,** listed only if distinctive: sweet, putrefactive, fruity, etc.

Figure 11.3 Growth characteristics of isolated microbial colonies

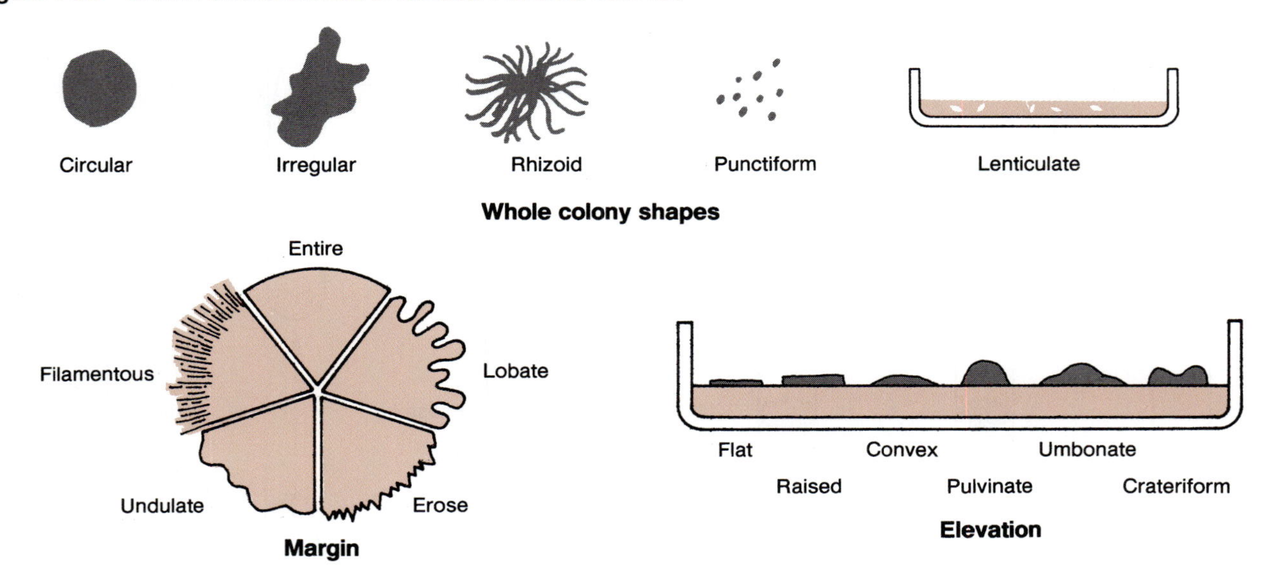

The terms utilized to characterize individual colonies are often mixed together. For example, your colony may be 2 mm in diameter, circular, with an entire margin. But it may also have a raised, opaque, dull, nonwater-soluble red center and a flat, translucent, glistening, unpigmented edge. The odor may be slight and indistinct.

Bergey's Manual utilizes far more terms than those listed here, but these get you started toward a mastery of microbiology's descriptive vocabulary.

PROCEDURES EXERCISE 11

Procedure 1

Designing and Performing an Experiment Concerning the Ubiquity of Microorganisms

Working in groups designated by **your instructor,** employ the scientific method of investigation to test a hypothesis concerning the ubiquity of microorganisms. Steps enumerated in table 11.1 are listed in this procedure to help you design and carry out your group experiment.

First Laboratory Session

a *Step 1.* **Recognize the situation:** Microorganisms are ubiquitous. *Step 2.* **Gather information:** Listen to **your instructor;** read your text, this exercise, and tables 11.1 and 11.2.

b *Step 3.* As a group, **formulate a testable hypothesis.** Be sure that it is narrow enough so that you can adequately test it with the media and materials provided. For example, your group might hypothesize that there are greater **numbers of microorganisms** and more **kinds of microorganisms** on the silverware in the school snack bar than on the plates under the food.

Check with other groups to be sure your hypotheses differ. Record your group's hypothesis in your laboratory report.

c *Step 4.* Individually, **make a prediction** concerning the outcome of your group investigation. Record the prediction in your laboratory report.

d *Step 5.* As a group, **design an experiment that tests your hypothesis.** Try to utilize the 3 Tryptic Soy Agar plates and 2 sterile, moist or dry swabs per person that are available to your group. Use table 11.2 as a guide for exposure areas and times. Record the experimental design (your plan of action) in your laboratory report. **Gather your samples.**

e *Step 6.* Include **a control for each variable** element of your experiment. **Your instructor** will incubate 2 Tryptic Soy Agar plates—one at 35° C and the other at room temperature until next laboratory session. These plates serve as media controls to test whether the cultures you obtain are from the environment or from contaminated media.

f How can you test the sterility of the moist swabs? Include a test of the swabs as one of your controls.

g Record the controls in your laboratory report with the experimental design. **Gather your controls.**

h **Invert** all plates and incubate them until the next laboratory session. Samples from warm areas such as the human body are incubated at 35° C. Room temperature incubation is more appropriate for samples from cooler environments such as soil, dust, and tabletops.

Second Laboratory Session

a Gather data by evaluating growth on your plates. **Estimate numbers of colonies and numbers of different kinds of colonies.** Record your estimations.

b Gather and record the estimates from other members of your group.

c *Step 7.* **Analyze your data and arrive at a conclusion.** Record it in your laboratory report.

d *Step 8.* Now that you have experience in the scientific method, **reevaluate your experimental design.** Enter suggestions for improvement in your laboratory report.

e *Step 9.* Rather than formulating **new hypotheses** of your own, investigate another group's experimental design, including controls. Does their experimental design sufficiently test the proposed hypothesis? Are their controls adequate? Enter your comments in your laboratory report.

f Now that the group work is completed, select any 3 colonies from any plates and describe each colony thoroughly. This exercise allows you to employ your microbiological vocabulary. Enter your descriptions in your laboratory report.

g After your group has recorded all observations, follow **your instructor's** directions for safe disposal of the plates.

Procedure 2

Describing Growth Patterns in Broths and on Agar Slants

First Laboratory Session

a Obtain 3 Tryptic Soy Broths and 3 Tryptic Soy Agar slants. Label 1 broth and 1 slant *"Mi. l.,"* label a second tube of each medium *"B. sub.,"* and label the third broth and slant *"E. f."* Add your name and the date to each label; loosen lids.

b Use your sterilized, cooled loop to transfer a little *Micrococcus luteus* into your appropriately labeled **broth.** Then, without flaming, use the same loop to draw a single stroke in a straight line up the center of the correctly labeled **slant's** surface without gouging the agar. Flame your loop.

c Repeat step (*b*) with *Bacillus subtilis* and your *"B. sub."* tubes.

d Repeat step (*b*) with *Enterococcus faecalis* and your *"E. f."* tubes.

e Incubate all 6 cultures at 35° C for 24 to 48 hours.

Second Laboratory Session

a **Without jostling the media,** take your Tryptic Soy Broth culture of *Micrococcus luteus* and describe its characteristic growth pattern. Enter your thorough description in your laboratory report.

b **Without jostling the media,** repeat step (*a*) using your broth culture of *Bacillus subtilis.*

c **Without jostling the media,** take your broth culture of *Enterococcus faecalis* and repeat step (*a*).

d Take your Tryptic Soy Agar slant culture of *Micrococcus luteus* and describe its characteristic growth pattern. Enter your thorough description in your laboratory report.

e Repeat step (*d*) with your slant culture of *Bacillus subtilis.*

f Repeat step (*d*) with your slant culture of *Enterococcus faecalis.*

g After your group has recorded all its observations, follow **your instructor's** directions for safe disposal of the tubes.

POSTTEST EXERCISE 11

Part 1

Matching

Each response may be used once, more than once, or not at all. All items relate to bacterial growth in **broth cultures.**

a. Turbidity
b. Flocculent
c. Ring
d. Pellicle
e. Sediment
f. Streamers

1. _____ Growth around the top rim
2. _____ Cloudiness
3. _____ Lines of growth floating down
4. _____ Heavy growth floating on the surface
5. _____ Growth floating in clumps
6. _____ A visible accumulation of cells on the bottom

Part 2

Matching

Each response may be used once, more than once, or not at all. All items relate to bacterial growth on an **agar slant.**

a. Spreading
b. Beaded
c. Arborescent
d. Rhizoidal
e. Filiform
f. Echinulate

1. _____ Smooth edges
2. _____ Small, individual colonies
3. _____ Covering the entire surface
4. _____ Rootlike
5. _____ Branching
6. _____ Spiny or saw-toothed edges

Part 3

Matching

Each response may be used once, more than once, or not at all. All items relate to the characteristics of isolated **colonies on agar plates.**

a. Punctiform
b. Entire
c. Cream
d. Erose
e. Water-soluble
f. Transparent
g. Putrefactive
h. Undulate
i. Raised
j. Opaque
k. Umbonate
l. Circular
m. Measured in mm or cm

1. _____ Size
2. _____ Whole colony shape
3. _____ Edge or margin
4. _____ Elevation
5. _____ Surface
6. _____ Pigmentation
7. _____ Opacity
8. _____ Odor

Part 4

True or False (Circle one, then correct every false statement.)

1. T F A colony is an individual bacterial cell.
2. T F The appearance of a colony reflects the biochemistry of the species that formed it, so all bacterial colonies look alike.
3. T F When utilizing the scientific method of investigation, you must first recognize a problem or situation and then gather information concerning it.
4. T F When the scientific method of investigation is employed, only experiments that prove the accuracy of the original hypothesis are useful.
5. T F A control is a test in which only one factor is different than in the experiment.
6. T F *Bergey's Manual of Systematic Bacteriology* is a four-volume sourcebook for bacterial taxonomy.
7. T F Conditions of incubation include temperature, pH, and salinity, but not the medium and its nutrients.
8. T F All bacteria form a pellicle when grown in broth.
9. T F Opaque means translucent.
10. T F Since agar media are largely water, a water-soluble microbial pigment tints the agar as well as the colony.

Laboratory Report **11**

Name ______________________

Section ____________

Ubiquity and Culture Characteristics

Procedure 1

Designing and Performing an Experiment Concerning the Ubiquity of Microorganisms

First Laboratory Session

(*b*) State your group's hypothesis concerning the ubiquity of microorganisms.

(*c*) State your own prediction concerning the outcome of your group's investigation.

(*d*) Record the design of your group's experiment.

(*f*) How will your group test the sterility of the moist swabs?

(*g*) What controls are you including in your experimental design?

Second Laboratory Session

(*a,b*) Organize your data. Identify each plate and record the number of colonies on it. Write "too numerous to count" (TNTC) if you observe over 300 colonies on a plate. If fewer than 30 colonies have developed on a plate, record the results as "no growth" (NG) or "too few to count" (TFTC), whichever is appropriate.

Discuss the quantity and diversity of colonies found on the plates prepared by your group.

(*c*) Examine your raw data. A valid analysis of data is essential to the scientific method of investigation. What can you conclude from your data?

What conclusions can you draw from your data regarding your hypothesis?

(*d*) How could you improve your group's experimental design?

(*e*) Examine another group's hypothesis, their experimental design, and their results. What is their hypothesis?

What do their data show?

Does their experimental design sufficiently test their hypothesis? __________ If not, how could it be improved?

Are their controls adequate? __________ If not, which ones are unnecessary (why), and/or what controls were necessary but omitted?

(*f*) Select any 3 isolated colonies and describe each one thoroughly.

Colony Characteristic	Colony Descriptions		
Colony no.:	1	2	3
Size			
Shape			
Margin			
Elevation			
Surface			
Pigmentation			
Color			
Water-solubility			
Opacity			
Odor (optional)			

Procedure 2

Describing Growth Patterns in Broths and on Agar Slants

(*a*) Draw and describe completely the growth pattern of *Micrococcus luteus* in broth.

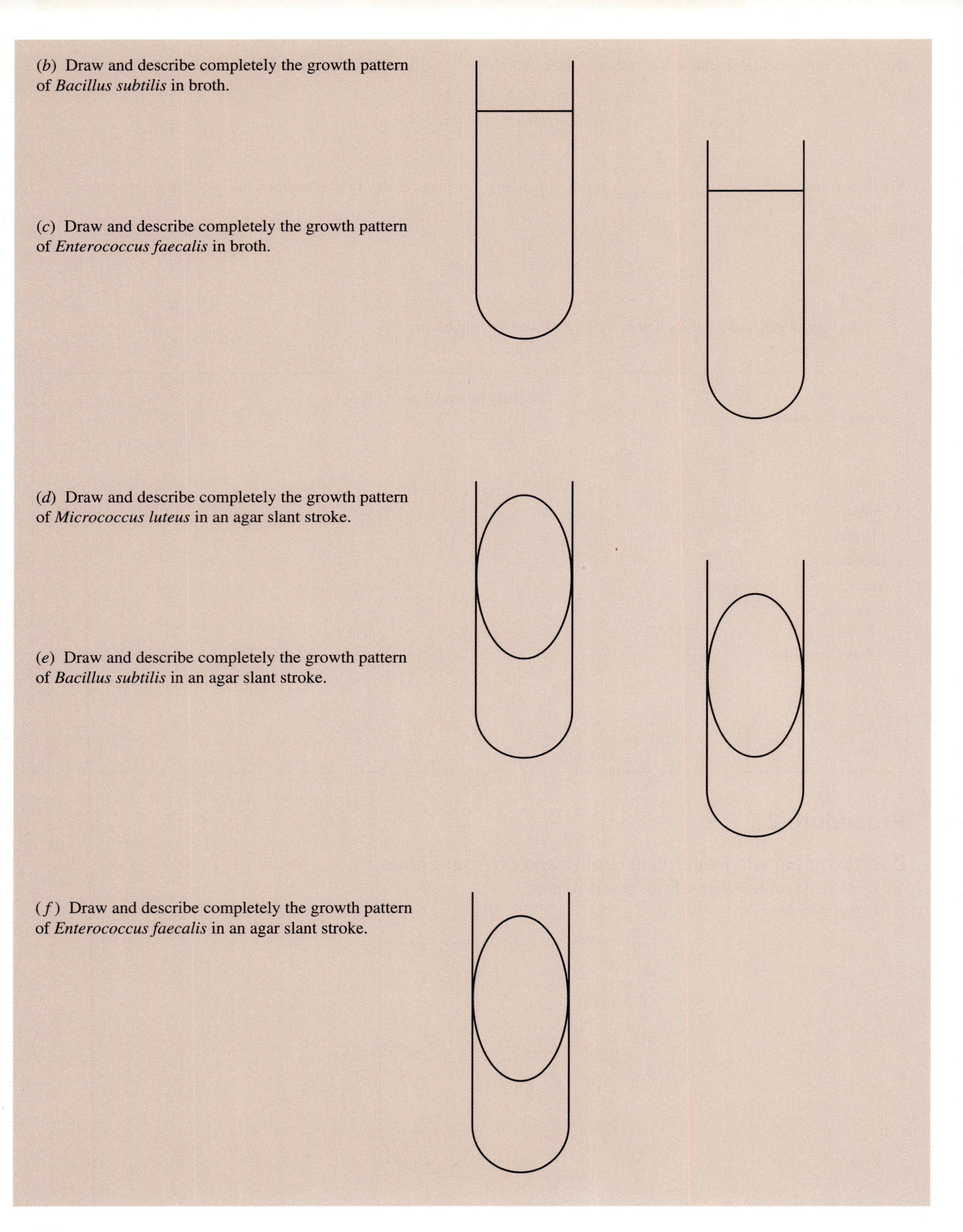

(*b*) Draw and describe completely the growth pattern of *Bacillus subtilis* in broth.

(*c*) Draw and describe completely the growth pattern of *Enterococcus faecalis* in broth.

(*d*) Draw and describe completely the growth pattern of *Micrococcus luteus* in an agar slant stroke.

(*e*) Draw and describe completely the growth pattern of *Bacillus subtilis* in an agar slant stroke.

(*f*) Draw and describe completely the growth pattern of *Enterococcus faecalis* in an agar slant stroke.

Exercise **12**

Pour-Plate Techniques

Necessary Skills

Aseptic transfer techniques, Gram stain

Materials

First Laboratory Session

Cultures (24-hour, 35° C, Tryptic Soy Broth, 1 per set, at least 5 ml each):
- *Micrococcus luteus*
- *Serratia marcescens*

Supplies (per group):
- 2 500 ml Beakers
- Thermometer
- Ring stand
- Screw cap bottle or other container holding 100 ml of sterile Nutrient Agar
- 3 Sterile Nutrient Agar pours (deeps), about 15 ml each
- 9 Sterile Petri dishes
- Sterile test tube with closure
- Test tube rack
- 2 Microscope slides (optional)
- 2 Sterile Pasteur pipets
- Bulb

Additional supplies:
- Beakers of disinfectant for contaminated Pasteur pipets and bulbs
- 30° C Incubator

Second Laboratory Session

Containers for the 6 Nutrient Agar plates poured by each group
2 Microscope slides per group (optional)
Gram stain reagents (optional)
Access to Quebec colony counter
Access to tally register (optional)

Primary Objective

Learn to pour adequate, sterile Petri dishes of agar medium and to prepare loop-dilution plates.

Other Objectives

1. List two common uses for poured plates.
2. Distinguish between a mixed culture and a pure culture.
3. Explain why a microbiologist needs pure cultures.
4. List and discuss the significance of Koch's postulates.
5. Define each of these microbiological terms and abbreviations: serial dilution, agar pour (deep), TNTC, TFTC.
6. Use the Quebec colony counter and a tally register to tabulate numbers of colonies on countable plates.
7. Find colonies throughout pour plates.

INTRODUCTION

In microbiological laboratories plates of liquefied agar media are poured either to suspend diluted bacteria or to prepare sterile, solid surfaces for inoculation. In this exercise you prepare sterile plates of media, and you practice a pour-plate dilution technique. Your sterile plates are utilized by the class for subsequent exercises.

Diluting bacteria in **pour plates** is one of the two most common techniques that separate individual bacteria from each other. The other frequently employed separation method is the **streak plate;** see Exercise 13, Streak-Plate and Spread-Plate Techniques.

A Loop-Dilution Series

In this exercise you isolate individual colonies of bacteria by preparing pour plates from a **loop-dilution series.** This is a **serial dilution** (each tube diluted from the previous tube) performed with your inoculating loop.

You prepare the inoculum for your loop-dilution series by mixing together two kinds of bacteria: ***Micrococcus luteus*** and ***Serratia marcescens.*** The mixture should hold approximately equal numbers of each species. Any species that is present in disproportionately low numbers may be diluted beyond the **"extinction point"** (where it no longer exists) before appearing on a countable plate.

Figure 12.1 Preparing a serial loop dilution and pouring plates

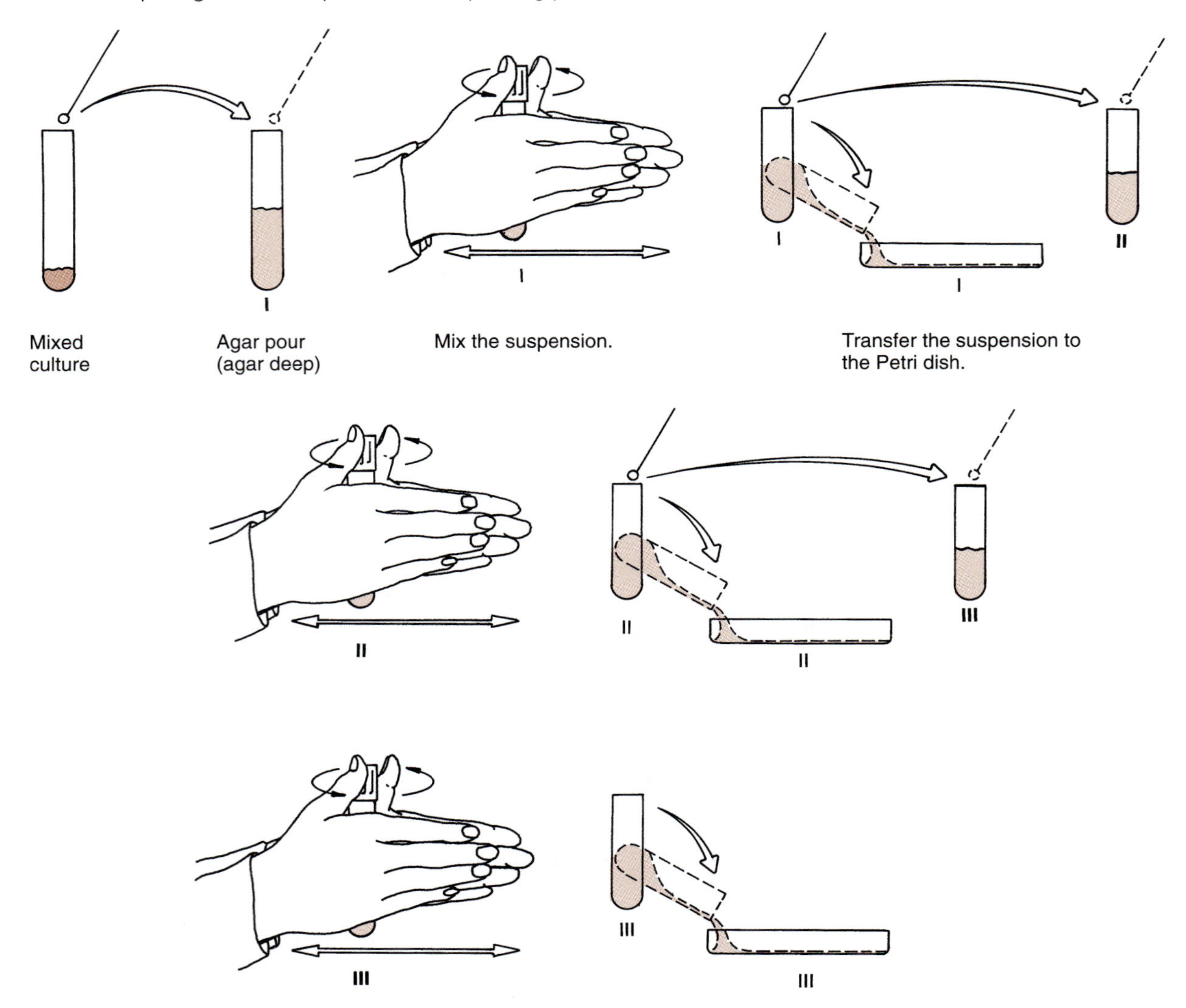

If you mixed accurately and maintained aseptic technique, you will observe two types of colonies and no contamination on your plates.

As shown in figure 12.1, you dilute your mixture by first transferring one loopful of it into an **agar pour** (a test tube containing about 15 ml of a liquefied agar medium; also called an **agar deep**). After mixing this suspension, you transfer a loop of it to the second tube of melted agar and pour the first tubeful into a Petri dish. This process is repeated. After mixing, you pour the second and third tubes into their Petri dishes.

Colony Growth

After incubation, one of the three plates usually is **countable;** that is, it has between 30 and 300 colonies. Why does a microbiologist only count the colonies on plates that develop 30 to 300 colonies? Statistics show that the appearance of fewer than 30 colonies may be a chance event. In addition, there is too much crowding for accurate counting if over 300 colonies develop.

Each colony grows from an isolated bacterial cell or a small cluster of similar bacteria. In microbiology laboratories measured dilutions, as opposed to loop dilutions, are frequently prepared. After incubation, colonies are counted to determine the number of bacteria per milliliter of the original solution. The bacterial load of water, milk, urine, or other fluids can be determined this way.

You do not prepare dilutions with specific amounts of liquid until later exercises of this laboratory manual. For now, count the numbers of colonies in countable plates to learn the technique and increase your powers of observation. Utilize the **Quebec colony counter** (a device that holds your Petri dish while providing glareless light and variable magnification) to help you count the isolated colonies. A **tally register,** which is a hand-held counting device, may also be available.

The number of colonies on countable plates is recorded. The growth on plates with fewer than 30 colonies is recorded as **"NG,"** no growth, or **"TFTC,"** too few to count, whichever is appropriate. If over 300 colonies have developed, **"TNTC,"** too numerous to count, is the usual notation.

Since the bacteria are evenly distributed in the agar, colonies develop throughout the medium, as illustrated in figure 12.2. You observe flattened colonies under the agar. Small, **lenticulate** (lens-shaped) colonies grow where bacteria are caught in the agar, and larger, "typical" colonies develop on the surface where they are not confined by the medium. Wherever colonies are crowded together, note that they are smaller since they must compete for nutrients.

Figure 12.2 Colonies develop throughout a pour plate.

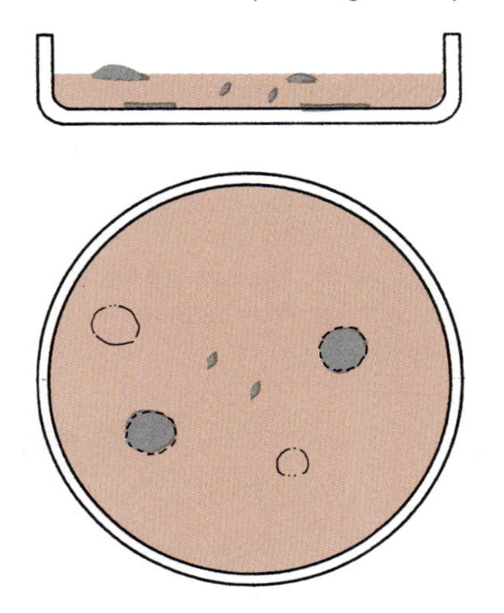

The Importance of Examining Pure Cultures

In samples taken from nature, bacteria are usually found in **mixed culture,** with different kinds of organisms growing together in communities. Normally, many different genera live in close association with each other; for instance, a sample of soil, pond water, human sputum, or feces contains a wide assortment of genera.

But to identify a bacterial species, a microbiologist generally needs to isolate the microbes in a **pure culture,** one containing organisms that are all descended from one bacterial cell. **Robert Koch,** a pioneer of medical microbiology, emphasized the importance of a pure culture.

Koch's postulates, illustrated in figure 12.3, define a procedure that determines which organism causes a specific disease. The postulates demand that the investigator isolate a pure culture of microbes. These organisms must be present in every case of the disease. The same germs must later be *recovered in pure culture* from a previously healthy experimental animal that was inoculated with them and then developed the disease.

Figure 12.3 Koch's postulates prove that a certain organism causes a specific disease.

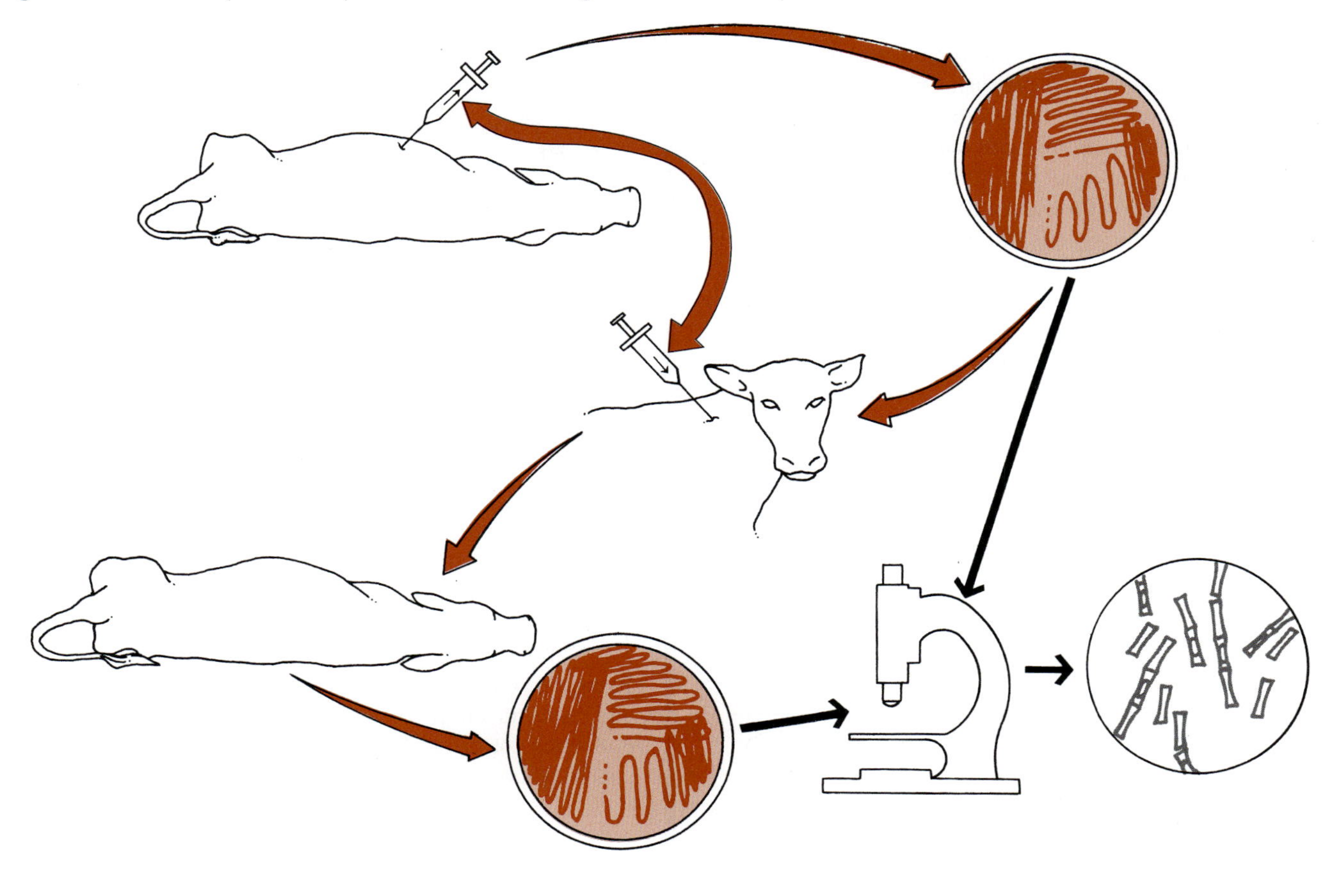

To study a bacterial species, the microbiologist uses pure cultures like the ones you will prepare with pour plates. The **physiological characteristics** (ability to use various substrates and produce specific substances) of a species can be determined only from a pure culture. Characterization of microscopic morphology requires pure culture, as does description of colony, broth culture, and slant stroke features.

PROCEDURES EXERCISE 12

Procedure 1

Pouring Petri Dishes of Sterile Nutrient Agar

> **Warning: Never** try to liquefy glass containers of agar media directly over a flame. The container is likely to explode.

First Laboratory Session

Note: The following instructions are written for right-handed students. If you are left-handed, make appropriate corrections.

a Place 1 screw cap bottle or other container holding 100 ml of sterile Nutrient Agar in a 500 ml beaker.

b Fill the beaker with hot water until the water is just above the level of the Nutrient Agar in the bottle.

c Loosen the bottle cap.

d Heat the beaker of water on a ring stand over your Bunsen burner. Keep the water boiling until the agar in the bottle is completely **liquefied** (about 15 minutes). Do not contaminate the medium with the boiling water.

e While the agar is liquefying, partially fill the other 500 ml beaker with water that is about 55° C.

f When the agar is completely liquefied, transfer the bottle to the 55° C beaker. **Temper** the agar. Do not let the temperature of the agar fall below about 45° C or it solidifies, and you must liquefy it again.

g Obtain 6 sterile Petri dishes. *Keep them closed to maintain their sterility.* Label the outside of the **bottoms** (small diameter halves) of the dishes with your name, the date, and the abbreviation "NA," for Nutrient Agar. Arrange the dishes near your Bunsen burner, as shown in the illustration.

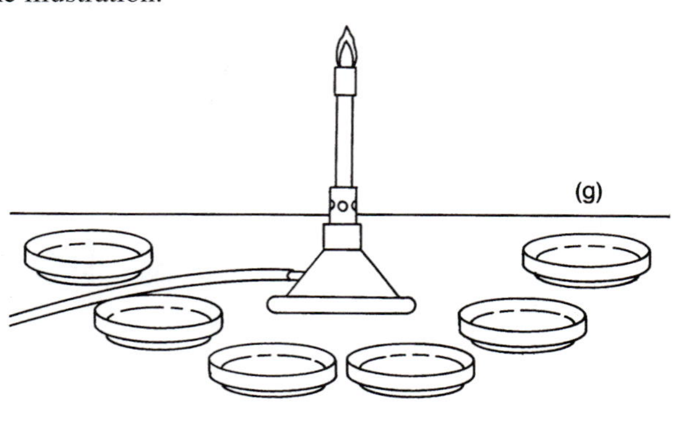

h When the agar has cooled to about 55° C, it is tempered and ready to be poured. **You must maintain the sterility of the agar and the plates while pouring.** *Read the next six paragraphs before pouring plates.*

i Dry the bottle. Remove the cap with the small finger of your right hand by holding the lid stationary while rotating the bottle. Keeping the lid in your curled finger, pass the bottle to your right hand and flame the lip of the bottle. Work with your hands near the flame and keep your nose away from it.

j With your left hand, lift the lid off the leftmost plate. With the bottle in your right hand, pour in enough agar to cover the bottom of the dish **almost** completely.

k To spread the medium, cover the dish with its lid, and then gently rock the plate or circumscribe a slow, 10 cm diameter circle with the dish on the bench top as shown in the illustration. **Do not splash the medium,** but be sure it spreads out to cover the bottom of the dish.

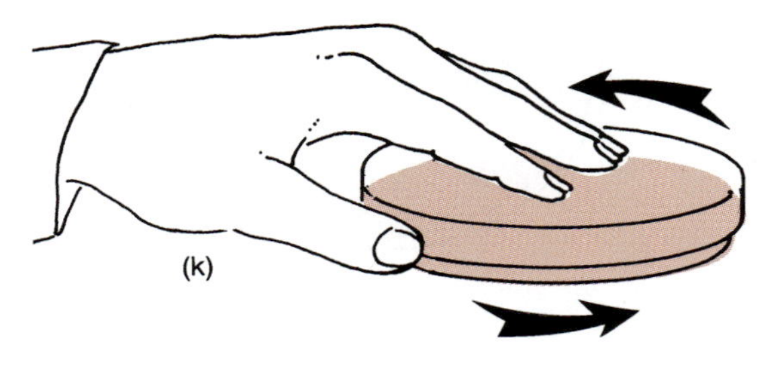

l **Keep the bottle tilted.** Do not hold it upright until you have poured the last plate. You do not want the agar to drip down the outside of the bottle and then into the next pour plate. Work near the flame.

m You are now ready to pour the second plate. Proceed as you did for the first one. To avoid scooping air into your bottle, go from left to right pouring plates. Finish pouring all the plates.

Now that you have read the procedure, go back to (i) and pour the plates.

n If there is Nutrient Agar left in your bottle, flame its lip, replace the cap, and return the bottle to the class supply. If you need a little more agar to complete your plates, reliquefy and temper some that is left from another group's work.

o Allow the Nutrient Agar in your plates to solidify. This generally takes less than 5 minutes. **Invert** the plates and incubate them in your drawer until the next class session.

Second Laboratory Session

Examine your pour plates. If they are free of contamination and have no lumps or bare spots, place them in the class collection for use in future laboratory sessions. Discard unusable plates as directed by **your instructor.**

Procedure 2

Preparing a Loop-Dilution Series

First Laboratory Session

This procedure consists of 3 steps:

1. Liquefy and temper 3 deeps.
2. Mix the bacteria.
3. Inoculate the deeps and prepare pour plates; see figure 12.1.

> **Warning: Never** try to liquefy an agar deep by holding it directly over a flame. The test tube is likely to explode.

Liquefy and Temper 3 Deeps

a **Liquefy** 3 Nutrient Agar deeps in a beaker of boiling water. Keep the water level just above the height of the Nutrient Agar in the deeps. **Temper** the liquefied deeps by keeping them in a beaker of 55° C water for about 10 minutes. Do not allow the temperature to drop below 45° C.

Mix the Bacteria

b Disperse the bacteria in the broth cultures of *Micrococcus luteus* and *Serratia marcescens.* Estimate the comparative density of microorganisms by observing the turbidity of each culture.

c Use a sterile Pasteur pipet and bulb to transfer a drop of the more concentrated culture into a sterile test tube. Close and replace the culture. Close the test tube and discard the contaminated pipet into a beaker of disinfectant. If you have contaminated the bulb, place it in a separate beaker of disinfectant. If it is not contaminated, use it again. Label the test tube "Mixed Culture."

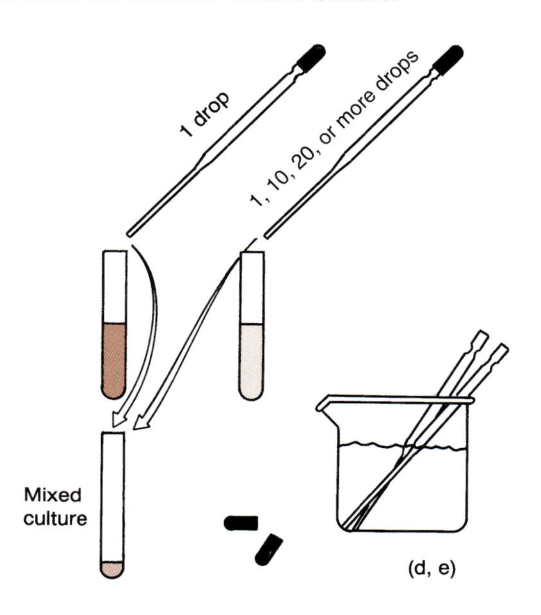

d As shown in the illustration, you want to transfer approximately equal numbers of bacteria from each genus into the "Mixed Culture" tube. Use turbidity as a guide. Estimate how many drops of the remaining culture contain about as many bacteria as 1 drop of the more concentrated culture.

e Attach a bulb to a sterile Pasteur pipet and transfer 1, 10, 20, or more drops of the next culture into the "Mixed Culture" test tube. Discard the Pasteur pipet into the beaker of disinfectant. If the bulb is contaminated, place it in the separate beaker of disinfectant.

Inoculate the Deeps and Prepare Pour Plates (see figure 12.1)

f Label the outer surface of the **bottoms** (small diameter halves) of 3 sterile, empty Petri plates "I," "II," and "III." Add your name, the date, and the words "Mixed Culture." **Your instructor** may suggest further labeling.

In the following steps, remember to work rapidly with the liquefied agars to prevent them from solidifying. Use sterile technique; remember to flame the necks of test tubes. Refer frequently to figure 12.1.

g Move 1 tempered, liquefied agar pour (deep) from the warm water beaker to your test tube rack. Resuspend the bacteria in the "Mixed Culture" tube. **Transfer 1 loopful of the mixture into the pour.**

h Suspend the bacteria in the pour by rapidly rolling the tube between your palms. This method distributes the organisms without producing unwanted bubbles in the agar.

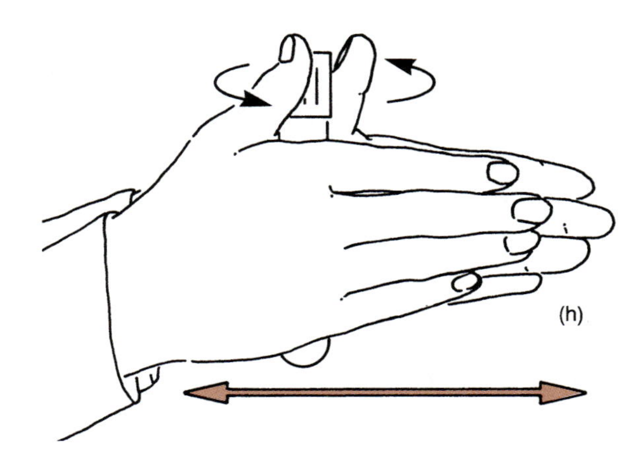

i Remove the next deep from the warm water beaker to your test tube rack. **Transfer 1 loopful from the bacteria-laden deep into the uninoculated one.**

j Pour the contents of the mixed deep into the appropriately labeled Petri dish. Cover the Petri dish. Gently rotate the dish clockwise and counterclockwise, back and forth, and from side to side, mixing and spreading the suspended bacteria. Do not splash.

k Return the empty test tube and its closure to the rack.

l Repeat steps (*h*) through (*k*).

m Repeat step (*h*). Then repeat steps (*j*) through (*k*).

n Discard the empty, contaminated test tubes as directed by **your instructor.**

o After about 5 minutes, when the Nutrient Agar in the plates has solidified, **invert** the plates and incubate them at 30° C until the next laboratory session.

p **Your instructor** may ask you to label a microscope slide and prepare a smear from each of the original broth cultures. Allow these smears to dry in your drawer until the next laboratory session.

Second Laboratory Session

a Examine plates "I," "II," and "III" of the loop-dilution series. Note flat, lenticulate, and surface colonies. Estimate whether one of the plates is "countable" (contains 30 to 300 colonies).

b If you have a countable plate, improve your observational skills by placing it on the Quebec colony counter and counting the colonies.

c Zero the tally register, if one is available, by turning the screw. Use the tally register to keep track of the number of colonies.

Do not move your head while using the Quebec colony counter. If you do, the location of the colonies in relation to the lines shifts and you must start again.

d Begin counting all colonies within the squares and across the top lines. Press the tally counter for each colony. Count colonies that lie on the top line and the preceding line, as shown in the illustration. At the end of a row, go down one row and start back across the plate, again counting colonies in each square and on the top and preceding lines. Continue counting until you finish the plate. Do not miss tiny, lenticulate colonies or flattened growth caught under the medium.

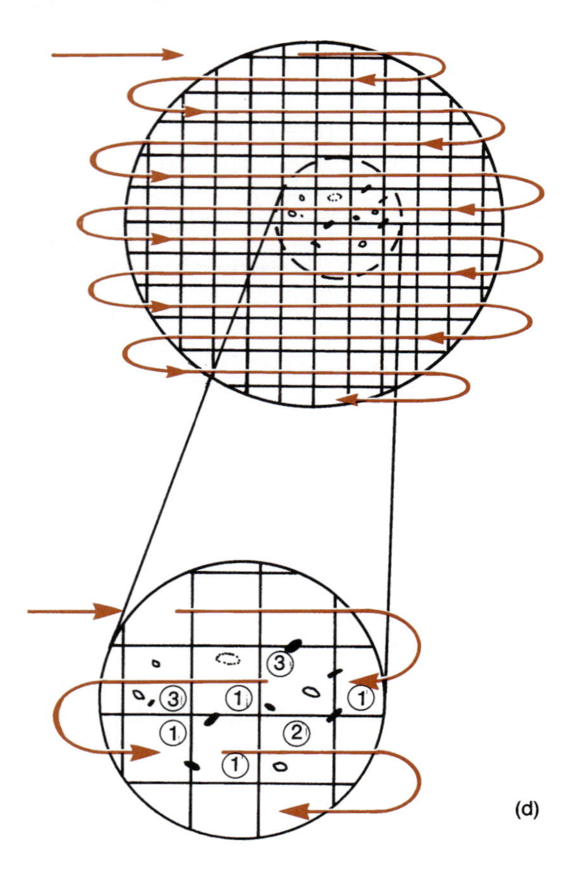

(d)

e Record the number of colonies in the laboratory report. **Repeat** the counting procedure and record your second total. Your second number should be within about ± 10% of the first. Are your counts reliable?

f The primary purpose of the loop dilution is to isolate individual colonies of bacteria from a mixed culture. Have you isolated colonies of both types of bacteria?

Brief descriptions of surface colonies of the two species in the "Mixed Culture" (Nutrient Agar, 48-hour, 30° C) are provided here. ***Micrococcus luteus*** colonies are circular, shiny or mat, opaque, and generally yellow with entire edges. ***Serratia marcescens*** forms circular, shiny, translucent colonies with entire edges. *Serratia marcescens* generally produces red pigment at 30° C. It produces more pigment at lower temperatures. Record your observations in the laboratory report.

g **Your instructor** may ask you to check your isolation technique by preparing smears of the different colony types. While the smears are drying, Gram stain the smears made during the last laboratory session from the original broth cultures. Gram stain the smears from your 2 colonies. Compare the bacteria.

Gram stain reactions that you will observe are briefly described here. **Micrococci** are gram-positive cocci. The micrococci form singles, pairs, short chains, cuboidal packets, and many tetrads. ***Serratia*** is a gram-negative rod.

Did you recover both genera? Record your observations in the laboratory report.

h After your group has finished working with the cultures, follow **your instructor's** directions for safe disposal of the plates.

POSTTEST EXERCISE 12

Part 1

True or False (Circle one, then correct every false statement.)

1. T F The two most common methods that separate individual bacteria from each other are pour plates and lawns.
2. T F In samples from nature, bacteria are usually found in mixed culture.
3. T F In a sample of human sputum or feces, you usually find a mixture of many different kinds of organisms.
4. T F Robert Koch was a pioneer of medical microbiology.

5. T F Koch's postulates are used to prove that a specific microorganism causes a specific disease.
6. T F To fulfill Koch's postulates, a researcher must cultivate a mixed culture of organisms from the blood of a diseased experimental animal.
7. T F The physiological characteristics of an organism are its Gram stain reactions.
8. T F In a serial dilution, each tube contains the same dilution as the previous one.
9. T F A countable plate has between 1 and 300 colonies.
10. T F In microbiology laboratories colonies are usually counted to determine the number of bacteria per milliliter of the original solution.
11. T F Lenticulate colonies are usually found on the surface of a pour plate.
12. T F In a pour plate, colonies develop throughout the medium.
13. T F Colonies are smaller in crowded areas of a pour plate than in areas where they are well separated.
14. T F If you are right-handed, go from left to right when pouring plates to avoid scooping air into the bottle of medium.
15. T F Pour the plates one at a time from a bottle of medium. Tilt the bottle to the upright position before each pour.

Part 2

Completion and Short Answer

1. Pour plates are commonly used to ________________ ____________________________ and to ____________ ____________________________ .
2. In a ________________ ________________ , all the bacteria are descended from one bacterial cell.
3. The physiological characteristics of an organism include its ___________________________ of substrates and its ______________________________ of substances.
4. An agar ________________________________ is a test tube with about 15 ml of agar medium in it.
5. A device that holds your Petri dish, while providing glareless light and variable magnification to help you count isolated colonies, is the _______________ _____________________ _____________________ .
6. Define the term *loop-dilution series.*
7. Define the following abbreviations.
   TF TC
   TNTC
   NG
8. Define the term *extinction point.*

Part 3

List Koch's postulates.

1.

2.

3.

4.

Laboratory Report **12**

Name ______________________________

Section ____________

Pour-Plate Techniques

Procedure 2

Preparing a Loop-Dilution Series

Second Laboratory Session

(*e*) Record the total number of colonies on your countable plate. _______ Count the plate again. What is the second count? _______

You can calculate the percent difference between the counts by applying the following formula:

$$\frac{\text{higher count} - \text{lower count}}{\text{lower count}} \times 100 = \%\ \text{difference}$$

Are your two numbers within 10% of each other? _______ If not, count the plate again. Your counts must be both accurate and repeatable.

(*f*) Have you isolated both colony types? _______

(*f,g*) If required by **your instructor,** describe the shape, Gram stain reaction, and arrangement of the bacteria in each smear from the original cultures.

Micrococcus luteus:

Shape ______________________________ Gram reaction ______________

Arrangement ______________________________

Serratia marcescens:

Shape ______________________________ Gram reaction ______________

Arrangement ______________________________

Describe your 2 surface colony types and, if required by **your instructor,** the microscopic appearance of bacteria from each of your surface colony types. Compare this information with the microscopic analysis you made of each genus in the inoculum or with information provided in Procedure 2. Use the information to make a tentative identification of the bacteria forming each of the 2 colony types.

First Colony

Probable identification ______________________________

Microscopic evaluation:

Shape ______________________________ Gram reaction ______________________________

Arrangement ______________________________

Colony description:

Second Colony

Probable identification ______________________________

Microscopic evaluation:

Shape ______________________________ Gram reaction ______________________________

Arrangement ______________________________

Colony description:

Exercise 13

Streak-Plate and Spread-Plate Techniques

Necessary Skills

Aseptic transfer techniques, Ubiquity and culture characteristics, Pour-plate techniques (optional)

Materials

First Laboratory Session

Cultures (1 per set):

Staphylococcus epidermidis (24-hour, 35° C, Tryptic Soy Broth)

Mixture: *Staphylococcus epidermidis* and *Escherichia coli* (both 24-hour, 35° C, Tryptic Soy Broths; freshly prepared mix in a ratio of approximately 5 parts staphylococci to 1 part *Escherichia*)

Supplies (per group):

2 Sterile cotton-tipped applicators

Empty, nonsterile test tube with closure

3 Petri dishes of sterile Tryptic Soy Agar (Student-prepared Nutrient Agar plates from Exercise 12, Pour-Plate Techniques, may be used.)

Beaker of disinfectant for contaminated swabs

Second Laboratory Session

Cultures (24-hour, 35° C, Tryptic Soy Broth, 1 per set):

Staphylococcus epidermidis

Supplies:

2 Petri dishes of sterile Tryptic Soy Agar or Nutrient Agar per group

1 Sterile cotton-tipped applicator per group

Beakers of disinfectant for contaminated swabs

Primary Objective

Streak bacterial plates, consistently obtaining well-isolated colonies. Prepare bacterial spread plates that completely cover the medium with even, confluent growth.

Other Objectives

1. Avoid contamination of streak plates and spread plates.
2. Define or describe: micromanipulation, concentration gradient, confluent growth, bacterial growth.
3. Calculate the number of bacteria that theoretically could result from the binary fission of a single, original cell that divides every 30 minutes for 12 hours.
4. List three uses of spread plates.

INTRODUCTION

In this exercise you learn two fundamental microbiological laboratory techniques. With the **streak-plate** technique, you separate bacterial cells and grow them into isolated colonies. Then you employ a pure culture of bacteria to prepare a **spread plate** in which bacterial growth completely covers the surface of an agar medium in a Petri dish. Both of these techniques are used thousands of times daily in research and medical laboratories worldwide.

As discussed in Exercise 12, Pour-Plate Techniques, bacteria in nature and in clinical samples such as sputum or feces exist in mixed cultures. To characterize these microorganisms, microbiologists traditionally separate the cells and grow them in pure culture.

Streak-Plate Technique

The streak-plate technique performs this separation. It is not a **micromanipulation** of single bacterial cells using fine needles under a microscope lens, although this difficult technique is possible. Instead, the streak-plate technique establishes a **concentration gradient** (decreasing quantity) of bacteria across the surface of a Petri dish of solid medium.

Streaking a plate involves drawing a loopful of organisms back and forth across the surface of a solid medium in a Petri dish until the microbes fall off the loop one at a

Figure 13.1 A quadrant method that efficiently separates bacteria

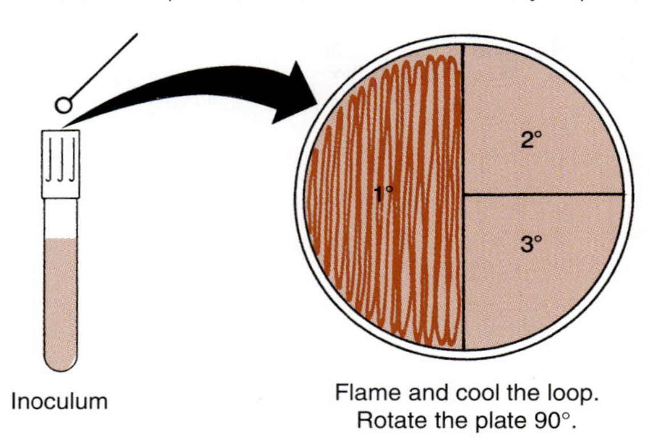

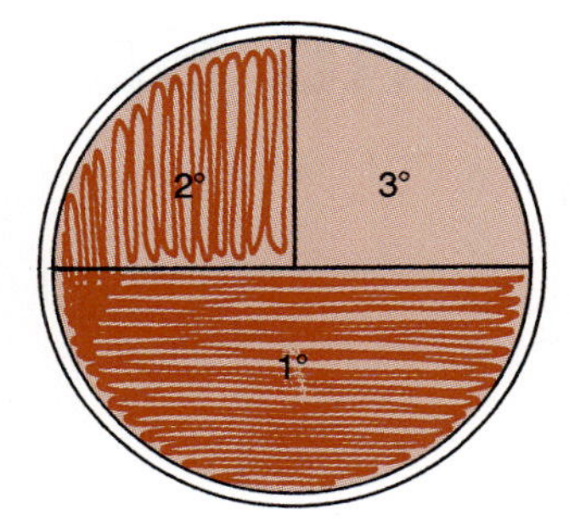

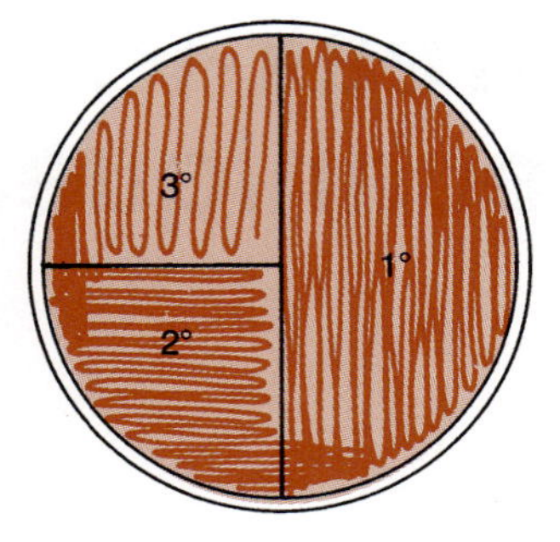

time. There are many patterns employed to draw the line. Each variation attempts to use as much of the agar as possible, thereby assuring separation of the cells.

One common streak-plate pattern, a **quadrant** method, is illustrated in figure 13.1. It is called a quadrant because the surface of the plate is approximately halved, then one-half is further divided into quarters.

In the quadrant method, a loopful of **inoculum** (organisms from the original culture) is spread over the primary area. Then the loop is flamed and cooled and the plate turned 90°. A few bacteria are spread into the secondary area from the primary sector. They are drawn back and forth over the quadrant. Again the loop is flamed and cooled, and the plate is turned. A few bacteria are drawn out from the secondary sector into the tertiary quadrant. Here they are dispersed further by the zigzag motion of the loop.

Quadrant methods of streaking efficiently isolate cells. Even the bacteria in stool specimens or other highly concentrated mixed cultures are separated by quadrant streaking.

In medical laboratories the specimen is often a mixed culture on a swab from the patient's throat, pus, feces, and so on. The cotton swab is covered with bacteria that must be separated from each other so that the microbiologist can identify pathogens. First, the primary section of a Petri plate is streaked with the swab, then the loop is employed to isolate cells.

In one procedure in this exercise, you simulate the clinical situation with a swab dipped into a mixed culture of ***Staphylococcus epidermidis*** and ***Escherichia coli.***

Colony Development

During incubation the progeny of a single cell, or a small group of similar cells, form a **colony,** a visible clump of growth on the surface of the agar. In the primary area, or wherever bacteria are crowded closely together on the surface of the growth medium, **confluent growth** (colonies flowing together) appears. In any sector where there is more space between the cells, they grow into larger colonies.

There are visible differences between the colonies of many species of bacteria. These colonial characteristics often provide the first hints that help microbiologists identify bacteria that are of particular interest. For more information on significant colony characteristics, refer to Exercise 11, Ubiquity and Culture Characteristics.

Although a single microbe of the type used in this exercise is only 1 or 2 μm in diameter, it divides into two cells by **binary fission** (simple splitting) within 20 or 30 minutes. As shown in figure 13.2, just two divisions per hour produce millions of cells overnight. This is **bacterial growth—**a multiplication of population numbers, not the enlargement of individual cells.

After incubation, plates are inspected. Pure cultures are prepared from well-isolated colonies, such as those that are illustrated in figure 13.3.

Spread-Plate Technique

Your group has a pure culture of *Staphylococcus epidermidis* to utilize for the second major technique in this exercise, the preparation of a bacterial spread plate (a "lawn" of bacteria).

Spread plates are employed by the microbiologist to test reactions of a pure culture when it is exposed to various substances. For instance, the susceptibility of bacteria to various **antiseptics** may be tested on spread plates. Lawns are also used to help a microbiologist evaluate the sensitivity of bacteria to **antibiotics. Bacteriophage** (viruses that infect specific strains of bacteria) are often grown on bacterial spread plates. The viruses show the microbiologist the **bacteriophage type** (subspecies) of a pure, bacterial culture.

You prepare both streak plates and spread plates in many of the following exercises of this manual. Both procedures are employed frequently in clinical and research laboratories.

Perfect the techniques now. Be highly critical when evaluating your results. Media for further practice will be available during the second laboratory session.

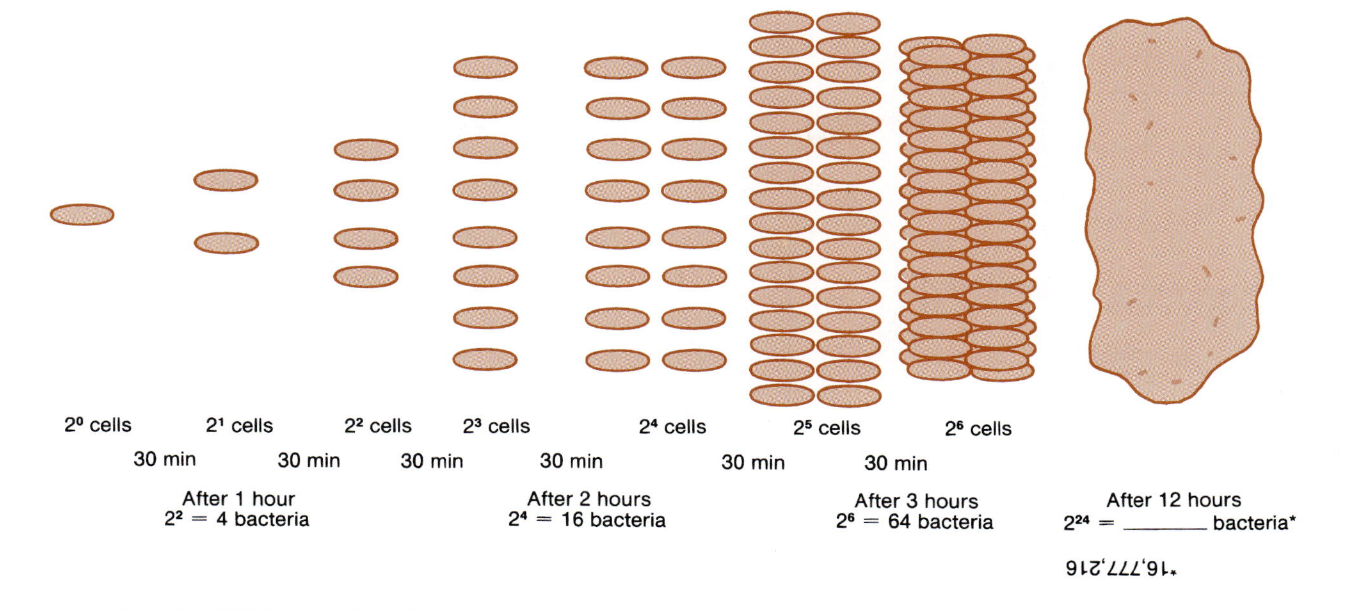

Figure 13.2 Bacteria double by binary fission

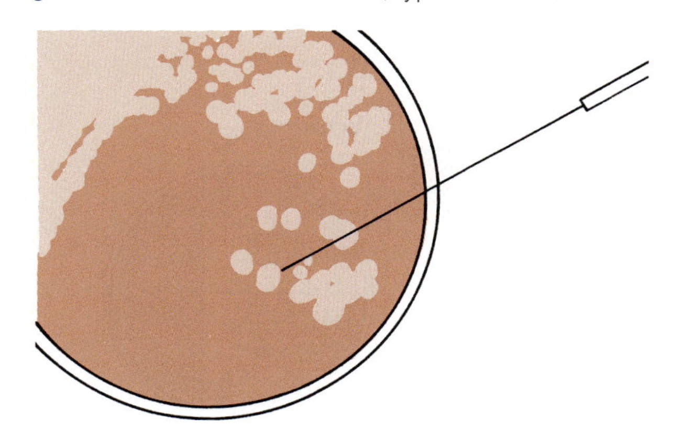

Figure 13.3 Select well-isolated, typical colonies

Table 13.1 Steps in Preparing a Streak Plate[1]

1. **Evaluate the concentration of bacteria** in the inoculum, then draw a "T" on the outside of the small diameter half of the plate. The mark divides the surface into primary, secondary, and tertiary sectors. The "T" is a guide. Abandon its use when you feel ready.
   a. **For heavy growth** (e.g., feces, colonies on solid media), make the primary area small; see figure 13.4a.
   b. **For light growth** (e.g., cerebrospinal fluid), utilize at least half the plate for the primary area.
2. **Turn the plate right side up;** place the primary sector to your left.
3. Using sterile technique, **obtain a loopful of inoculum.** Close the culture tube and replace it in your rack.
4. With your left hand, tilt the lid of the Petri dish open. Use the lid as a "germ guard" between your face and the plate. Starting at the high point distal to the "T," **glide your loop very gently and rapidly back and forth across the primary sector many times.** Fill the primary area with close together lines. Use the entire surface area; see figure 13.4b.
5. Close the lid on your Petri dish. **Flame your loop.** Turn the dish ¼ turn; see figure 13.4c.
6. Tilt the lid of your dish open, using it to shield your face and the surface of the medium. **Cool your loop** by
   a. Waiting (with the dish closed);
   b. Stabbing the loop into a sterile place in the agar; or
   c. Dragging the loop through some of the sterile condensed water on the inside of the Petri dish lid.
7. Drawing closely spaced lines, streak your loop gently across the agar in the primary area, **carry organisms into the secondary sector.** Zigzag back and forth about 6 times. Then proceed to **fill in the second area;** see figure 13.4d. *Do not overlap the primary area after the first 6 strokes.* You are not trying to spread the bacteria evenly. You want to separate the bacteria until they fall off your loop one at a time.
8. Close the lid on your Petri dish. **Flame your loop.** Turn the dish ¼ turn.
9. Tilt the lid of your dish open. **Cool your loop.**
10. **Draw bacteria from the secondary area into the tertiary quadrant. Fill the area,** spacing your lines a little farther apart than before; see figure 13.4e.
11. Close the dish; **flame your loop. Invert the plate** for incubation.

[1]Instructions are for right-handed students. Make appropriate corrections if you are left-handed.

PROCEDURES EXERCISE 13

Procedure 1

A Dry Run (Practice without Microorganisms) of Streak-Plate Technique

Note: All directions are given for right-handed students. If you are left-handed, make appropriate corrections.

a To prepare for your dry run of the streak-plate technique, trace a Petri dish onto a piece of paper. Use your tracing for a plate, your pencil for an inoculation loop, an empty tube with a closure for a culture, and work next to your unlit Bunsen burner.

b Using your tracing, practice the steps in preparing a streak plate, which are explained in table 13.1 and illustrated in figure 13.4.

Figure 13.4 Preparing a streak plate

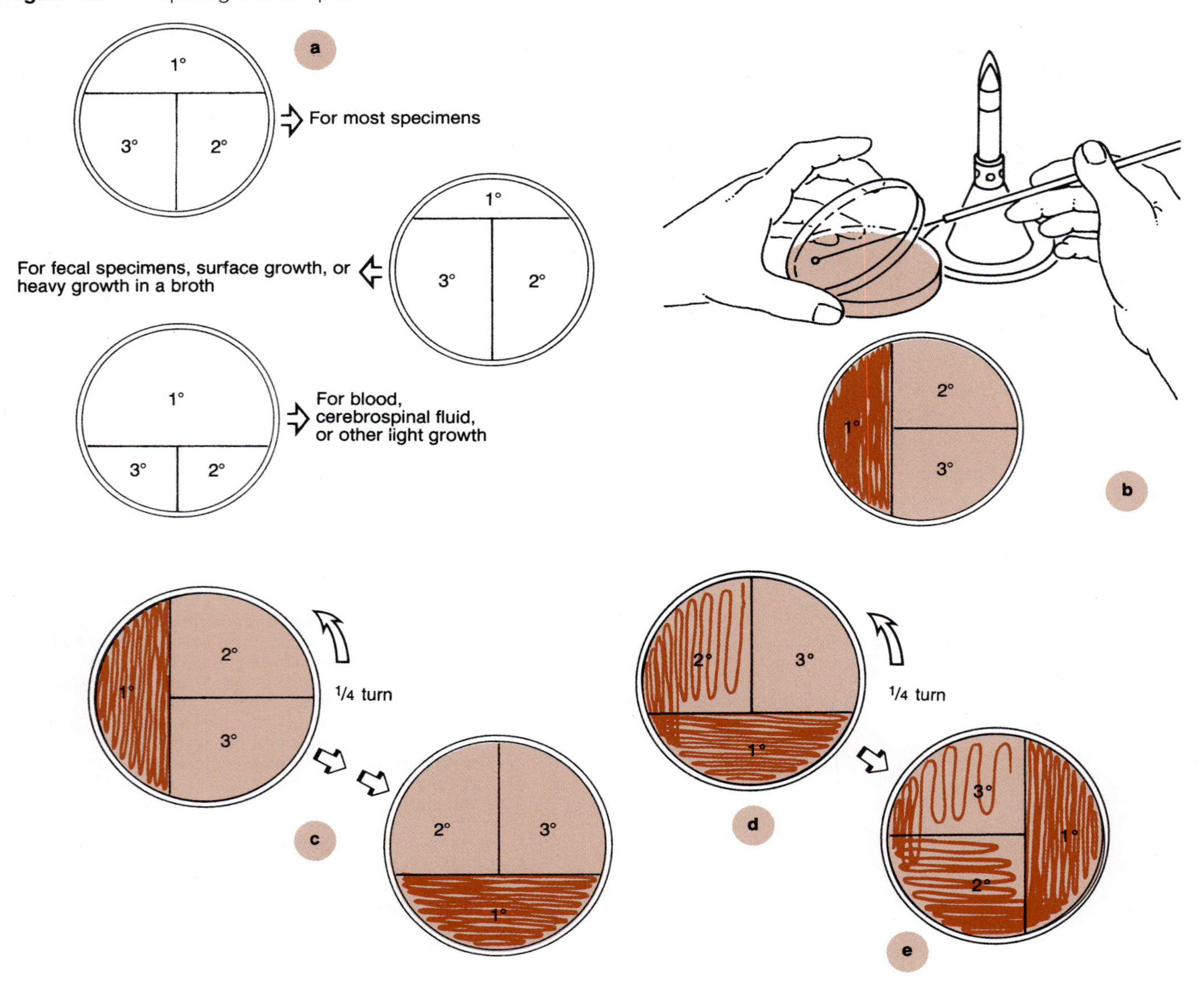

c Repeat the dry run until you are comfortable with the procedure. Remember, to avoid contamination, work quickly, keeping your plates and tubes as close to the flame as possible. Keep your work away from your nose.

Procedure 2

Preparing a Streak Plate

First Laboratory Session

a Obtain a plate of Tryptic Soy Agar or Nutrient Agar. Label the back of the plate with your name, the date, the name of the culture, and the temperature of incubation (35° C). **Your instructor** may suggest further labeling.

b Utilizing a well-suspended broth culture of *Staphylococcus epidermidis,* prepare a streak plate employing the protocol described in table 13.1.

c Incubate the inverted plate at 35° C until the next laboratory session.

Second Laboratory Session

a Examine your streak plate for well-isolated colonies and any evidence of contamination. Contaminating colonies generally look different from the other colonies. They often grow off of the line of streak rather than on it.

b In your laboratory report, evaluate your work. Remember, adequate isolation of colonies may occur in the primary, secondary, or tertiary sections. Discard your plate as directed by **your instructor.**

c Employ table 13.1 and figure 13.4 as aids in preparing another streak plate of *Staphylococcus epidermidis.* If possible, improve your technique.

Third Laboratory Session

a Examine your plate for well-isolated colonies and any evidence of contamination.

b In your laboratory report, evaluate your work. Discard your plate appropriately.

Procedure 3

Preparing a Streak Plate from a Swab of a Mixed Culture

First Laboratory Session

a Label the back of a sterile Tryptic Soy Agar or Nutrient Agar plate with your name, the date, the words "1° Area Swab, Mixed Culture, *Staphylococcus epidermidis* and *Escherichia coli,*" and the temperature of incubation (35° C). **Your instructor** may suggest further labeling. Draw a guide on the back of the plate.

b Using sterile technique, dip a **cotton-tipped applicator** into a mixed broth culture of *Staphylococcus epidermidis* and *Escherichia coli.* Press the cotton end of the swab against the inside rim of the tube, as shown in the illustration. Ring the inside of the tube two times to press excess liquid out of the swab. Close the culture tube and return it to your rack.

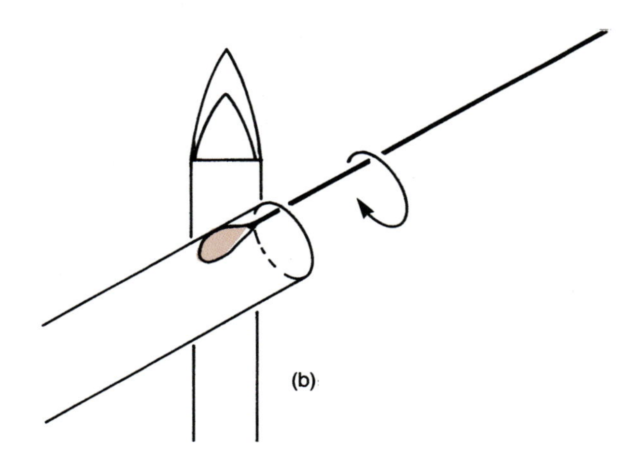

c Working directly beneath the flame, with your face back, away from your work, quickly use the damp swab to fill the primary area of your plate with closely spaced zigzag lines.

d Close the plate. Rotate it one-quarter turn. Discard the swab as directed by **your instructor.**

e Use a flamed and cooled **inoculating loop** to complete the secondary and tertiary sections of your streak for isolation of colonies.

f Incubate your inverted plate at 35° C until the next laboratory session.

Second Laboratory Session

a Examine your plate for well-isolated colonies of both *Staphylococcus epidermidis* and *Escherichia coli.* Use colonies of *Staphylococcus epidermidis* from Procedure 2 as a reference. Also watch for contamination. Discard your plate as directed by **your instructor.**

b In your laboratory report, evaluate your work.

Procedure 4

Preparing a Spread Plate

First Laboratory Session

a Label the back of a sterile Tryptic Soy Agar or Nutrient Agar plate with your name, the date, the name of the culture, *Staphylococcus epidermidis,* the temperature of incubation (35° C), and the word "Lawn." **Your instructor** may suggest further labeling.

b Using sterile technique, dip a cotton-tipped applicator into the broth culture of *Staphylococcus epidermidis;* ream the swab twice. Close the tube and replace it in your rack. You do **not** dip into it again.

c As shown in the illustration, use the swab to fill three-quarters of the plate with closely spaced zigzag lines. To avoid contamination, work quickly and near the flame, keeping your head back, away from the plate. Leave the right fourth of the plate (the part of the surface that is under the wrist of your inoculating hand) empty so that you do not have to hold your hand over the medium. This helps prevent contamination.

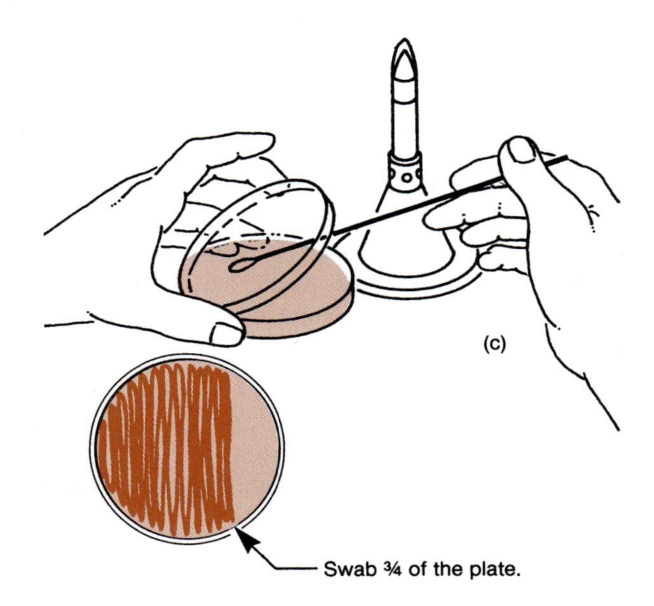

d Rotate the plate one-quarter turn. Rotate the applicator then swab the surface of the medium thoroughly, but stop before you need to bend your wrist and reach over the edge.

e Give the plate a final one-quarter turn. Swab it 1 more time with your rotated applicator. You have now swabbed the plate 3 times and turned it twice, as shown in the illustration. Your plate should be entirely covered with the pure culture of bacteria.

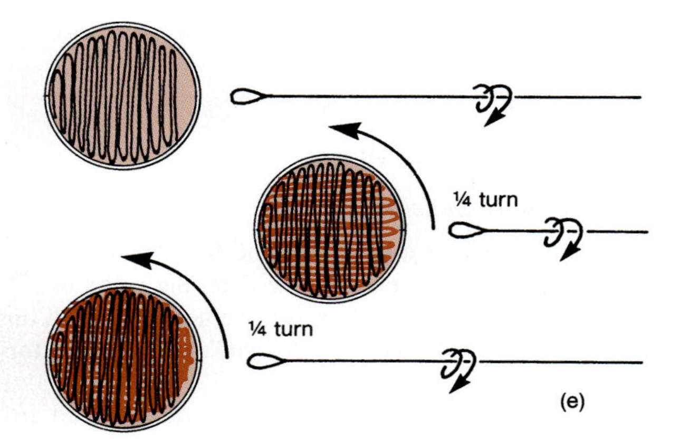

f Discard the swab as directed by **your instructor.** Invert your plate. Incubate it at 35° C until the next laboratory session.

Second Laboratory Session

a Examine your spread plate for complete coverage of the medium. Also watch for contamination. Evaluate your work in your laboratory report. Discard your plate as directed by **your instructor.**

b Using a broth culture of *Staphylococcus epidermidis,* a sterile plate of Tryptic Soy Agar, and a sterile swab, prepare an improved spread plate, if possible. Label the plate, invert it, and incubate it at 35° C until the next laboratory session.

Third Laboratory Session

a Examine your spread plate for complete coverage of the medium and any evidence of contamination.

b Complete the plate evaluation in the laboratory report. Discard your plate appropriately.

POSTTEST EXERCISE 13

Part 1

True or False (Circle one, then correct every false statement.)

1. T F To prepare a streak plate, spread bacteria evenly over the entire surface of a plate of medium.
2. T F When preparing a bacterial spread plate, swab a pure culture of bacteria evenly over the entire surface of an agar medium in a Petri dish.
3. T F A bacterial concentration gradient is a well-dispersed mixture of bacteria, thoroughly suspended in a broth medium.
4. T F Streaking a plate involves drawing a loopful of organisms back and forth across a solid medium in a Petri dish until the microbes fall off the loop one at a time.
5. T F The progeny of a single cell may form a colony, a visible clump of growth, on the surface of an agar medium.
6. T F The area of confluent growth is where colonies overlap and grow together.
7. T F All bacterial colonies look alike. Only staining reveals differences between genera.
8. T F Bacteriophage are special bacteria that are grown in spread plates.
9. T F Bacterial sensitivity to antibiotics is often tested on a spread plate.
10. T F In a dry run of a laboratory procedure, the microbiologist practices without using test organisms.
11. T F While streaking the primary area of a streak plate, place the Petri plate lid on the laboratory bench.
12. T F Be sure to flame your swab after preparing a spread plate.

Part 2

Completion

1. Bacterial cells can be separated by ________________ ____________________ using fine needles under a microscope lens.
2. To streak a plate for isolation of colonies, place one loopful of ____________________________ on the surface of the medium and use the quadrant-streaking method to dilute the bacteria.
3. In a medical laboratory, the bacteriological specimen is often a mixed culture from the patient's ______________ , ______________ , ______________ , and so on.
4. Bacteria divided by ______________ ______________ , simple splitting.
5. ________________ ________________ is an increase in bacterial population, not the size of individual cells.

Laboratory Report **13**

Name ____________________

Section ____________

Streak-Plate and Spread-Plate Techniques

Procedure 2

Preparing a Streak Plate

Second Laboratory Session

(*a*) Describe a well-isolated colony of *Staphylococcus epidermidis.*

Is there evidence of contamination among the colonies on your streak plate? __________

(*b*) Are you satisfied with your streak plate? __________ If not, how can you improve your streak-plate technique?

Third Laboratory Session

(*b*) Evaluate your streak-plate technique.

Procedure 3

Preparing a Streak Plate from a Swab of a Mixed Culture

Second Laboratory Session

(*b*) Did well-isolated colonies of both *Staphylococcus epidermidis* and *Escherichia coli* develop on the streak plate you prepared using a swab from a mixed culture? __________
Is there contamination? __________
Are you satisfied with your technique? __________ If not, how can you improve it?

Procedure 4

Preparing a Spread Plate

Second Laboratory Session

(*a*) Did you achieve complete coverage of the medium with a confluent growth of bacteria? ___________ There should be no holes and no ragged edges. If you are not completely satisfied with the outer edge of your plate, try circling it with the inoculating swab next time.
Is there any evidence of contamination? ___________
If you are not completely satisfied with your spread plate, how can you improve your technique?

Third Laboratory Session

(*b*) Evaluate your spread plate.

Exercise 14

Hand Washing

Attention: Do *not* wash your hands before this exercise.

Necessary Skills

Aseptic transfer techniques, Streak-plate techniques

Materials

Supplies (per group):
- 4 Petri plates containing Tryptic Soy Agar or Nutrient Agar
- 4 Sterile cotton-tipped applicators
- About 1 ml sterile water in a capped test tube
- Sterile hand brush

Additional supplies (per class):
- Soap or other cleanser (several containers or bars of whatever kind is utilized by your teaching facility)
- Paper towels
- Several containers of Betadine antiseptic hand cleanser per class (other antiseptic cleansing solutions such as pHisoHex with hexachlorophene may be used)
- Hand lotion dispenser
- Petroleum jelly
- Receptacle for used brushes
- Containers of disinfectant for contaminated swabs
- At least 5 Tryptic Soy Agar or Nutrient Agar plates for class controls
- At least 5 moist, sterile cotton-tipped applicators for preparing class controls
- Sterile hand brush for class control

Primary Objective

Demonstrate the cleansing power of hand washing and fingernail scrubbing with soap, a hand brush, and antiseptic solution.

Other Objectives

1. Distinguish between transient and resident microbial flora.
2. Define the terms *soap* and *detergent.*
3. Diagram saponification.
4. Explain how soap cleans.
5. Discuss the reason that soap is not an effective cleansing agent when used with hard water.
6. Evaluate the effectiveness of scrubbing the hands and fingernails with a wet, soapy brush to remove microorganisms.
7. Demonstrate the hand-cleansing capability of an antiseptic detergent solution.

INTRODUCTION

In the microbiology laboratory, you cleanse your hands to protect your microbial cultures, yourself, and the community. Generally, you begin a laboratory session by washing your hands with soap or other detergent to remove **transient** (temporary) microorganisms that might contaminate your cultures. The last thing you do before leaving the microbiology laboratory is thoroughly wash, perhaps also disinfect, your hands to remove pathogens.

As you demonstrate in this exercise, **resident** (long-term) flora that generally inhabit hair follicles and sebaceous glands of your skin remain even after thoroughly washing, scrubbing, and disinfecting your hands.

The Chemistry of Soap

An increased use of **detergent** (any cleansing solution, including soap) to remove bacteria embedded in fats and oils has increased human life expectancy. But what is a "soap" and how does it work? How effectively do various hand-washing techniques and solutions remove microbes? You answer these questions as you read this introduction and perform the experiment in this exercise.

Figure 14.1 Soap

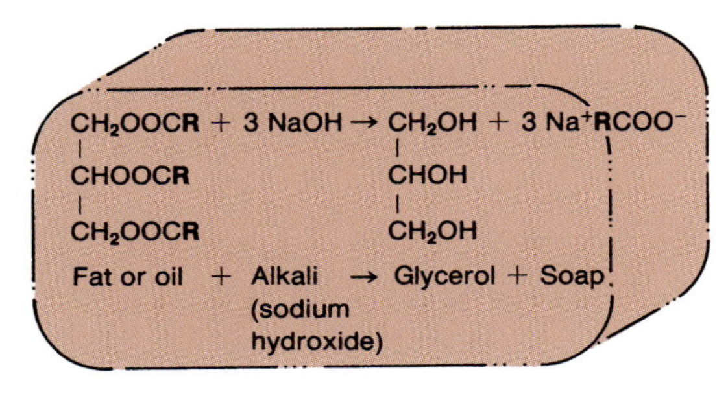

The chemistry of detergents is often complex. Their formulation keeps many chemists employed by large companies such as Colgate-Palmolive. But making a soap (one type of detergent) is fairly straightforward.

In a process called **saponification,** a **soap** is formed when fats are heated in the presence of a base such as sodium hydroxide **(NaOH).** As shown in figure 14.1, the sodium salt **(Na^+RCOO^-)** of the acid **($RCOO^-$)** is the **"soap."**

Generally, oily substances do not mix with water. While water can remove most dirt, it is ineffective in eliminating bacteria suspended in fats, hydrocarbon chains, and other water-insoluble substances.

But notice in figure 14.1 that each soap molecule has a long **hydrocarbon chain (R)** attached to a negatively charged **carboxylate ion (COO^-)** and its associated **sodium (Na^+) ion.** In greasy (often bacteria-laden) water, the hydrocarbon end of the soap molecule is soluble in oils, and the carboxylate end is soluble in water. A chemical link forms between oil and water. Then the oily ends of the molecules dissolve together. Now the greasy oil is chemically combined with water; it can be carried away in the liquid. This is how soap cleans.

When soap is used with **hard water** (water containing mineral salts), magnesium and calcium salts react with soapy oil droplets, forming a lighter-than-water precipitate. This scum, with its load of grease and bacteria, adheres to whatever is being washed. Because of soapy scum formation, other detergents are frequently better cleansers than soap when utilized with hard water.

The Effectiveness of Hand Washing

In this exercise you perform an experiment designed to determine if microbes on your hands are removed when you wash with hot soapy water and then scrub with disinfectant solution. You also evaluate the diversity of microorganisms that remains after each cleansing procedure. For a discussion of the antiseptic you employ, see Exercise 19, Disinfectants and Antiseptics.

Since there are nearly as many conventions for hand washing as there are microbiology laboratories, commercial food establishments, medical centers, and parents, no single method is advocated here. For the sake of uniformity, however, all those who wash their hands for this experiment are asked to employ the method outlined in Procedure 1.

The experiment you perform is illustrated in figure 14.2.

PROCEDURES EXERCISE 14

Attention: Do **not** wash your hands before this exercise.

Procedure 1

The Effectiveness of Hand Washing; see figure 14.2

First Laboratory Session

Warning: Remove jewelry from your hands and store it in a safe place during this experiment.

a Label a plate of Tryptic Soy Agar or Nutrient Agar with your name and the date. Add "#1" and "Unwashed Control." Dampen a sterile cotton swab in the test tube of sterile H_2O. As shown in figure 14.2, swab a portion of the palm and small finger of your unwashed, nondominant hand.

b With rolling, overlapping strokes, streak the swab across the top third of a plate of Tryptic Soy Agar or Nutrient Agar. Discard the swab as directed by **your instructor.** As in figure 14.2, press the tip of the sampled finger and its nail into an unswabbed portion of the agar.

c With a sterile loop, complete the secondary and tertiary portions of a 3-part streak for isolation of colonies. **Do not** streak over the fingertip impression.

d Incubate the plate at 35° C until the next laboratory session.

e In your laboratory report, discuss the function of this plate in your experiment, and why you utilized sterile H_2O.

f Label your second plate of agar medium with your name and the date. Add "#2" and "Wash." Have a sterile swab and several towels ready.

g Wash your hands for 20 seconds with hot soapy water, then rinse your hands. Do **not** dry your hands. Use your dominant hand to turn off the water with the towels, then pick up the sterile swab. As in figure 14.2, swab the ring finger and part of the palm of your nondominant hand.

h Repeat steps (*b*) through (*d*).

i Label your third plate of agar medium adding "#3" and "Wash and Brush." Have a sterile swab, a sterile unwrapped hand brush, and several towels ready.

j Wash your hands for 20 seconds with hot soapy water and a sterile hand brush, then spend 20 seconds using the wet, soapy brush on the fingernails of your nondominant hand. Rinse your hands and the brush thoroughly with running water. Do **not** dry your hands.

Figure 14.2 Testing the effectiveness of hand washing

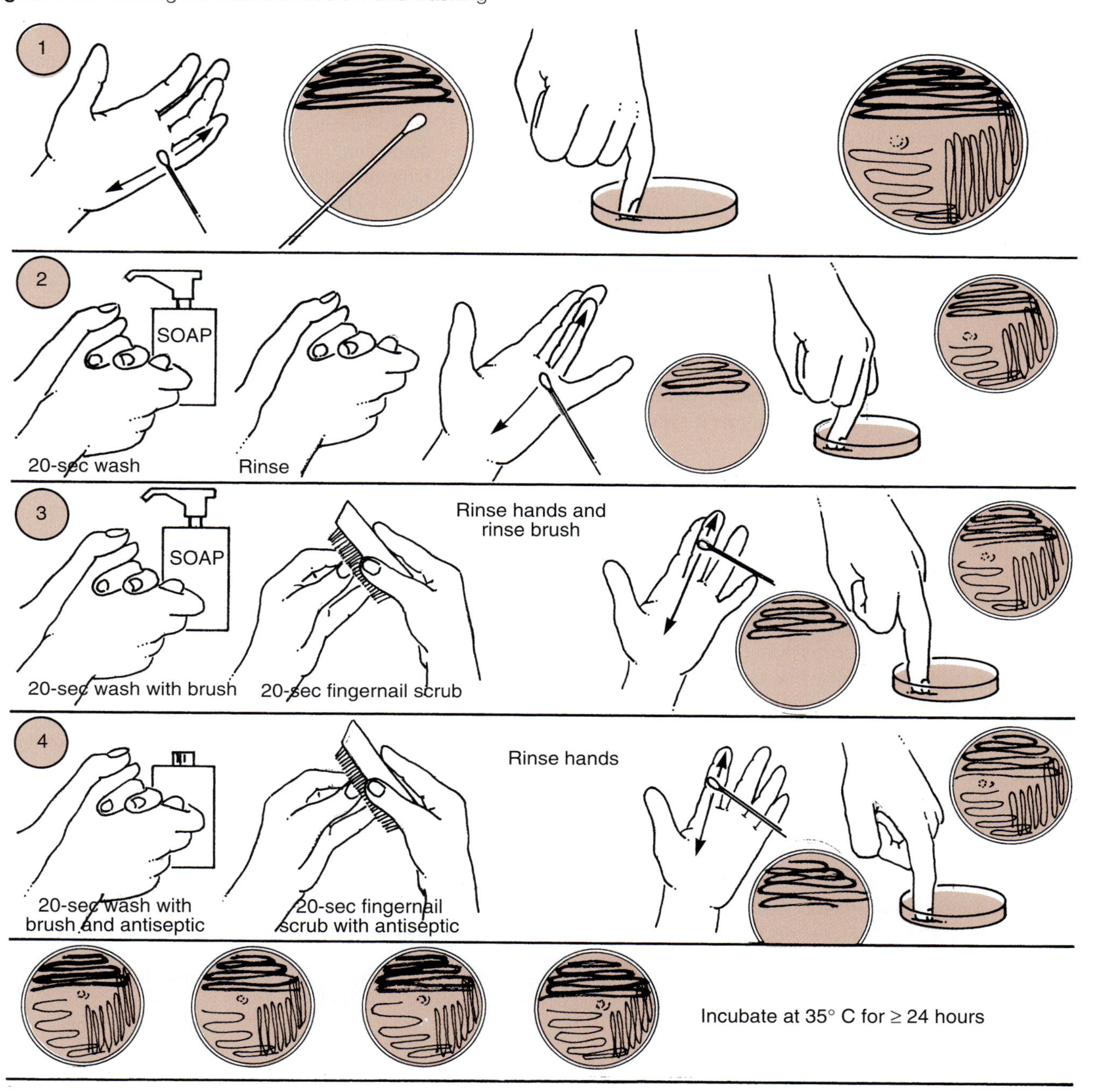

k Use your dominant hand to turn off the water with the towels and to manipulate the sterile swab. Place the brush, bristles up, on a towel; you will use the brush again.

l As shown in figure 14.2, swab the middle finger and part of the palm of your nondominant hand. Repeat steps (*b*) through (*d*).

m Label your fourth plate of agar medium adding "#4" and "Antiseptic and Brush." Have a sterile swab, your hand brush, and several towels ready.

n Wash your hands for 20 seconds with hot water, your hand brush, and an antiseptic cleansing solution (Betadine, pHisoHex with hexachlorophene, etc.). Spend the next 20 seconds using the brush and cleansing solution on the fingernails of your nondominant hand. Rinse your hands. Do **not** dry your hands.

o Use your dominant hand to turn off the water with the towels and to manipulate the sterile swab. Discard the brush in an appropriate manner.

p As shown in figure 14.2, swab the index finger and part of the palm of your nondominant hand.

q Repeat steps (*b*) through (*d*).

r If you wish, apply hand lotion or petroleum jelly to your hands. If you have removed jewelry for this experiment, remember to retrieve it.

s **Your instructor,** or assigned groups will prepare appropriate controls for this experiment. The primary area of each control plate is swabbed or pressed with an agent employed in the experiment. The swab is discarded appropriately. The secondary and tertiary sectors of each plate are streaked with a loop for isolation of colonies. Controls are incubated at 35° C until the next laboratory session.

Prepare and adequately label a control plate for each of the following variables: each soap or detergent utilized, a sterile hand brush, each container of Betadine or other disinfectant, water, and towel. Add other controls if appropriate.

Second Laboratory Session

a Examine your Petri plates. In your laboratory report, estimate how much washing your hands with hot soapy water changed their microbial population. Then estimate the amount of change produced by washing 1 minute and brushing.

b How much of the microbial load is removed by the full complement of hot soapy water, brushing, rinsing, and then disinfecting the hands while scrubbing them?

c Also, for each phase of this experiment, as represented by your 4 Petri dishes, estimate the change in the microbial load carried by your fingertips and under your fingernails.

d Compare the approximate numbers of different kinds of colonies on each plate.

e What conclusions can you draw concerning the numbers and kinds of microorganisms you find on your hands after various cleansing procedures?

f Examine the class controls. Comment on the microbial load of the various cleansers, disinfectants, and other agents employed in this exercise. Does the growth on any of the control plates alter your conclusions concerning hand washing? If so, how?

g After the class has recorded all observations, safely discard the plates as directed by **your instructor.**

POSTTEST EXERCISE 14

Part 1

True or False (Circle one, then correct every false statement.)

1. T F Any cleansing solution, including soap, is a detergent.
2. T F Saponification is a highly complex chemical process, but soap making is fairly straightforward.
3. T F Soaps are formed when fats are heated in the presence of an acid such as hydrochloric acid.
4. T F During saponification a sodium salt of an acid is produced.
5. T F The carboxylate end of a soap molecule is soluble in hydrocarbons, the oily end in water.
6. T F Soap forms a chemical link between oil and water.
7. T F Soap cleans by chemically linking oil to grease.
8. T F The soapy scum that forms in oily, hard water removes bacteria when it sinks to the bottom of the liquid, weighted down by its load of heavy mineral salts.
9. T F Thoroughly scrubbing the hands and fingernails with hot soapy water for 1 minute removes all bacteria from the area cleansed.
10. T F Scrubbing the hands with antiseptic solution and a brush for 40 seconds always kills all bacteria on the cleansed area.
11. T F Washing hands usually reduces their microbial load of transient flora.
12. T F Hand washing by medical personnel helps reduce the patient's risk of **nosocomial** (hospital-acquired) infection.

Part 2

Completion and Short Answer

1. In the space at the left, diagram saponification, labeling each molecule.
2. ________________________ microorganisms are on your body for only a short time, while ________________________ flora live there for years.
3. ____________________ and ____________________ salts in greasy, hard water form a lighter-than-water scum.
4. The ________________________ end of a soap molecule is soluble in oils, and the ________________________ end is soluble in water.

Laboratory Report **14**

Name ______________________________

Section ______________

Hand Washing

Procedure 1

The Effectiveness of Hand Washing

First Laboratory Session

(*e*) Why do you prepare the first plate before washing your hands, and why is this the only swab you moisten before use?

Second Laboratory Session

(*a,b*) Compare the numbers and the diversity of colonies on your 4 plates. What reduction (or increase) in numbers do you observe with washing, scrubbing, and disinfecting? Discuss your results.

(*b*) Before swabbing your hand for the fourth plate, what was the total time you spent washing with hot soapy water? __________ What was the total time spent scrubbing? __________ What was the total time spent disinfecting? __________
(*c*) Discuss the fingertip and fingernail impressions on your Petri plates. Compare the numbers and, if possible, the diversity of colonies in these impressions.

(*f*) Examine your class control plates. Is there evidence that any of the agents carried a heavy microbial load that may have altered the results of your hand-washing experiment? _________ If so, which one(s)?

Discuss the effects of contaminated soaps, disinfectants, towels, and so forth on hand washing.

Exercise 15

Quantifying Bacteria

Necessary Skills

Aseptic transfer techniques, Pour-plate techniques

Materials

First Laboratory Session

Culture (24-hour, 35° C, Tryptic Soy Broth at least 8 ml per group):
Escherichia coli
Supplies (per group):
Procedure 1
3 Sterile 99 ml water blanks
Container of boiling water to hold 4 deeps
4 Tryptic Soy Agar deeps
4 Empty, sterile Petri dishes
4 Sterile 1.0 or 1.1 ml pipets
Pipettor
Access to a 55° C water bath
Racks to hold class deeps in the water bath
Procedure 2
Wash bottle of purified water
Kimwipes® or other lint-free tissue
Container of disinfectant
Clean cuvette
Nonabrasive holder for cuvette
At least 15 ml of sterile Tryptic Soy Broth
6 Sterile test tubes
2 Sterile 10 ml pipets
4 Sterile 5 ml pipets
Pipettor
Access to a spectrophotometer

Second Laboratory Session

Access to a Quebec colony counter (optional)
Access to a tally register (optional)

Primary Objective

Quantify bacteria.

Other Objectives

1. Give reasons a microbiologist might wish to ascertain the number of bacteria per milliliter in a liquid.
2. Explain how it is possible to distinguish living bacterial cells from dead ones.
3. Identify two methods of growing, then enumerating, living bacterial cells.
4. Differentiate a single cell from a colony-forming unit.
5. Discuss preparation of bacterial dilutions. Explain why dilutions are necessary. Calculate test tube and plate dilutions.
6. Calculate the number of bacteria per milliliter of an original solution given the number of colonies on a countable plate and the dilution of the plate.
7. Compare and contrast two methods of preparing bacterial dilution plates.
8. Select examples of circumstances under which the turbidity of a bacterial solution might remain unchanged while cell numbers fluctuate.
9. Perform and evaluate a plate-count determination of the numbers of living microorganisms in a solution.
10. Prepare controls and a serial dilution of bacteria.
11. Perform and evaluate a turbidometric assay of bacterial density employing a spectrophotometer, a serial dilution of bacteria, and appropriate controls.
12. Construct and evaluate a standard curve of optical density versus the number of cells in a suspension.

INTRODUCTION

In this exercise you determine the number of microorganisms per milliliter of a solution. Such evaluations of microbial population size are employed frequently by microbiologists in public health, industry, and medicine.

Public health microbiologists measure microbial contamination levels to ascertain whether water is **potable** (drinkable), milk products are free of excessive contamination, and foods are safe to eat. In brewing, antibiotic production, and other industrial processes, microbial concentrations are monitored. To evaluate the efficacy of a chemotherapeutic agent, a physician may need to know the number of bacteria per milliliter surviving in a patient's body fluids while treatment for infection is under way.

These investigators require information on the concentration of **living** cells in liquids.

Counting Viable Microorganisms

How is it possible to distinguish living bacterial cells from dead ones? Through a microscope lens, they often look alike. But given a favorable set of **incubation conditions** (time, temperature, nutrients, pH, oxygen tension, etc.) the living cell is capable of reproduction. Dead cells do not reproduce or form colonies.

When microbes are very dilute, as in a city water supply, it is often convenient to quantify the microbial content of the fluid by simply passing a known amount of the liquid through a **filter.** The filter is then incubated on a Petri dish of nutrient medium. Colonies that develop on the filter are counted. With a little mathematical manipulation, the number of colonies tells the investigator how many bacteria per milliliter are in the water and are able to survive the incubation conditions.

In this exercise you employ **dilution plate counts;** see figure 15.1. This method of quantification is especially useful for examining more highly contaminated liquids, microbe-laden fluids that could clog a filter.

There are two common techniques for preparing dilution plates. In the **pour-plate** technique, a sample from an accurate dilution of bacteria is pipetted into a Petri dish, then agar medium is poured over the liquid and mixed. In the **spread-plate** technique, generally 0.1 ml of the diluted sample is pipetted onto the surface of a solidified agar medium in a Petri dish. The liquid is spread over the medium with a sterilized, bent rod. To improve the accuracy of both methods, multiple plates of each dilution are prepared.

Both spread plates and pour plates are incubated. Then colonies are counted, and the microbiologist, by applying mathematical calculations, estimates the number of viable bacteria per milliliter of the original solution.

Sometimes bacteria clump. A colony may grow from several bacteria clustered together. For this reason, microbiologists often say colonies develop from **colony-forming units** rather than from single cells.

Water-Blank Dilutions

Dilutions are generally prepared in 9 or 99 ml **water blanks** (containers holding measured amounts of water). When you pipet 1 ml of bacterial suspension into 9 ml of water, you create a **1:10** (called a "one in ten") dilution. This dilution may be written 10^{-1}. It holds 1 ml of suspension in a total of 10 ml of fluid. What dilution is created when you add 1 ml of suspension to 99 ml of water? __________ *

Figure 15.1 Dilution plate counts—Procedure 1

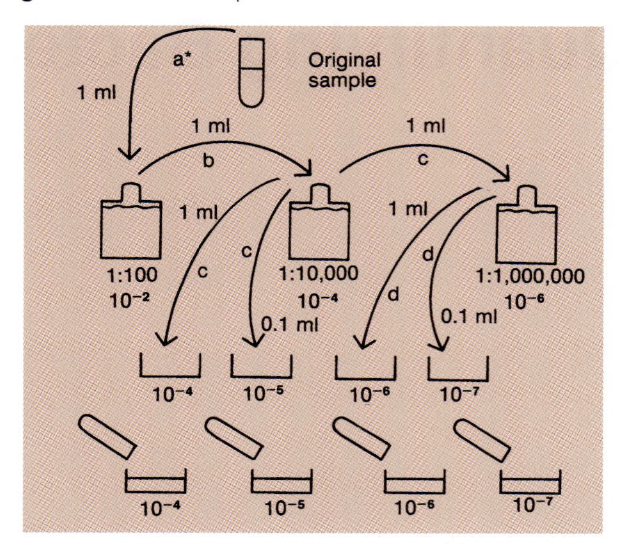

*Each letter, a–d, represents a separate, sterile pipet.

You may need to dilute further. Adding 1 ml from a 1:10 dilution to a second 9 ml water blank produces a 10^{-2} (1:100) dilution of the original bacteria.** What dilution do you create by adding 1 ml of a 1:100 suspension to a 99 ml water blank?***

Why does a microbiologist interested in counting viable bacteria need to prepare dilutions? Because measured amounts of liquid from each dilution are added to agar media and then allowed to grow. An acceptable number of colonies must develop on a plate before the microbiologist can utilize the results.

Only plates with 30 to 300 colonies are considered **countable.** When fewer than 30 colonies appear, the results are not as repeatable as they need to be for scientific investigations. If more than 300 colonies are crowded onto a plate, one cluster of bacteria often hides another, making counts inaccurate.

Calculations

To find how many bacteria are in the original solution, multiply the number of colonies on a countable plate times the inverse of that plate's dilution:

bacteria per ml in the original solution

$$= \textbf{number of colonies on a plate} \times \frac{\textbf{1}}{\textbf{dilution of plate}}$$

*1:100 or $\mathbf{10^{-2}}$.

**Each milliliter in the 10^{-1} dilution contains 1/10th as many bacteria as the original solution. You now dilute a milliliter 10 times more. As you can see, each milliliter of the new dilution holds 1/10th of 1/10th of the original concentration of bacteria. In mathematics the word "of" means "times"; therefore $10^{-1} \times 10^{-1} = \mathbf{10^{-2}}$ (also written 1/10 × 1/10 = 1/100).

***$\mathbf{10^{-4}}$ (1:10,000) because $10^{-2} \times 10^{-2} = 10^{-4}$ (also written 1/100 × 1/100 = 1/10,000).

For example, if there are 79 colonies on a 1:1,000,000 plate, then in the original solution there are

$$79 \times \frac{1}{1{:}1{,}000{,}000}$$

bacteria that are able to grow in your set of incubation conditions. That is:

Bacteria per ml in the original solution
= 79 × 1,000,000 = 79,000,000

Dilutions of Plates

You have read how to calculate the dilution of a water blank, but how do you compute the dilution of a plate? Examine figure 15.1 for examples and remember two rules:

1. If you add 1 ml of a suspension to a plate, retain the dilution of the water blank for the plate.
2. If you add 0.1 ml (1/10 ml) of a suspension to a plate, multiply the dilution of the water blank by 1/10 to calculate the dilution of the plate.

Quantifying Living and Dead Bacteria

When you are interested in determining the total concentration of both living and dead microorganisms in a solution, you can measure **cell mass.**

Cell mass can be estimated with **dry weight.** This method is most useful with filamentous microorganisms such as fungi.

Mass is also indicated by the ability of a bacterial suspension to scatter light. We generally see a broth become **turbid** (cloudy) as its microbial population increases. Remember, turbidity reflects mass, not numbers.

This fact is based on the observation that small microbes may enlarge, scattering more light, without increasing the number of cells in their population. Conversely, large cells that are rapidly multiplying may simply divide, greatly increasing microbial numbers without appreciably altering total cell volume.

Performing a Turbidometric Assay

In this exercise you perform a **turbidometric assay** of bacterial density. That is, you measure the turbidity of bacterial dilutions with a spectrophotometer and use these measurements to estimate the numbers of bacteria in a solution.

Instructions for using a common spectrophotometer, the Bausch and Lomb Spectronic 20, are given in table 15.1; the instrument is illustrated in figure 15.2.

You prepare controls and a **serial dilution** of bacteria as in figure 15.3. (A serial dilution is the progressive dilution of a sample in a series of tubes.) You then employ the spectrophotometer to measure the **optical density** (amount of light scattered and absorbed by cells in a solution) of the bacterial suspension in each tube.

Table 15.1 Using the Bausch and Lomb Spectronic 20

1. Connect the Spectronic 20 to an electrical outlet. Many spectrophotometers need to warm up for about 20 minutes before use. Turn on the instrument by rotating the **power/zero control knob** clockwise; see figure 15.2. Turn the **wavelength control knob** to 686 nm. Wait.
2. With the **sample holder** empty and closed, rotate the **power/zero control knob** until the needle on the galvanometer reads **"0% Transmittance."** To read the dial accurately, move your head until the needle is at eye level and directly in front of its mirrored reflection. You have now "zeroed" the spectrophotometer. Simply read the number on digital models.
3. Align the index lines of the sample holder and a cuvette containing uninoculated medium. Insert the cuvette and close the lid of the sample holder.
4. Rotate the **light control knob** until the needle on the galvanometer reads **"100% Transmittance."** You have now "standardized" the spectrophotometer.
5. Repeat steps 2 through 4 until the readings remain consistent. After measuring the **absorbance** of a few, well-suspended samples, repeat steps 2 through 4 again. The Spectronic 20 is more accurate when it is frequently zeroed and standardized.

Figure 15.2 The Bausch and Lomb Spectronic 20

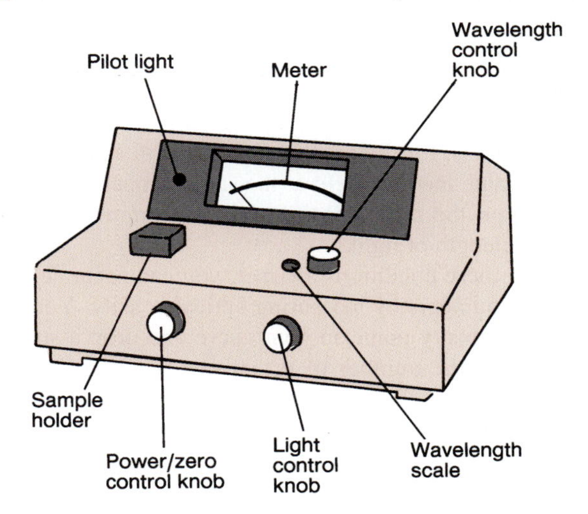

Graphing Optical Density versus Number of Cells

When measuring turbidity, you graph the data placing optical density on the vertical *Y* axis and the dilution of each bacterial suspension on the horizontal *X* axis. The resulting optical density versus dilution graph is located in your laboratory report for Procedure 2.

Figure 15.3 A serial dilution—Procedure 2

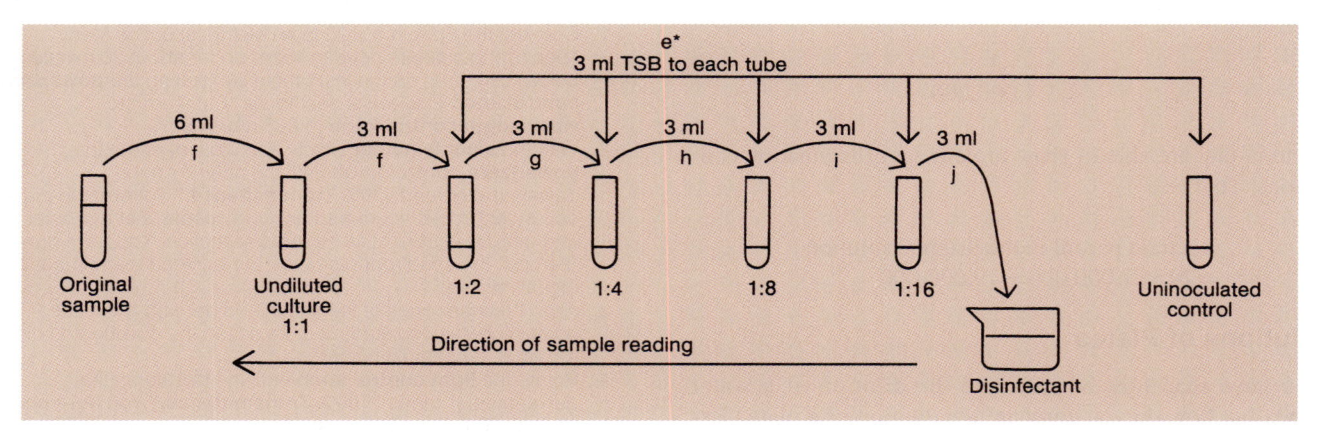

*Each letter, e–j, represents a separate, sterile pipet.

Your plate-count analysis tells you how many bacteria per milliliter of the original solution survived under your set of conditions of incubation. Given this information, you calculate the number of bacteria per milliliter in each test tube of your dilution series. Then you fill in the number of organisms per milliliter of each dilution on the *X* axis of the optical density versus dilution graph.

By linking the data points on your graph, you construct a **standard curve of the optical density versus the number of cells.** This graph is a tool that you can employ to estimate cell concentration any time the same conditions are applied. That is, for the graph to be valid, all samples must be in the same medium held at the same temperature. They must contain the same species and be examined with the same wavelength of light.

When these conditions are met, you can accurately estimate cell numbers by measuring optical density. You find the optical density point on your curve and drop a straight line down to the number of organisms per milliliter. By serving as a substitute for plate-count dilution techniques, such an estimate saves time, space, and money.

PROCEDURES EXERCISE 15

Procedure 1

Plate Count

First Laboratory Session

a Examine figure 15.1 before embarking upon this procedure. Have all sterile materials handy; label your 3 dilution blanks and the bottoms of your 4 empty, sterile Petri dishes as in the figure.

b Bring a container of water holding your 4 Tryptic Soy Agar deeps to a boil. The water level should be above the level of the medium. To avoid contamination, keep the water level below the bottom rims of the test tube closures.

c Liquefy the deeps—liquefaction takes approximately 10 minutes in boiling water—then temper (cool) them in a 55° C water bath.

d Resuspend *Escherichia coli* in the original culture by vortexing. With a sterile pipet and pipettor, transfer 1 ml from the original culture into your 10^{-2} dilution blank. Discard the pipet in an appropriate manner.

e Tighten the lid of your first dilution. Shake the 10^{-2} dilution with 25 rapid, full-arm swings, circumscribing at least a 0.5 m arc with the bottle.

f With a pipettor and a fresh, sterile pipet, transfer 1 ml of fluid from the 10^{-2} bottle to the 10^{-4} bottle. Discard the pipet appropriately; tighten the 10^{-4} dilution lid; shake as before.

g With a third sterile pipet, make all your transfers from your 10^{-4} dilution. Deliver 1.0 ml to the 10^{-4} plate, 0.1 ml to the 10^{-5} plate, and 1.0 ml to the 10^{-6} dilution bottle. Discard the pipet appropriately; tighten the 10^{-6} dilution bottle lid; shake as before.

h With a fourth sterile pipet, make both your transfers from your 10^{-6} dilution. Deliver 1.0 ml to the 10^{-6} plate and 0.1 ml to the 10^{-7} plate.

i Using sterile technique, flame the lip of a tempered deep. Pour the medium over the contents of 1 plate. Gently, without splashing, mix the liquids by rotating the Petri dish first in one direction, then the other, and back and forth, and from side to side.

j Repeat step (*i*) until all 4 plates are poured and mixed.

k After the medium cools and hardens, invert the Petri dishes and incubate them at 35° C until the next laboratory session.

Safely discard your dilution bottles as directed by **your instructor.** Save your original culture of *Escherichia coli.* You use it again in Procedure 2.

Second Laboratory Session

a Inspect each pour plate. Count the colonies on any plates that have between 30 and 300 colonies. Use a Quebec colony counter and a tally register if they are available. In your laboratory report, record the plate dilution and the number of colonies on any countable plate.

b Calculate the number of bacteria per milliliter in the original bacterial solution. If you have several countable plates, and therefore several numbers to use, average your results. In your laboratory report, record the number of bacteria per milliliter in your original solution.

c If none of your plates were "countable," record the results of another group that employed the same original suspension of bacteria, giving credit in your laboratory report for the source of your information.

Procedure 2

Turbidimetry; see figure 15.3

First Laboratory Session

a Turn on the spectrophotometer and allow it to warm up for at least 20 minutes before use. **Your instructor** may demonstrate calibration of the spectrophotometer. Unless you receive other instructions, follow the directions in table 15.1, and see figure 15.2. Set the wavelength to 686 nm, if possible (550 nm, if not).

b Label 6 test tubes with the dilutions shown in figure 15.3. With a sterile 10 ml pipet, transfer 3 ml of **Tryptic Soy Broth** into the tubes labeled **"1:2," "1:4," "1:8," "1:16,"** and **"Uninoculated Control."** Dispose of the pipet appropriately.

c Use the same original suspension of *Escherichia coli* that you employed in Procedure 1. Resuspend the culture thoroughly by vortexing.

d With a pipettor and a sterile 10 ml pipet, transfer 6 ml from the **original culture** into your test tube labeled **"Undiluted Culture, 1:1."** Use the same pipet to transfer 3 ml from the **"1:1"** tube to the **"1:2"** tube. Dispose of the pipet in an appropriate manner.

e With a pipettor and a sterile 5 ml pipet, mix the contents of your **"1:2"** tube. Transfer 3 ml of the mixture into your **"1:4"** tube. Dispose of the pipet appropriately.

f With a pipettor and a second sterile 5 ml pipet, mix the contents of your **"1:4"** tube. Transfer 3 ml of the mixture into your **"1:8"** tube. Dispose of the pipet appropriately.

g With a pipettor and your third sterile 5 ml pipet, mix the contents of your **"1:8"** tube. Transfer 3 ml of the mixture into your **"1:16"** tube. Dispose of the pipet appropriately.

h With a pipettor and your fourth sterile 5 ml pipet, mix the contents of your **"1:16"** tube. Transfer 3 ml from the **1:16** dilution into a container of **disinfectant.** Dispose of your pipet appropriately. Congratulations! You have now prepared a serial dilution of *Escherichia coli.*

i With the **power/zero control knob,** adjust the spectrophotometer to **"0% Transmittance."** Wipe your cuvette with lint-free tissue. Pour the "Uninoculated Control" into the cuvette. Wipe again if necessary.

j Align the index line on the cuvette with the index mark on the instrument. Insert the cuvette, which holds your uninoculated control, into the spectrophotometer. Close the lid of the sample holder.

k With the **light control knob,** adjust the instrument to read **"100% Transmittance."** Remove the sample; close the lid of the sample holder; adjust to "0% Transmittance" again if necessary.

l When the "0%" and "100% Transmittance" readings are steady, pour the sample back into the "Uninoculated Control" tube and replace it in the cuvette with your resuspended 1:16 dilution. Wipe the cuvette; insert the cuvette, aligning the index marks; close the lid of the sample holder. Read the absorbance scale and record the reading in your laboratory report. Absorbance reflects optical density.

m Remove the cuvette from the spectrophotometer. Pour the sample back into its appropriate test tube. After resuspending the bacteria, pour the next dilution into the cuvette. Since you are progressing from least concentrated to most concentrated bacterial samples, it is not necessary to rinse your cuvette with purified water between each sample. Read the absorbance scale and record the reading in your laboratory report.

n Measure the absorbance of all your dilutions. After every second sample, rinse the cuvette well with purified water. Avoid contaminating your hands. Collect the rinse water in a container of disinfectant. Wipe the cuvette. Rezero the transmittance scale; then, with your "Uninoculated Control," adjust the instrument to read "100% Transmittance." **Your instructor** will discuss decontamination of cuvettes and tissues.

o Enter all test readings in your laboratory report.

p In your laboratory report, construct a curve of optical density versus bacterial dilution.

q In your laboratory report, explain why this experiment would be impossible to perform without an uninoculated control.

Second Laboratory Session

a After completing Procedure 1, calculate the number of bacteria per milliliter in each tube of your spectrophotometery dilution series.

b Complete your graph by calculating and recording the number of organisms per milliliter in each dilution.

c In your laboratory report, explain the usefulness of the completed graph.

POSTTEST EXERCISE 15

Part 1

Matching

Each response may be used once, more than once, or not at all.

a. Quantification of viable microorganisms only
b. Quantification of both viable and nonliving microorganisms

1. ____ Filtration and cultivation of colonies on the filter
2. ____ Spread-plate colony count
3. ____ Pour-plate colony count
4. ____ Dry weight measurement
5. ____ Measurement of optical density

Part 2

True or False (Circle one, then correct every false statement.)

1. T F Estimations of microbial population size are useful for industrial microbiologists but are of little value to the medical or public health microbiologist.
2. T F Only a living colony-forming unit forms a colony.
3. T F Filtration is most useful as a tool in the enumeration of microbiota from heavily populated liquid samples.
4. T F A colony-forming unit may be a single cell or a clump of bacteria.
5. T F If you pour 15 ml of agar medium into a Petri dish that contains 1 ml of a 10^{-6} bacterial suspension, the dilution of the plate is still recorded as 10^{-6}.
6. T F If you pour any suitable amount of agar medium into a Petri dish that contains 0.1 ml of a 10^{-6} bacterial suspension, the dilution of the plate is recorded as 10^{-7}.
7. T F If 38 colonies develop on a 10^{-6} dilution plate, it seems likely that between 0 and 5 colonies might grow on a 10^{-5} plate prepared from the same original suspension of bacteria.
8. T F A single standard curve of optical density versus number of cells is a convenient tool. By simply measuring the turbidity of any solution—milk, water, urine, and so on—and dropping a line from the curve to the *X* axis, you discover the number of bacteria in the solution.
9. T F Calculate the number of bacteria per milliliter in each test tube of a dilution series by multiplying the number of bacteria per milliliter in the original solution times the dilution of the tube. For example, if there are 1,100,000 bacteria per milliliter in the original solution, there are $1{,}100{,}000 \times 1{:}16 = 69{,}000$ bacteria per milliliter (rounded to two significant figures) in the 1:16 dilution.

Part 3

Completion and Short Answer

1. Across the bottom of this page, illustrate how to prepare 10^{-4}, 10^{-5}, 10^{-6}, and 10^{-7} plates. Label all dilution blanks and plates. Indicate the amount of fluid in each transfer and where to employ fresh pipets.
2. You expect (more, fewer) colonies on a 10^{-4} plate than on a 10^{-6} plate prepared from the same, original bacterial suspension. (Circle one.)
3. To calibrate a spectrophotometer, adjust the power/zero control knob to "0% ________________." Then adjust the light control knob to "100% ______________________" after inserting your ______________________ into the sample holder. Repeat this process frequently.

Laboratory Report **15**

Name ______________________________

Section ______________

Quantifying Bacteria

Procedure 1

Plate Count

Second Laboratory Session

(*a*) Record the plate dilution and the number of colonies on any countable plates.

Plate Dilution	Number of Colonies
________	________
________	________

(*b,c*) Calculate the number of bacteria per milliliter in the original bacterial solution, averaging results if necessary. Record your calculations.

Procedure 2

Turbidimetry

First Laboratory Session

(*l,m,o*) Record your absorbance readings.

1:16 ____________________

1:8 ____________________

1:4 ____________________

1:2 ____________________

1:1 ____________________

(*p*) Enter your readings in the following graph. Connect your data points constructing a curve of optical density (absorbance) versus bacterial dilution.

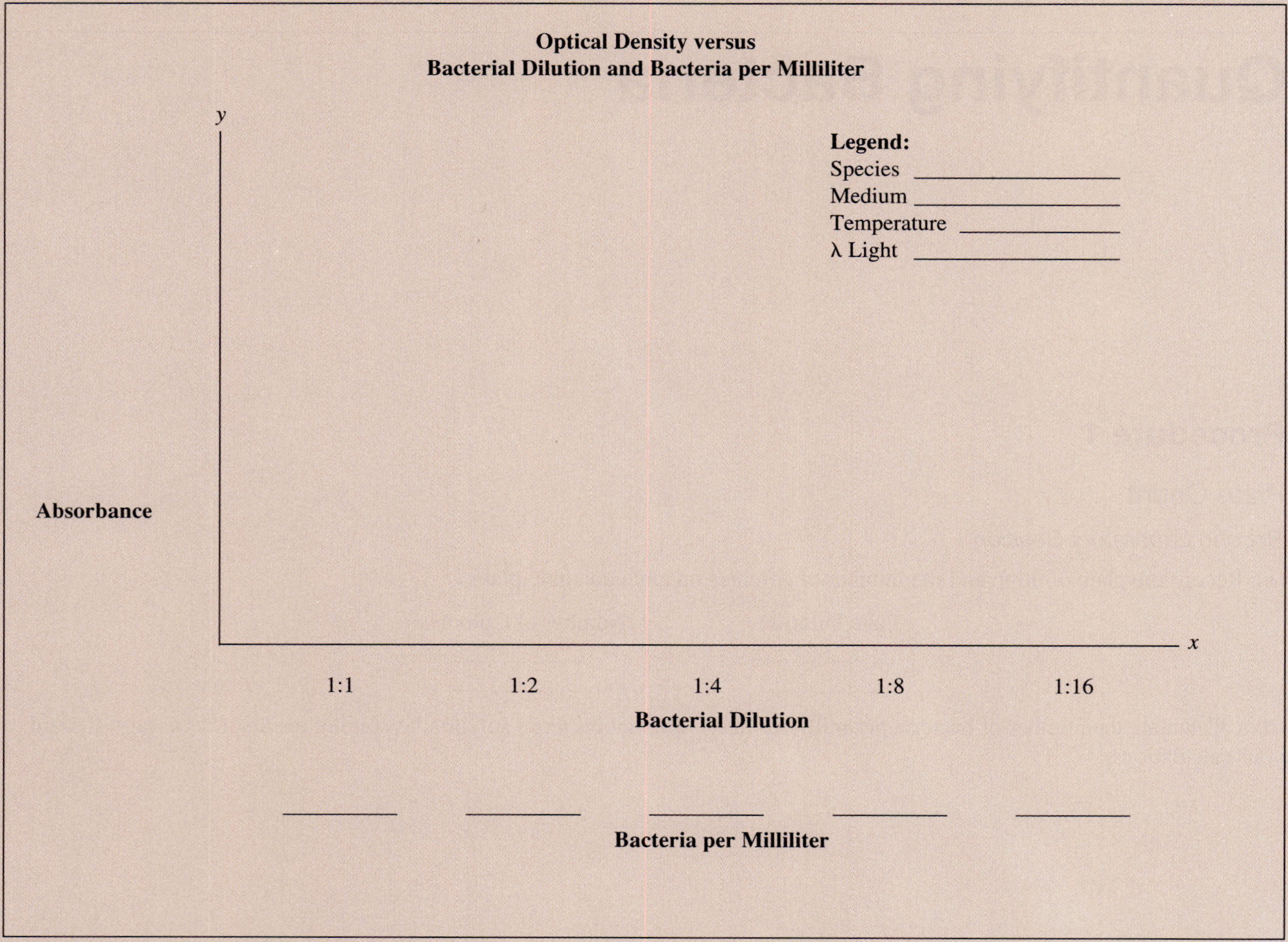

Optical Density versus
Bacterial Dilution and Bacteria per Milliliter

y

Legend:
Species ________________
Medium ________________
Temperature ______________
λ Light ________________

Absorbance

x

1:1 1:2 1:4 1:8 1:16

Bacterial Dilution

________ ________ ________ ________ ________

Bacteria per Milliliter

(*q*) Explain why this experiment would be impossible to perform without an uninoculated control.

Second Laboratory Session

(*a*) With data from Procedure 1, calculate the number of bacteria per milliliter in each test tube of your spectrophotometry dilution series. Round off all numbers to 2 significant figures.

1:1 ____________________
1:2 ____________________
1:4 ____________________
1:8 ____________________
1:16 ____________________

(*b*) Complete the graph you began in the first laboratory session of this procedure by filling in your data showing the numbers of bacteria per milliliter in dilution series test tubes.
(*c*) Describe how employing this graph might save a microbiologist time, laboratory space, and expense.

Exercise **16**

Anaerobic Culture: Oxygen Requirements

Necessary Skills

Aseptic transfer techniques

Materials

Cultures (48-hour, 35° C, 1 per set):
- *Micrococcus luteus* (Tryptic Soy Broth)
- *Staphylococcus epidermidis* (Tryptic Soy Broth)
- *Clostridium sporogenes* (Cooked Meat Broth or fresh Fluid Thioglycollate Medium)

Supplies:
- 6 Tryptic Soy Agar deeps per group
- 1 500 ml Beaker
- 3 Fluid Thioglycollate Medium tubes per group
- 3 Tryptic Soy Agar plates per group
- Water bath, 50° C, with test tube racks to hold 6 tubes per group
- Candle jar(s) to hold 1 plate per group
- 1 Short candle per jar
- ½ Petri dish per jar
- Petroleum jelly
- Matches
- Water-dampened filter paper or paper towel
- GasPak® (BBL) jars(s) to hold 1 plate per group
- Anaerobic indicator strip(s) (1 per GasPak® jar)
- Hydrogen + carbon dioxide generator envelope(s) (1 per GasPak® jar)
- 10 ml Pipet
- Pipettor
- Scissors
- Container of water (at least 10 ml per GasPak® jar)

Warning: Hydrogen gas is flammable. Observe appropriate precautions when handling gas-generating envelopes.

Primary Objective

Utilize various anaerobic culture methods to cultivate microorganisms, then compare and analyze the results.

Other Objectives

1. Recount the discovery of anaerobes.
2. List five groupings of microorganisms according to their oxygen tolerance.
3. Name at least one example of a pathogenic genus in each group and list a disease it causes.
4. Identify: H_2O_2, O_2^-.
5. Explain the terms *oxidation reduction potential* and *reduced environment.*
6. Prepare and evaluate the growth in agar shake cultures, stab cultures, Fluid Thioglycollate Medium cultures, candle jar cultures, and GasPak® cultures of organisms with various oxygen tolerances.
7. Appraise the redox potential of a GasPak® anaerobic system and tubes of Fluid Thioglycollate Medium.

INTRODUCTION

Do all living things require oxygen? Pasteur, studying clostridia, observed bacteria swimming in the center of a wet mount preparation. He saw that they slowed and then died at the edge of the coverslip where they contacted air. Correctly, he surmised that oxygen killed the microbes. Pasteur coined the term ***anaerobe*** to describe an organism that is poisoned by oxygen. In this exercise you cultivate anaerobes and examine the oxygen requirements of microorganisms.

Oxygen Requirements

You might anticipate that because humans and other familiar animals require oxygen, microbes would, too. But in reality—since liquid environments cover much of the earth,

Table 16.1 Bacterial Oxygen Tolerance and Requirements

Group	Catalase[1]	Superoxide Dismutase[2]	Comments	Pathogenic Example	Disease
Aerobes	Yes	Yes	Require O_2 for aerobic respiration	*Bordetella pertussis*	Whooping cough
Microaerophiles	Yes	Yes	Grow best with decreased O_2	*Neisseria gonorrhoeae*	Gonorrhea
Facultative anaerobes	Yes	Yes	Aerobic and anaerobic respiration; most also ferment	*Staphylococcus aureus*	Food poisoning, pimples, boils, wound infections
Aerotolerants	No	Yes	Do not require O_2 but are not killed by it	*Streptococcus pyogenes*	Septic sore throat
Anaerobes	No	No	Some tolerate reduced O_2 levels, others are killed quickly by free oxygen	*Clostridium perfringens*	Food poisoning, gas gangrene

[1] $2H_2O_2 \xrightarrow{\text{catalase}} 2H_2O + O_2$
(hydrogen peroxide) → (water) (oxygen)

[2] $2O_2^- + 2H^+ \xrightarrow{\text{superoxide dismutase}} O_2 + H_2O_2$
(superoxide) (hydrogen ion) → (oxygen) (hydrogen peroxide)

and oxygen is not highly soluble in water—organisms that require this gas are at a disadvantage in nature. Microorganisms need water; only a few actually require oxygen.

In the microbial world, there is great variation in the amount of oxygen that organisms require or can tolerate; see table 16.1. Five oxygen requirement groupings are generally recognized: **aerobes, microaerophiles, facultative anaerobes, aerotolerants,** and **anaerobes.**

Oxidation Reduction Potential

In this exercise you cultivate anaerobes by reducing the **oxidation reduction potential** of their environment. The oxidation reduction potential is a means of expressing a compound's affinity for electrons. This affinity is compared with the attraction of H_2, as in H_2O, for electrons.

Chemists often refer to an environment as either **oxidized** or **reduced.** A highly **oxidized environment,** air for example, has a great affinity for electrons; it has a high oxidation reduction potential. When, in an environment, there is a greater proportion of hydrogen than in H_2O, that environment has a negative oxidation reduction potential. Thus an electron-rich environment is a **reduced environment.** Another name for the oxidation reduction potential is the **redox potential.**

Cultivating Anaerobes

Microbiologists use a variety of methods to cultivate anaerobes. In this exercise you create an environment with a reduced oxidation reduction potential by boiling media to drive out dissolved oxygen; see figure 16.1. For the **agar shake culture,** you inoculate bacteria into the liquefied deep. To prepare a **stab culture,** you inoculate a freshly boiled, then solidified, deep. What prevents anaerobes from growing near the top surface of the medium in a stab or agar shake culture?*

You also utilize a **GasPak® (BBL) anaerobic system** to cultivate anaerobes; see figure 16.2. Cultures are placed in the jar along with an envelope containing chemicals that release hydrogen and carbon dioxide when activated by water. Water is added to the envelope; the jar is then sealed. In the presence of the palladium catalyst, hydrogen reacts with oxygen forming water. This reaction removes free oxygen. A haze of water droplets forms on the inside walls of the jar, and the lid above the catalyst chamber warms slightly as the reaction progresses.

But how does the microbiologist know that the jar is providing a sufficiently reduced environment? To evaluate the oxidation reduction potential in the jar, the scientist always includes a **redox potential indicator** (a device that shows whether free oxygen is present).

There are dyes that are colorless in a reduced environment. **Litmus, resazurin,** and **methylene blue,** for example, show color only when in an oxidized environment. These dyes are excellent redox potential indicators. One standard GasPak® anaerobic indicator is a methylene blue-saturated pad in a sealed, peel-apart foil package.

In this exercise you evaluate the growth of an aerobe, a facultative anaerobe, and an anaerobe in **Fluid Thioglycollate Medium.** Reducing compounds such as **sodium thioglycollate** are incorporated into certain media so that the medium itself provides a reduced environment. Sodium thioglycollate binds free oxygen. Fluid Thioglycollate Medium also contains the redox potential indicator **resazurin,** which turns pink in an oxidized environment.

*A small amount of oxygen dissolves in water, and since agar media are largely water, oxygen dissolves into the medium where it contacts air. Anaerobes cannot grow in the presence of this dissolved oxygen.

Figure 16.1 Boiling reduces agar for a stab culture and an agar shake culture.

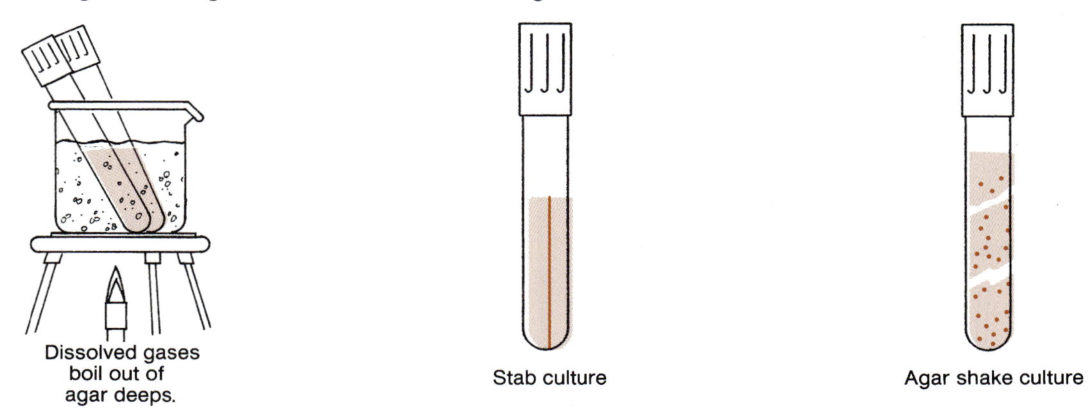

Figure 16.2 The GasPak® (BBL) anaerobic system

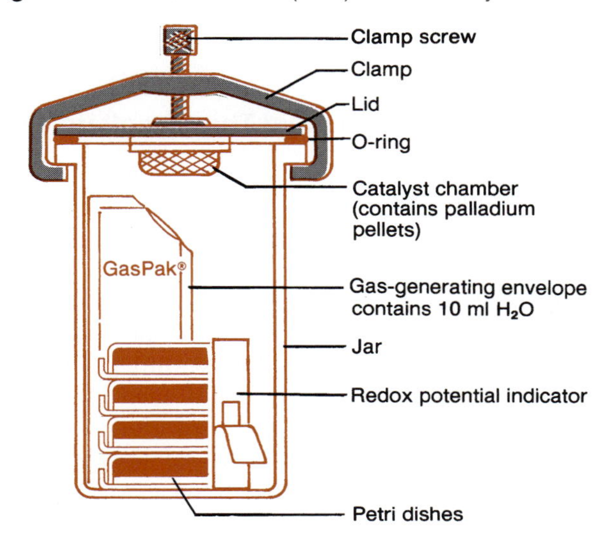

Figure 16.3 Candle jar incubation

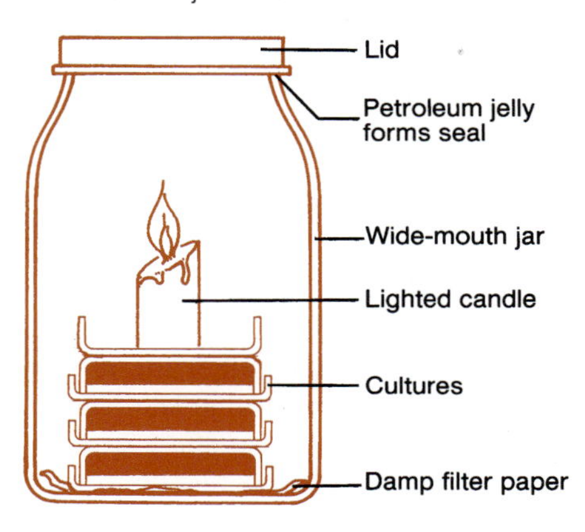

Here you cultivate the anaerobe ***Clostridium sporogenes.*** Pathogenic or opportunistic anaerobes and the diseases they cause include ***Clostridium perfringens,*** gas gangrene and food poisoning; ***Clostridium tetani,*** tetanus; ***Clostridium botulinum,*** botulism; ***Bacteroides fragilis,*** peritonitis; and ***Treponema pallidum,*** syphilis.

The Candle Jar

A **candle jar** is a wide-mouthed jar into which the microbiologist places a damp piece of paper, cultures, and a lighted candle, as shown in figure 16.3. The apparatus is then tightly sealed; after a few minutes the flame goes out. The humid atmosphere that remains in the jar provides higher CO_2 and lower oxygen concentrations than are normally found in air.

The usual levels of these gases in air are approximately 21% O_2 and 0.3% CO_2. In a candle jar, the amounts are altered to about 16% O_2 and 4% CO_2. Some **fastidious** (delicate, requiring special nutrients and/or growth conditions) organisms grow best in this environment. Pathogens that are generally cultured in a candle jar include ***Neisseria gonorrhoeae,*** which causes gonorrhea, and ***Haemophilus influenzae,*** frequently the etiologic agent of meningitis and otitis media in infants.

In this experiment you attempt to cultivate an aerobe, a facultative anaerobe, and an anaerobe in a candle jar. Which one(s) do you predict will grow?

PROCEDURES EXERCISE 16

Procedure 1

Preparing Agar Shake Cultures and Stab Cultures

First Laboratory Session

a Obtain 6 Tryptic Soy Agar deeps. Loosen any screw caps.

b As in the illustration, liquefy them in a beaker of boiling water. The water should reach just above the level of the medium in the tubes but not contact the tube lids. Leave the deeps in boiling water for 10 minutes to drive out dissolved oxygen.

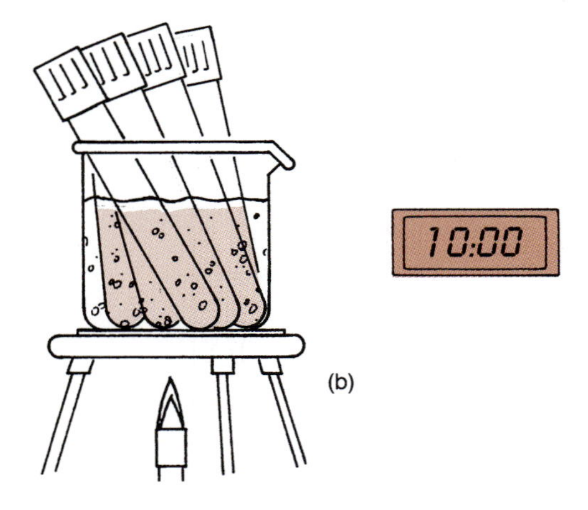

c Temper the liquefied media in 3 of the tubes in a 50° C water bath. Allow the other 3 to solidify.

d To prepare an **agar shake culture,** obtain 1 liquefied, reduced, and tempered tube of Tryptic Soy Agar.

e Resuspend cells in the culture of *Micrococcus luteus,* and then transfer 1 loopful deep into the tempered medium. Do not introduce air bubbles into the deep.

f Flame the lip of the tube, then cap it; flame your loop.

g As in the illustration, roll the warm tube between your palms to disperse the bacteria throughout the medium without introducing oxygen. Label the tube with your name, the name of the bacteria, the date, and the type of culture, "Shake."

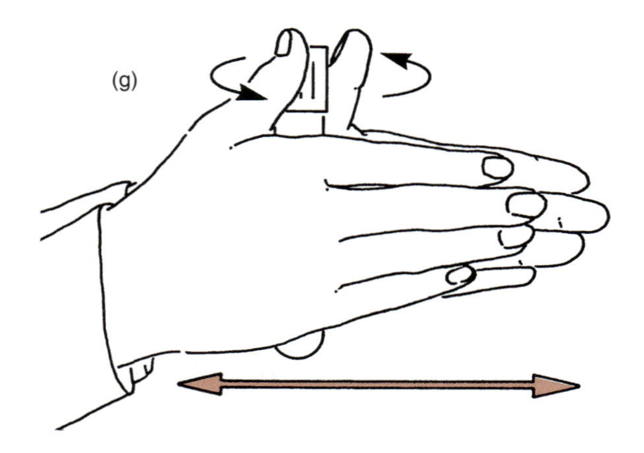

h Repeat steps (*d*) through (*g*) with *Staphylococcus epidermidis* and then with *Clostridium sporogenes.*

Note whether the medium in any of your 3 tubes has solidified before you were able to complete preparation of an agar shake culture. If so, boil the tube again, then temper the medium and reinoculate. In the laboratory report, explain why you must reinoculate the medium.

i To prepare a **stab culture,** obtain 1 liquefied, reduced, cooled, and solidified tube of Tryptic Soy Agar.

j Resuspend cells in the culture of *Micrococcus luteus.* Transfer 1 inoculating needle of them into the agar deep with a clean, single stab to the bottom of the test tube.

Withdraw the needle slowly, allowing the reduced medium to surround the microbes left behind.

k Label the tube with your name, the name of the bacteria, the date, and the type of culture, "Stab."

l Repeat steps (*i*) through (*k*) with *Staphylococcus epidermidis* and then with *Clostridium sporogenes.*

m Loosen lids of the screw cap tubes. Incubate your stab cultures and agar shake cultures at 35° C until the next laboratory session.

Second Laboratory Session

a Examine your 3 agar shake cultures and 3 stab cultures. In the laboratory report, illustrate and discuss the growth pattern of each culture. After the group has examined the tubes, dispose of the cultures appropriately.

Procedure 2

Culturing Bacteria in Fluid Thioglycollate Medium

First Laboratory Session

a Obtain 3 tubes of Fluid Thioglycollate Medium. Loosen any screw caps.

b Unless the medium is fresh and less than 30% in each tube is oxidized (indicated by pink, oxidized resazurin at the surface), heat the tubes in a beaker of boiling water. Leave the broths in boiling water for 10 minutes to drive out dissolved oxygen.

c Cool the liquefied media in the 3 tubes.

d To prepare a Fluid Thioglycollate Medium culture, obtain 1 reduced, cooled tube of the medium.

e Resuspend bacteria in the class culture of *Micrococcus luteus.*

f Transfer 1 loopful of the bacteria deep into the medium. Label the tube with your name, the name of the bacteria, the date, and the type of culture, "Thio."

g Repeat step (*f*) with *Staphylococcus epidermidis* and then with *Clostridium sporogenes.*

h Incubate the 3 Fluid Thioglycollate Medium cultures at 35° C until the next laboratory session.

Second Laboratory Session

a Examine your 3 Fluid Thioglycollate Medium cultures. In your laboratory report, discuss the growth pattern of each culture. After the group has examined the tubes, dispose of the cultures appropriately.

Procedure 3

Comparing the Effects of Aerobic, Anaerobic, and Candle Jar Incubation

First Laboratory Session

a Obtain 3 plates of Tryptic Soy Agar.

b As illustrated, divide the bottom of each plate into 3 sections. Label 1 section of each plate for each of your 3 bacterial cultures: *Micrococcus luteus, Staphylococcus epidermidis,* and *Clostridium sporogenes.* Label each plate with your name and the date. Do not write on the lids of the plates.

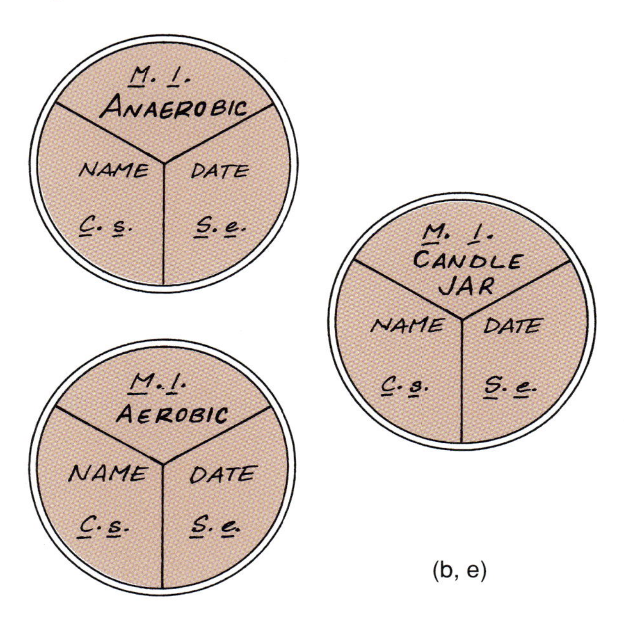

(b, e)

c With your sterilized loop, draw a single, short line of *Micrococcus luteus* in the appropriate triangle of 1 plate. Resterilize the loop; similarly inoculate the next plate. Resterilize and repeat the procedure in the third plate.

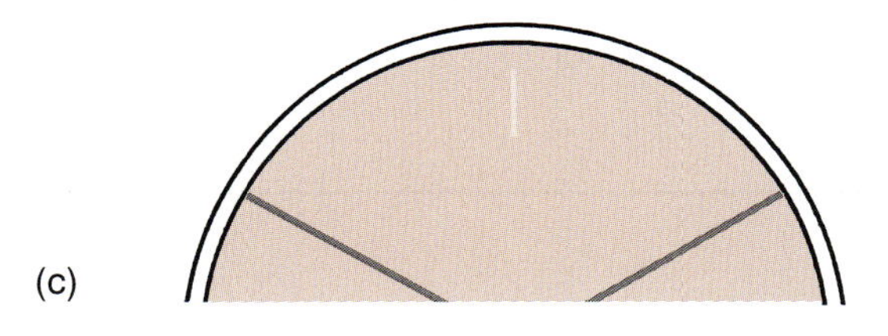

(c)

d Repeat step (*c*) with *Staphylococcus epidermidis* and then with *Clostridium sporogenes.*

e Label 1 plate "Anaerobic"; place it in a GasPak® jar for incubation. Label 1 plate "Candle Jar"; place it in a candle jar that has a water-dampened paper towel or filter paper lining its bottom surface. Label 1 plate "Aerobic"; place it in your 35° C incubator until the next laboratory session.

f Watch as **your instructor,** or an assigned group, prepares a GasPak® anaerobic system as shown in figure 16.2. The preparer checks the condition of the O-ring and jar rim and makes sure there is palladium in the catalyst chamber. Then, after adding 10 ml of H_2O to a gas-generating envelope and opening the redox potential indicator strip, the person preparing the jar clamps down the lid.

g Note in your laboratory report the color of the redox potential indicator strip when it is first opened and again after about 10 minutes.

h **Your instructor** will incubate the jar(s) at 35° C until the next laboratory session. If possible, check the jar again after it has been sealed for at least 1 hour. Note any changes in the jar and indicator strip.

i Watch as **your instructor,** or an assigned group, prepares a candle jar. The preparer checks the condition of the jar rim and lid, then smears the rim with a thin layer of petroleum jelly (to ensure a complete seal). After using melted wax to attach a short candle to a Petri dish, the preparer places the candle in the jar, lights the candle, and finally seals the jar.

j Note in your laboratory report about how long it takes before the candle is extinguished.

k **Your instructor** will incubate the jar(s) at 35° C until the next laboratory session.

Second Laboratory Session

a Before opening the GasPak® anaerobic system jar, observe the redox potential indicator strip. Did the system maintain anaerobic conditions? Examine your 3 plates. In your laboratory report, illustrate and discuss the growth pattern of each culture and how it relates to the form of incubation: aerobic, anaerobic, or candle jar. After the group has examined the plates, dispose of the cultures appropriately.

POSTTEST EXERCISE 16

Part 1

Matching

Each response may be used once, more than once, or not at all. Some statements may require more than one response. Use the definitions and examples that are given in this exercise.

a. Aerobes
b. Facultative anaerobes
c. Aerotolerants
d. Microaerophiles
e. Anaerobes

1. ____ Must have oxygen to live
2. ____ Utilize aerobic respiration but require increased CO_2 and decreased O_2 concentrations compared with those usually found in air
3. ____ Do not utilize oxygen and are not killed by it
4. ____ Utilize aerobic and anaerobic respiration and often also produce enzymes that carry out fermentation

5. ____ Do not produce the enzyme catalase nor do they manufacture superoxide dismutase
6. ____ Produce superoxide dismutase
7. ____ Include *Neisseria*
8. ____ Include *Staphylococcus*
9. ____ Include *Clostridium*
10. ____ Include *Streptococcus*

Part 2

True or False (Circle one, then correct every false statement.)

1. T F Pasteur discovered anaerobes, microorganisms that are poisoned by oxygen.
2. T F Since there is oxygen in the air that surrounds earth, being an aerobe is a great advantage for microorganisms in nature.
3. T F Fermentation does not require oxygen.
4. T F Fermentation releases some but not all of the energy in organic compounds.
5. T F The enzyme that catalyzes the reaction $2H_2O_2 \rightarrow 2H_2O + O_2$ is superoxide dismutase.
6. T F Bacteria that are exposed to air as they grow are generally in a highly reduced environment.
7. T F A redox potential is the opposite of an oxidation reduction potential; when one is high the other is low.
8. T F Anaerobes grow in reduced environments.
9. T F After incubation, the sealed redox potential indicator envelope is removed from a GasPak® jar and opened.
10. T F Fluid Thioglycollate Medium is pink in the bottom of the tube where the resazurin is highly reduced.
11. T F A candle jar contains the best mix of gases available for culturing anaerobes.
12. T F *Neisseria gonorrhoeae* and *Haemophilus influenzae* are generally incubated in a candle jar.
13. T F To prepare an agar shake culture, boil the medium, solidify it, and then inoculate it with a clean stab to the bottom of the tube.
14. T F Loosen the caps of screw cap tubes before incubating oxygen-utilizing bacteria in them.
15. T F Spread a thick layer of petroleum jelly around the O-ring of a GasPak® jar before sealing the lid.
16. T F After inoculating a tube of Fluid Thioglycollate Medium, shake the tube thoroughly to aerate and separate the bacteria.

Part 3

Completion and Short Answer

1. List the specific epithet of three pathogenic clostridia, and name a disease caused by each.

   Specific epithet   Disease

   a.

   b.

   c.
2. Name two anaerobic cultivation methods that utilize freshly boiled, then cooled, agar deeps.

   a.

   b.
3. Draw a GasPak® anaerobic system; label at least eight of its parts.

4. Three dyes that serve microbiologists as redox potential indicators are ____________________, ____________________, and ____________________.
5. The percentages of the gases O_2 and CO_2 in air and in a candle jar are about:

Percentages of Gases

Gas	% in Air	% in Candle Jar
O_2	a.	b.
CO_2	c.	d.

Laboratory Report **16**

Name ______________________________

Section ______________

Anaerobic Culture: Oxygen Requirements

Procedure 1

Preparing Agar Shake Cultures and Stab Cultures

First Laboratory Session

(*h*) If your agar shake culture medium solidifies before you have dispersed the bacteria in it, what must you do and why?

Second Laboratory Session

(*a*) Illustrate the results of your shake cultures and stab cultures of the 3 assigned bacteria.

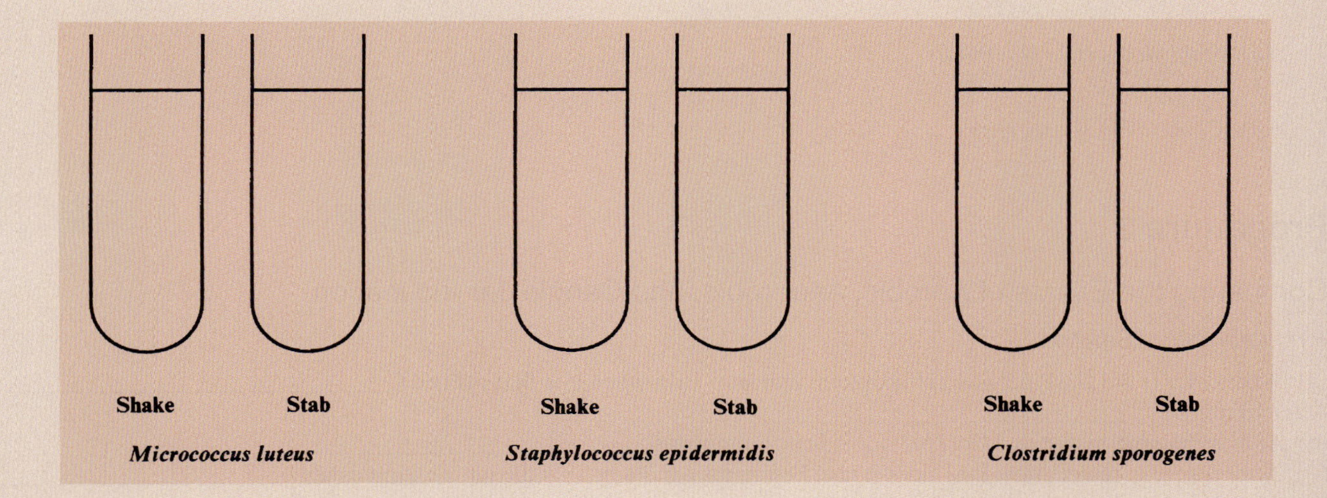

Comment on how the growth pattern you see in each tube relates to the description of micrococci as aerobic, staphylococci as facultative, and clostridia as anaerobic.

Shake culture, *Micrococcus luteus*

Stab culture, *Micrococcus luteus*

Shake culture, *Staphylococcus epidermidis*

Stab culture, *Staphylococcus epidermidis*

Shake culture, *Clostridium sporogenes*

Stab culture, *Clostridium sporogenes*

Procedure 2

Culturing Bacteria in Fluid Thioglycollate Medium

Second Laboratory Session

(*a*) Discuss the growth patterns in each of the 3 Fluid Thioglycollate Medium cultures.

(i) *Micrococcus luteus*

(ii) *Staphylococcus epidermidis*

(iii) *Clostridium sporogenes*

Procedure 3

Comparing the Effects of Aerobic, Anaerobic, and Candle Jar Incubation

First Laboratory Session

(*g*) What color is the GasPak® anaerobic system indicator strip when it is first opened? ______________________________
After 10 minutes? ______________________________

(*h*) After remaining in a sealed GasPak® jar at least 60 minutes? ______________________________

(*j*) After the jar is sealed, how long does it take before the candle in a candle jar is extinguished? ______________________
Why check the candle?

Second Laboratory Session

(*a*) Did the GasPak® anaerobic system jar maintain anaerobic conditions? ______________________________

In the following chart, compare the results of your aerobic, candle jar, and GasPak® anaerobe jar incubations. How well does each genus grow in each environment?

The Effects of Oxygen (and CO_2) Concentration on Growth

Genus	Environment		
	Aerobic	**Candle Jar**	**GasPak® Jar**
Micrococcus			
Staphylococcus			
Clostridium			

Exercise **17**

Bacteriophage

Necessary Skills

Knowledge of the metric system of measurement, Aseptic transfer techniques, Quantifying bacteria

Materials

First Laboratory Session

Cultures (Procedures 1 and 2):
- *Escherichia coli* strain B (24-hour, 35° C, Tryptic Soy Broth, at least 8 ml per group)
- Bacteriophage T2 (approximately 1:100 suspension in Tryptic Soy Broth, appropriate dilution to be determined before class, at least 0.7 ml per group)

Cultures (Procedure 3), 24-hour, 35° C, Tryptic Soy Broth, labeled "Unknown____"):
- *Escherichia coli* strain B
- *Enterobacter aerogenes*
- *Escherichia coli* (not strain B)

Supplies:

Procedure 1
- 1 Container per group holding 40 ml of Tryptic Soy Broth
- 8 Sterile test tubes per group
- 8 Test tubes, each containing 2.5 ml of Soft Agar (Tryptic Soy Broth + 0.7% agar), per group
- 1 500 ml Beaker per group
- 8 Tryptic Soy Agar plates per group
- Water bath, 50° C, with test tube racks for class Soft Agar tubes
- At least 9 sterile 1 ml pipets per group
- 1 Sterile 5 ml pipet per group
- At least 1 sterile 10 or 25 ml pipet per group
- Pipettors

Procedure 2
- Control and 10-fold serial dilution of phage from Procedure 1
- 1 Container per group holding 40 ml of Tryptic Soy Broth
- 8 Sterile test tubes per group
- At least 1 sterile 10 or 25 ml pipet per group
- At least 1 sterile 2 or 5 ml pipet per group
- At least 1 sterile 1 ml pipet per group
- Pipettors

Procedure 3 (per group)
- 3 Tryptic Soy Agar plates
- 3 Sterile cotton-tipped applicators
- 2 Sterile Pasteur pipets
- Bulb
- 1 Tryptic Soy Broth

Second Laboratory Session

Access to a Quebec colony counter (optional)
Access to a tally register (optional)

Primary Objective

Quantify bacteriophage.

Other Objectives

1. Describe a virus.
2. Indicate in what way viruses seem to go back and forth between life and a nonliving existence.
3. Identify the characteristics utilized in viral classification.
4. Define: bacteriophage, tissue culture, PFU.
5. Prepare and interpret the plaque assay method of bacteriophage quantification.
6. Quantify the phage in a liquid by preparing and interpreting a broth-clearing assay.
7. Use phage to identify a bacterial unknown.

INTRODUCTION

A **virus** is a noncellular, obligate intracellular parasite. Generally, a virus consists of a **capsid** (protein coat) surrounding **nucleic acid,** either RNA or DNA but not both. An **envelope** encloses the capsid of some viruses.

The virus seems to fluctuate back and forth between life and a nonliving existence. Outside a **permissive host,** the virus is inert, unable to take in nutrients and metabolize them to form energy or viral substance. Once inside a host cell, nucleic acid from a single viral parasite may direct the host activities to produce several hundred new viruses within less than 30 minutes.

For classification, each type of virus is first assigned to one of three large divisions based on its usual host. Generally, the hosts are either animals, plants, or bacteria. There is some overlap, since viruses that infect plants occasionally invade insects and are carried about among the vegetation. Further classification is by nucleic acid, RNA or DNA; size; shape; and presence or lack of an envelope.

Laboratory Cultivation of Plant, Animal, and Bacterial Viruses

Since a virus requires a living, susceptible host cell in which to replicate, the laboratory study of plant and animal viruses is often inconvenient and expensive. Living host plants or animals, including embryonated eggs, are sometimes required.

Many plant and animal viruses can be grown in **tissue culture** (cultivation of selected cells from a multicellular organism) on the surfaces of Petri plates, narrow tubes, or revolving bottles. Tissue cultures (also called **cell cultures**) are much easier to handle in the laboratory than whole, live plants or animals.

Bacteriophage (viruses that infect bacteria), often called **phage,** are by far the most convenient viruses for laboratory investigations. Phage are rapidly replicated in their bacterial hosts. The phage that infect ***Escherichia coli,*** **coliphage,** have been extensively characterized.

How can you see the effects of coliphage on *Escherichia coli?* In the laboratory a semisolid **overlay** of soft, nutrient medium can be inoculated with *Escherichia coli.* If the overlay is poured onto an agar medium, then incubated appropriately, a confluent layer (lawn) of bacteria grows on the plate. To see the effects of bacteriophage, add coliphage to the bacteria in your Soft Agar overlay before pouring. If the coliphage **lyse** (burst) cells while *Escherichia coli* is multiplying in your overlay, **plaques** (holes, cleared areas) develop in the lawn of bacteria.

Quantifying Bacteriophage

Your first experiment, a **plaque assay** (test for plaque formation) is designed to discover the number of **plaque-forming units (PFUs)** in a viral solution. A plaque roughly corresponds to a bacterial colony since it generally indicates the place where one agent multiplies until there is macroscopic evidence of its existence. A plaque-forming unit most often develops from a single **virion** (infectious, viral particle).

To be more specific, you **assay** (make an analysis of) the approximate number of phage in a sample by preparing bacteriophage **titrations** (determinations of the concentration of an agent by measuring a given reaction).

In Exercise 15, Quantifying Bacteria, you set up a serial dilution of bacteria. Here you prepare a serial dilution of viruses and grow them in a relatively constant supply of bacteria. The **titer** (concentration) of virions in the original coliphage sample equals the inverse of the plate dilution times the number of plaques on a countable plate; see figure 17.1.

Your plaque assay establishes the number of coliphage plaque-forming units present on a solid surface, but how can you tell whether virulent coliphage are present in a liquid culture of *Escherichia?* Grow susceptible *Escherichia coli* for 24 hours at 35° C in a nutrient fluid. If the broth is clear the next day, it is likely that the bacteria were lysed by coliphage. As in the plaque assay, clearing of bacteria generally indicates the presence of bacteriophage.

In your second experiment, you perform a **broth-clearing assay;** see figure 17.2. To quantify virions, you utilize broth cultures of bacteria that are infected with a sequential dilution of phage.

In your broth-clearing assay, the greatest dilution of (least concentrated) phage to completely clear a broth by lysing the bacteria is called the **end-point dilution.** The titer of phage in the original coliphage sample equals the inverse of the end-point tube's dilution.

Bacteriophage Typing

Epidemiologists searching for the origins of an epidemic try to identify the agent that is causing the disease and trace the epidemic back to a common source. For remarkably detailed identification of ubiquitous bacteria including ***Pseudomonas aeruginosa, Staphylococcus aureus,*** or ***Salmonella* spp., bacteriophage typing** is performed. In bacteriophage typing, which is also called **phage typing,** extremely host-specific phages that can be purchased from catalogs are dropped onto a spread plate of a pathogen that has already been identified to the species level. Later, the microbiologist notes which phages lysed the test bacteria. Similarly, in this exercise, you utilize a bacteriophage to identify a bacterial unknown.

PROCEDURES EXERCISE 17

Procedure 1

Phage Titration by Plaque Assay

First Laboratory Session

a Examine figure 17.1 thoroughly before continuing with this procedure.

b Liquefy 8 tubes of Soft Agar in a boiling water bath. Cool the tubes to 50° C.

Figure 17.1 Phage titration by plaque assay*

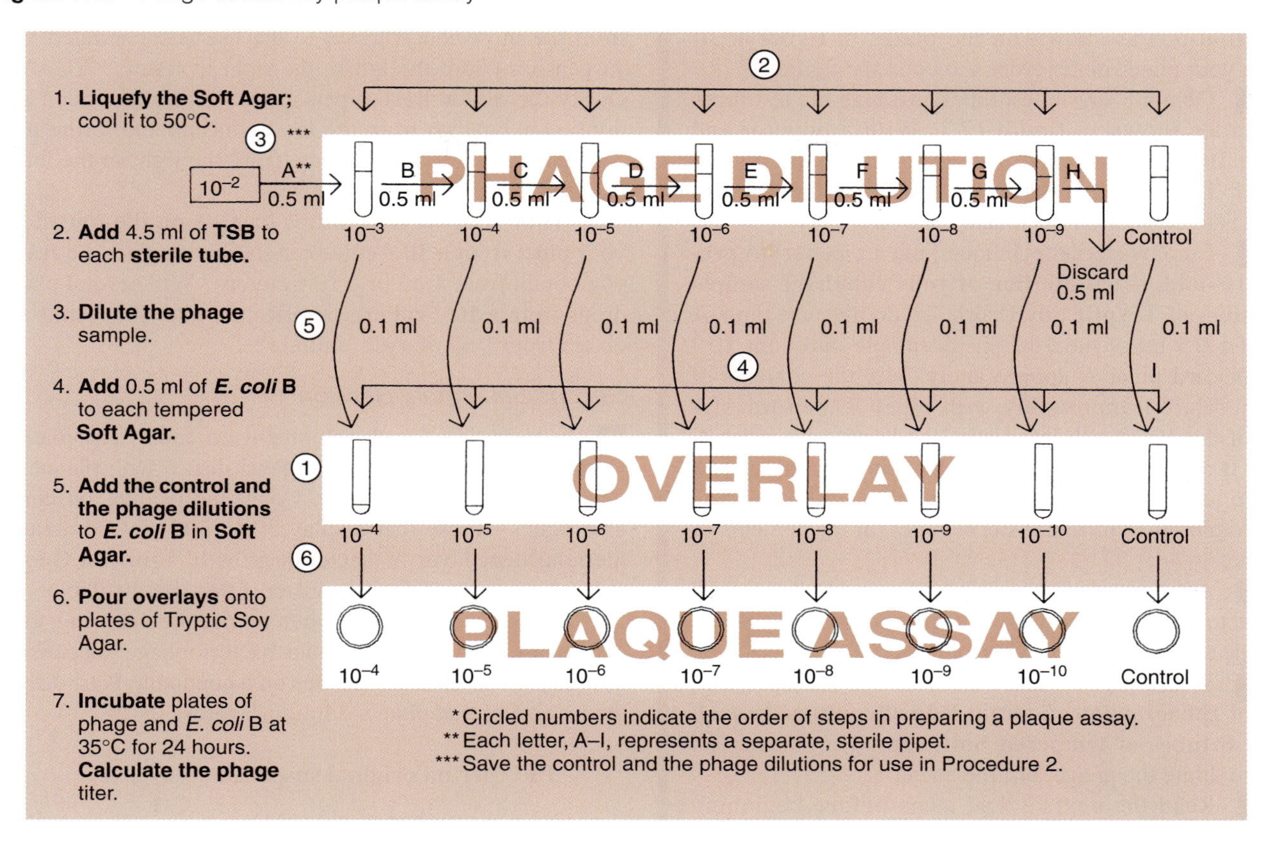

Figure 17.2 Phage titration by broth-clearing assay*

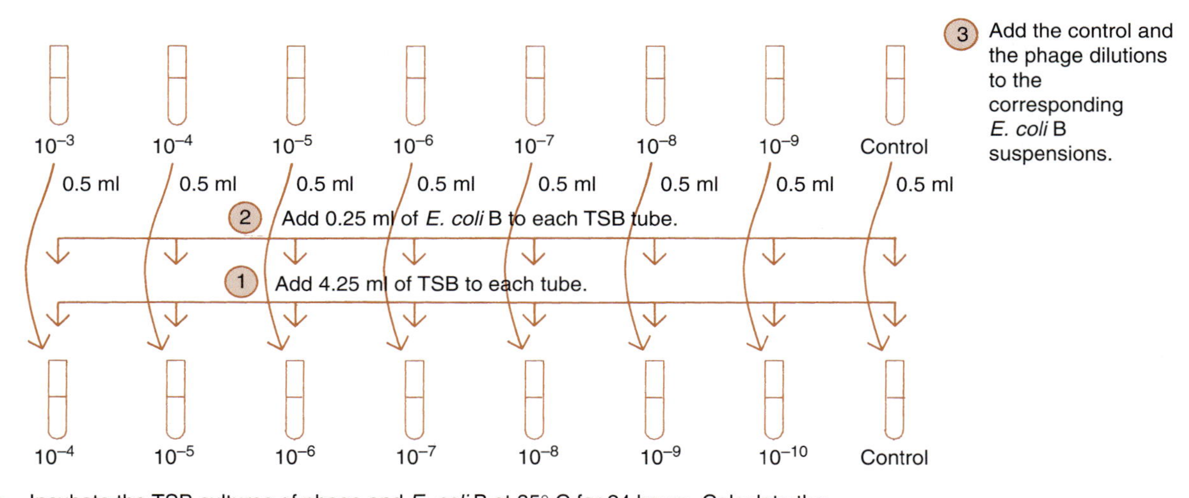

c Label 8 Petri plates of Tryptic Soy Agar with your group identification, the date, and the word "Control" or the dilutions as shown on the plates in figure 17.1. Arrange your plates in the order shown in the figure.

d Obtain 8 sterile test tubes. Arrange the test tubes in the order shown in figure 17.1. With a pipettor and a sterile 10 or 25 ml pipet, use aseptic technique to add 4.5 ml of Tryptic Soy Broth to each tube. Set the last tube aside. It is to be your uninoculated control.

e Employ sterile technique (and a pipettor) to **prepare a 10–fold, serial dilution of your coliphage sample in the tubes of Tryptic Soy Broth.** To do this, use pipet A to transfer 0.5 ml of the 10^{-2} phage sample into your 10^{-3} broth. Discard pipet A appropriately. Mix the contents of the 10^{-3} dilution thoroughly with pipet B then transfer 0.5 ml of the 10^{-3} broth into your 10^{-4} broth with pipet B. Discard B appropriately. Mix your 10^{-4} broth with pipet C. Continue to dilute the phage in your tubes of Tryptic Soy Broth, employing a different sterile pipet for the contents of each tube.

f After preparing your 10^{-9} dilution, discard 0.5 ml from the tube as directed by **your instructor.** Do **not** add any phage to your virus-free control tube.

g With a pipettor and a sterile 2 or 5 ml pipet, **add 0.5 ml of *Escherichia coli* strain B broth culture to each of your 8 tubes of tempered Soft Agar.** (In this experiment you dilute the phage, **not** the bacteria.)

h Read the next several steps **before** beginning your transfers. **You will add 0.1 ml from your control tube and each of your phage dilutions to the *Escherichia coli* in your tempered tubes of Soft Agar.** Remember that the Soft Agar must remain warm or it solidifies, ruining your experiment.

i **Your instructor** may ask you to prepare and pour the Soft Agars 1 at a time, utilizing 8 sterile 1 ml pipets. However, if you can keep the Soft Agar at 50° C, it is possible to inoculate all your Soft Agars with one sterile 1 ml pipet. To do this, inoculate the control into its Soft Agar first. Then, with the same pipet, quickly make transfers into corresponding Soft Agars from the most dilute (10^{-9}) to the most concentrated (10^{-3}) phage tubes.

j After the phage transfer, roll a test tube containing a Soft Agar overlay between your palms to distribute the organisms. Avoid forming bubbles. Use sterile technique to **pour the overlay onto the corresponding Petri plate of Tryptic Soy Agar.** Rock and swirl the plate gently to distribute the Soft Agar overlay evenly. Without splashing, rotate the plate clockwise and counterclockwise, back and forth. Discard the contaminated tube in an appropriate manner; do not lay it on your bench top.

k Mix and pour each of the Soft Agar overlays in a similar fashion. Save your serial dilution of phage. You use it again in Procedure 2.

l After the overlays have solidified, incubate your inverted plates of phage and *Escherichia coli* strain B at 35° C until the next laboratory session.

m In your laboratory report, explain why it is necessary to use a new sterile pipet for each transfer as you go from the most concentrated to the least concentrated viral suspension (from the left to the right in figure 17.1). Also clarify the reason fresh pipets are not required at each new dilution as you go from the least concentrated to the most concentrated viral suspensions (from the right to the left in the figure).

Hint: Will several drops carried on the outside of your pipet from a 10^{-3} culture significantly alter the results you obtain from 1 ml of a 10^{-4} culture? Will several excess drops from a 10^{-4} culture significantly alter the results you obtain from 1 ml of a 10^{-3} culture?

Second Laboratory Session

a Examine plaque development on each Petri plate in your phage titration. Compare Petri dishes with plaques to your virus-free control plate. Count the plaques on countable plates—those that display 30 to 300 cleared areas in the confluent layer of bacterial growth. Utilize a Quebec colony counter and a tally register if they are available.

b To find the number of plaque-forming units (PFUs) in 1 ml of the original bacteriophage suspension, multiply the number of plaques on a countable Petri dish by the inverse of that plate's dilution. That is:

$$\textbf{PFU per ml original suspension} = \textbf{number of plaques} \times \frac{\textbf{1}}{\textbf{plate dilution}}$$

c Compare your results with those of other groups. Record your results in your laboratory report.

Procedure 2

Phage Titration by Broth-Clearing Assay

First Laboratory Session

a As in figure 17.2, employ a sterile 10 or 25 ml pipet to add 4.25 ml of Tryptic Soy Broth to each of 8 sterile test tubes. Employ a pipettor and a sterile 2 or 5 ml pipet to transfer 0.25 ml of *Escherichia coli* strain B into each of your Tryptic Soy Broth tubes.

b From Procedure 1, obtain your serial dilution of phage and the control tube. With a pipettor and a sterile 1 ml pipet, add 0.5 ml from the control tube to the control Tryptic Soy Broth.

c The same pipet may be employed for your remaining transfers if you work with the most dilute virus suspension first, then progress to the most concentrated virus. In other words, transfer 0.5 ml from the 10^{-9} viral dilution into its corresponding broth. Then transfer liquid from the 10^{-8} viral dilution to its broth, from the 10^{-7} dilution, the 10^{-6}, 10^{-5}, and so on.

d Incubate the Tryptic Soy Broth control and your cultures of phage and *Escherichia coli* strain B at 35° C until the next laboratory session. Discard your serial dilution of phage and the control tube appropriately.

Second Laboratory Session

a Examine your broth-clearing assay. In your laboratory report, note the dilution of the least concentrated virus sample that completely clears an *Escherichia coli* strain B broth culture. This is the end-point dilution.

b In your laboratory report, calculate the titer of the phage in your original coliphage sample. The titer equals the inverse of the end-point tube's dilution.

Procedure 3

Phage Identification of a Bacterial Unknown

First Laboratory Session

a Obtain a broth culture of unknown A, a plate of Tryptic Soy Agar, and a container of sterile, cotton-tipped applicators. Label the plate with your group identification, the date, and "Unknown A." As in the illustration, draw a line down the center of the plate. Then draw 2 well-separated, dime-sized circles, one on each side of the line. Label 1 circle "T2" and the other circle "TSB."

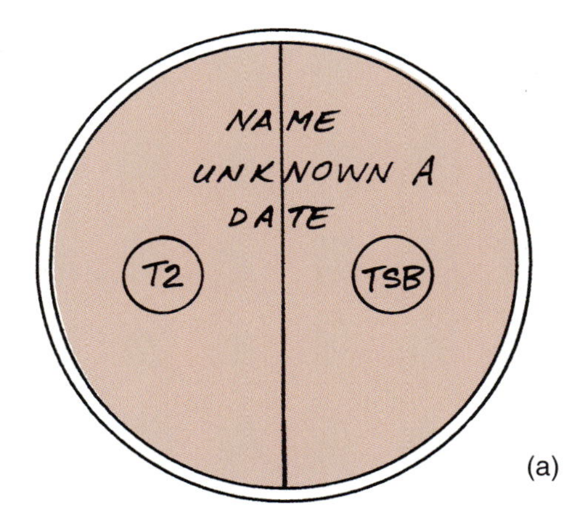

(a)

b Using the method described in Exercise 13, Streak-Plate and Spread-Plate Techniques, prepare a spread plate for confluent growth of the appropriate unknown. Discard your contaminated cotton-tipped applicator as directed by your instructor.

c Repeat steps (*a*) and (*b*) with unknowns B and C, labeling the 2 additional plates appropriately.

d With a sterile Pasteur pipet, transfer a small drop of bacteriophage T2 to the T2 circle on each plate. Be careful to avoid touching the bacterial lawns and cross contaminating plates. Discard your pipet appropriately.

e With a new sterile pipet, carefully transfer a small drop of Tryptic Soy Broth onto each of the 3 "TSB" circles. Discard your pipet appropriately.

f After the drops have dissolved into the bacterial lawn, invert your plates and incubate them at 35° C until the next laboratory session.

g In your laboratory report, explain why scientists include controls in their experiments and why you employed Tryptic Soy Broth as a control. See the materials list at the beginning of this exercise and also review Exercise 11, Ubiquity and Culture Characteristics, for help with this explanation.

Second Laboratory Session

a Examine your spread plates of bacterial unknowns. In your laboratory report, briefly describe the appearance of the TSB and T2 spots on each plate. Which unknown is *Escherichia coli* strain B?

POSTTEST EXERCISE 17

True or False (Circle one, then correct every false statement.)

1. T F Viruses are defined as "cellular" because they replicate inside of living host cells.
2. T F Cultivation of viruses in tissue culture is generally as quick, inexpensive, and convenient as growing bacteria in nutrient media.
3. T F Coliphage are common microbiological laboratory "weeds" that infect all human pathogens.
4. T F A viral plaque is a hole in a confluent layer of permissive host cells.
5. T F A plaque-forming unit represents one bacterial cell.
6. T F A plaque assay is designed to discover the number of bacteria in a viral solution.
7. T F To prepare a coliphage plaque assay, add *Escherichia coli* strain B to Soft Agar overlays, add dilutions of a virulent coliphage, pour the overlays onto Tryptic Soy Agar in Petri dishes, incubate appropriately, then examine for liquefaction of the Soft Agar.
8. T F The titer (concentration) of virions in your original coliphage sample equals the number of plaques on a countable plate times the inverse of that plate's dilution.
9. T F In a broth-clearing assay, the titer of phage in the original coliphage sample equals the end-point tube's dilution.
10. T F Viruses are classified by host, nucleic acid, size, shape, and presence or absence of an envelope.
11. T F Bacteria are used to treat virus strains (with ice packs and tiny Ace bandages) in bacteriophage typing.
12. T F Host-specific phages can distinguish one strain of a single bacterial species from another strain of the same species.

Laboratory Report **17**

Name ______________________________

Section ______________

Bacteriophage

Procedure 1

Phage Titration by Plaque Assay

First Laboratory Session

(*m*) Explain why it is necessary to use a new sterile pipet for each transfer as you move from left to right in figure 17.1.

Explain why it is **not** necessary to employ a new sterile pipet for each transfer as you go from right to left in figure 17.1.

Second Laboratory Session

(*b,c*) Did you have any countable plates? _____ If not, why?

If so, how many plaques developed on each countable plate?

Record the number of plaque-forming units in the original bacteriophage suspension. Average your plate counts if necessary. Show your calculations.

Compare the results of your plaque assay to the data collected by at least 2 other groups. Fill in the following blanks with the number of plaque-forming units per milliliter of the original viral suspension.

My Group ________________________ Group A ________________________ Group B ________________________

If possible, explain any discrepancies between the figures in the preceding list.

Procedure 2

Phage Titration by Broth-Clearing Assay

Second Laboratory Session

(*a*) Note the dilution of the least concentrated virus sample that completely clears an *Escherichia coli* strain B broth culture. ______________________ What is this dilution called? ____

(*b*) What is the titer of the phage in your original coliphage sample? __________________

Procedure 3

Phage Identification of a Bacterial Unknown

First Laboratory Session

(*g*) Why do scientists include controls in their experiments, and why is TSB an appropriate control for your bacteriophage T2 identification of a bacterial unknown?

Second Laboratory Session

(*a*) Briefly describe the appearance of TSB and T2 spots on each plate. Which unknown is *Escherichia coli* strain B? ______

Temperature and pH

Necessary Skills

Microscopy, Aseptic transfer techniques, Endospore staining, Methods of scientific investigation

Materials

Cultures (24- to 48-hour, Tryptic Soy Broth, 1 per set):
- *Bacillus stearothermophilus* (55° C)
- *Escherichia coli* (35° C)
- *Pseudomonas fluorescens* (30° C)
- *Lactobacillus delbrueckii* subspecies *bulgaricus* (formerly *Lactobacillus bulgaricus*) (35° C)
- *Saccharomyces cerevisiae* (30° C)

Equipment (per class):
- Refrigerator
- 35° C Incubator
- 55° C Incubator
- Thermometer
- Contaminated pipet receptacles

Supplies (per group):

Procedure 1
- 4 Tryptic Soy Agar plates

Procedure 2
- 2 Tryptic Soy Agar plates
- 2 Sterile test tubes
- 400 ml Beaker
- Thermometer
- Pipettor
- Sterile 5 ml pipets
- Hot plate **or** Bunsen burner and ring stand with wire gauze
- Endospore stain reagents in dropper bottles
  - Malachite green
  - Safranin
- Other staining supplies and equipment

Procedure 3
- 16 Tryptic Soy Broths (4 with each of the following pH values: 3, 5, 7, 9)
- Reference books that discuss fungi and/or bacteria, including *Bergey's Manual*

Primary Objective

Investigate the effects of temperature and pH variation on microbial growth.

Other Objectives

1. List five physical factors that alter the growth patterns of microorganisms and explain their effects upon the optimum temperature range of a species.
2. Compare and contrast the natural environments of psychrophiles, mesophiles, and thermophiles.
3. Compare the temperature requirements and tolerances of thermophiles and thermodurics.
4. Specify times and temperatures for pasteurization and flash pasteurization.
5. Measure and evaluate the parameters of temperature maximum, minimum, and range.
6. Define: thermal death time, thermal death point, pH, buffer.
7. Propose, evaluate, perform, and appraise the results of an experiment that establishes thermal death time.
8. Summarize the characteristics of neutralophiles, acidodurics, acidophiles, and alkalophiles.
9. Discuss how microorganisms are able to change the pH of their own environment.
10. Explain the buffering action of dibasic potassium phosphate (K_2HPO_4) and monobasic potassium phosphate (KH_2PO_4).

INTRODUCTION

Temperature and pH, along with osmotic pressure, oxygen tension, and radiation, are vital **physical factors** that alter the growth patterns of microorganisms. Lacking **homeostatic mechanisms** (systems that maintain an equilibrium), microbes are directly affected by changes in their environment. In this exercise you examine how temperature and pH influence growth and death among representative members of the microbial world.

Figure 18.1 Temperature parameters

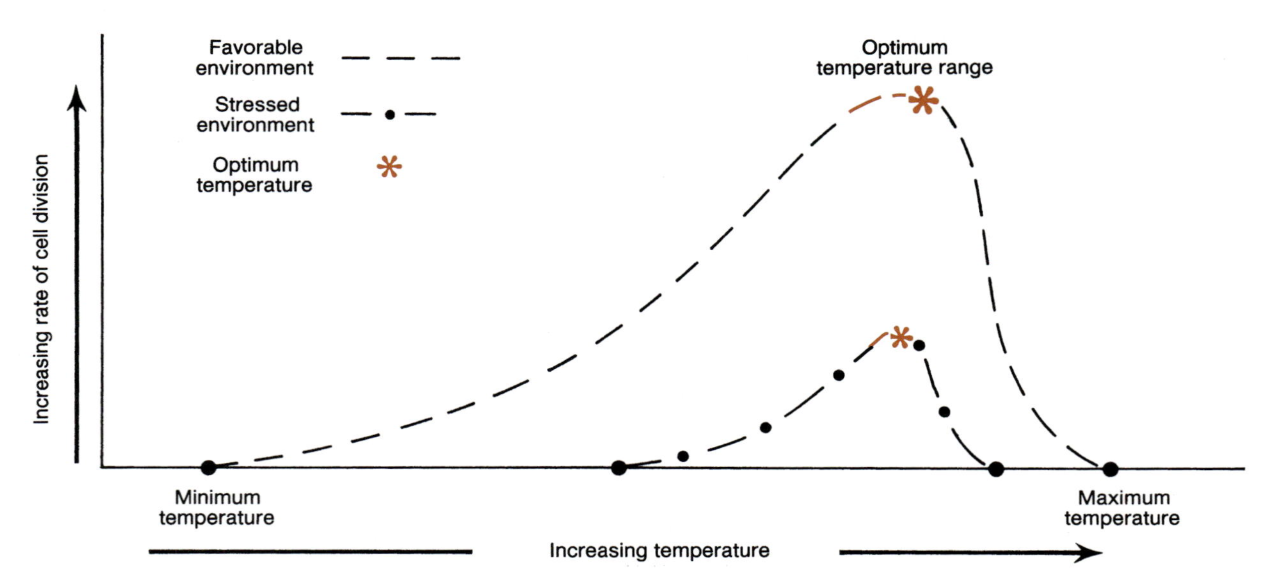

Most microorganisms are well adapted to survival in their usual habitats. Microbial exoenzymes and membranes are stable and function optimally when surrounded by a range of physical factors such as the temperature, pH, osmotic pressure, and oxygen concentration that they normally encounter.

Temperature

Each microbial species has its own **optimum** (most favorable) **temperature** and **optimum temperature range** (the span of temperatures where it multiplies most rapidly); see figure 18.1.

A species' optimum temperature range may be either narrow or broad. Breadth depends not only on the genetic potential of the microbes but also upon all of the other physical and chemical parameters that may be stressing the culture. For example, in an acid environment, the optimum temperature range narrows among organisms that grow optimally in neutral pH surroundings. Stress in one parameter narrows the range of another.

Above and below the optimum temperature range lie the **temperature maximum** and **minimum.** These numbers define limits beyond which multiplication of the species stops.

Temperature Groupings

Microbiologists divide microorganisms into three temperature groupings. **Psychrophiles** multiply best at temperatures ranging from below 0° C to about 20° C. They grow in a refrigerator, causing food spoilage. Where in nature do you expect to find psychrophiles?

Mesophiles abound in temperatures from about 20° C to 40° C. Human pathogens are mesophiles, as are the normal flora that inhabit our bodies. Most topsoils in temperate climates are rich in mesophiles.

Thermophiles flourish in high temperatures. They grow in hot springs, in desert soils, and in spas, thriving in hot environments above 40° C.

Organisms that survive but do not multiply at high temperatures are called **thermodurics.** Endospore-formers are generally thermoduric, since the endospores of most sporing species endure boiling for at least 10 minutes.

These and other thermodurics tolerate **pasteurization** (63° C, 30 minutes) and **flash pasteurization** (72° C, 15 seconds). The pasteurization processes kill vegetative cells of pathogens. However, some of the surviving thermodurics are able to spoil pasteurized liquid when temperatures fall back into the species' optimal ranges.

The Effects of Temperature on Bacterial Growth

In this exercise you examine the effects of temperature on growth, and you design your own experiment to discover the thermal death time of selected species.

Thermal death time is the *time* required to kill all the members of a species, and their endospores, at a given temperature. Most often that temperature is 100° C.

Thermal death point is the *temperature* required to kill all the members of a species, and their endospores, within 10 minutes.

pH

In this exercise you investigate the effects of another physical factor, **acidity** (and **alkalinity**), on microbial growth. Acidity is measured by the logarithmic **pH** scale.

The **pH** scale runs from 0 to 14, as shown in figure 18.2. Low numbers represent **acids** and indicate excess H^+ ions. The high numbers, representing **alkaline solutions (bases),** have excess OH^- ions or other ions that attract H^+.

Figure 18.2 The pH scale

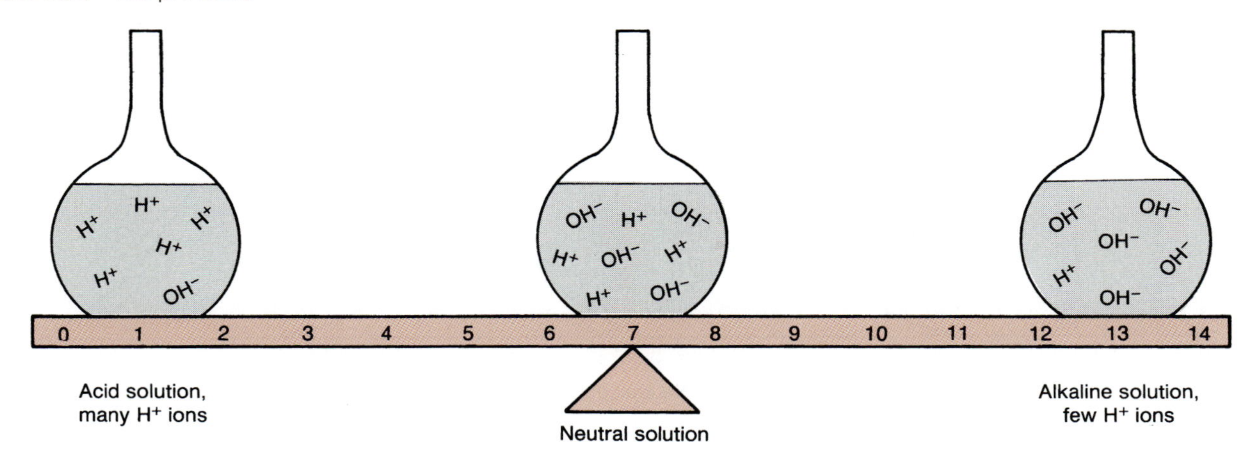

Table 18.1 Approximate pH Values of Some Common Substances

Approximate pH	Example	Approximate pH	Example
0		7	Distilled H_2O
1	0.1N HCl		Human blood
	Volcanic soils		Spinal fluid
	Lime juice		Saliva
2	Lemon juice		Eggs
	Acid mine drainage	8	Seawater
3	Vinegar		0.1N Sodium bicarbonate
	Grapefruit	9	Very alkaline soil
	Cherries		Soapy water
	Apples	10	Milk of magnesia
	Oranges	11	Household ammonia
	Wines	12	0.01N Sodium hydroxide
	Soft drinks	13	Oven cleaner
4	Tomatoes	14	N Sodium hydroxide
	Beers		
5	Breads		
	Beans		
	Peas		
6	Corn		
	Butter		
	Milk		

Water (H_2O) consists of $H^+ + OH^-$. This liquid has a neutral pH because the concentration of H^+ ions equals that of OH^- ions. The concentration of H^+ ions in water is $\mathbf{10^{-7}}$ moles/liter.

The pH of a solution may be defined as the negative log to the base 10 of the concentration of H^+ ions in the solution.* Water's pH equals:

$$-(\log_{10}\mathbf{10^{-7}})$$

that is, $-(\mathbf{-7})$ or simply **7.**

Because the pH scale represents logarithmic numbers, each whole number is 10 times more acid (or less alkaline) than the number above it. For example, stomach acid, at pH 2, is 10 times more acid than a carbonated soft drink with a pH of 3. Household ammonia, pH 11, is 1,000 times more alkaline than seawater with its pH of about 8. The pH values of some common compounds are shown in table 18.1.

*Refer to a college chemistry text for a discussion of the relationship between hydronium ion (H_3O^+) concentration and pH.

pH Tolerance of Microorganisms

Most bacteria are **neutralophiles;** that is, their optimum pH range is about 6 to 8.5. Many fungi and some bacteria are **acidoduric (acidotolerant),** able to survive but not multiply in acid. A greater proportion of fungi than bacteria are **acidophiles,** capable of growth in acid environments. **Alkalophiles (alkalinophiles)** are also represented among both the fungi and bacteria. Classification by pH preference is shown in figure 18.3.

Even some normal flora of the human body prefer pH values outside the neutral range. While inhabitants of the

Figure 18.3 Classification of microbes by pH preference

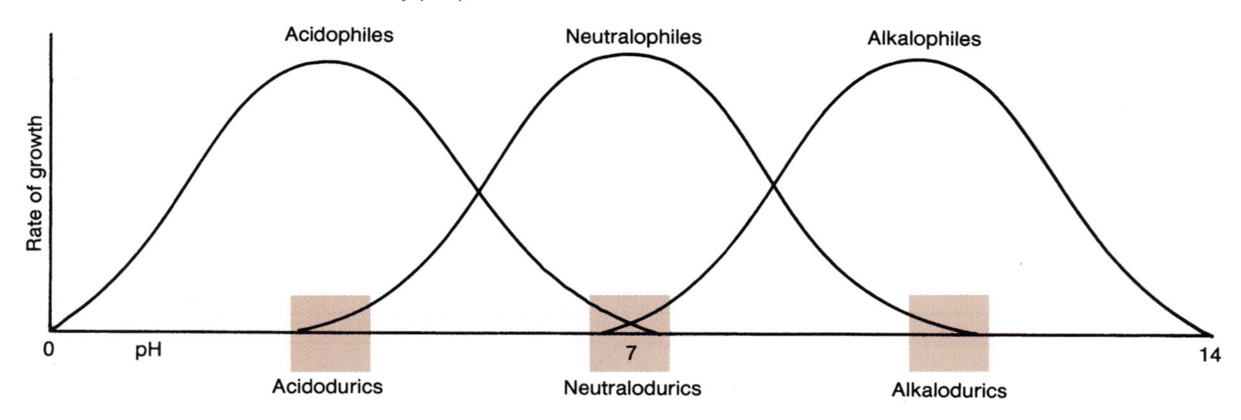

respiratory tract (saliva pH 6.5 to 7.5) are generally neutralophiles, flora of the human skin, with its pH of approximately 5, must tolerate acid. Human gastric contents range from pH 1 to 3, yet bacteria proliferate on the mucosal layer of the stomach wall.

Normal flora of the intestinal tract are likely to survive a broad range of pH values, similar to the range they tolerate in their usual habitat. The pH range of human feces is about 4.5 to 8.5, yet the material contains abundant microorganisms.

Buffers

This pH tolerance is not universal; wide fluctuations in pH kill many microorganisms. To retard a toxic buildup of acid or alkaline products, **buffers** are included in most culture media. A buffer is a compound that slows pH changes; buffers maintain an equilibrium.

In complex media **amino acids, peptides,** and **proteins** are natural buffers. In synthetic media two phosphate salts that together neutralize acids and bases at pHs near 7 are often included. These are (1) **monobasic potassium phosphate, KH_2PO_4,** the salt of a weak acid; it donates H^+ and becomes a weak base and (2) **dibasic potassium phosphate, K_2HPO_4,** the salt of a weak base; it binds excess H^+ and becomes a weak acid.

Bacteria Alter pH

As illustrated in figure 18.4, microbes can change the pH of their own environment, overcoming the moderating influence of natural buffers.

In unrefrigerated, raw milk, for example, bacteria ferment the sugar lactose, producing acid. This acid kills many of the original bacteria that generated it. Acidophilic fungi and bacteria then attack amino acids, peptides, and proteins in the milk, releasing amines and other alkaline compounds. The resulting elevation of pH inhibits growth of the acidophiles that created it.

In this exercise you investigate the effects of a broad range of pH values on representative bacteria and fungi. Pay particular attention to adequate labeling of your work since you share your results with your group.

Figure 18.4 Microbes change the pH of unrefrigerated raw milk

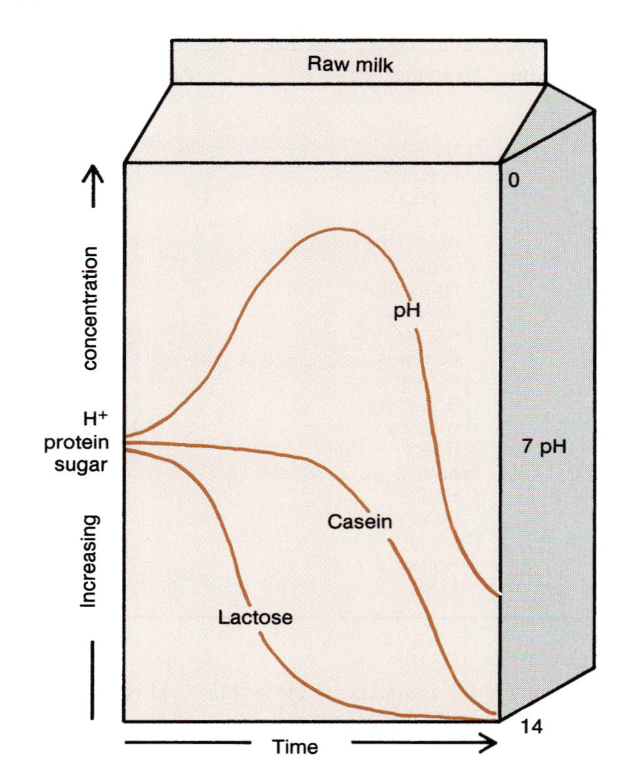

PROCEDURES EXERCISE 18

Procedure 1

How Temperature Alters Growth

First Laboratory Session

a Locate 4° C (refrigerator), ambient ("surrounding," meaning room temperature), 35° C, and 55° C incubation areas. If other temperatures for incubation are available, **your instructor** will direct you to them. In your laboratory report, note the ambient temperature of your laboratory.

b Gather 1 plate of Tryptic Soy Agar for every temperature of incubation. Draw a "Y" on the bottom of each plate. Label 1 section for each of your 3 cultures: *Bacillus stearothermophilus, Escherichia coli,* and *Pseudomonas fluorescens.* Also add your name and the date.

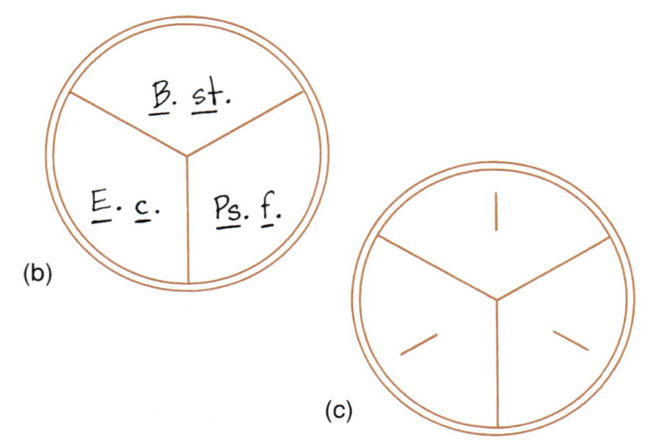

c Use your sterile loop to inoculate each section with a short, straight line of the appropriate culture, as in the illustration. Invert your plates and incubate them at the designated temperatures until the next laboratory session.

Second Laboratory Session

a In your laboratory report, evaluate the quantity of growth produced by each culture at each temperature. To check the repeatability of your results, compare them with those of another group. Discuss temperature tolerance. Classify each organism as psychrophilic, mesophilic, or thermophilic.

Procedure 2

Thermal Death Time

First Laboratory Session

Thermal death time is the amount of time required to kill bacteria and their spores at a given temperature.

a As a group, plan an experiment to test the thermal death time of *Bacillus stearothermophilus.* Use 100° C as the test temperature.

Utilize only the supplies and equipment provided for your class; these include sterile test tubes and plates of Tryptic Soy Agar. You also have access to a beaker, heat source, thermometer, incubators, and so on.

b Refer to Exercise 11, Ubiquity and Culture Characteristics, to review the scientific method of investigation. *The simplest experiments are often the most accurate and revealing.* Briefly, the scientific method requires you to

1. Recognize a problem;
2. Gather information;
3. Formulate a testable hypothesis;
4. Make a prediction;
5. Design and perform experiments, gathering data;
6. Include controls;
7. Arrive at a conclusion by evaluating your data;
8. Reevaluate experimental design; and
9. Formulate another hypothesis and test it.

c Record your experimental design in your laboratory report. Record and explain the controls that you include. Perform an endospore stain to confirm the presence of endospores in the culture.

d Incubate labeled plates. Dispose of all contaminated items in an appropriate manner.

Second Laboratory Session

a Examine the test results of your experiment. Did your group discover the thermal death time for *Bacillus stearothermophilus?* If not, reevaluate your experimental design. Enter suggestions for improvement in your laboratory report.

b In your laboratory report, consider the medical significance of other endospore-forming bacteria that survive boiling and can infect wounds.

Procedure 3

Investigating pH Preferences of Microorganisms

First Laboratory Session

a Select 1 culture—a bacterial species, *Pseudomonas fluorescens, Escherichia coli,* or *Lactobacillus delbrueckii* subspecies *bulgaricus;* or the yeast, *Saccharomyces cerevisiae.* You perform this procedure utilizing your culture. Then you evaluate results from your entire group. Be sure that all 4 organisms are cultivated by your group.

b Label each broth clearly with the name of your organism, the pH, the date, and your name. Inoculate your culture into 4 Tryptic Soy Broths, 1 with each of the following pH values: 3, 5, 7, 9. Dispense 2 loops of thoroughly resuspended organisms into each broth.

c Incubate *Escherichia coli* and *Lactobacillus delbrueckii* subspecies *bulgaricus* at 35° C, the fungus and *Pseudomonas fluorescens* at 30° C. Examine them during the next laboratory session.

d In *Bergey's Manual,* this laboratory manual, or other microbiological reference works, identify at least 1 common habitat for each of the bacteria. List these environments in your laboratory report. Employ this information to predict the optimum pH for each of the bacteria.

e In this laboratory manual, your text, or other microbiology books, find at least 1 common habitat for the yeast. List this environment in your laboratory report, and use the information to predict the optimum pH for the fungus.

f In your laboratory report, explain why you do not investigate the common habitats of *Saccharomyces* by reading about this yeast in *Bergey's Manual.*

Second Laboratory Session

a If possible, thoroughly resuspend the microbes in each broth. Evaluate your own species, assigning each tube a value of either 0 for no growth or + to + + + + from least to most growth. Enter your results in your laboratory report.

b Examine each of the other 3 genera. The maximum amount of growth for 1 species is considerably different from the maximum amount of growth for another species. Therefore evaluate each species separately. Evaluate the growth at each pH from 0 or + (least growth for that species) to + + + + (maximum growth for that species).

c Compare your results with those of your laboratory partners. Enter the results in your laboratory report.

d For each culture, tabulate and graph the pH response.

e Estimate pH optimum, minimum, and maximum values. Classify each species as neutralophilic, acidophilic, or alkalophilic. If possible, identify species which are acidoduric, neutraloduric, or alkaloduric.

Do not dispose of the pH culture tubes until **everyone** in your group has completed recording all the results.

POSTTEST EXERCISE 18

Part 1

Matching

Each response may be used once, more than once, or not at all. Pick the temperature classification that contains microorganisms most likely to be found in each environment.

a. Psychrophile c. Mesophile
b. Thermophile

1. ____ In the soupy, backyard hot tub of a decadent microbiology professor
2. ____ In a cat's ear
3. ____ Rotting a forgotten, refrigerated cucumber
4. ____ On the runners of Santa's sleigh at the North Pole
5. ____ Recycling the elements contained in a deceased lizard, Death Valley, California, in August, temperature 55° C

Part 2

True or False (Circle one, then correct every false statement.)

1. T F Temperature is a physical factor that alters the growth patterns of microorganisms.
2. T F Psychrophiles grow on snow and in cold seawater at the bottom of the oceans.
3. T F Endospore-formers are thermophilic, not thermoduric.
4. T F Below a species' temperature minimum, the microorganisms thrive.
5. T F Above a species' temperature maximum, multiplication of the microbes ceases.
6. T F The human stomach, with its pH of around 2, is especially well suited for the growth of alkalophiles.
7. T F The pH scale runs from 0 to 14 with low numbers indicating a dearth of H^+ ions.
8. T F The pH of water is defined as $-(\log_{10}10^{-7})$, which equals 7.
9. T F Vinegar with a pH of about 3 is 100 times more acid than cheese with a pH of about 5.
10. T F Most bacteria are neutralophiles.
11. T F Most fungi are alkalophiles, but a few are acidophiles.
12. T F Microorganisms often alter the pH of their surroundings.
13. T F Buffers speed pH changes.
14. T F Microbiologists often use the two phosphate salts K_2HPO_4 and KH_2PO_4 to buffer synthetic media.
15. T F Acidoduric microorganisms are more likely to be found in human saliva than in human gastric contents.

Part 3

Completion

1. Pasteurization is performed at ____° C for ____ minutes; flash pasteurization requires ____° C for ___ seconds.
2. The span of temperatures where a species multiplies fastest is the ____________ ____________ ____________.
3. The time required to kill all bacteria and their endospores at a given temperature is the ____________ ____________ ____________.
4. The temperature required to kill all bacteria in 10 minutes is the ________________ ____________ ____________.

Laboratory Report **18**

Name ______________________________

Section ______________

Temperature and pH

Procedure 1

How Temperature Alters Growth

First Laboratory Session

(*a*) What is the ambient temperature in your laboratory? ______

Second Laboratory Session

(*a*) Discuss the amount of growth each species produced at each temperature.

Temperature and Growth

Species	**4° C**	**Ambient**	**35° C**	**55° C**
Bacillus stearothermophilus				
Escherichia coli				
Pseudomonas fluorescens				

How do your results compare with those of another group?

Which microbe has the widest range of temperature tolerance? ____________________ Which has the narrowest range of temperature tolerance? ____________________

If possible classify each species as psychrophilic, mesophilic, or thermophilic.

Bacillus stearothermophilus ____________________

Escherichia coli ____________________

Pseudomonas fluorescens ____________________

Procedure 2

Thermal Death Time

First Laboratory Session

(*a*) State your group's hypothesis concerning the thermal death time of *Bacillus stearothermophilus.*

(*c*) Record your group's experimental design. Discuss the controls.

Second Laboratory Session

(*a*) Collect data and evaluate your group's experimental results.

If your group discovered the thermal death time for *Bacillus stearothermophilus,* what is it? ___________________________
If you did not discover the thermal death time, what changes do you suggest for your group's experimental design?

(*b*) Discuss the medical significance of an endospore-forming species such as *Clostridium perfringens*—an etiologic agent of gas gangrene—that survives boiling and can infect wounds.

Procedure 3

Investigating pH Preferences of Microorganisms

First Laboratory Session

(*a*) What is the genus of your culture? ________________________

(*e, f*) For each of the species utilized in this exercise, list a common environment. Predict the optimal pH for each species.

pH Prediction for Microbiota

Microorganism	Common Environment	Predicted pH Optimum
Pseudomonas fluorescens		
Escherichia coli		
Lactobacillus delbrueckii subspecies *bulgaricus*		
Saccharomyces cerevisiae		

(*f*) Why not look up the yeast *Saccharomyces cerevisiae* in *Bergey's Manual?*

Second Laboratory Session

(*a,b,c,d*) After resuspending the microbes as much as possible, evaluate the growth in your set of pH culture tubes. Then evaluate and discuss the growth in the tubes of your laboratory partners. Record your results in the following table and graph. For no growth enter 0. To indicate least to most growth **for each species,** enter + to ++++. **Each species is evaluated separately.**

Microbial Growth Response to pH Variation

Microorganism	pH 3	pH 5	pH 7	pH 9
Pseudomonas fluorescens				
Escherichia coli				
Lactobacillus delbrueckii subspecies *bulgaricus*				
Saccharomyces cerevisiae				

Using a different color line for each microbial genus, graph the growth responses to pH. This gives you a linear representation of the material in the previous table.

++++

+++

++

+

0

3 5 7 9

Growth **pH**

Microbial Responses to pH Variation

Legend: (Indicate color of line)

______ *Pseudomonas fluorescens*

______ *Lactobacillus delbrueckii* subspecies *bulgaricus*

______ *Escherichia coli*

______ *Saccharomyces cerevisiae*

Exercise **19**

Disinfectants and Antiseptics

Necessary Skills

Aseptic transfer techniques, Spread-plate techniques

Materials

Cultures (dilutions of 24-hour, 35° C, Tryptic Soy Broths, 1 per set):
- *Staphylococcus epidermidis* (1:10 dilution)
- *Escherichia coli* (1:100 dilution)

First Laboratory Session

Supplies:

Procedure 1 (per group)
- Water Agar plate (1.5% agar in purified H_2O)
- Forceps
- 4 Nonsterile filter paper disks
- Bottles of the following dyes:
  - Crystal violet
  - Safranin
  - Methylene blue
  - Carbolfuchsin
- Access to a refrigerator

Procedure 2 (per group)
- 2 Mueller Hinton Medium plates
- 2 Sterile cotton-tipped applicators
- At least 6 containers of various antiseptics and/or disinfectants from the class supply. **Students may provide antimicrobials they wish to test.**
- Forceps
- Petri dish containing at least 6 sterile filter paper disks
- Alcohol for flaming

Second Laboratory Session

Procedures 1 and 2
- Metric rulers

Primary Objective

Evaluate the effects of antiseptics and disinfectants on bacteria.

Other Objectives

1. List six factors that influence the efficiency of antimicrobials.
2. Distinguish between disinfectants and antiseptics.
3. Define: bacteriostatic, bactericidal, phenol coefficient.
4. Select features of an ideal disinfectant or antiseptic.
5. Compare and contrast characteristics of common types of antimicrobials.
6. Demonstrate that various molecules diffuse through agar solutions at different rates.
7. Show and evaluate the bacteriostatic effects of a variety of antiseptics and/or disinfectants.

INTRODUCTION

In this exercise you examine the effects of liquid, **antimicrobial agents** on bacteria. Many factors influence the efficiency of antimicrobials. Performance depends upon degree of microbial contamination, sensitivity of the population, and concentration of the antimicrobial. Duration and temperature of exposure influence effectiveness. In addition to these factors, organic matter such as feces, blood, or sputum may chemically bind with and inactivate the antimicrobial.

Antimicrobial Applications

Manufacturers of antimicrobials design their products for specific applications. **Germicides** (chemicals lethal to pathogens) that are meant to be used on inanimate surfaces such as floors, workbenches, and large equipment are called **disinfectants.** Some disinfectants are **bactericidal** (lethal for bacteria).

Table 19.1 Some Characteristics of an Ideal Disinfectant

1. **Highly effective,** even when dilute; fast acting; broad range of activity over wide temperature range
2. **Nontoxic;** not teratogenic, carcinogenic, or irritating when applied topically or inhaled; safe if ingested
3. **Inexpensive**
4. **Stable** diluted, concentrated, or dehydrated; no loss of activity when transported or stored in powdered form without refrigeration
5. **Soluble** and active in water and oils; low surface tension, penetrates into crevices
6. **Colorless, odorless, deodorant, detergent**
7. **Harmless** to wood, metal, glass, plastic, and other materials
8. **Biodegradable**

Other solutions, the **antiseptics,** are appropriate for use on living tissue such as the human skin or throat. Antiseptics are often **bacteriostatic** (able to prevent multiplication but not kill bacteria).

To indicate the effectiveness of their **proprietary** (patented) disinfectants, manufacturers often print the **phenol coefficient (PC)** of the product on the bottle label. The phenol coefficient is a number that indicates the killing power of a disinfectant as compared with that of phenol.

In the 1800s **Joseph Lister** introduced dilute phenol, formerly called "carbolic acid," as a disinfectant for surgical dressings. Because his disinfection procedure dramatically reduced postsurgical infection and death, other surgeons soon accepted it. Phenol, in various dilutions, was the first disinfectant and antiseptic to gain favor in Western medicine. For this reason, other solutions are now compared with it.

Selecting an Antimicrobial Agent

The phenol coefficient is not the only characteristic of a disinfectant that microbiologists consider when selecting an appropriate germicidal agent. No disinfectant or antiseptic is perfect—but if one were, it would have characteristics including those listed in table 19.1.

Besides phenolics, medical professionals may choose from a variety of antimicrobial solutions. Some of the most common types and a few of their salient characteristics are listed in table 19.2.

PROCEDURES EXERCISE 19

Procedure 1

Variation in Dye Diffusion Rates

First Laboratory Session

This procedure is performed by pairs of groups.

a Each group obtains a Petri plate of Water Agar. Write the date and your group identification on the bottom of the plate. Draw a cross on the bottom of the plate. Label each quadrant: 1 for crystal violet, the second for safranin, the third for methylene blue, and the last for carbolfuchsin.

b With a forceps, pick up a nonsterile, filter paper disk and dip it halfway into the liquid in an open bottle of crystal violet dye. Allow the solution to rise up into the disk by capillary action.

c Place the stained disk in the appropriately labeled quadrant of your plate. Touch the disk with the tips of your forceps to improve its contact with, and adhesion to, the surface of the medium. Wipe your forceps clean on a paper towel.

d Repeat the staining and positioning of disks with each of the other dyes: safranin, methylene blue, and finally carbolfuchsin.

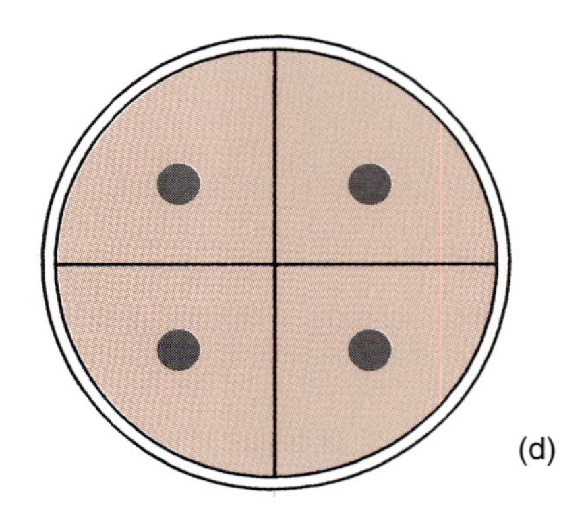

e Incubate your inverted Water Agar Petri plate until the next laboratory session. One group's plate is incubated at 35° C, the other's plate is incubated in the refrigerator (approximately 4° C). Add the incubation temperature to the label of each plate.

f In your laboratory report, explain why it is not necessary to observe sterile technique when utilizing Water Agar and dyes. (Hint: Note the formula for Water Agar in the materials section of this exercise.)

Second Laboratory Session

a Examine the diffusion of dye molecules into the Water Agar. The intensity of color around each disk is proportional to the number of dye molecules present in the agar.

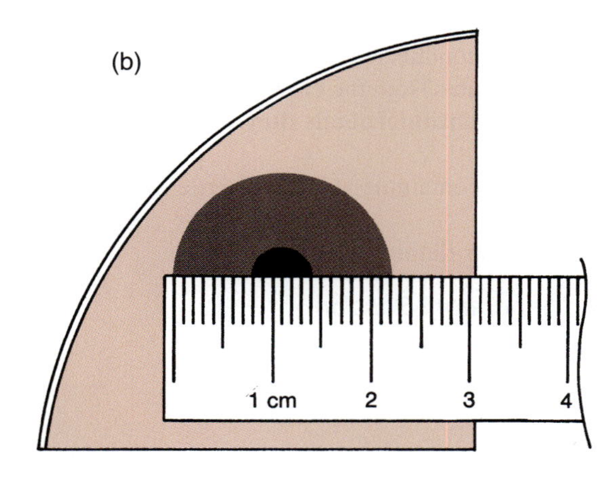

Table 19.2 Antimicrobial Solutions

Antimicrobial	Examples	Action	Effectiveness	Selected Applications
Alcohols	Ethanol and isopropanol	Denature proteins, dissolve lipids, and dehydrate molecules	Kill vegetative cells but not endospores	Disinfect instruments and cleanse skin
Alkylating agents	Formaldehyde	Inactivates proteins and nucleic acids	Kills vegetative cells and endospores	Embalming and vaccines
	Glutaraldehyde	Inactivates proteins and nucleic acids	Kills vegetative cells and endospores	Antiseptic (Cidex®)
Halogens	Iodine	Inactivates proteins	Kills vegetative cells and some endospores	Antiseptic; surgical preparation (Betadine®)
	Chlorine	Oxidizing agent	Kills vegetative cells and endospores	Disinfection for water, dairies, restaurants
Phenolics	Cresols	Denature proteins and alter membranes	Kill vegetative cells but not endospores	Preservatives
	Hexachlorophene	Denatures proteins and alters membranes	Kills vegetative cells but not endospores	Antiseptic
Quaternary ammonium compounds	Zephiran, Roccal, and other cationic detergents	Denature proteins and alter membranes	Kill most vegetative cells; do not kill *Mycobacterium tuberculosis, Pseudomonas aeruginosa,* or endospores; do not destroy unenveloped viruses	Sanitization in restaurants, laboratories, and industrial settings

b Measure the diameter of each colored area and enter the measurements in your laboratory report. In the report, note any conclusions you can draw relating temperature of incubation to the diameter of diffusion zones. What other factors influence diffusion rates of molecules moving through Water Agar?

Procedure 2

Disk Testing for Bacteriostasis

First Laboratory Session

a Obtain 2 plates of Mueller Hinton Medium.

b Label one of the plates *Staphylococcus epidermidis* and the other *Escherichia coli.* Add your group identification and today's date.

c With a sterile cotton-tipped applicator, prepare a spread plate of the thoroughly resuspended 1:10 dilution of *Staphylococcus epidermidis* on the appropriate plate. Discard the swab in an approved fashion.

d With a fresh, sterile cotton-tipped applicator, prepare a spread plate of the thoroughly resuspended 1:100 dilution of *Escherichia coli* on the other plate. Discard the swab correctly.

e Obtain 6 antiseptic and/or disinfectant solutions from the class supply. Also get a forceps, alcohol for flaming, and a container of sterile filter paper disks.

f With care to avoid injury to yourself, alcohol-flame sterilize the forceps.

g Touching only the disk you are taking, remove 1 sterile filter paper disk from the Petri dish with your sterilized forceps. Close the dish immediately.

h Dip the disk halfway into 1 antimicrobial solution, allowing the liquid to rise into the disk by capillary action.

i Place the disk onto the lawn of *Staphylococcus epidermidis.* With the tips of your forceps, touch the surface of the disk to improve its contact with, and adhesion to, the medium.

j Label the bottom of the plate with an abbreviation for the name of the solution.

k Resterilize your forceps and, placing the disks as far apart as possible, repeat steps (*g*) through (*j*) until you have challenged *Staphylococcus epidermidis* with all 6 solutions. Sterilize your forceps between solutions and when you have completed your transfers.

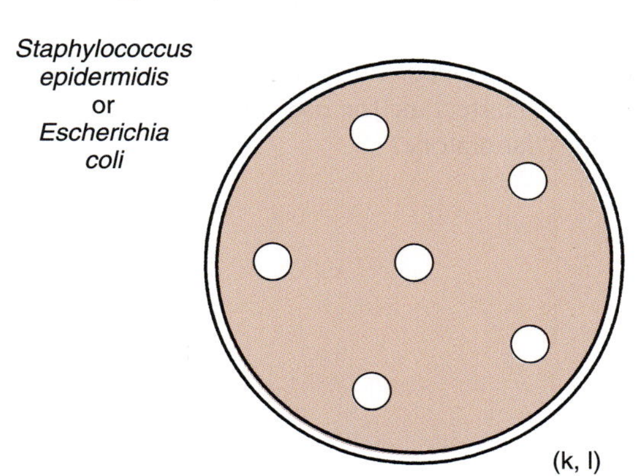

l Repeat steps (*f*) through (*k*), utilizing the same antimicrobial solutions but placing the disks onto your lawn of *Escherichia coli.*

m In your laboratory report, list the abbreviation, the name, and if possible the concentration of the active ingredients for each antimicrobial agent you employ.

n Return the antimicrobials and remaining sterile disks to the class set.

o Incubate your labeled, inverted plates at 35° C until the next laboratory session.

Second Laboratory Session

a As in the illustration, measure the diameter, or double the radius, of the zone of inhibition of bacterial growth around each disk. Enter your measurements in the laboratory report.

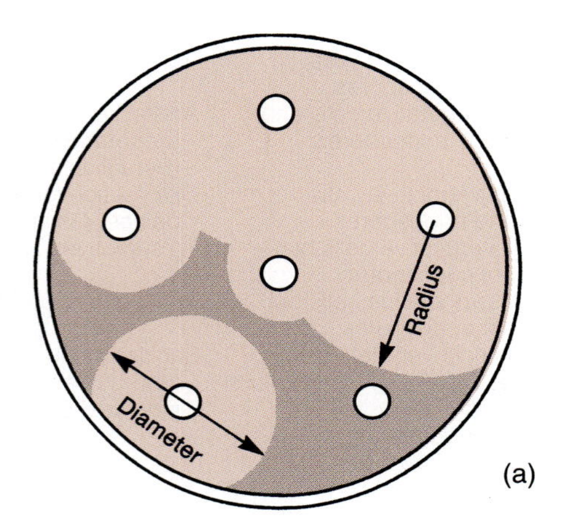

(a)

b Antimicrobial disks are said to test bacteriostatic but not necessarily bactericidal effects. Discuss this distinction in your laboratory report.

c Compare the effects of each antimicrobial agent on gram-positive *Staphylococcus epidermidis* and gram-negative *Escherichia coli.*

d The size of the zone of inhibition depends partially upon the sensitivity of the organism. But remember what you demonstrated in Procedure 1 of this exercise—not all solutions have the same rate of diffusion in an agar medium.

With these facts in mind, and utilizing information from table 19.2, discuss in your laboratory report the extent to which it is appropriate to utilize the results of your disk procedure. For instance, can your findings help you select an adequate disinfectant for use on workbenches in your microbiology laboratory?

POSTTEST EXERCISE 19

Part 1

Matching

Choose the one best response. Use each response only once.

a. Germicides
b. Bactericides
c. Bacteriostatic agents
d. Disinfectants
e. Antiseptics

1. ____ Solutions that prevent an increase in population size of bacteria but do not kill the organisms
2. ____ Chemicals that kill all bacteria and bacterial endospores
3. ____ Chemicals that kill or inhibit the growth of microbes and are designed to be used on inanimate objects
4. ____ Solutions that kill viral, bacterial, and protozoan pathogens
5. ____ Chemicals that generally inhibit the growth of microbes and are designed for use on skin or other tissue

Part 2

True or False (Circle one, then correct every false statement.)

1. T F Since the performance of a disinfectant depends in part upon the degree of microbial contamination, it is faster to disinfect a clean workbench surface than a soiled one.
2. T F The ability of an antimicrobial to kill or inhibit microorganisms depends partly upon the concentration of active ingredients in the solution.
3. T F Organic substances such as blood, sputum, and feces bind to many antimicrobial agents, increasing their activity.
4. T F Disinfecting a walk-in cold storage room generally takes longer than disinfecting an equally contaminated, 22° C area.
5. T F The phenol coefficient compares the bactericidal capabilities of test solutions to those of household ammonia.
6. T F In this exercise you perform the phenol coefficient test.
7. T F Since disinfection is instantaneous, it is a good idea to wipe disinfectant solution off your bench top immediately so that no residue is left on the surface.

Part 3

Short Answer

1. List characteristics that you might hope to find in an ideal disinfectant.

2. Tabulate 5 major types of antimicrobial agents. Indicate the effectiveness of each type.
   a.
   b.
   c.
   d.
   e.

Laboratory Report **19**

Name ______________________

Section ____________

Disinfectants and Antiseptics

Procedure 1

Variation in Dye Diffusion Rates

First Laboratory Session

(*f*) Considering the formula for Water Agar and the use of dyes, rather than bacterial cultures, in this procedure, why is it unnecessary to observe sterile technique here?

Second Laboratory Session

(*b*) Draw your dye diffusion plate and that of another group. With colored pencils, indicate the intensity and diameter of the tinted zone around each dye-impregnated disk.

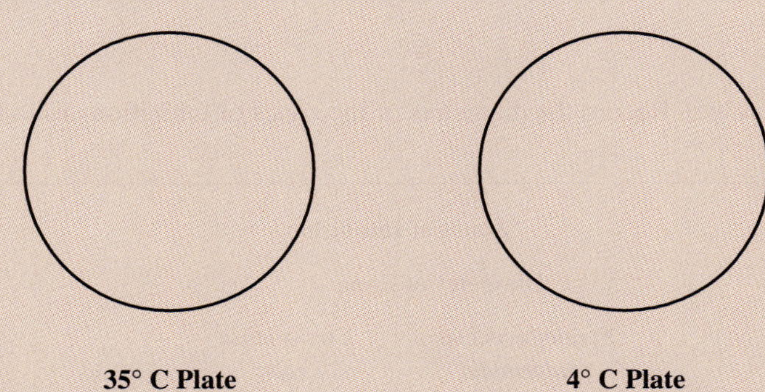

35° C Plate **4° C Plate**

What conclusions can you draw relating temperature of incubation to the diameter of diffusion zones?

You can see that not all dyes have the same diffusion rate in Water Agar. Some tinted areas are larger than others, indicating that the dye has spread farther. List factors other than temperature that account for the variation in diffusion rate that is demonstrated by different dyes.

Procedure 2

Disk Testing for Bacteriostasis

First Laboratory Session

(*m*) In the following chart, list the abbreviation, the name, and if possible the concentration of the active ingredients for each antimicrobial agent you employ in this experiment.

Antimicrobials Employed

Abbreviation	Name	Active Ingredients and Their Concentration
______		
______		
______		
______		
______		
______		

Second Laboratory Session

(*a*) List the name of each antimicrobial. Record the diameters of the zones of inhibition around disks on both spread plates.

Zones of Inhibition

Antimicrobial	Diameter of Zone	
	Staphylococcus epidermidis	***Escherichia coli***
______________________	______	______
______________________	______	______
______________________	______	______
______________________	______	______
______________________	______	______
______________________	______	______

(*b*) From these test results, can you be sure whether the bacteria are dead or have only been prevented from multiplying? ______ Explain your answer.

(*c*) Evaluate your zone of inhibition data. Compare the effects of each antimicrobial agent on *Staphylococcus epidermidis* and *Escherichia coli.*

(*d*) Is it valid to utilize the results of your disk procedure to help select a laboratory disinfectant? ______ Discuss the reasons for your answer.

Exercise 20

Antibiotic Disk Sensitivity Testing

Necessary Skills

Microscopy, Knowledge of the metric system of measurement, Aseptic transfer techniques, Gram staining, Spread-plate techniques, Culture characteristics

Materials

First Laboratory Session

Cultures (isolated colonies on 24-hour, 35° C, Tryptic Soy Agar streak plates, 1 plate of each species per set):
- *Staphylococcus epidermidis*
- *Escherichia coli*

Supplies (per group):
- At least 2 sterile Pasteur pipets
- Bulb for the pipets
- 2 Sterile cotton-tipped applicators
- 2 Mueller Hinton Medium plates, 25 ml of medium per 100 mm plate
- Forceps
- Alcohol for flaming
- 4 small, capped test tubes, each containing 2.5 ml of sterile 0.85% NaCl aq. (test tube size to match the McFarland standards)

Additional supplies:
- Access to an antibiotic disk dispenser (optional)
- Group access to about 8 varieties of high potency chemotherapeutic disks
- 1 McFarland No. 0.5 turbidity standard per group*

*To prepare the McFarland No. 0.5 turbidity standard, add 0.16 ml of 1% barium chloride aq. to 33.16 ml of 1% (0.36N) sulfuric acid. Dispense approximately 2.5 ml per screw cap test tube. In the sealed tubes, the standard remains stable for about 1 month.

Second Laboratory Session

Supplies:
- Metric rulers
- Microscope slides
- Gram stain reagents

Primary Objective

Prepare and evaluate a modified Kirby-Bauer antibiotic disk sensitivity test.

Other Objectives

1. Define: chemotherapeutic drugs, antibiotics.
2. List the three genera that produce most antibiotics and one antibiotic originally isolated from each of the genera.
3. Explain why it is no longer sufficient to define antibiotics as those antimicrobial chemicals that are produced by bacteria and fungi.
4. Summarize and give an example of how most antibiotics accomplish selective toxicity.
5. Distinguish between broad- and narrow-spectrum antibiotics.
6. Identify and evaluate a contribution of Kirby and Bauer to the field of antibiotic sensitivity testing.
7. List eight variables in the antibiotic disk diffusion test.
8. Identify the factor that the Kirby-Bauer antibiotic disk diffusion test is designed to evaluate.
9. Use the "R," "I," and "S" abbreviations listed in a zone diameter interpretation table to evaluate chemotherapeutic drugs' effects.
10. Examine colonies within a zone of inhibition and classify them as resistant mutants or contaminants.
11. Diagram a minimum inhibitory concentration test.

INTRODUCTION

Chemotherapeutic drugs are chemicals used to treat disease. They include the **antibiotics,** a group of compounds originally produced by metabolic reactions of bacteria and fungi, which kill or inhibit the multiplication of other microbes. Most antibiotics are produced by the two bacterial genera ***Streptomyces*** and ***Bacillus*** and by the fungal genus ***Penicillium.*** Among the hundreds of antibiotics these genera produce are streptomycin, bacitracin, and penicillin, respectively.

Other chemotherapeutic drugs, including the sulfonamides, are produced in chemical laboratories and are not considered to be antibiotics.

However, the distinction between artificial synthesis and biosynthesis is breaking down. Many antibiotics, originally isolated from living microbes, are now manufactured or altered by organic chemists to improve activity and minimize toxicity.

Chemotherapeutics

As **Paul Ehrlich** discovered in the late 1800s, only chemotherapeutics that display **selective toxicity** are useful in treating human disease. That is, the agents must harm the germs more than the patient.

Generally, an antibiotic accomplishes selective toxicity by blocking a **biochemical pathway** (series of enzymatic reactions) that is required by the bacterium but absent or nonessential for the patient. For example, penicillin prevents the formation of new linkages in some bacterial cell walls. Can interrupting cell wall synthesis harm human cells? Why or why not?*

Not every antimicrobial kills all bacteria. Some chemotherapeutics are **broad-spectrum**—effective against a wide range of bacteria, both gram-positive and gram-negative. Others have a **narrow spectrum** of activity—few species are killed or inhibited by these drugs.

Throughout the world, pathogenic bacteria are becoming more resistant to antibiotics. Some of the reasons for resistance are listed in table 20.1.

Kirby-Bauer Antibiotic Disk Diffusion Test

In this exercise you perform a modified version of a rapid, inexpensive, simple test that answers the question, "Which chemotherapeutic drugs stop the spread of this particular strain of bacteria?" It is the **antibiotic disk diffusion test,** standardized by **Kirby** and **Bauer** for use in diagnostic laboratories. Although automated, computerized sensitivity test modules are available, the Kirby-Bauer procedure is still accepted and widely utilized worldwide.

The test is standardized to avoid variation. Every laboratory that correctly performs Kirby-Bauer antibiotic disk diffusion tests provides the same accurate results.

*Hint: Do human cells have cell walls?

Table 20.1 Development of Chemotherapeutic Resistance among Pathogens

1. The pathogens may lack the targeted biochemical pathway. By employing an alternative pathway, they escape the effects of the chemotherapeutic agent.
2. Microbial enzymes may destroy the chemotherapeutic.
3. Some pathogenic bacteria contain plasmids that provide resistance to many chemotherapeutics. Treatment with any one of these antimicrobials selects for multiresistant genera by killing sensitive microbiota while leaving invaders containing resistance factors unharmed.
4. Undertreatment of an infectious disease with antibiotics selects for the survival of only the most resistant strains. When treatment is stopped too soon, resistant bacteria proliferate.
5. Resistant strains of pathogenic species survive the worldwide overuse of chemotherapeutic drugs. They become the dominant representatives of their kind. Susceptible strains are being eradicated.

The procedure is designed to evaluate one variable, the **sensitivity** (susceptibility) of a pathogen to assorted chemotherapeutics. To interpret whether a bacterial strain is sensitive to a certain antibiotic, all other variables must be held constant.

Variables that can alter results of an antibiotic disk diffusion test include concentration and rate of diffusion of the chemotherapeutic in each disk, density of bacterial growth, thickness of medium, reactions between antimicrobial and medium, viscosity of medium (broth or agar preparation), and temperature and duration of incubation.

Performing a Disk Diffusion Test

To perform a disk diffusion test, a microbiologist emulsifies isolated colonies of the pathogenic bacteria in saline, creating a suspension whose **turbidity** (cloudiness) matches the **McFarland** No. 0.5 turbidity standard. (See the materials section in this exercise.) A Petri dish filled 4 mm deep with Mueller Hinton Medium is **seeded** with the suspension. That is, the bacteria are swabbed evenly over the surface of the plate.

After the surface dries, commercially available filter paper disks impregnated with known concentrations of antimicrobials are pressed onto the lawn of bacteria. The dish is incubated at 35° C for 16 to 18 hours.

The antimicrobials diffuse into the surrounding medium. Bacteria grow up to the edge of some disks. The contents of other disks kill or inhibit the growth of microbes, creating a clear **zone of inhibition** in the lawn of bacterial growth; see figure 20.1.

The microbiologist compares the diameter of a zone of inhibition around one disk with the standardized measurements given for that antimicrobial's zone; see table 20.2. If there is little or no inhibition, the bacteria fall into the **"R," resistant,** category—indicating that amounts of this particular chemotherapeutic that can be made available in human tissue do not stop the growth of the germ. A wider zone may indicate the category **"S,"** meaning that the microbe is **sensitive** (or **susceptible**) to the drug. An intermediate reading, **"I,"** is **inconclusive.**

Figure 20.1 In an antibiotic disk diffusion test, the zone of inhibition around each disk reflects the resistance (R) or sensitivity (S) of the bacteria to the chemotherapeutic agent in the disk. "I" indicates an intermediate reaction.

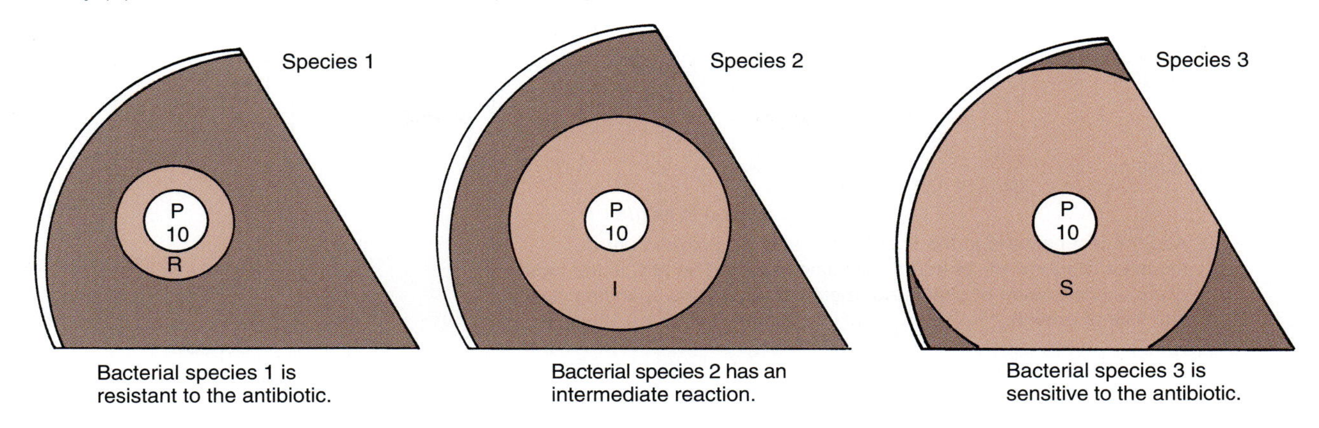

Bacterial species 1 is resistant to the antibiotic.

Bacterial species 2 has an intermediate reaction.

Bacterial species 3 is sensitive to the antibiotic.

Table 20.2 Evaluation of Zones of Inhibition[1]

Antimicrobic	Content	Zone Diameter to Nearest mm		
		Resistant mm or Less	Intermediate mm Range	Susceptible mm or More
Amikacin	30 μg	14	15–16	17
Ampicillin (gram-negative and enterococci)	10 μg	11	12–13	14
Ampicillin (staphylococci)	10 μg	20	21–28	29
Bacitracin	10 Units	8	9–12	13
Carbenicillin (Enterobacteriaceae including *Escherichia*)	100 μg	17	18–22	23
Carbenicillin (*Pseudomonas aeruginosa*)	100 μg	13	14–16	17
Cefamandole and cefoxitin	30 μg	14	15–17	18
Cefotaxime	30 μg	14	15–22	23
Cephalothin	30 μg	14	15–17	18
Chloramphenicol	30 μg	12	13–17	18
Clindamycin and lincomycin	2 μg	14	15–16	17
Colistin	10 μg	8	9–10	11
Erythromycin	15 μg	13	14–17	18
Gentamicin	10 μg	12	13–14	15
Kanamycin	30 μg	13	14–17	18
Methicillin (Results may be applied to cloxacillin, dicloxacillin, nafcillin, and oxacillin.)	5 μg	9	10–13	14
Nalidixic Acid (urinary tract infection organisms only)	30 μg	13	14–18	19
Neomycin	30 μg	12	13–16	17
Novobiocin	30 μg	17	18–21	22
Penicillin G (staphylococci)	10 Units	20	21–28	29
Penicillin G (other microorganisms)	10 Units	11	12–21	22
Polymyxin B	300 Units	8	9–11	12
Rifampin (*Neisseria meningitidis*)	5 μg	24	–	25
Streptomycin	10 μg	11	12–14	15
Sulfisoxazole (Results may be applied to all sulfonamides.)	300 μg	12	13–16	17
Tetracycline (Results may be applied to all tetracyclines.)	30 μg	14	15–18	19
Trimethoprim	5 μg	10	11–15	16
Trimethoprim/ sulfamethoxazole	1.25 μg 23.75 μg	10	11–15	16
Vancomycin	30 μg	9	10–11	12

Adapted from the *DIFCO Manual,* 10th ed. DIFCO Laboratories, Inc., Detroit, MI, 1984. For further information refer to the discussion of sensitivity disks in the *DIFCO Manual.*

[1] These diameter sizes fulfill World Health Organization (WHO) requirements and "Performance Standards for Antimicrobic Disc Susceptibility Tests" published by the National Committee for Clinical Laboratory Standards (NCCLS).

Figure 20.2 Evaluating the minimum inhibitory concentration (MIC) of a chemotherapeutic agent

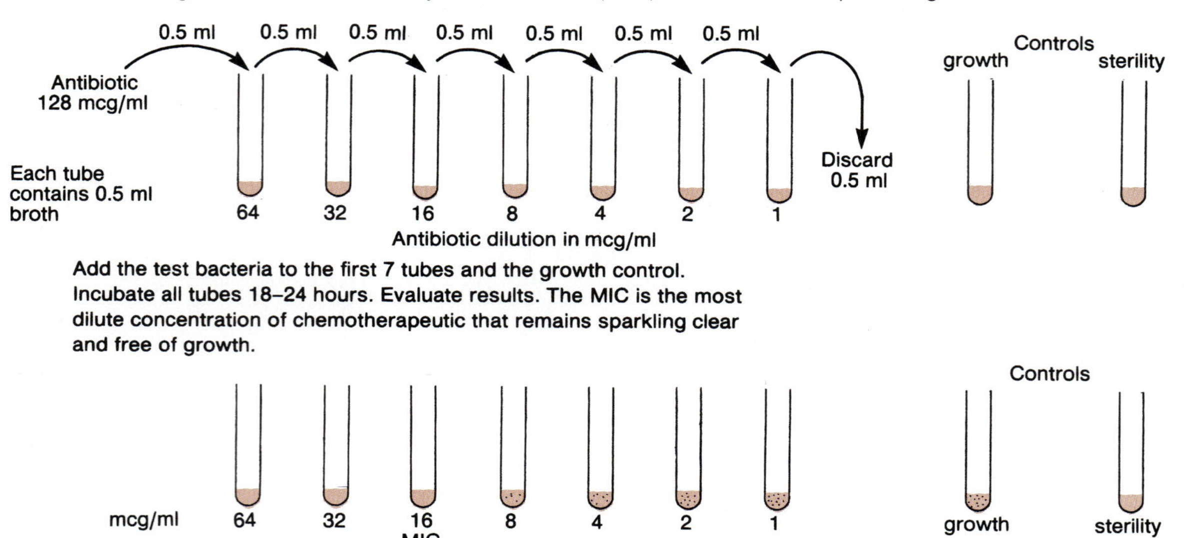

Sometimes colonies grow within a zone of inhibition. The microbiologist Gram stains this growth and examines both the slides and the colonies to see if they represent contaminating microbes or mutant, **resistant strains** of the pathogen. What effect could resistant strains of the pathogen have if the patient is treated with this chemotherapeutic?

Minimum Inhibitory Concentration

Sometimes a physician needs quantitative results that specify the minimum antibiotic concentration that inhibits a particular microbe.

To provide this information, the microbiologist prepares a two-fold dilution series of the antibiotic in liquid medium, adds a measured suspension of the test bacteria to each tube, and incubates the microbes in the dilution tubes; see figure 20.2. The least concentrated amount of chemotherapeutic that completely inhibits growth of the pathogen is the **minimum inhibitory concentration.**

By utilizing a microdilution tray and a microplate reader, a device that measures light absorbance through the sample in each well of the tray, diagnostic laboratories that perform many tests automate the minimum inhibitory concentration procedure.

In this exercise you prepare antibiotic disk diffusion plates of the gram-positive bacterium ***Staphylococcus epidermidis*** and the gram-negative species ***Escherichia coli.*** You perform a modification of the Kirby-Bauer method suitable for the teaching laboratory and evaluate the zones of inhibition that develop on your plates.

PROCEDURES EXERCISE 20

Procedure 1

The Antibiotic Disk Diffusion Test

First Laboratory Session

a In your laboratory report, enter complete descriptions of typical, well-isolated colonies of both *Staphylococcus epidermidis* and *Escherichia coli.*

b Maintaining aseptic technique, use your flamed and cooled inoculating loop to emulsify a few colonies of *Staphylococcus epidermidis* in 1 tube containing 5 ml of sterile saline. (While the Kirby-Bauer method specifies utilizing well-isolated colonies, sharing plates with your laboratory partners may necessitate use of confluent growth areas.)

c Now match the turbidity in your bacterial suspension to the McFarland standard. To do this, thoroughly resuspend the particles in a McFarland No. 0.5 dilution series test tube and the cells in your capped bacterial suspension. Hold the tubes up to a light source or in front of a black line and compare turbidity in the 2 test tubes.

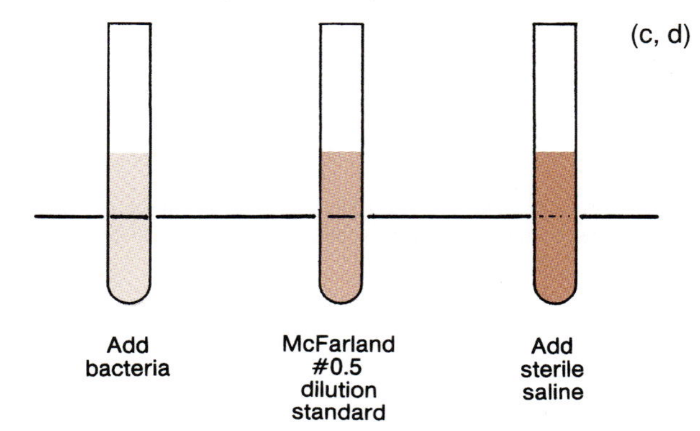

d If turbidity does not match, utilize sterile technique to add Pasteur pipet drops of sterile saline **or** loopfuls of *Staphylococcus epidermidis* to your bacterial suspension. Close the suspension and completely mix the contents. Examine the 2 tubes again. Repeat this procedure until turbidity is equalized in the 2 tubes.

e Prepare a spread plate of the diluted *Staphylococcus epidermidis* on an accurately labeled dish of Mueller Hinton Medium.

f Dispose of the cotton-tipped applicator appropriately.

g Allow the *Staphylococcus epidermidis* suspension to dry onto the agar surface for at least 5 minutes before applying disks.

h During your wait, repeat the suspension and seeding processes utilizing a sterile test tube of saline, colonies of *Escherichia coli,* a sterile cotton swab, and a second adequately labeled plate of Mueller Hinton Medium.

i If an automatic disk dispenser is available for your use, continue with steps (*j*) and (*k*); if not, go to step (*l*).

j **Your instructor** will demonstrate how to use the dispenser.

k Have your alcohol-flamed sterilized forceps ready. Utilizing the dispenser, apply chemotherapeutic disks onto your plate of *Staphylococcus epidermidis.* If any disks roll out of place, use the sterile forceps to return them immediately to a well-distributed pattern.

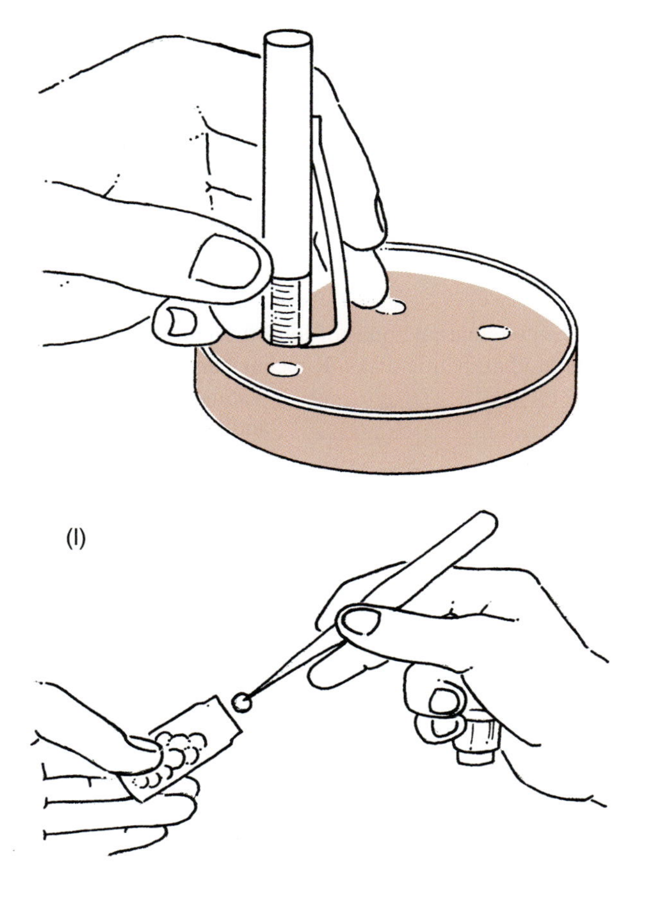

(l)

l If an automatic disk dispenser is unavailable, utilize cartridge dispensers or your alcohol-flamed forceps to place approximately 8 different kinds of antibiotic disks onto the plate of *Staphylococcus epidermidis.* Space the disks well apart.

m Whichever method you utilized to place the disks, touch each disk with the sterilized tips of your forceps to ensure complete contact with, and adhesion to, the surface of the agar. Alcohol-flame sterilize your forceps.

n Repeat the placement of chemotherapeutic disks utilizing your spread plate of *Escherichia coli.*

o Incubate inverted plates at 35° C for 16 to 18 hours. Measure the zones of inhibition immediately after incubation, if possible. If not, refrigerate your inverted plates until the next laboratory session.

p Examine the chemotherapeutic disk cartridges. In your laboratory report, record the symbol on each disk; list the agent and concentration represented by each symbol.

Second Laboratory Session

a Measure the zone of inhibition that surrounds each chemotherapeutic disk on each spread plate. Record the zone diameters in your laboratory report.

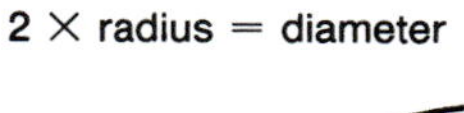

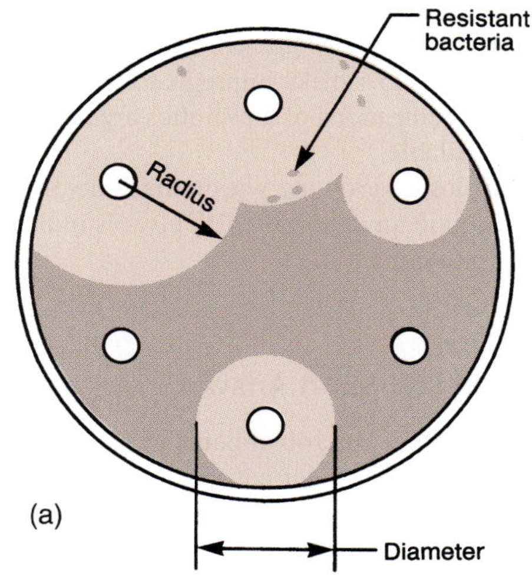

(a)

b Utilizing the chart in table 20.2, indicate whether the reactions of your laboratory strains of *Staphylococcus epidermidis* and *Escherichia coli* were susceptible, intermediate, or resistant to each antibiotic.

c If colonies are growing within a zone of inhibition, examine them. In your laboratory report, enter a complete description of an antibiotic-resistant colony.

d Prepare a Gram stain of the resistant bacteria, another of the seeded bacteria. Examine the stains.

e In your laboratory report, record your evaluation of the resistant growth. Is it caused by an antibiotic-resistant, mutant strain of the seeded bacteria or by a contaminant?

POSTTEST EXERCISE 20

Part 1

True or False (Circle one, then correct every false statement.)

1. T F Antibiotics are chemicals produced by, or originally produced by, microbes.
2. T F Organic chemists produce artificial antibiotics, but these are inferior to the natural products, demonstrating less activity and more toxicity.
3. T F Antibiotics are designed to block a biochemical pathway required by the patient but not by the invading pathogenic bacteria.
4. T F Broad-spectrum antibiotics are effective against both gram-positive and gram-negative bacteria, while narrow-spectrum antibiotics kill or inhibit relatively few species.
5. T F The Kirby-Bauer method measures whether specific antibiotics are able to kill or inhibit the growth of individual bacterial strains.
6. T F Microbiologists compare the color of a bacterial suspension to the McFarland No. 0.5 standard, ignoring turbidity.
7. T F The Mueller Hinton antibiotic sensitivity test, accepted worldwide as a clinical standard, utilizes Kirby-Bauer Medium.
8. T F Filter paper disks impregnated with specific concentrations of antibiotics are commercially available.
9. T F Colonies growing within the zone of inhibition around an antibiotic disk always indicate contamination.

Part 2

Completion and Short Answer

1. Name the three microbial genera that produce the most antibiotics useful to humans.
   a.
   b.
   c.
2. Name a group of chemotherapeutic drugs that are not antibiotics. ______________________
3. ______________________ discovered the principle of ______________________ ______________________, which specifies that a chemotherapeutic must be more toxic to the parasite than to the patient.
4. List some of the reasons bacteria are becoming more resistant to antibiotics.
5. What variables can alter the size of a zone of inhibition around an antibiotic disk?
6. When evaluating zones of inhibition around chemotherapeutic disks, R means bacteria are ______________________, S stands for ______________________, and I indicates ______________________ results.
7. Diagram a test for the minimum inhibitory concentration of an antibiotic.

Laboratory Report **20**

Name ______________________________

Section ______________

Antibiotic Disk Sensitivity Testing

Procedure 1

The Antibiotic Disk Diffusion Test

First Laboratory Session

(*a*) Describe typical colonies of *Staphylococcus epidermidis* and *Escherichia coli.*
Staphylococcus epidermidis:

Escherichia coli:

(*p*) In the following chart, record the symbols printed on various chemotherapeutic disks. List the agent and the concentration represented by each symbol.

Symbols on Chemotherapeutic Disks

Symbol	Chemotherapeutic	Concentration
1.		
2.		
3.		
4.		
5.		
6.		
7.		
8.		

Second Laboratory Session

(*a,b*) In the following chart, record the diameter of the zone of inhibition that surrounds each disk.

You may lack information to assign an R, I, or S evaluation for each entry in the following chart. Wherever you have sufficient information, decide whether the bacteria are resistant, intermediate, or sensitive to a chemotherapeutic, and place an × in the appropriate column of the chart.

Chemotherapeutic Efficacy

Chemotherapeutic	**Species**							
	Staphylococcus epidermidis				***Escherichia coli***			
	Zone Diameter	**R**	**I**	**S**	**Zone Diameter**	**R**	**I**	**S**
1. ________	______	__	__	__	______	__	__	__
2. ________	______	__	__	__	______	__	__	__
3. ________	______	__	__	__	______	__	__	__
4. ________	______	__	__	__	______	__	__	__
5. ________	______	__	__	__	______	__	__	__
6. ________	______	__	__	__	______	__	__	__
7. ________	______	__	__	__	______	__	__	__
8. ________	______	__	__	__	______	__	__	__

(*c*) If there are colonies growing within the zone of inhibition around a chemotherapeutic disk, describe a typical one.

(*d*) Describe the morphology and Gram stain reaction of the resistant cells growing within the zone of inhibition.

Describe the morphology and Gram stain reaction of the seeded cells on the same plate.

(*e*) Considering your descriptions of both the seeded and the resistant microbes, do you think that the antibiotic-resistant bacteria represent a contaminant or a mutant strain of the seeded bacteria? ____________________
Explain your answer.

Selected Microbial Enzymes: Amylase, Lipase, Catalase, Cytochrome Oxidase, Cysteine Desulfhydrase

Necessary Skills

Aseptic transfer techniques, Preparing culture media

Materials
First Laboratory Session

Cultures (24- to 48-hour, 35° C, Tryptic Soy Broth, 1 per set):
- *Bacillus subtilis* (Procedures 1 and 2)
- *Escherichia coli* (Procedures 1, 2, 3, and 4)
- *Enterococcus faecalis* (Procedure 3)
- *Pseudomonas aeruginosa* (Procedure 3)
- *Proteus vulgaris* (Procedure 4)

Supplies:
- Access to the *DIFCO Manual,* 10th ed.
- Access to *Bergey's Manual*

Procedure 1
- 1 Starch Agar plate (1% soluble starch, 0.3% beef extract, 1.2% agar, aq.) per group

Procedure 2
- 1 Egg Yolk Agar plate (4 parts sterile Dextrose Agar plus 1 part 5% egg yolk emulsified in sterile, purified water)

Procedure 3
- 1 Tryptic Soy Agar plate per group
- Access to 3% hydrogen peroxide in dropper bottles

Procedure 4
- 2 Triple Sugar Iron Agar slants

Second Laboratory Session

Supplies:
- Access to the *DIFCO Manual*

Procedure 1
- Iodine solution in dropper bottles

Procedure 2
- Access to a flat container of ice and/or a refrigerator

Procedure 3
- Access to 3% hydrogen peroxide in dropper bottles
- 2 DIFCO oxidase differentiation disks (or other oxidase testing method) per group
- Forceps
- 2 Microscope slides per group
- Purified water, deionized and/or distilled, several drops per group
- Toothpicks or wooden applicator sticks
- Alcohol for flaming

Procedure 4
- Access to a labeled, uninoculated control tube of Triple Sugar Iron Agar

Primary Objective

Perform and interpret tests for bacterial production of amylase, lipase, catalase, cytochrome oxidase, and cysteine desulfhydrase.

Other Objectives

1. Discuss the work of microbial physiologists.
2. Consider preparing a media file.
3. Explain what information is contained in the genetic code of a bacterial cell.
4. Differentiate four types of information a microbiologist uses to identify bacteria.
5. Define: exoenzyme, endoenzyme, hydrolyze, oxidase, desulfhydrase.
6. Tabulate the molecule degraded and the product(s) produced by each of these enzymes: amylase, lipase, and catalase.
7. Describe the value of exoenzymes to bacteria.
8. Diagram the reaction catalyzed by lipase, by catalase, by cysteine desulfhydrase.
9. Name two bacterial genera that tolerate oxygen but do not produce catalase.
10. Indicate the function of cytochrome oxidase in a bacterial cell.
11. Explain how ferrous sulfide is used to indicate bacterial production of hydrogen sulfide.

INTRODUCTION

In this exercise you begin to examine bacterial **physiology,** the biochemistry of cells. Microbial physiologists determine which substances bacterial species utilize and which products they produce. To a microbiologist, the biochemistry of a species is its virtual fingerprint, helping to identify the organisms and differentiate them from all other species.

Preparing your own **media file** helps you organize your study of microbial physiology. Table 21.1 shows you what to include on each card in your file. Add cards for the media that you use as you progress through this microbiology course.

Enzymes and Identification of Microbes

Which reactions a species can perform is determined by its genetic makeup. A microbe inherits deoxyribonucleic acid (DNA) from its parent cell. The DNA carries instructions, the genetic code, which includes a pattern for all the **enzymes** (biological **catalysts**) that the microorganism can produce.

Enzymes are substances that alter the rate of chemical reactions, generally making them go faster. Enzymes located inside cells are called **endoenzymes** or **intracellular enzymes.** Those excreted outside of the cell are **exoenzymes** or **extracellular enzymes.**

A microbiologist analyzes enzymatic reactions and **morphological** (size, shape, arrangement, staining reactions) data to begin identification of bacteria. In addition to physiological and morphological tests, **immunological** (antigen, antibody) testing is required to identify some bacterial species.

Nucleic acid probes are also available to help clinicians identify selected microbes. Some probes are radioactive, others are chemically treated to form visible spots when they contact the complementary nucleic acids of the target species in a specimen.

Amylase

Amylases are exoenzymes that **hydrolyze starch.** That is, amylases split starch by adding the H^+ and OH^- ions of H_2O. **Starch** is a **polysaccharide** (a large molecule consisting of repeated monosaccharide units), a many-branched polymer of the **monosaccharide** (simple sugar) glucose.

Starch, a rich source of carbon and energy, abounds in nature where it is a common reserve carbohydrate supplying the nutritional needs of plant roots and seeds. But the polysaccharide is too large to cross bacterial semipermeable cell membranes. Amylases cleave large starch molecules into monosaccharide and disaccharide units small enough to enter into a bacterial cell. There the sugars are degraded by endoenzymes, releasing energy and carbon. Microbes that do not produce the exoenzyme amylase can starve while surrounded by starch.

Table 21.1 Preparing a Media File Card

1. Name the medium.
2. State its intended use.
3. List its significant ingredients.
4. Give the names and significant ingredients of the reagent(s) you add, if any.
5. Draw colored sketches of positive and negative results.
6. State the recommended duration of incubation.
7. List other notes, if necessary.

Figure 21.1 Starch Agar plate flooded with iodine solution

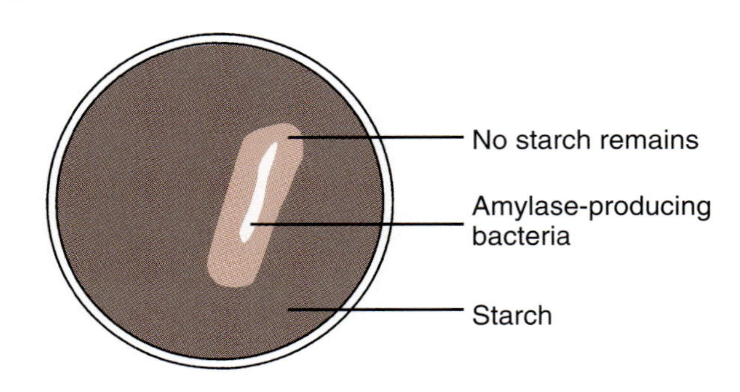

In this exercise you streak bacteria onto an agar medium containing starch. After incubating the plate, you analyze the amylase-producing capabilities of the microbes by flooding the medium with iodine solution, as shown in figure 21.1.

Starch remaining in your medium turns brownish-blue in the presence of iodine. Wherever bacterial amylase has degraded the starch into soluble sugar units, no color develops. By examining the Starch Agar plate, you determine which bacteria produce amylase. These "amylase-positive" species are surrounded by clear areas.

Lipase

Lipid hydrolysis is also characteristic of some bacteria. Lipids include fats (solid at room temperatures) and oils (liquid at these temperatures), lecithins, sterols, and other substances that are non-water-soluble but extractable by nonpolar solvents. These solvents include alcohols, acetone, chloroform, ether, and benzene.

Lipids are found in every kind of cell membrane, and they are common plant and animal nutrient storage compounds. It is advantageous for microorganisms living in close association with plants and animals to produce hydrolytic **lipases,** thus sharing the fatty feast; see figure 21.2.

Lipases cleave large fat molecules. The smaller units enter microbes and are degraded by bacterial endoenzymes that release the rich, lipoidal supply of carbon and energy.

Figure 21.2 Hydrolysis of a triglyceride by bacterial lipase

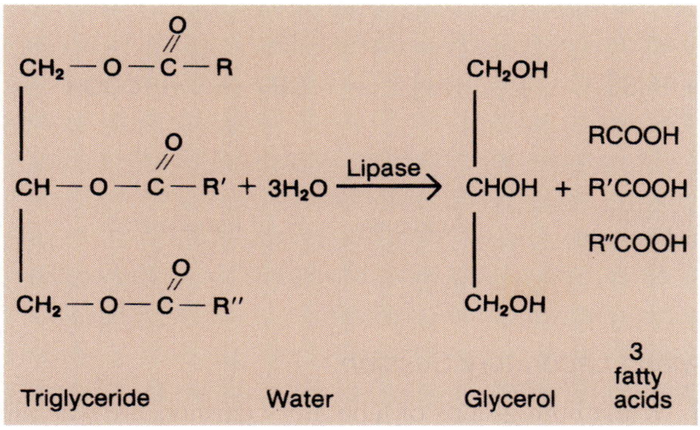

$$\begin{array}{l} CH_2-O-\overset{O}{\overset{//}{C}}-R \\ | \\ CH-O-\overset{O}{\overset{//}{C}}-R' + 3H_2O \xrightarrow{\text{Lipase}} \\ | \\ CH_2-O-\overset{O}{\overset{//}{C}}-R'' \end{array} \quad \begin{array}{l} CH_2OH \\ | \\ CHOH \\ | \\ CH_2OH \end{array} + \begin{array}{l} RCOOH \\ R'COOH \\ R''COOH \end{array}$$

Triglyceride Water Glycerol 3 fatty acids

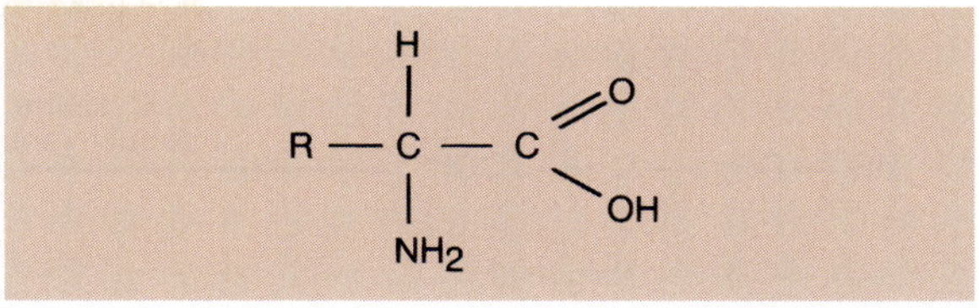

One virulence factor of ***Staphylococcus aureus,*** the pathogenic bacterium most often responsible for pimples, boils, and carbuncles, is lipase production. How does staphylococcal lipase help the bacterium invade tissue?*

In addition to infecting human tissue, *Staphylococcus aureus* is the most common bacterial cause of food poisoning. What fatty foods, frequently stored at room temperature, might harbor this facultatively anaerobic, non-endospore-producing, lipase-producing species?

Catalase

Catalase is an enzyme that breaks hydrogen peroxide into water and oxygen:

$$\begin{array}{ccccc} & \textbf{catalase} & & & \\ 2H_2O_2 & \rightarrow & 2H_2O & + & O_2\uparrow \\ \begin{pmatrix}\text{hydrogen}\\ \text{peroxide}\end{pmatrix} & \rightarrow & (\text{water}) & + & \begin{pmatrix}\text{oxygen}\\ \text{gas}\end{pmatrix} \end{array}$$

Hydrogen peroxide is toxic to cells. But it is also a common by-product of metabolic reactions that take place in the presence of water and oxygen. Therefore most organisms that are able to survive in an atmosphere containing oxygen produce enzymes to degrade the peroxide. ***Streptococcus, Enterococcus,*** and ***Lactobacillus*** are the most frequently encountered exceptions to this rule. These genera do not produce catalase. Nor is the enzyme produced by anaerobic bacteria.

In this exercise you practice several methods of evaluating microbial catalase production.

Cytochrome Oxidase

You also examine bacteria for the production of **cytochrome oxidase.** By an odd twist of chemical terminology, an **oxidase** is an enzyme that reduces (adds electrons to) oxygen; thus, it is an **oxygen reductase.** Cytochrome oxidase is the final link in the **electron transfer system** that provides **adenosine triphosphate** (**ATP,** an energy-rich molecule) during aerobic respiration. This enzyme receives electrons and passes them to oxygen, forming water.

Many bacteria that live in the presence of oxygen produce cytochrome oxidase, but some do not. In this exercise you perform a biochemical test to differentiate these microbial groups. Among the gram-negative rods, for example, ***Pseudomonas,*** and bacteria closely related to it, are oxidase producers. Members of the family **Enterobacteriaceae** are not.

Can you name the genus and specific epithet of several common inhabitants of the human intestine that are members of the Enterobacteriaceae? If not, check *Bergey's Manual* or your microbiology text.

Figure 21.3 An amino acid

$$\begin{array}{c} H \\ | \\ R - C - C \begin{array}{l} {}^{/\!/O} \\ {}_{\backslash OH} \end{array} \\ | \\ NH_2 \end{array}$$

Cysteine Desulfhydrase

To identify and differentiate bacteria, especially gram-negative, facultatively anaerobic rods, a microbiologist often determines whether the microorganisms produce enzymes that **degrade** (make less complex by splitting off groups) amino acids.

Twenty different amino acids occur naturally in proteins. Each amino acid molecule has a **carboxyl group** (—COOH) and an **amino group** (—NH_2) attached to its first carbon atom, as shown in figure 21.3. Also attached to the **alpha-carbon** (first carbon) is an **R group** (a radical, a side group of atoms). Each different amino acid has a unique R group.

Bacteria that produce the enzyme **cysteine desulfhydrase** are able to strip the amino acid **cysteine** of both its **sulfhydryl (—SH)** and amino groups. The reaction yields **hydrogen sulfide** (H_2S), **ammonia** (NH_3), and **pyruvic acid,** as shown in figure 21.4.

To discover whether a test culture produces the enzyme, a microbiologist checks for **hydrogen sulfide** (H_2S). Popularly known as rotten egg gas, the end product H_2S has a distinctly unpleasant and memorable odor. But identification of bacteria by the "sniff" method has not been standardized. Investigators utilize another chemical property of H_2S to determine the presence of this malodorous gas.

Hydrogen sulfide reacts with **heavy metals** such as lead or iron to form a visible, **black precipitate.** This reaction involves metal ions, not bacterial enzymes. But if

*Pathogenic *Staphylococcus aureus* hydrolyzes phospholipid in cell membranes and utilizes the metabolites for growth. This ability, along with other virulence factors, enables staphylococci to colonize skin and subcutaneous tissues, forming abscesses.

Figure 21.4 Cysteine desulfhydrase

$$HS{-}CH_2{-}\underset{NH_2}{\overset{H}{C}}{-}COOH \xrightarrow{\text{Cysteine desulfhydrase}} H_2S\uparrow + NH_3 + CH_3{-}\overset{O}{\overset{\|}{C}}{-}COOH$$

Cysteine → Hydrogen sulfide gas + Ammonia + Pyruvic acid

bacteria produce hydrogen sulfide in the presence of a heavy metal, then visible evidence—the black precipitate—accumulates.

The medium you utilize in this test, **Triple Sugar Iron Agar,** contains a plentiful supply of cysteine. The agar also contains charged particles of iron. When bacteria growing in Triple Sugar Iron Agar are able to release H_2S from amino acids, **ferrous sulfide** (FeS) blackens the medium.

PROCEDURES EXERCISE 21

Procedure 1

Amylase Testing

First Laboratory Session

a Groups of students work in pairs to perform this procedure. Each group obtains a **Starch Agar plate** and labels it with the group identification, the date, and the word "Starch." One group labels its plate *Bacillus subtilis;* the other group labels its plate *Escherichia coli.*

b Use aseptic technique and your inoculating loop to streak a short line of the appropriate bacteria near one edge of your plate, as in the illustration.

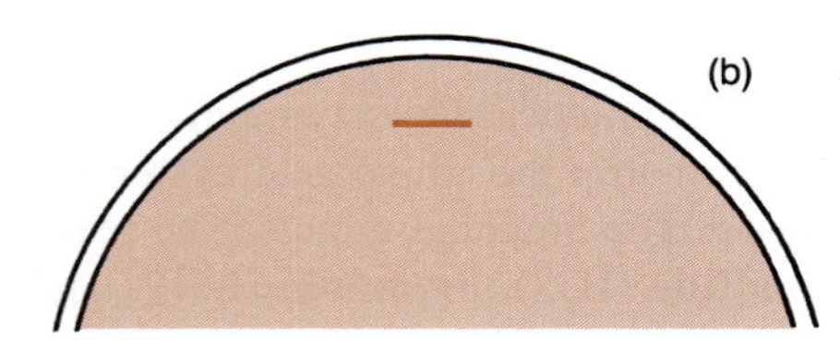

c Invert your plates and incubate them together at 35° C for at least 48 hours or until the next laboratory session.

d After examining the formula for Starch Agar that is given in the materials section of this exercise, explain in your laboratory report why **growth of bacteria** on a Starch Agar plate is **not sufficient evidence** to indicate starch hydrolysis.

Second Laboratory Session

a While both groups of laboratory partners are watching, flood the **Starch Agar plates** with **iodine.**

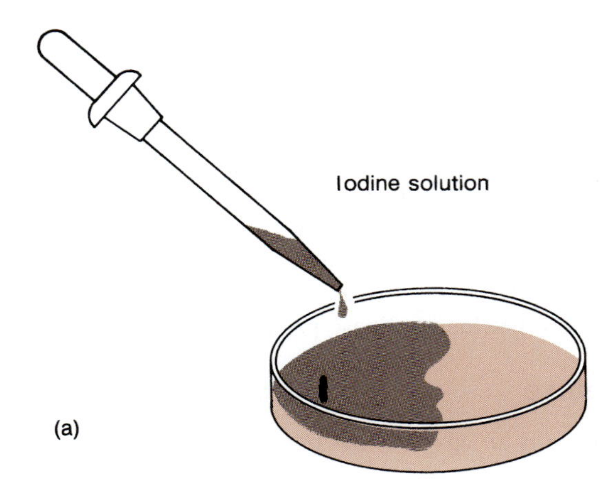

b In your laboratory report, note the length of time that elapses before the dark purple to brown color, indicating the presence of starch, develops. This color fades, so make your observations promptly.

Use caution to avoid dripping bacteria-laden iodine solution out of the plate while making your observations.

c In your report illustrate the lines of bacterial growth on your plates and any clear zones around the growth. The clear zones indicate areas where bacterial amylase has destroyed starch. Which species produced amylase?

d After all partners have recorded their observations, discard the Starch Agar plates in an appropriate manner.

Procedure 2

Lipase Testing

First Laboratory Session

Groups of students work in pairs to perform this procedure.

a Each group obtains an **Egg Yolk Agar plate** and labels it with the group identification, the date, and the abbreviation "EYA." One group labels its plate *Bacillus subtilis;* the other group labels its plate *Escherichia coli.*

b To spot inoculate your labeled, lipid-enriched medium, place a loopful of either *Bacillus subtilis* or *Escherichia coli* on an appropriately labeled spot near one edge.

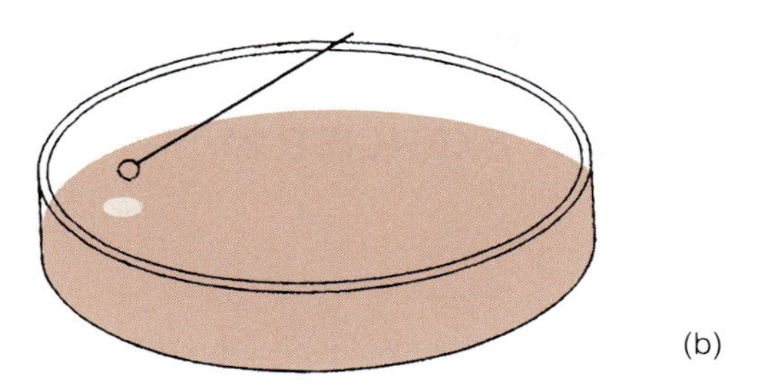

(b)

c Invert your Petri plate and incubate it at 35° C for at least 48 hours or until the next laboratory session.

Second Laboratory Session

a Carefully examine your plate of **Egg Yolk Agar** and that of your partner group. Is there an area of clearing around either spot inoculation? If you have any difficulty interpreting the results, place your plate on ice for 10 minutes or in the refrigerator for 30 minutes, then observe it again. Lipids become more opaque when cooled.

b In your laboratory report, illustrate and discuss any evidence of lipase production that you observe.

c After all partners have recorded their observations, the lipid-enriched medium can be discarded in an appropriate manner.

Procedure 3

Catalase and Cytochrome Oxidase Testing

First Laboratory Session

a Obtain a plate of **Tryptic Soy Agar.** Label the plate with your group identification, the date, and the words "Tryptic Soy Agar." Inoculate the plate with short streaks of *Enterococcus faecalis, Pseudomonas aeruginosa,* and *Escherichia coli,* as in the illustration.

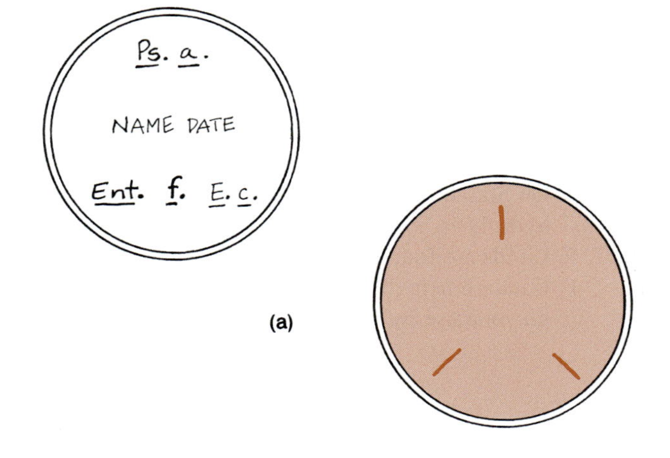

(a)

b Incubate the plate at 35° C no longer than 48 hours. *It is best to perform the catalase test and the cytochrome oxidase test on young bacteria.*

The next few steps outline a procedure for quickly testing the **catalase** reaction of bacteria in a broth culture. Unfortunately, the procedure contaminates the culture. Therefore delay this test until all who share your set of class broth cultures have completed their work with *Escherichia* and *Enterococcus faecalis.*

c Check with everyone who shares your set of cultures and, when all have completed their work with *Escherichia,* gather the group around the broth.

d One student adds a dropperful of **3% hydrogen peroxide** to the **broth culture** of *Escherichia,* then mixes the liquids gently. Watch closely. In your laboratory report, note the reaction or the lack of a reaction.

e Now 1 student from the group adds a dropperful of **3% hydrogen peroxide** to the *Enterococcus faecalis* **broth culture** and gently mixes the liquids. Again, watch closely. In your laboratory report, evaluate the reaction or the lack of a reaction that you observe.

Second Laboratory Session

Catalase Testing

a As in the illustration, place 2 small drops of **3% hydrogen peroxide** on opposite ends of a microscope slide.

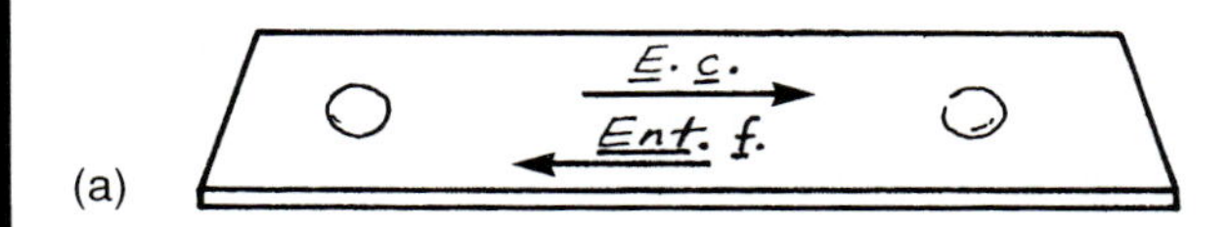

(a)

b Use sterile technique to transfer a loopful of the *Escherichia coli* **growth from the Tryptic Soy Agar plate** streak that you prepared during the last laboratory session into 1 of the drops.

c Stir the bacterial suspension with the loop.

d In your laboratory report, describe the catalase-positive reaction and identify the gaseous content of the bubbles that form.

e Flame and cool your loop; mix a loopful of *Enterococcus faecalis* from your **Tryptic Soy Agar plate** streak into the other drop of **hydrogen peroxide.**

f In your report describe the catalase-negative reaction. Without contaminating your fingers, discard the slide appropriately.

Cytochrome Oxidase Testing

There are a number of reliable methods available that test for **cytochrome oxidase** production. **Your instructor** may ask you to follow the procedure listed here or may demonstrate some other protocol.

g Place a small drop of purified water on a clean microscope slide.

h Alcohol-flame sterilize your forceps and use them to deliver an **oxidase disk** onto the drop of water.

i With a wooden applicator stick or toothpick, smear a little *Pseudomonas aeruginosa* from the streak on your **Tryptic Soy Agar plate** onto the damp disk. Dispose of the applicator in an appropriate manner.

j Repeat steps (*g*), (*h*), and (*i*), placing a bit of *Escherichia coli* from the streak on your Tryptic Soy Agar plate onto a second, damp oxidase disk.

k Observe the disks for up to 20 minutes. If the dab of **bacterial growth** on a disk turns pink to maroon and then almost black within 20 minutes, this color development indicates that the species produces cytochrome oxidase.

l In your laboratory report, note how long it takes for color to develop.

m After the group has completed its observation of catalase and cytochrome oxidase testing, discard your plate in an appropriate fashion.

n In the *DIFCO Manual,* 10th edition, or other reference work, find the name of at least 1 form of the reagent used for oxidase testing. Enter the name in your laboratory report.

Procedure 4

Cysteine Desulfhydrase Production

First Laboratory Session

a Gather an inoculating needle, 2 test tubes of **Triple Sugar Iron Agar,** and cultures of *Proteus vulgaris* and *Escherichia coli.*

b Label each tube with your group identification, the date, and the type of medium. Label 1 tube to receive *Proteus vulgaris,* the other to receive *Escherichia coli.*

c Flame sterilize your needle and part of its handle. Cool the needle; use aseptic technique to obtain an inoculum of *Proteus vulgaris* on the needle. Then, going straight in and coming straight back out again, stab the appropriately labeled Triple Sugar Iron Agar to the bottom of the tube. Avoid touching the inner walls of the tube with your needle handle.

d After resterilizing your needle and part of its handle, inoculate *Escherichia coli* into the other test tube of Triple Sugar Iron Agar.

e In your laboratory report, note the color of the Triple Sugar Iron Agar.

Incubate both Triple Sugar Iron Agar cultures at 35° C for at least 24 hours until the next laboratory session.

Second Laboratory Session

a Examine your **Triple Sugar Iron Agar** cultures. The formation of a black precipitate that is not present in the uninoculated control indicates that the bacteria produced H_2S.

b In your laboratory report, note cysteine desulfhydrase production abilities of each species. Illustrate the reaction, correctly labeling the visible precipitate.

c After all students in your group have completed this protocol, discard your Triple Sugar Iron Agar cultures appropriately.

POSTTEST EXERCISE 21

Part 1

Matching

Each response may be used once, more than once, or not at all. Match the enzyme with one of its substrates and/or the molecules it produces.

a. Amylases
b. Lipases
c. Catalase
d. Cytochrome oxidase
e. Cysteine desulfhydrase

1. ____ Hydrogen peroxide
2. ____ Glycerol and fatty acids
3. ____ Fats, oils, phospholipids, and other molecules soluble in nonpolar solvents
4. ____ Starch
5. ____ Hydrogen sulfide gas

Part 2

True or False (Circle one, then correct every false statement.)

1. T F Microbial physiology is the study of the size, shape, arrangement, and staining reactions of bacteria.
2. T F Along with other information, DNA carries the genetic code for all the enzymes that a microorganism produces.
3. T F Some exoenzymes break down molecules that are too large to cross a cell membrane.
4. T F Endoenzymes speed reactions outside of cells.
5. T F Amylases hydrolyze starch by adding the Na^+ and Cl^- ions of H_2O.
6. T F Starch is a large, many-branched polymer of the monosaccharide glucose.
7. T F Any bacteria that grow on a plate of Starch Agar are amylase producers.
8. T F To analyze the reactions of bacteria growing on Starch Agar, flood the plate with 3% hydrogen peroxide.
9. T F On Starch Agar a darkened area indicates starch hydrolysis.
10. T F Lipids are found in all cell membranes.
11. T F Catalase helps *Staphylococcus aureus* poison mayonnaise by attacking fat molecules.

12. T F Egg yolk contains the phospholipid lecithin (choline phosphoglyceride), an emulsifying agent. About two-thirds of the dry weight of egg yolk is lipid.
13. T F Hydrogen peroxide is toxic to cells, but it is also a common by-product of reactions that accompany aerobic respiration.
14. T F *Enterococcus, Streptococcus,* and *Lactobacillus* tolerate oxygen and produce catalase.
15. T F An oxidase is an enzyme that oxidizes compounds.
16. T F Cytochrome oxidase is produced by *Pseudomonas* but not by members of the Enterobacteriaceae.
17. T F All bacteria that utilize desulfhydrases produce hydrogen sulfide, rotten egg gas.
18. T F Cysteine desulfhydrase yields hydrogen sulfide, hydrochloric acid, and pyruvic acid.
19. T F The black precipitate that may form in a test tube containing Triple Sugar Iron Agar consists of H_2S.
20. T F Probes are pieces of nucleic acid that scientists sometimes use to identify microbes.

Part 3

Short Answer

1. Illustrate the hydrolysis of a triglyceride by a lipase.

2. Give the formula for the splitting of hydrogen peroxide by catalase.

3. Diagram the cysteine desulfhydrase reaction.

4. Tabulate the significant ingredients of Starch Agar, Egg Yolk Agar, and Triple Sugar Iron Agar.

Significant Ingredients

Medium	**Ingredients**
Starch Agar	
Egg Yolk Agar	
Triple Sugar Iron Agar	

Laboratory Report **21**

Name ______________________________

Section ______________

Selected Microbial Enzymes: Amylase, Lipase, Catalase, Cytochrome Oxidase, Cysteine Desulfhydrase

Procedure 1

Amylase Testing

First Laboratory Session

(*d*) List the ingredients in Starch Agar per 1,000 ml of purified water; see the materials section of this exercise.

Explain how it is possible that bacteria may grow heavily on Starch Agar but not necessarily produce amylase.

Second Laboratory Session

(*b*) How long after flooding your Starch Agar plate with iodine solution did a dark color appear, indicating the presence of starch? ______________________

(*c*) Illustrate the lines of bacterial growth and the dark and clear areas that surround them.

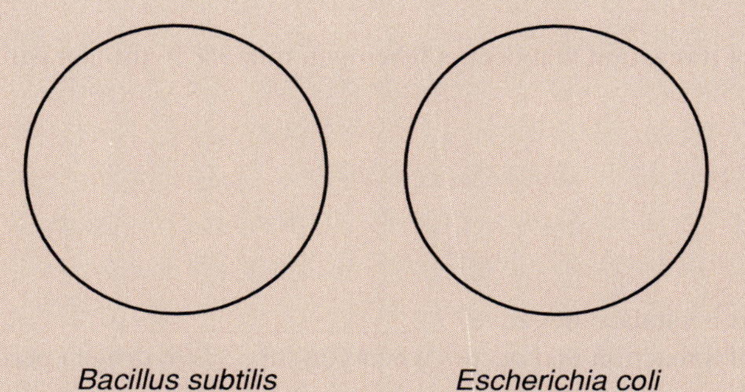

Starch Agar

Which species produced amylase? ______________________
Is it amylase-positive? ________

Procedure 2

Lipase Testing

Second Laboratory Session

(*a*) To interpret the results of your lipase production test, is it necessary to ice your Egg Yolk Agar? ________

(*b*) Illustrate the spots of bacterial growth and any area of clearing on your Egg Yolk Agar.

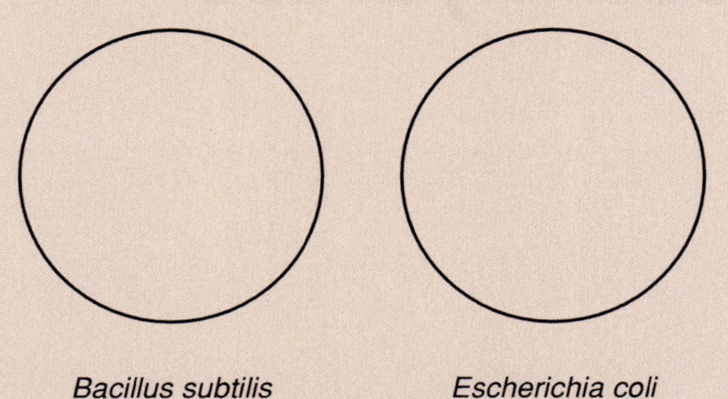

Bacillus subtilis *Escherichia coli*

Egg Yolk Agar

Discuss the evidence of lipase production. Is the reaction as easily interpreted as the amylase reaction?

Which species produced lipase? ____________________

Is it lipase-positive? _____

Procedure 3

Catalase and Cytochrome Oxidase Testing

First Laboratory Session

(*d*) Describe the reaction or lack of a reaction that occurs when you mix 3% hydrogen peroxide into a broth culture of *Escherichia coli.*

Is the culture catalase-positive or is it catalase-negative? ____________________

(*e*) Describe the reaction or lack of a reaction that occurs when you mix 3% hydrogen peroxide into the broth culture of *Enterococcus faecalis.*

Is the culture catalase-positive or is it catalase-negative? ____________________

Second Laboratory Session

(*d*) Describe the appearance of a positive catalase reaction when solid, visible bacterial growth is mixed into 3% hydrogen peroxide.

Genus ___________________________

What gas is contained in the bubbles? ______________________

If a classmate contaminates the hydrogen peroxide with *Escherichia coli,* what liquid remains in the bottle? _____

(*f*) Describe the appearance of a negative catalase reaction when solid, visible bacterial growth is mixed into 3% hydrogen peroxide.

Genus ___________________________

(*l*) How long did it take for your smear of *Pseudomonas aeruginosa* on the oxidase differentiation disk to complete the color change, indicating production of cytochrome oxidase? _______________________

After 20 minutes what color was the spot of *Escherichia coli* on its dampened oxidase differentiation disk? _____

(*n*) Name 1 form of the reagent used for oxidase testing. ____

Procedure 4

Cysteine Desulfhydrase Production

First Laboratory Session

(*e*) Describe the color of the unincubated tube of Triple Sugar Iron Agar. _________________________

Second Laboratory Session

(*b*) Describe the appearance of a positive test for bacterial cysteine desulfhydrase production in Triple Sugar Iron Agar.

Which species produced cysteine desulfhydrase? _____________________

Name the compound that forms the visible precipitate. ________________________

Exercise **22**

Enzymatic Attacks on Nitrogen-Containing Substrates: Urea Hydrolysis, Nitrate Reduction, and Proteolysis of Gelatin

Necessary Skills

Aseptic transfer techniques, Knowledge of the pH scale, Microbial enzymatic reactions, Microbial oxygen requirements

Materials

First Laboratory Session

Cultures (24-hour, 35° C, heavy growth on Tryptic Soy Agar slants, 1 of each per set):
- *Escherichia coli*
- *Proteus vulgaris*
- *Pseudomonas aeruginosa*
- *Staphylococcus epidermidis*

Supplies (per group):
- 3 Nitrate Broths
- 2 capped, 0.5 ml Urea Broths
- 2 Tryptic Soy Gelatin tubes (no more than 3 ml per tube of 6.0% gelatin in Tryptic Soy Broth)
- Access to the *DIFCO Manual*

Second Laboratory Session

Supplies (per class):
- Zinc powder
- Toothpicks
- Nitrite A reagent (0.8 g of sulfanilic acid/100 ml of 5N acetic acid) in dropper bottles
- Nitrite B reagent (0.6 g of dimethyl-alpha-naphthylamine/100 ml of 5N acetic acid) in dropper bottles
- Access to refrigeration or ice in a beaker

> **Warning:** Dimethyl-alpha-naphthylamine is closely related to carcinogenic compounds. If it comes in contact with your hands, wash them immediately.

Primary Objective

Understand, perform, and evaluate tests for urea hydrolysis, nitrate reduction, and proteolysis of gelatin.

Other Objectives

1. Explain the reactions catalyzed by urease, nitrate reductase, nitrite reductase, and gelatinase.
2. Define reduction.
3. Distinguish between *Proteus* and *Salmonella* or *Shigella* on the basis of pathogenicity and urease production.
4. Explain why nitrite may not be present in an appropriately incubated Nitrate Broth culture.
5. Explain the function of zinc in a test for nitrate reduction.

INTRODUCTION

In this exercise you examine the ability of bacteria to alter the nitrogen-containing molecules **urea, nitrate,** and **gelatin.** The enzymes that catalyze these reactions are unrelated to each other.

Microbiologists frequently test bacterial cultures for production of **urease, nitrate reductase,** and **gelatinase.** Evidence of these microbial enzymes helps to identify and differentiate bacteria, especially the gram-negative, intestinal rods.

Urea Hydrolysis

Urea is a common metabolic waste product that is toxic to most living organisms. The enzyme **urease** catalyzes the breakdown of urea into ammonia and carbon dioxide as shown here:

$$O{=}C\begin{matrix}NH_2\\NH_2\end{matrix} + H_2O \xrightarrow{\textbf{urease}} 2NH_3 + CO_2\uparrow$$

(urea) + (water) → (ammonia) + (carbon dioxide)

Testing for Urease Production

Clinicians, investigating the cause of severe diarrhea or dysentery, routinely check suspect fecal bacteria for urease production. Manufacture of the enzyme distinguishes ***Proteus*** from ***Salmonella*** or ***Shigella.*** All three genera are gram-negative, facultatively anaerobic, lactose nonfermenting rods. However, *Proteus,* a member of the normal intestinal flora of humans, produces the exoenzyme urease. *Salmonella* and *Shigella,* intestinal pathogens, do not.

In this exercise you perform a test for urease production. The test medium, **Urea Broth,** contains urea and the dye **phenol red,** a pH indicator. If the bacteria produce urease, there will be a positive test reaction, and the color of the medium will change from yellow to hot pink in response to the alkalinity of ammonia.

Nitrate Reduction

Many **facultatively anaerobic** (generally able to utilize an electron transport system to manufacture adenosine triphosphate with or without oxygen) bacteria produce **nitrate reductase.** This enzyme **reduces** (adds electrons to or removes oxygen atoms from) **nitrate (NO_3^-).** By utilizing nitrate reductase, facultative anaerobes can substitute NO_3^- for O_2 as the final electron acceptor during their cellular respiration.

$$\underset{\text{(nitrate)}}{NO_3^-} + 2e^- + 2H^+ \xrightarrow[\text{(anaerobic conditions)}]{\textbf{nitrate reductase}} \underset{\text{(nitrite)}}{NO_2^-} + \underset{\text{(water)}}{H_2O}$$

Testing for Nitrate Reduction

To perform the **nitrate reduction test,** cultivate a pure culture of bacteria in a **Nitrate Broth,** a medium that contains large amounts of **nitrate.** After sufficient incubation, test the broth for the presence of **nitrite.**

If you find nitrites, the bacteria have proven themselves positive for **nitrate reductase** production. But what does an absence of nitrite indicate? If there is no nitrite in the broth, there are at least three explanations for its absence.

1. The bacteria do not reduce nitrate.
2. The bacteria reduce nitrate to nitrite (NO_2^-). With **nitrite reductase,** they further reduce nitrite to **ammonia (NH_3).**
3. In a process called **denitrification,** bacteria such as ***Pseudomonas aeruginosa*** reduce nitrate to nitrite, and then they reduce the nitrite to **nitrogen gas** ($N_2\uparrow$).

As you can see from explanations 2 and 3, simply testing a Nitrate Broth culture for the presence of nitrite is not enough to reveal the presence of nitrate reductase activity. Bacterial enzymes may have further reduced the nitrite. Yet, the reagents commonly utilized in this procedure, **sulfanilic acid** and **dimethyl-alpha-naphthylamine,** only produce red pigment when they react with nitrite.

Figure 22.1 Bacterial reduction of nitrate

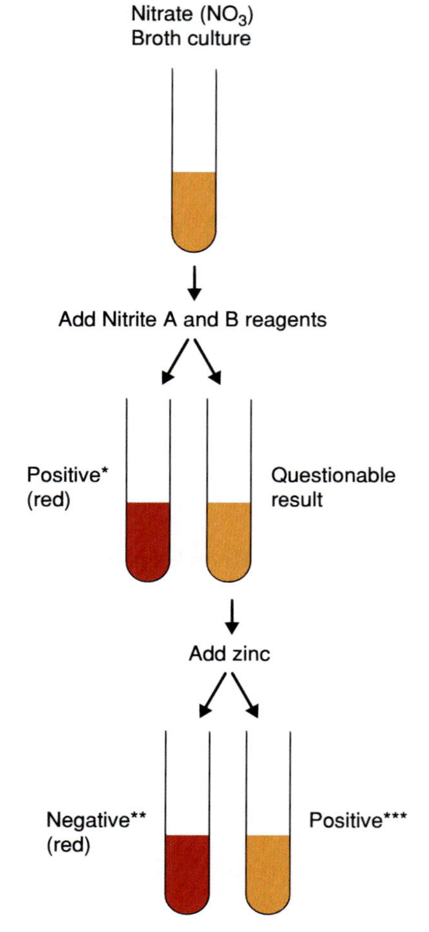

*Bacteria reduced NO_3 to NO_2 (nitrite). This culture is positive for NO_3 reduction.
**Bacteria did not reduce NO_3 to NO_2. This culture is negative for NO_3 reduction.
***Bacteria reduced NO_3 to NH_3 (ammonia) or N_2 (nitrogen gas). This culture is positive for NO_3 reduction.

As shown in figure 22.1, microbiologists utilize **zinc,** an inorganic catalyst that reduces nitrate, yielding nitrite, to confirm the test for nitrate reduction. Table 22.1 lists the steps in a test for bacterial reduction of nitrate.

Gelatin Hydrolysis

Gelatinase production is one of the many characteristics microbiologists use to identify species of bacteria. One reliable test for the enzyme involves **gelatin liquefaction** (formation of a liquid). Liquefaction occurs when bacterial gelatinase attacks and removes gelatin that was solidifying a medium.

When gelatin is dissolved in most liquids, it forms a colloidal suspension that is in the **gel** (semisolid) state below 25° C and the **sol** (liquid) condition above that temperature. Since heat liquefies gelatin, remember that immediately

Table 22.1 A Test for the Bacterial Reduction of Nitrate

I. Incubate a pure culture of bacteria in a **Nitrate Broth.**
II. Add reagents **Nitrite A** (sulfanilic acid) and **Nitrite B** (dimethyl-alpha-naphthylamine). These chemicals react with **nitrite,** producing a red color.
   A. Red: Culture positive for nitrate reduction.
   B. Not red: No nitrite present. **Continue testing.** Add **zinc.**
      1. Red: Culture negative for nitrate reduction by bacteria. The zinc reduced nitrate to nitrite.
      2. Not red: Culture positive for nitrate reduction by bacteria. No nitrate remained in the broth.

after incubation at 35° C, the sol state of a warm **Tryptic Soy Gelatin** culture can be confused with the results of bacterial liquefaction. Therefore always chill your incubated gelatin cultures before reporting gelatin liquefaction.

PROCEDURES EXERCISE 22

Procedure 1

Urea Hydrolysis

First Laboratory Session

a Obtain 2 tubes of **Urea Broth** and Tryptic Soy Agar slant cultures of *Escherichia coli* and *Proteus vulgaris.*

b Label the 2 broths with your group identification, the date, and the word "Urea." Label 1 tube for *Escherichia coli,* the other to receive *Proteus vulgaris.*

c Inoculate each broth with the appropriate bacteria. Incubate the cultures from 8 to 48 hours at 35° C.

Second Laboratory Session

a Examine your **Urea Broth** cultures. In your laboratory report, record the color you observe in each tube and the significance of each color.

Procedure 2

Nitrate and Nitrite Reduction

First Laboratory Session

Warning: Nitrite B (dimethyl-alpha-naphthylamine) is closely related to carcinogenic compounds. Avoid contact with this reagent. If it does contaminate your hands, wash them immediately.

a Note that **nitrate and nitrite reductions** are **anaerobic** reactions. Bacteria generally begin reduction in the bottom of the tube of Nitrate Broth after they deplete oxygen dissolved in the liquid. Avoid stirring or otherwise introducing oxygen into the broth.

b Label 3 **Nitrate Broths** with your group identification, the date, and the word "Nitrate." Label 1 for *Proteus vulgaris,* 1 to receive *Pseudomonas aeruginosa,* and the third for *Staphylococcus epidermidis.*

c With your flamed, sterilized, and cooled inoculating loop, transfer *Proteus vulgaris, Pseudomonas aeruginosa,* and *Staphylococcus epidermidis* each into the appropriately labeled Nitrate Broth.

d Incubate the cultures at 35° C for at least 48 hours.

Second Laboratory Session

a When all in your group are ready to test the cultures for nitrate and nitrite reductions, check each culture for **bacterial growth.** If there is no growth, record NG. Test cultures that show signs of growth. Begin the 2-step protocol outlined in table 22.1 and shown in figure 22.1.

b To the **Nitrate Broth** culture of *Proteus vulgaris,* add 10 to 15 drops of **Nitrite A reagent.** To the same broth, add an equal amount of **Nitrite B reagent.**

c In your laboratory report, record and interpret the color that appears in the culture within 15 minutes.

d If the culture does **not** become red within 15 minutes, use a toothpick to add a very small amount of **zinc powder** to the broth.

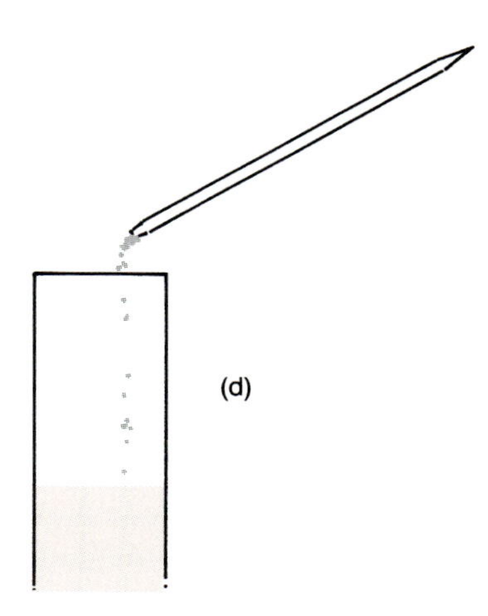

e In your laboratory report, record and interpret the color that appears in your Nitrate Broth culture within 15 minutes.

f Utilizing *Pseudomonas aeruginosa,* repeat steps (*b*) through (*e*).

g With *Staphylococcus epidermidis,* repeat steps (*b*) through (*e*).

Procedure 3

Proteolysis of Gelatin

First Laboratory Session

a Label 2 tubes containing **Tryptic Soy Gelatin** with your group identification, the date, and the abbreviation "TSG." Label 1 tube for *Escherichia coli,* the other to receive *Pseudomonas aeruginosa.*

b Inoculate each gelatin tube with the appropriate bacteria. Incubate the cultures at least 48 hours at 35° C.

Second Laboratory Session

a Place your gelatin culture tubes upright in a container of ice and water for at least 10 minutes or upright in a refrigerator for at least 30 minutes.

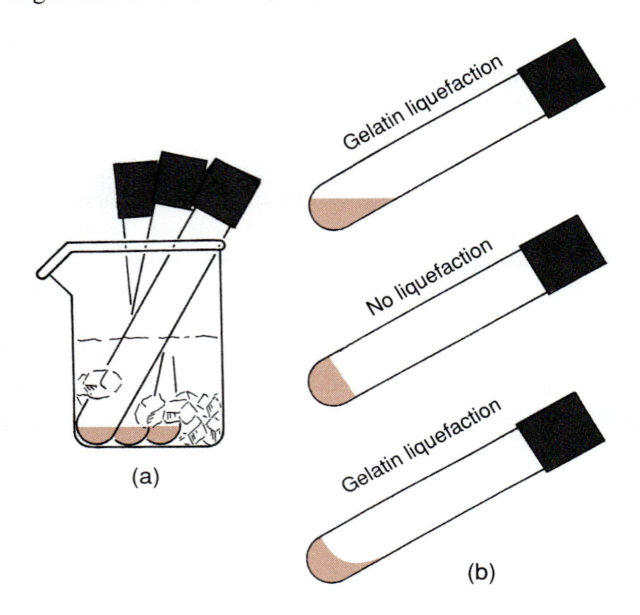

b Examine each tube for evidence of liquefaction, as in the illustration. Enter your results in your laboratory report.

POSTTEST EXERCISE 22

Part 1

Matching

Each response may be used once, more than once, or not at all. Some statements may require more than one response.

a. Nitrate reductase
b. Nitrite reductase
c. Urease
d. None of these

1. ____ $NO_3^- \rightarrow NO_2^-$
2. ____ Removes electrons from NO_2^-
3. ____ $NO_2^- \rightarrow NO_3^-$
4. ____ $CO(NH_2)_2 + H_2O \rightarrow 2NH_3 + CO_2\uparrow$
5. ____ Involved in denitrification

Part 2

True or False (Circle one, then correct every false statement.)

1. T F Urea is toxic to most cells, but urease producers can degrade this nitrogen-containing compound.
2. T F Urea is a protein.
3. T F Facultatively anaerobic bacteria generally do not produce nitrate reductase.
4. T F Urease catalyzes the reduction of NO_3^-; nitrate reductase catalyzes the hydrolysis of NO_2^-.
5. T F *Proteus* is urease positive; *Salmonella* and *Shigella* are urease negative.
6. T F *Proteus* is generally nonpathogenic in the human intestine, but it is a common cause of urinary tract infections.
7. T F If nitrite is missing from an incubated culture tube of Nitrate Broth, it may indicate that the bacteria did not produce nitrate reductase.
8. T F Denitrification is the complete oxidation of NO_3^- to form nitrogen gas.
9. T F In a Nitrate Broth culture, if zinc reduces nitrate to nitrite, then the bacteria in the tube did not.
10. T F Gelatin suspensions are in the gel state below 25° C.

Part 3

Short Answer

1. Illustrate or outline a complete test for the presence of nitrate reductase.

Laboratory Report **22**

Name ____________________

Section ____________

Enzymatic Attacks on Nitrogen-Containing Substrates: Urea Hydrolysis, Nitrate Reduction, and Proteolysis of Gelatin

Procedure 1

Urea Hydrolysis

Second Laboratory Session

(*a*) In the following chart, record your observations and interpretations of color development in the Urea Broth cultures.

Urea Broth Results

Bacteria	Phenyl Red pH Indicator Color	Significance, Urease Positive	Significance, Urease Negative
Proteus vulgaris			
Escherichia coli			

Procedure 2

Nitrate and Nitrite Reduction

Second Laboratory Session

(*a, c, e, f, g*) Enter your observations and their significance in the following chart. In the significance column, record NG for no growth or note which of the following results is (are) indicated:

1. Bacterial nitrate reduction
2. Bacterial nitrite reduction
3. No bacterial reduction
4. A questionable result
5. Other—explain

For each of the 3 species tabulated here, include at least 1 entry (1, 2, 3, 4, and/or 5 from the preceding list) concerning the significance of the color you observe. If you enter the number "5," explain your result.

Nitrate and Nitrite Reduction

Culture	Add Nitrite A and B		Add Zinc (if required)	
	Color	Significance	Color	Significance
Proteus vulgaris				
Pseudomonas aeruginosa				
Staphylococcus epidermidis				

Procedure 3

Proteolysis of Gelatin

Second Laboratory Session

(*b*) After cooling your Tryptic Soy Gelatin cultures, examine them and illustrate your findings.

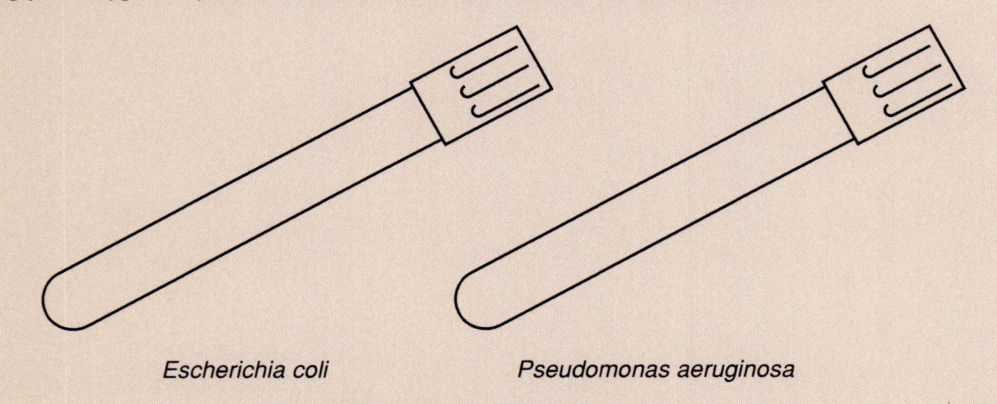

Escherichia coli *Pseudomonas aeruginosa*

Gelatin Liquefaction

Which species produce(s) gelatinase? ____

Exercise 23

Carbohydrate Fermentation

Necessary Skills

Aseptic transfer techniques, Knowledge of the pH scale, Microbial oxygen requirements

Materials

Cultures (24-hour, 35° C, Tryptic Soy Broth, 1 per set):
- *Escherichia coli*
- *Proteus vulgaris*
- *Enterococcus faecalis*

Supplies (per group):
- 3 Phenol Red Dextrose (Glucose) Broths with Durham tubes
- 3 Phenol Red Lactose Broths with Durham tubes
- 3 Phenol Red Saccharose (Sucrose) Broths with Durham tubes

Additional supplies:
- 1 Set of 3 uninoculated control broths per class. The set contains 1 of each broth: Phenol Red Dextrose, Lactose, and Saccharose.

Several *DIFCO Manuals* per class

Labeled bottle of dehydrated Phenol Red Carbohydrate Broth Base or any of the specific Phenol Red Broths

Primary Objective

Perform and interpret tests for microbial fermentation of carbohydrates.

Other Objectives

1. Define: catabolism, monosaccharide, disaccharide, polysaccharide, permease.
2. List the significant ingredients of a Phenol Red Carbohydrate Broth.
3. Give examples of monosaccharides, disaccharides, and polysaccharides.

INTRODUCTION

In this exercise you investigate the abilities of bacteria to **catabolize** (metabolize, releasing energy and simpler carbon compounds) carbohydrates. These compounds are common sources of nutrients and energy for plants and animals as well as for most microorganisms.

As the term **carbohydrate** implies, this type of molecule contains carbon plus the elements hydrogen and oxygen in the correct proportion to form water. Carbohydrates often have the formula $(CH_2O)_n$.

Generally, scientists further categorize carbohydrates as **monosaccharide, disaccharide,** or **polysaccharide**—having one, two, or many sugar units; see figure 23.1.

Enzymes That Attack Carbohydrates

To help identify an unknown bacterial species, microbiologists generally investigate which carbohydrates the microbes utilize and what end products the microbial metabolic reactions produce.

The dissimilation of a single carbohydrate often involves an assortment of enzymes. Enzymes that act outside of the cell that produces them **(exoenzymes)** are involved, as are enzymes that act inside the cell **(endoenzymes).**

Both exoenzymes and endoenzymes are often required because, while monosaccharides can enter bacterial cells by diffusion, larger units are generally degraded by exoenzymes or brought across the cell membrane by specific **permeases** (transfer enzymes). Once inside a cell, the sugar is catabolized only if the microorganism produces specific enzymes to degrade it.

Catabolism may be oxidative or fermentative, as is demonstrated in Exercise 16, Anaerobic Culture: Oxygen Requirements. During **fermentation,** organic compounds donate electrons, and other carbon-containing molecules are the final acceptors of electrons. Fermentations yield energy and molecular building blocks. Table 23.1 lists six common microbial fermentations.

Figure 23.1 Representative carbohydrates

Typical monosaccharides

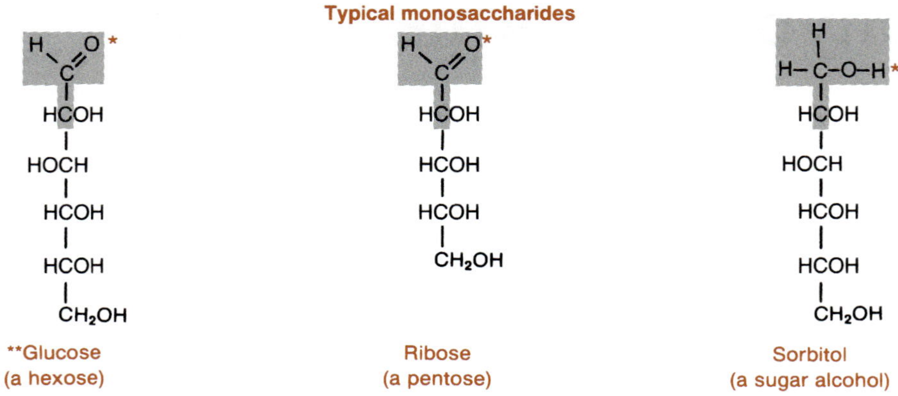

Monosaccharides:
Hexoses ($C_6H_{12}O_6$): Glucose (dextrose), fructose (levulose), galactose, mannose
Pentoses ($C_5H_{10}O_5$): Ribose, arabinose, xylose, ribulose
Sugar alcohols (*The carbonyl [C = O] group of monosaccharides is reduced to C — O — H): Sorbitol, glycerol, inositol, mannitol

A typical disaccharide, sucrose

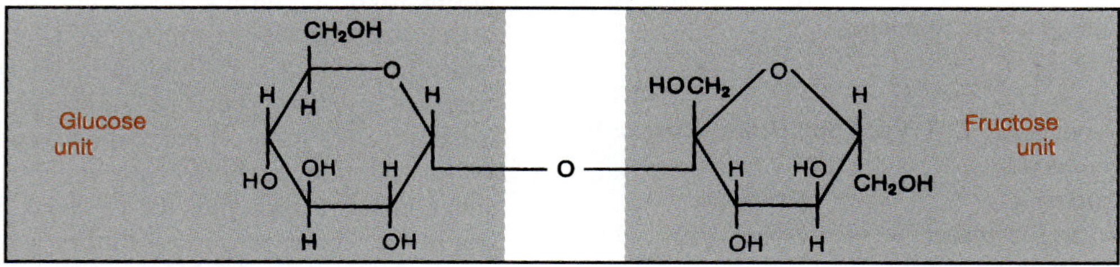

Disaccharides: Sucrose, maltose, lactose

Polysaccharides: Starch, glycogen, dextrans, agar

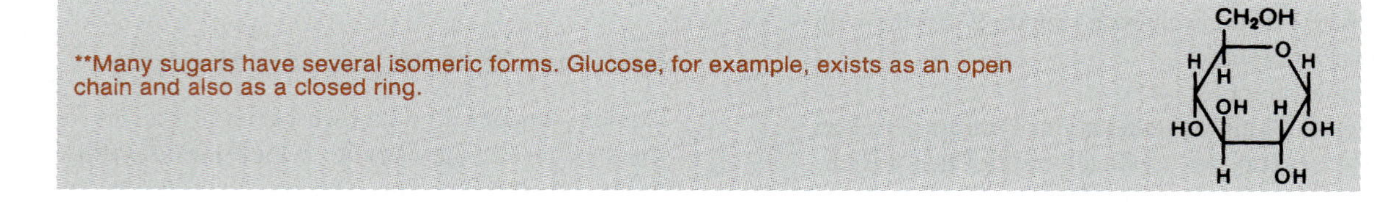

**Many sugars have several isomeric forms. Glucose, for example, exists as an open chain and also as a closed ring.

Table 23.1 Representative Microbial Fermentations of Pyruvic Acid

Microorganisms	Primary End Products
Clostridium	Acetone, butanol, butyric acid, CO_2, isopropanol
Enterobacter	Butanediol, CO_2, ethanol, formic acid, H_2, lactic acid
Escherichia	Acetic acid, CO_2, ethanol, H_2, lactic acid, succinic acid
Lactobacillus, Enterococcus, and *Streptococcus*	Lactic acid
Propionibacterium	Acetic acid, CO_2, H_2, propionic acid
Saccharomyces	CO_2, ethanol

Testing for Carbohydrate Fermentation

In this exercise you inoculate **Phenol Red Carbohydrate Broths** with bacteria. After the microbes have grown in the carbohydrate broths, you examine the tubes for evidence of fermentation. During fermentation, most bacteria produce acids and many produce gases. The ingredients of a Phenol Red Carbohydrate Broth enable you to observe evidence of the end products of fermentation.

A single type of carbohydrate is provided as the sole, fermentable nutrient in each broth. For instance, the only sugar available to fermenters growing in a Phenol Red Dextrose Broth is dextrose. Therefore, only bacteria capable of fermenting dextrose will produce acid in this broth.

Figure 23.2 Carbohydrate fermentation

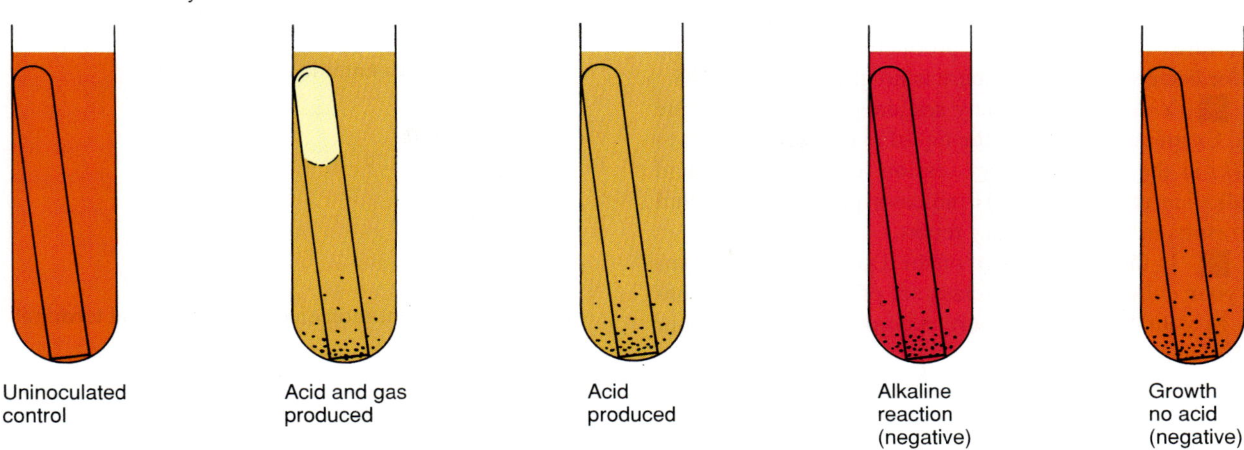

Nonfermentable nutrients are also available in the carbohydrate broth; peptone and beef extract are included to support growth.

To indicate whether acid-generating fermentation has occurred, each carbohydrate broth contains **phenol red,** a pH indicator. The tube of carbohydrate broth also contains a **Durham tube,** a small, inverted vial that collects gas; see figure 23.2.

The manufacturer of the medium suggests that when the tubes of carbohydrate broth are stored for longer than a few days, they should be placed in a boiling water bath, then cooled before use. If you are unsure about the reasoning behind this recommendation, check the *DIFCO Manual.*

PROCEDURES EXERCISE 23

Procedure 1

Carbohydrate Fermentation

First Laboratory Session

a Carbohydrate fermentation is an anaerobic reaction. Avoid stirring or otherwise introducing oxygen into the broth. Before inoculating the broths, check each Durham tube for gas. There should not be any empty space in the top of the vial.

b Label 3 **Phenol Red Dextrose Broths with Durham tubes,** 1 broth for each of your 3 bacterial cultures: *Escherichia coli, Proteus vulgaris,* and *Enterococcus faecalis.* Add your group identification, the date, and the word "Dextrose" to each label.

c Resuspend your stock culture broths. Using aseptic technique to avoid cross-contamination, transfer a loopful of each stock culture—*Escherichia coli, Proteus vulgaris,* and *Enterococcus faecalis*—into the appropriately labeled carbohydrate broth.

d Label 3 **Phenol Red Lactose Broths with Durham tubes,** 1 broth for each of your 3 bacterial cultures. Add your group identification, the date, and the word "Lactose" to each label.

e Repeat step *(c).*

f Label 3 **Phenol Red Saccharose Broths with Durham tubes.**

g Repeat step *(c).*

h Incubate your 9 fermentation broths at 35° C for 24 or 48 hours as directed by **your instructor.**

i **Your instructor,** or an assigned group, will prepare a class control set of uninoculated Phenol Red Carbohydrate Media. For the control set, gather 1 dextrose, 1 lactose, and 1 saccharose broth. Check the Durham tubes for gas. Label the tubes appropriately and incubate the set at 35° C for the duration of the experiment.

j The *DIFCO Manual* states that some investigators suggest including 0.5% of the substrate carbohydrate for 24-hour cultures, but 1.0% of the fermentable substance in broths that are to be incubated 48 hours. In your laboratory report, explain why it is a good idea to include a greater percentage of carbohydrate when increasing incubation duration.

k In the *DIFCO Manual,* discover the pH of an uninoculated Phenol Red Carbohydrate Broth. Also in the *DIFCO Manual,* investigate pH indicators to find the acid color and the alkali color of phenol red. Record your findings in your laboratory report.

l Some laboratories store refrigerated Phenol Red Carbohydrate Broths for months. During this time, gases at the air-broth interface may dissolve into the broth. Frequently, a small amount of gas is caught in the Durham tube of a sterile, uninoculated, but long-refrigerated, control tube incubated for 2 days at 35° C. In your laboratory report, explain this phenomenon.

Second Laboratory Session

a When all members of your group are ready to examine the cultures for acid and gas production, check each test tube for bacterial growth. If there is no growth, record NG.

b You do not add any reagents to a carbohydrate broth. Examine the color of the liquid in each tube. Refer to step *(k)* in the first laboratory session to discover the acid and alkaline colors of the pH indicator phenol red. Record your results in your laboratory report.

c Is there gas in any of the Durham tubes? Compare your cultures with the class set of uninoculated controls. In your laboratory report, record and interpret your results.

d Two hypothetical fermentation results are presented in your laboratory report; evaluate them.

POSTTEST EXERCISE 23

Part 1

True or False (Circle one, then correct every false statement.)

1. T F Catabolism is the energy-requiring, enzymatic breakdown of substrate molecules.
2. T F Carbohydrates are among the most common sources of nutrients for plants and animals but not for microorganisms.
3. T F A polysaccharide is composed of only one or two sugar units.
4. T F The dissimilation of polysaccharides is performed by endoenzymes exclusively.
5. T F The acid color of a Phenol Red Carbohydrate Broth is red, the alkali color is yellow.
6. T F After "dextrose-positive" bacteria exhaust the glucose in a Phenol Red Dextrose Broth, they may attack proteins in the nutrient supply, deaminating this new substrate and raising the pH.
7. T F All bacteria that grow in a Phenol Red Dextrose Broth ferment dextrose.
8. T F At the air-liquid interface, small amounts of oxygen dissolve into a Phenol Red Carbohydrate Broth.
9. T F Saccharose is the same as lactose.
10. T F An inverted vial, the Durham tube, is placed in each carbohydrate broth to collect acid.

Part 2

Completion and Short Answer

1. List seven fermentable carbohydrates.
   a.
   b.
   c.
   d.
   e.
   f.
   g.
2. During fermentation electron donors and terminal receivers are ________________ molecules; in respiration, electron donors are ________________ substances, and the final receivers are ________________ molecules.
3. List five acids, three neutral pH substances, and two gases produced by microorganisms during fermentation.
   acids:
   a.
   b.
   c.
   d.
   e.
   neutral pH substances:
   a.
   b.
   c.
   gases:
   a.
   b.
4. Boiling helps drive dissolved ________________ out of a culture medium.

Laboratory Report **23**

Name ______________________________

Section ______________

Carbohydrate Fermentation

Procedure 1

Carbohydrate Fermentation

First Laboratory Session

(*j*) List the contents of a Phenol Red Carbohydrate Broth.

With the following facts in mind, explain why a higher percentage of fermentable carbohydrate should be included in Phenol Red Carbohydrate Broths that are destined for longer incubation.

(i) Bacteria growing surrounded by a mixture of protein and carbohydrate substrates generally attack fermentable carbohydrates first (assuming that a sufficient complement of enzymes is available to the species).

(ii) Only after exhausting the saccharide supply do the microbes degrade proteinaceous material.

(iii) Protein dissimilation often releases amines. These compounds act as weak bases in water solutions.

(*k*) What is the pH of an uninoculated Phenol Red Carbohydrate Broth? (Check the *DIFCO Manual* or a bottle label for this information.) ______________________

What is the acid color of phenol red pH indicator? ______________________

What is the alkali color of phenol red? ______________________

(*l*) Explain why gas collects in the Durham tube of an **uninoculated** fermentation broth that was stored in a refrigerator for months, then incubated 48 hours at 35° C. If necessary, refer to a chemistry text to investigate the comparative solubility of gases in cold and warm liquid.

Second Laboratory Session

(*a, b, c*) In the following chart, record the changes you observe in your Phenol Red Carbohydrate Broth cultures. **Your instructor** may recommend a method of recording results. Two common methods of notation follow.

Methods of Notation

Condition	**Notation**	
	Either	**Or**
No growth	No growth	NG
Alkaline or no change	Negative	Alk
		NC
Acid	Positive or weak positive	A
Acid and gas	Positive with gas	AG

Carbohydrate Fermentations

Culture	**Substrate**		
	Dextrose	**Lactose**	**Saccharose**
Escherichia coli			
Proteus vulgaris			
Enterococcus faecalis			

(*d*) Explain the degradations of carbohydrates and proteins that are occurring in the following tubes of inoculated, incubated Phenol Red Dextrose Broth.

(i) The broth becomes turbid and yellow within 2 hours. Within 24 hours it is beginning to redden at the surface. After 48 hours the entire broth is orange.

(ii) The broth becomes turbid within 24 hours. After 48 hours it is dark red.

Exercise 24

The IMViC Tests: A Set of Biochemical Reactions

Necessary Skills

Aseptic transfer techniques, Knowledge of the pH scale, Fermentation, Enzymes

Materials

First Laboratory Session

Cultures (24-hour, 35° C Tryptic Soy Agar slants, 1 per set):
- *Escherichia coli*
- *Enterobacter aerogenes*

Supplies (per group):
- 2 Test tubes, each containing at least 5 ml of 1% tryptone, aq.
- 2 Test tubes, each containing at least 5 ml of MR-VP Medium
- 2 Simmons Citrate Agar slants

Supplies (per class), uninoculated controls:
- Tryptone, 1% aq.
- MR-VP Medium
- Simmons Citrate Agar slant

Second Laboratory Session

Supplies:

Reagents in dropper bottles:

> **Warning:** Kovacs' reagent and alpha-naphthol are potentially hazardous. They should be handled with care in a well-ventilated room.

- Kovacs' reagent
- Methyl red pH indicator
- Barritt's reagents
  - Barritt's solution A: alpha-naphthol, 5% in 95% ethanol
  - Barritt's solution B: potassium hydroxide, 40% aq.

Equipment (per group):
- 2 Clean, nonsterile, capped test tubes
- 2 Clean, nonsterile 1 or 5 ml pipets
- Pipettor

Class-incubated, uninoculated controls:
- Tryptone, 1% aq.
- MR-VP Medium
- Simmons Citrate Agar slant

Equipment (per class):
- Clean, nonsterile, capped test tube
- Clean, nonsterile 1 or 5 ml pipet
- Pipettor

Primary Objective

Perform and evaluate each of the IMViC tests.

Other Objectives

1. Correlate the medium employed, its significant ingredients, and the reagent added for each of the following tests: indole, methyl red, Voges-Proskauer.
2. Name the medium employed in the Simmons Citrate test. List the significant ingredients of the medium.
3. Judge the significance of the IMViC reactions when testing for fecal contamination of milk or water.
4. Define the term *coliform.*
5. Contrast the positive and negative test result for each of the IMViC tests.
6. Evaluate the importance of incubation duration for the indole, methyl red, and Voges-Proskauer tests.

Table 24.1 IMViC Reactions

Bacteria	Test			
	Indole	**Methyl Red**	**Voges-Proskauer**	**Simmons Citrate**
Escherichia coli	Positive	Positive	Negative	Negative
Enterobacter aerogenes	Negative	Negative	Positive	Positive

INTRODUCTION

In this exercise you perform and interpret the **IMViC** set of tests. The acronym "IMViC" is a mnemonic aid for remembering the order and names of four tests:

- **I** **I**ndole
- **M** **M**ethyl red
- **Vi** **V**oges-**P**roskauer (The "i" is for ease of pronunciation.)
- **C** **C**itrate, Simmons

Microbiologists have developed groups of biochemical tests that they use to differentiate morphologically and physiologically similar organisms. One such group is the IMViC set. It is primarily used to separate **coliforms.** Coliforms are defined as gram-negative, aerobic or facultatively anaerobic, non-endospore-forming, rod-shaped bacteria that produce acid and gas from lactose within 48 hours at 35° C.

Bergey's Manual of Determinative Bacteriology places the coliforms and their relatives in the family **Enterobacteriaceae,** commonly called the **"enterics."** Among the enterics are many species that normally reside in human and other animal intestines, including the common sewage indicator organism ***Escherichia coli.***

Using the IMViC tests, ***Escherichia coli*** is easily differentiated from other morphologically and physiologically similar coliforms, including ***Enterobacter aerogenes;*** see table 24.1.

This differentiation is important in the testing of water, milk products, and sewage effluent, where fecal contamination is monitored. While *Escherichia* is always associated with feces, *Enterobacter* is not. *Enterobacter* may be found in feces, but it is also common in soil and on decayed vegetation that is free of fecal contamination.

In this exercise you perform the IMViC tests to differentiate enterics. Beyond simply performing the IMViC tests (see figure 24.1), you should understand the basic principles of how each test works.

Indole

Coliforms can be differentiated by their ability or inability to produce the enzyme **tryptophanase.** This enzyme hydrolizes the amino acid **tryptophan** into indole, pyruvic acid, and ammonia as shown in figure 24.2. When evaluating the IMViC set of tests, you look for the production of **indole** by bacteria.

The medium employed in the indole test is **1% tryptone in water.** Tryptone, rich in tryptophan, is a pancreatic digest of the milk protein casein.*

Indole itself can be further degraded by some bacteria. Therefore, to avoid a false-negative reaction, the tryptone culture must be tested for the presence of indole after no more than 4 or 5 days of incubation.

To discover whether indole has been produced in a tryptone culture, add **Kovacs' reagent,** an acid solution of para-dimethylaminobenzaldehyde in alcohol. In the positive reaction, Kovacs' reagent combines with indole, forming a red layer that floats on the surface of the medium. In a negative response, Kovacs' reagent does not become red.

Methyl Red and Voges-Proskauer

Both the methyl red and Voges-Proskauer tests are performed on a culture grown in **MR-VP Medium.** The significant ingredient of the MR-VP Medium is **dextrose** (glucose).

Most coliforms growing in an MR-VP Medium initially convert some dextrose to acids. Then certain genera go on producing more acid in a **"mixed acid fermentation."** Other genera utilize the dextrose but yield products with a more neutral pH such as those end products that are formed in the **"butanediol fermentation."**

Microbiologists look for evidence of these different reactions to help identify the coliforms. Typical enteric responses are shown in figure 24.3.

As you can see from figure 24.3, timing is important for interpretation of the methyl red and Voges-Proskauer tests. Two to five days of incubation are recommended for the methyl red test; the Voges-Proskauer test is also inaccurate before at least 48 hours of incubation.

How does a microbiologist perform two different tests on one culture grown in a single MR-VP Medium? After incubation, part of the culture is pipetted into a clean test tube and reserved for the Voges-Proskauer test.

Methyl red pH indicator is added to the remaining culture. This indicator turns red at pH 4.4 and below. It becomes yellow at pH 6.2 and above. In a positive methyl red test, the reagent remains red, indicating that large amounts of acid were produced by bacteria growing in the MR-VP

*To prepare a **pancreatic digest** of casein, a microbiologist adds pancreas enzymes to casein. The enzymes hydrolyze (digest) the milk protein, breaking it down into amino acids.

Figure 24.1 How to perform the IMViC tests

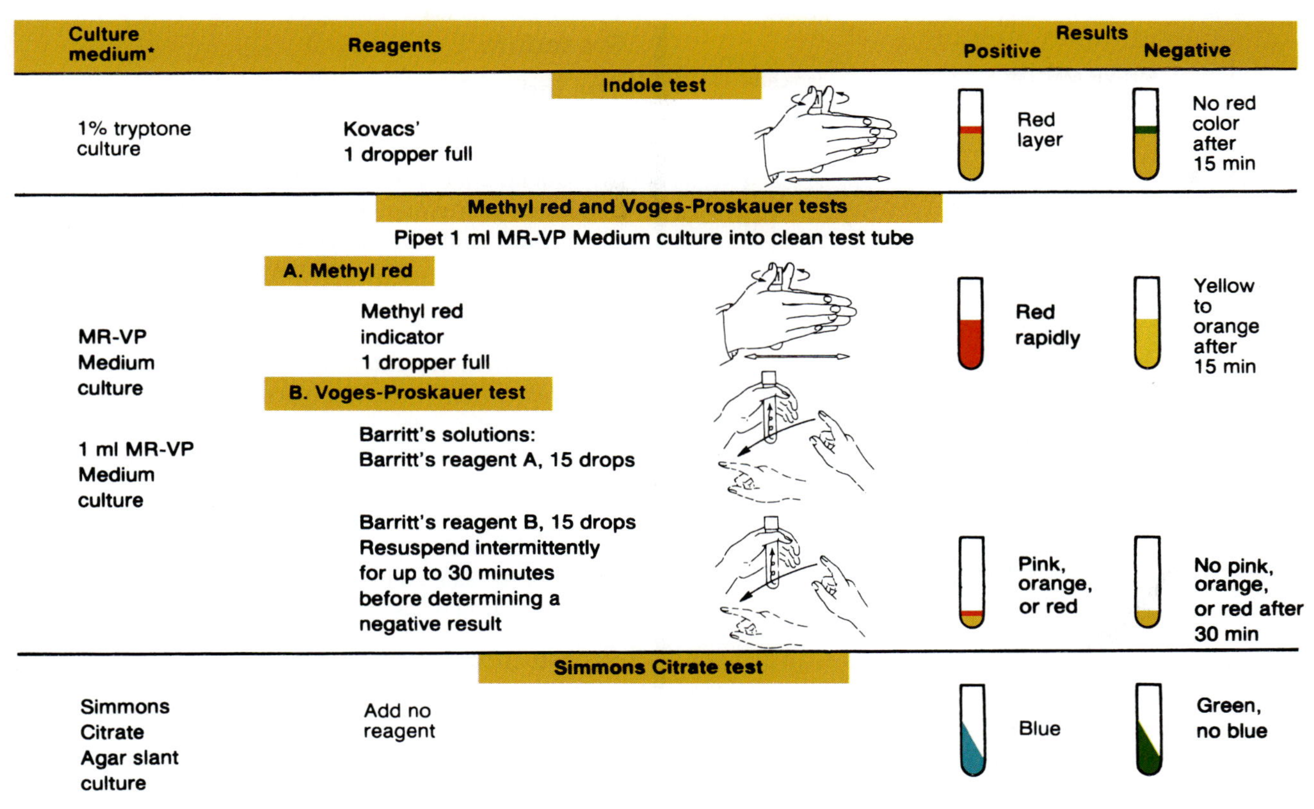

Figure 24.2 Indole production

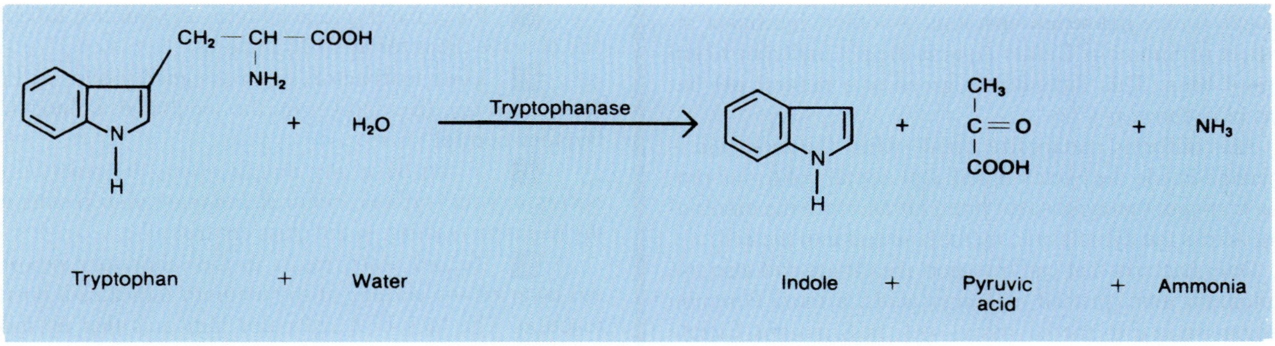

$$\text{Tryptophan} + H_2O \xrightarrow{\text{Tryptophanase}} \text{Indole} + CH_3COCOOH + NH_3$$

Tryptophan + Water → Indole + Pyruvic acid + Ammonia

Medium. The negative reaction is orange or yellow, indicating less acid production.

The **Voges-Proskauer** test is performed on the part of the culture that was pipetted into a clean test tube. This test checks for the presence of **acetyl methyl carbinol** (acetoin), a precursor of 2,3–butanediol.

In this exercise you check to see whether bacteria in the MR-VP Medium are capable of a butanediol fermentation. To do this, you add **Barritt's reagent A** (alpha-naphthol) and **Barritt's reagent B** (potassium hydroxide) to the incubated culture.

Figure 24.3 pH of the mixed acid and butanediol fermentations

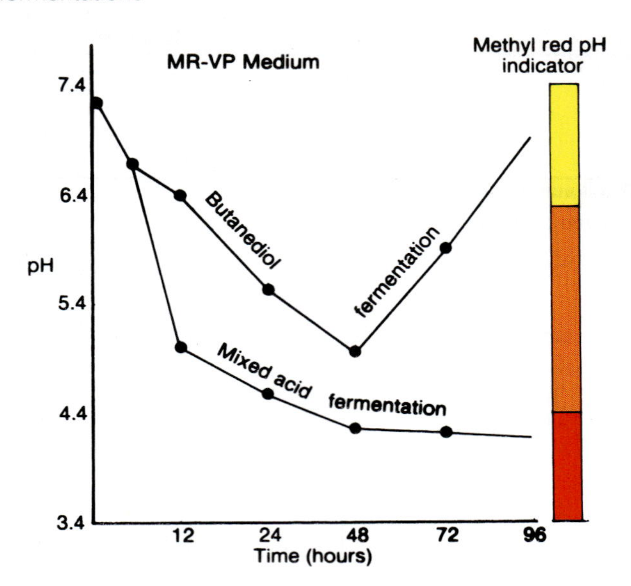

Barritt's reagents react with acetyl methyl carbinol and oxygen to form a red-colored positive test. This reaction may take up to 30 minutes, and it only colors the air-broth interface of the medium unless the broth culture is repeatedly shaken to supply sufficient oxygen. In a negative test, the bacteria have not produced acetyl methyl carbinol, so no red color appears.

Citrate

Simmons Citrate Agar is a synthetic medium. It contains mineral salts, the only carbon source is **citrate,** and the sole nitrogen source is **ammonium ions.** Another significant ingredient of Simmons Citrate Agar is the pH indicator **brom thymol blue.** This indicator is green at a neutral pH but turns blue above pH 7.6.

All coliforms metabolize citrate when the molecule is generated inside the bacterial cell. But not all coliforms produce **transport enzymes** that bring citrate from the environment across the cytoplasmic membrane and into the cell.

Only bacteria that can transport the citrate and also use ammonium ions (instead of amino acids, a more common nitrogen source in media) grow well in Simmons Citrate Agar. Can you think of a reason why amino acids are omitted from a medium that seeks to supply citrate as the sole source of carbon?

The microbiologist does not add reagents to a Simmons Citrate Agar to read the reaction. In a positive citrate test, heavy growth of coliforms generally turns the brom thymol blue indicator in the medium to a deep Prussian blue. This occurs because growing bacteria produce alkaline products from the ammonium salts in the medium. In a negative reaction, there is little or no growth, and the color of the agar remains unchanged after 48 hours of incubation.

PROCEDURES EXERCISE 24

Procedure 1

The Indole Test

First Laboratory Session

a Obtain 2 tubes of the medium **1% tryptone.** Label the tubes with your group identification, the date, and the word "Tryptone." Add *"Escherichia coli"* to the label of 1 tube, *"Enterobacter aerogenes"* to the label of the other.

b Obtain class cultures of *Escherichia coli* and *Enterobacter aerogenes.* Using aseptic technique, inoculate the appropriately labeled tryptone medium with *Escherichia coli.* Sterilize the inoculating loop. Inoculate the other tube of tryptone medium with *Enterobacter aerogenes.* Incubate the 2 tryptone cultures at 35° C for 48 to 96 hours.

c **Your instructor,** or an assigned group, will incubate a labeled, uninoculated test tube of 1% tryptone at 35° C for the same duration as the class cultures. This tube will serve the class as an incubated, uninoculated control.

Second Laboratory Session

> **Warning:** Kovacs' reagent and alpha-naphthol are potentially hazardous. They should be handled with care in a well-ventilated room.

a Add a dropperful of **Kovacs' reagent** to each **tryptone culture.**

b Roll each tube between your palms to mix the reagent through the culture.

c Place the tryptone cultures in your test tube rack and observe them for up to 15 minutes.

d **Your instructor,** or an assigned group, will perform steps (*a*), (*b*), and (*c*) on the incubated, uninoculated tryptone control.

e If you have any doubts about the reaction that occurs in your tryptone cultures, compare your results with the negative reaction in the class control tube.

f Record your results in the laboratory report. A red layer on the surface of the medium indicates a positive reaction. The lack of a red color indicates that no indole was produced. Discard your tubes into the appropriate receptacle.

Procedure 2

The Methyl Red and Voges-Proskauer Tests

First Laboratory Session

a Obtain 2 **MR-VP Medium** broths and label each with your group identification, the name of the medium, the date, and the name of the bacterium—*Escherichia coli* on 1 tube, *Enterobacter aerogenes* on the other.

b Using aseptic technique, inoculate the appropriately labeled MR-VP Medium with *Escherichia coli*. Sterilize the inoculating loop. Inoculate the other broth with *Enterobacter aerogenes*.

c Incubate the 2 MR-VP Medium broth cultures for at least 48 hours at 35° C.

d **Your instructor,** or an assigned group, will incubate a labeled, uninoculated test tube of MR-VP Medium at 35° C for the same duration as the class cultures. This tube will serve the class as an incubated, uninoculated control.

Second Laboratory Session

a Obtain 2 clean test tubes. Label 1 "VP *Escherichia coli*," the other "VP *Enterobacter aerogenes*."

b Using a pipettor and a pipet, transfer 1 ml from the MR-VP Medium culture of *Escherichia coli* into the appropriately labeled test tube. With a new, clean pipet, transfer 1 ml from the MR-VP Medium culture of *Enterobacter aerogenes* into the other empty test tube.

c **Your instructor,** or an assigned group, will perform steps (*a*) and (*b*) on the incubated, uninoculated MR-VP Medium control.

Methyl Red Testing

After removing 1 ml from each tube, perform the **methyl red test** on the 2 **original MR-VP Medium cultures.**

d Add 1 dropperful of the pH indicator **methyl red** to your *Escherichia coli* **MR-VP Medium culture.**

e Roll the tube between your palms to mix the contents.

f Observe your methyl red test culture. In a positive test, a red color appears, indicating substantial acid production; negative tests are yellow to orange.

g Repeat steps (*d*), (*e*), and (*f*) with your MR-VP Medium culture of *Enterobacter aerogenes*.

h **Your instructor,** or an assigned group, will perform steps (*d*), (*e*), and (*f*) on the incubated, uninoculated MR-VP Medium control and label the tube "Methyl Red Test Negative Control."

i If you have any doubts about the reaction that occurs in your methyl red test, compare your results with the negative reaction in the class control tube.

j Enter the results of your methyl red tests in your laboratory report. Discard your MR-VP Medium cultures into the appropriate receptacle.

Voges-Proskauer Testing

Perform the **Voges-Proskauer test** on the 2 **"VP"** test tubes **containing 1 ml each of the MR-VP Medium cultures.**

k Add 15 drops of **Barritt's reagent A** to the "VP" tube of *Escherichia coli*. Cap the tube and vortex it.

l Add 15 drops of modified **Barritt's reagent B** to the "VP" tube of *Escherichia coli*. Cap the tube and vortex it.

m Vortex the tube intermittently (to provide oxygen for the reaction) and observe the broth for up to 30 minutes. Any reddening indicates a positive reaction.

n Repeat steps (*k*), (*l*), and (*m*) using the "VP" test tube containing *Enterobacter aerogenes*.

o **Your instructor,** or an assigned group, will perform steps (*k*), (*l*), and (*m*) on 1 ml of broth from the incubated, uninoculated MR-VP Medium control and label the tube "Voges-Proskauer Test Negative Control."

p If you have any doubts about the reaction that occurs in your Voges-Proskauer test, compare your results with the negative reaction in the class control tube.

q Record the results of your Voges-Proskauer tests in the laboratory report. Discard your "VP" tubes into the appropriate receptacle.

Procedure 3

The Citrate Test

First Laboratory Session

a Obtain 2 **Simmons Citrate Agar** slants and label each with your group identification, the name of the medium, the date, and the name of the bacterium—*Escherichia coli* on 1 tube, *Enterobacter aerogenes* on the other.

b Using aseptic technique and an inoculating needle, inoculate the appropriately labeled Simmons Citrate Agar slant with *Escherichia coli* by stabbing into the medium, then zigzagging the needle back and forth across the surface of the slant. Sterilize the needle.

c Inoculate your other tube of Simmons Citrate Agar with *Enterobacter aerogenes,* again performing a "stab and streak" inoculation.

d Incubate the 2 Simmons Citrate Agar slant cultures for 48 hours at 35° C.

e **Your instructor,** or an assigned group, will label and incubate a tube containing Simmons Citrate Agar for the class set of uninoculated controls.

Second Laboratory Session

a Examine your 2 **Simmons Citrate Agar slant cultures,** comparing them with the class uninoculated control. **Bacterial growth** on the slant, usually accompanied by a change in the **color** of the medium from green to blue, indicates a positive test result.

In a negative test, there may appear to be slight growth where bacteria were transferred onto the Simmons Citrate Agar, but no blue color develops.

b Record the results of your Simmons Citrate Agar tests in the laboratory report. Discard your citrate cultures into the appropriate receptacle.

POSTTEST EXERCISE 24

Part 1

Fill in the following charts.

IMViC Chart A

Test	Medium	Significant Ingredients
Indole		
Methyl red		
Voges-Proskauer		
Simmons Citrate		

IMViC Chart B

Test	Reagents (Names and Ingredients)	Positive Reaction	Negative Reaction
Indole			
Methyl red			
Voges-Proskauer			
Simmons Citrate			

Part 2

True or False (Circle one, then correct every false statement.)

1. T F Coliforms are defined as gram-negative, aerobic or facultatively anaerobic, non-endospore-forming, coccus-shaped bacteria that produce acid and gas from lactose within 48 hours at 35° C.
2. T F *Bergey's Manual of Determinative Bacteriology* places the coliforms in the family Enterobacteriaceae.
3. T F *Escherichia coli* is a common sewage indicator organism.
4. T F *Enterobacter aerogenes* is a common sewage indicator organism.
5. T F Tryptophanase hydrolizes tryptophan into pyruvic acid and citric acid.
6. T F To perform the indole test, a tryptone broth may be used since this medium is rich in tryptophan.
7. T F Para-dimethylaminobenzaldehyde is the methyl red reagent.
8. T F Another name for dextrose is glucose.
9. T F If a coliform culture performs the butanediol fermentation in an MR-VP Medium, the results of the properly performed Voges-Proskauer test will be positive.
10. T F In a positive Simmons Citrate Agar test, bacteria lower the pH of the medium.

Part 3

Completion

1. Bacteria in the family Enterobacteriaceae are commonly called ______________________________ .
2. Three fluids that are routinely tested for fecal contamination are __________ , __________ , and __________ __________ .
3. Coliforms that lower the pH of an MR-VP medium below 4.5 have generally performed a __________ __________ fermentation.
4. The Voges-Proskauer test monitors the production of __________ __________ __________ by an MR-VP Medium culture.
5. Coliforms that yield a positive test reaction on a Simmons Citrate Agar must be able to transport __________ across their cell membranes and also be able to utilize __________ __________ as their sole source of nitrogen.

Laboratory Report **24**

Name ______________________________

Section ______________

The IMViC Tests: A Set of Biochemical Reactions

Procedures 1, 2, 3

The IMViC Tests

Fill in the following chart with reactions that you observe.

IMViC Reactions

	Indole	Methyl Red	Voges-Proskauer	Simmons Citrate
Escherichia coli				
Enterobacter aerogenes				

If your results differ from those mentioned in table 24.1, list factors that may be causing the incongruities.

Exercise 25

Selective and Differential Media

Necessary Skills

Aseptic transfer techniques, Streak-plate techniques, Knowledge of culture characteristics, Controlling microbial growth

Recommended Reading

Antibiotic Disk Sensitivity Testing, Chemical controls

Materials

First Laboratory Session

Cultures (48-hour, Tryptic Soy Broth, 1 per set):
- *Staphylococcus epidermidis* (35° C)
- *Escherichia coli* (35° C)
- *Enterobacter aerogenes* (35° C)
- *Pseudomonas fluorescens* (30° C)

Supplies (per group):
- 2 Columbia CNA Agar plates
- 4 Levine EMB Agar plates
- 2 Pseudomonas Agar F plates
- 2 Tryptic Soy Agar plates
- Access to the *DIFCO Manual*

Class demonstration:
- 3 Blood Agar plates, each streaked for isolation of colonies with 1 of the following species, then incubated 35° C, 24 hours. Seal the plates before class.
- *Streptococcus pyogenes* or other beta-hemolytic bacteria
- *Streptococcus pneumoniae* or other alpha-hemolytic bacteria
- *Staphylococcus epidermidis* or other nonhemolytic bacteria
- Access to a Quebec colony counter (optional)

Second Laboratory Session

Supplies:

Access to uninoculated control plates:
- Columbia CNA Agar
- Levine EMB Agar
- Pseudomonas Agar F
- Tryptic Soy Agar

Access to a darkened room is helpful.
1 box, or other eye shield, per ultraviolet lamp
Thin gloves near ultraviolet lamps (optional)
Shielded ultraviolet lamp

Primary Objective

Distinguish between bacteria by utilizing all-purpose, selective, and differential media.

Other Objectives

1. Define: all-purpose medium, selective medium, differential medium.
2. Demonstrate the advantages of a medium that is both selective and differential.
3. Explain why there is no universal medium.
4. Name five all-purpose media.
5. List six compounds utilized to inhibit the growth of unwanted organisms in selective media.
6. Demonstrate and discuss the salient characteristics of each of these media: Columbia CNA Agar, Levine EMB Agar, and Pseudomonas Agar.
7. Summarize adequate procedure for obtaining a pure culture from the growth of an isolated colony on a selective medium.
8. Distinguish between the appearance of selected gram-negative bacterial colonies growing on Levine EMB Agar.
9. Observe and describe alpha, beta, and "gamma" hemolysis on Blood Agar.
10. Examine fluorescent pigment.

INTRODUCTION

In this exercise you cultivate bacteria on an **all-purpose medium,** one designed to grow a broad range of genera. You examine colonies on **selective media.** These agars are representative of those that chemically inhibit some genera but not others. You also inoculate and then evaluate **differential media,** plates that allow the investigator to distinguish easily between the growth characteristics of various species. Note that selective, differential, and all-purpose media can be either agars or broths.

One medium may fall into several classification groupings. To save time and materials, microbiologists employ many media that are both selective and differential. A few nutrient solutions including Blood Agar, a medium commonly employed in clinical microbiology, are classified both as all-purpose and as differential. In this exercise you examine demonstration cultures grown on Blood Agar.

All-Purpose Media

From your work in the section of this laboratory manual entitled Influencing and Controlling Microbial Growth, you have probably already concluded that no medium is **"universal,"** able to support the growth of all microorganisms. Any set of nutrients inhibits some microbes while giving a selective advantage to others. In addition, every set of **conditions of incubation** (temperature, duration, humidity, pressure, etc.) favors some genera while preventing the multiplication of others.

But some nutrient solutions are classified as **all-purpose media.** These include Nutrient Agar, Tryptic Soy Agar, Blood Agar, Brain Heart Infusion Agar, Sabouraud Dextrose Agar (for fungi), and many others. Each all-purpose medium supports the growth of a wide variety of normal body flora, pathogens, and soil microbiota.

Selective Media

When concocting nutrient solutions for the cultivation of microbes, scientists often include antimicrobial chemicals. The concentration of these molecules is just high enough to inhibit growth of unwanted microbiota while allowing sought-after organisms to multiply. Dyes, antibiotics, sodium chloride, dextrose, quaternary ammonium compounds, sodium azide, and other inhibitors are found in various **selective media.**

In this exercise you inoculate both gram-positive and gram-negative bacteria onto **Columbia CNA Agar.** This blood-red, selective medium is designed to isolate gram-positive cocci including **staphylococci** and **streptococci** from specimens that may also contain gram-negative bacteria.

Cultivation of samples on Columbia CNA Agar assures the clinician that gram-positive bacteria that may be the etiologic agents of disease will not be overgrown by gram-negative bacteria and subsequently overlooked.

Some bacteria **hemolyze** (lyse erythrocytes) Columbia CNA Agar, and others do not. For a description of bacterial hemolysis, see the following section, Differential Media.

An isolated colony on a selective medium is never considered a **pure culture,** the descendants of one cell. The colistin and nalidixic acid in Columbia CNA Agar, for example, only slow the growth of gram-negatives; they do not kill all the suppressed bacteria. Therefore any colony may cover or touch viable, but invisible, contaminating organisms. The ability to retard, rather than stop or kill, undesirable growth is characteristic of selective media.

To ensure purity, the deft microbiologist who needs a pure culture streaks a colony from a selective medium onto an all-purpose medium. The wily investigator then utilizes well-isolated growth from the all-purpose medium for all further inoculations.

Differential Media

Differential media simplify the task of distinguishing between bacteria. Some differential media alter the appearance of colonies. Those colonies producing specific enzymes take on a different color than those that lack these enzymes. With other differential media, the medium itself changes color or opacity in response to microbial enzymatic reactions.

Correct interpretation of the reactions displayed on differential media saves the investigator both time and effort when identifying microorganisms.

Some media are both **differential and all-purpose. Blood Agar,** a rich nutrient solution plus 5% to 10% animal blood, is an all-purpose medium. It supports the growth of a wide variety of microbes. Since the hemolytic reactions of **streptococci, staphylococci,** and other bacteria growing on Blood Agar are fairly predictable and repeatable, this medium is also valued for its differential capabilities.

When lysis of erythrocytes in the medium is complete **(beta hemolysis),** a clear area surrounds the colony on Blood Agar. Colonies secreting enzymes that only partially destroy red blood cells **(alpha hemolysis)** are surrounded by a drab, greenish zone. If bacterial enzymes do not attack the red blood cells, there is no hemolysis. Microbiologists sometimes describe nonhemolytic colonies as **gamma-hemolytic.** What color is Blood Agar around a gamma-hemolytic colony? See figure 25.1.

Levine EMB Agar, which you utilize in this exercise, is a **selective and differential medium.** Dyes in the agar inhibit the growth of most gram-positive bacteria. Astute microbiologists also employ Levine EMB Agar to help differentiate gram-negative rods. On it, ***Enterobacter aerogenes*** colonies are a pink-to-buff color around darkened

Figure 25.1 Hemolytic reactions on Blood Agar

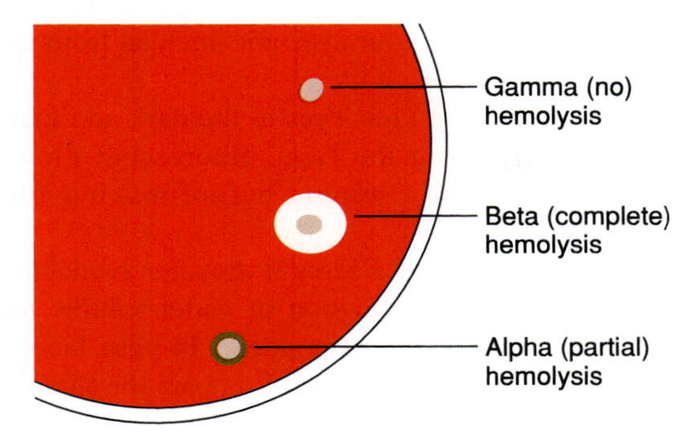

centers; ***Escherichia coli*** colonies develop a green metallic sheen; and ***Pseudomonas aeruginosa*** colonies are colorless, as are those of the intestinal pathogens ***Salmonella*** and ***Shigella.***

Colonies of all these species are nearly colorless on Tryptic Soy Agar. How does growth on Levine EMB Agar manage to differentiate them so well? The Levine EMB Agar contains the disaccharide lactose and the dyes eosin and methylene blue. Many gram-negative, nonpathogenic, intestinal rods produce enzymes that ferment lactose. This fermentation yields acid that reacts with the dyes to color fermenting colonies. Colonies of nonfermenters remain colorless.

Look up **MacConkey Agar** in the *DIFCO Manual.* Compare and contrast MacConkey Agar and Levine EMB Agar.

Another **differential medium** you inoculate in this exercise is **Pseudomonas Agar F.** This medium enhances bacterial generation of **fluorescein,** a fluorescent pigment that glows yellow-green. The medium also inhibits formation of **pyocyanin,** a blue-green, nonfluorescent pigment. Pseudomonas Agar F is generally used in conjunction with another medium, **Pseudomonas Agar P,** which has the opposite effects on bacterial enzymes that control production of these two pigments.

The two media help differentiate species of ***Pseudomonas,*** a genus known for its manufacture of various pigments. The pathogen ***Pseudomonas aeruginosa,*** for example, is commonly known as the "blue-green pus" bacterium.

Fluorescein, made by many species of the genus *Pseudomonas,* is a fluorescent pigment, hidden from view when examined with visible light. If you cannot see the color, how do you evaluate whether bacteria growing on Pseudomonas Agar F have produced it?

To perform this seemingly impossible feat, you employ ultraviolet radiation. The radiation excites electrons in the fluorescent pigment. Excited electrons glow and you observe their pale light.

When exposed to ultraviolet radiation, fluorescein in the medium, the colonies, or both, glows yellow. What characteristic of the dye molecules determines whether you find the pigment in colonies or in the medium? Check Exercise 11, Ubiquity and Culture Characteristics, if you need help answering this question.

PROCEDURES EXERCISE 25

Procedure 1

Tryptic Soy Agar, an All-Purpose Medium; and Columbia CNA Agar, a Selective and Differential Medium

First Laboratory Session

a Label 2 plates of **Tryptic Soy Agar** with your group identification, the date, and the name of the medium. Label 1 plate to receive *Escherichia coli,* the other to receive *Staphylococcus epidermidis.*

b With the appropriate culture, streak each plate for isolation of colonies.

c Label 2 plates of **Columbia CNA Agar** with your group identification, the date, and the name of the medium. Label 1 plate to receive *Escherichia coli,* the other to receive *Staphylococcus epidermidis.*

d With the appropriate culture, streak each plate for isolation of colonies.

e Incubate the 4 inverted plates at 35° C until the next laboratory session.

Second Laboratory Session

a Examine your 2 **Tryptic Soy Agar** cultures. In your laboratory report, describe well-isolated colonies of *Escherichia coli* and *Staphylococcus epidermidis* growing on Tryptic Soy Agar.

b Examine your 2 **Columbia CNA Agar** cultures. Compare them with the uninoculated control plates. In your laboratory report, describe the amount of growth that has developed on each plate of Columbia CNA Agar.

c In your laboratory report, evaluate your results. Do your findings support the claim that Tryptic Soy Agar is an all-purpose medium, and Columbia CNA Agar is a selective medium?

Procedure 2

Levine EMB Agar, a Selective and Differential Medium; and Pseudomonas Agar F, a Differential Medium

First Laboratory Session

a Label 4 plates of **Levine EMB Agar** with your group identification, the date, and the name of the medium. Label 1 plate to receive *Escherichia coli,* the second for *Staphylococcus epidermidis,* the third for *Enterobacter aerogenes,* and the fourth for *Pseudomonas fluorescens.* Mark the *Pseudomonas fluorescens* plate "30° C."

b With the appropriate culture, streak each plate for isolation of colonies.

c Label 2 plates of **Pseudomonas Agar F** with your group identification, the date, and the name of the medium. Label 1 plate to receive a single-line inoculation of *Escherichia coli,* the other to receive a single line of *Pseudomonas fluorescens.* Mark the *Pseudomonas fluorescens* plate "30° C."

d With the appropriate culture, draw a single, straight line across the center of each plate of Pseudomonas Agar F. **Do not** streak for isolation of colonies.

e Incubate the 6 inverted plates until the next laboratory session. All grow best at 35° C except *Pseudomonas fluorescens,* which flourishes at 30° C.

Second Laboratory Session

a Examine your Petri dishes of **Levine EMB Agar.** Compare them with the uninoculated control plates. In your laboratory report, note the amount of growth that has developed on each plate.

b If possible, record your description of the appearance of **well-isolated colonies** of *Escherichia coli, Staphylococcus epidermidis, Enterobacter aerogenes,* and *Pseudomonas fluorescens* on Levine EMB Agar.

c In your laboratory report, discuss whether your results support the claim that Levine EMB Agar is both selective and differential.

d Many ultraviolet lamps require a warm-up period of at least 15 minutes. **Your instructor** will inform you if your lamp must be turned on ahead of time.

e Examine your Petri dishes of **Pseudomonas Agar F** using visible, laboratory lighting. Compare them with the uninoculated control plate. In your laboratory report, describe evidence that you observe of water-soluble and non-water-soluble pigment production by either of the species.

> **Warning:** Ultraviolet radiation damages unprotected eyes. **Do not look** directly into the rays of the lamp.

> **Warning:** Overexposure of unprotected skin to ultraviolet radiation causes burns. **Do not expose** your ungloved hands to the rays of the lamp for more than a few seconds.

f Place your uncovered plates of Pseudomonas Agar F under an ultraviolet lamp. It is helpful if the lamp is in a dark room. Be sure your eyes are shielded from the radiation.

g To accustom your eyes to the dark and increase your chances of seeing the faint, fluorescent glow, try closing your eyes for a minute before making further observations.

h Use gloves and glasses if they are available. Examine both plates for production of water-soluble and/or non-water-soluble fluorescent pigment. In your laboratory report, describe your observations. Compare the fluorescent pigment-producing capabilities of *Pseudomonas fluorescens* with those of *Escherichia coli.*

Procedure 3

Demonstrating Hemolytic Reactions on Blood Agar

a Examine the 3 sealed **Blood Agar cultures.** Hemolysis is easier to see if you hold the plate of Blood Agar up to a light and let the light come through the medium. Employ a Quebec colony counter if one is available. Examine the area immediately surrounding an isolated, typical colony. Hemolysis is generally visible as a thin "halo" around a hemolytic colony.

b In your laboratory report, describe alpha-, beta-, and gamma-hemolytic reactions. Identify which species produced each type of hemolysis.

POSTTEST EXERCISE 25

Part 1

Matching

Each response may be used once, more than once, or not at all. Some statements may require more than one response.

a. Selective medium
b. Differential medium
c. All-purpose medium
d. None of these

1. _____ Columbia CNA Agar
2. _____ Blood Agar
3. _____ Levine EMB Agar
4. _____ Pseudomonas Agar F
5. _____ Tryptic Soy Agar

Part 2

True or False (Circle one, then correct every false statement.)

1. T F An all-purpose medium is one designed to allow the microbiologist to distinguish easily between growth characteristics of microbes.
2. T F There is no universal medium.
3. T F Gram-positive cocci, not gram-negative rods, are likely to grow well on Columbia CNA Agar.
4. T F An isolated colony on a selective medium is **not** considered a pure culture.
5. T F Microbiologists utilize dyes, antibiotics, sodium chloride, dextrose, quaternary ammonium compounds, sodium azide, and other inhibitors in all-purpose media.
6. T F Beta hemolysis is complete destruction of erythrocytes, alpha hemolysis is partial destruction, and the term *gamma hemolysis* signifies no lysis at all.
7. T F Fluorescein is a blue-green, nonfluorescent pigment.
8. T F *Pseudomonas aeruginosa* is commonly known as the "fluorescein pus" organism.
9. T F Ultraviolet radiation excites electrons in fluorescent pigment.

Part 3

Completion and Short Answer

1. A(n) ________________________ medium is designed to grow a broad range of genera.
2. A(n) ________________________ medium chemically inhibits some genera but not others.
3. List four conditions of incubation.
   a.
   b.
   c.
   d.
4. Name four all-purpose media.
   a.
   b.
   c.
   d.
5. Explain how to isolate a pure culture from growth on a selective medium.

6. The colonies on some differential media are distinct from each other. On other differential media, ________________ changes in response to the reactions of various bacterial enzymes.
7. Devise a test that will tell you whether *Staphylococcus epidermidis* is killed or simply inhibited by Levine EMB Agar.

Laboratory Report **25**

Name ______________________________

Section ______________

Selective and Differential Media

Procedure 1

Tryptic Soy Agar, an All-Purpose Medium; and Columbia CNA Agar, a Selective and Differential Medium

Second Laboratory Session

(*a*) Describe the amount of growth that has developed on each plate of **Tryptic Soy Agar.** Describe a **typical, well-isolated colony** of each species growing on Tryptic Soy Agar. Include colony size, shape, margin, elevation, surface characteristics, opacity, and pigmentation, including color and water solubility of any pigment you observe.

Escherichia coli:

Staphylococcus epidermidis:

(*b*) Describe the amount of growth that has developed on each plate of **Columbia CNA Agar.** If possible, describe a **typical, well-isolated colony** of each species growing on Columbia CNA Agar.

Escherichia coli:

Staphylococcus epidermidis:

(*c*) Do your findings support the claim that Tryptic Soy Agar is an all-purpose medium, and Columbia CNA Agar is a selective medium? ________ Explain your answer.

Procedure 2

Levine EMB Agar, a Selective and Differential Medium; and Pseudomonas Agar F, a Differential Medium

Second Laboratory Session

(*a, b*) Describe the amount of growth that has developed on each plate of Levine EMB Agar. If possible, describe a **typical, well-isolated colony** of each species growing on Levine EMB Agar.

Escherichia coli:

Enterobacter aerogenes:

Pseudomonas fluorescens:

Staphylococcus epidermidis:

(*c*) Do your results support the claim that Levine EMB Agar is both selective and differential? ________ If so, how?

(*e*) Did either species produce pigment that you can observe under **visible light?** ________ Describe color and water solubility of any bacterial pigment that you observe under visible light on Pseudomonas Agar F.

Escherichia coli:

Pseudomonas fluorescens:

(*h*) Did either species produce pigment that you can observe under **ultraviolet radiation?** ________ Describe color and water solubility of any bacterial pigment that you observe under ultraviolet radiation on Pseudomonas Agar F.

Escherichia coli:

Pseudomonas fluorescens:

Procedure 3

Demonstrating Hemolytic Reactions on Blood Agar

(*b*) In the following chart, describe alpha-, beta-, and gamma- (non-) hemolytic reactions. Identify which species produces each type of hemolysis.

Bacterial Hemolysis

Species	Type of Hemolysis	Description of Hemolysis
1.		
2.		
3.		

Exercise 26

Multiple-Test Media and Systems

Necessary Skills

Aseptic transfer techniques, Gram staining, Streak-plate techniques, Knowledge of microbial metabolic reactions

Materials

Warning: Handle your unknown with care; it may be a pathogen.

First Laboratory Session

Pure cultures (1 per group, 24-hour, 35° C, Tryptic Soy Agar plates streaked for isolation of colonies):
- Unknown Enterobacteriaceae (The dishes are numbered. The **instructor** receives a key to the unknowns.)

Supplies (per group):
- Triple Sugar Iron Agar slant
- API 20E® test strip*
- Serological tube containing 5 ml of sterile, 0.85% (w/v), saline
- Wash bottle containing tap water
- API 20E® incubation tray with cover
- Test tube containing 5 ml of sterile mineral oil
- Sterile Pasteur pipet
- Bulb
- Access to the *DIFCO Manual*

Additional supplies (optional):
- 1 Sterile oxidase differentiation disk per unknown
- 1 Sterile wooden applicator stick for oxidase test per unknown
- Forceps
- Alcohol for flaming forceps
- 2 Microscope slides per unknown
- Gram stain reagents

Second Laboratory Session

Supplies (per class, reagents in dropper bottles):
- Ferric chloride, 10% aq.
- Kovacs' reagent
- Barritt's reagents A and B
- Nitrate reduction reagents, Nitrite A and Nitrite B
- Zinc powder
- Toothpicks
- Hydrogen peroxide, 1.5% aq.
- API 20E® Analytical Profile Index (optional)

Primary Objective

Inoculate, incubate, and interpret the reactions in a multitest medium and a miniaturized multitest system.

Other Objectives

1. Explain why correct identification of microbial species is crucial for research and medical microbiology.
2. List the significant ingredients of Triple Sugar Iron Agar.
3. Illustrate, label, and analyze all reactions that may occur in Triple Sugar Iron Agar.
4. Inoculate and evaluate the reactions in a Triple Sugar Iron Agar.
5. List three multitest media (in addition to Triple Sugar Iron Agar) and contrast the biochemical tests performed in each.
6. Inoculate, incubate, and appraise the reactions in an API 20E® strip.
7. Utilize an API 20E® strip, with either a differential chart or a profile index number and an API 20E® *Index,* to identify an unknown culture of bacteria.
8. Define: triplates, "quads."
9. Evaluate the necessity of performing a Gram stain and an oxidase test in conjunction with API 20E® testing.

*Available from bioMériex Vitek, Inc., 595 Anglum Dr., Hazelwood, MO 63042.

Table 26.1 Ingredients (per liter) of Triple Sugar Iron Agar

Beef extract	3.0 g
Yeast extract	3.0 g
Peptone	15.0 g
Proteose peptone	5.0 g
Dextrose	1.0 g
Lactose	10.0 g
Sucrose	10.0 g
Ferrous sulfate	0.2 g
Sodium chloride	5.0 g
Sodium thiosulfate	0.3 g
Agar	12.0 g
Phenol red	0.024 g
dH_2O, q.s.	1,000.0 ml

From the *DIFCO Manual,* 10th ed. DIFCO Laboratories, Inc., Detroit, MI. 1984, p. 1019.

Table 26.2 Representative Multitest Media

Medium	Intended Use	Biochemical and Physiological Reactions
SIM Medium	*Salmonella* and *Shigella*	Hydrogen sulfide production, indole formation, motility
Kligler Iron Agar	Gram-negative, enteric bacilli	Fermentation of glucose and lactose, carbon dioxide and hydrogen sulfide production
Litmus Milk	Many bacteria	Fermentation of lactose, carbon dioxide gas production, degradation of casein, coagulation, reduction

INTRODUCTION

As you have discovered, investigators utilize the morphological, biochemical, nucleic acid, and immunological characteristics of microorganisms to identify species. Correct identification of microbiota is crucial for repeatability in scientific endeavors and for initiation of appropriate treatment and control in the clinical setting.

In earlier exercises, you have inoculated, incubated, and interpreted reactions on media that allow you to differentiate bacteria. But, generally, each test tube or Petri plate yielded information concerning only a single enzyme or a closely related group of enzymes. In this exercise you examine media, including **Triple Sugar Iron Agar,** that test for a variety of enzymes. You also survey systems, including the **API 20E®,** that are miniaturized sets of compartmentalized biochemical tests.

Multitest Media

Triple Sugar Iron Agar is a multitest medium that contains **0.1%** *dextrose* and **1.0%** each of *sucrose* and *lactose*. It also contains *ferrous sulfate* and *phenol red* pH indicator in a nutrient solution rich in protein; see table 26.1. From your previous work, recall which biochemical reactions cause microbes growing in media such as Triple Sugar Iron Agar to produce acid. Which enzymatic pathways might yield alkaline products?

Triple Sugar Iron Agar is designed to help differentiate gram-negative bacteria. The intestinal pathogens ***Salmonella*** and ***Shigella*** ferment dextrose but not sucrose or lactose. The most common gram-negative, nonpathogenic fecal rods do not share this fermentation pattern; see figure 26.1. Based on your knowledge of biochemical testing procedures, why is ferrous sulfate included in Triple Sugar Iron Agar? For help with this question, see the *DIFCO Manual* or Exercise 21, Selected Microbial Enzymes.

A tube of Triple Sugar Iron Agar is inoculated with a stab to the bottom of the butt; this puncture is followed by a zigzag streak covering the surface of the slant. After a few hours of incubation, bacteria that ferment dextrose (glucose) drop the pH of the entire medium below 6.8; see figure 26.2. In response to the presence of acid, phenol red colors all of the agar yellow. The Triple Sugar Iron Agar butt, with comparatively few bacteria present, may remain yellow for days.

But, since there is very little glucose present, only 0.1%, the teeming bacteria on the slant soon exhaust the supply. Deprived of dextrose, they attack lactose and/or sucrose if they are enzymatically equipped for these fermentations.

Microbes that consume all available sugars, and other bacteria that lack biochemical pathways for fermentation may use other enzymes to degrade the abundant peptones in Triple Sugar Iron Agar. This assault releases amines and additional products that raise the pH of the slant. What color is phenol red in the medium when acid-yielding lactose and/or sucrose fermentations occur; when peptone deaminations occur? See figure 26.2.

Two reactions involving gases are also evident in a Triple Sugar Iron Agar. If the bacteria produce hydrogen sulfide ($H_2S\uparrow$) gas, a black precipitate of iron sulfide forms. Species that release large amounts of primarily carbon dioxide ($CO_2\uparrow$) gas form visible bubbles that are often caught on the bottom or sides of the medium. Carbon dioxide gas may even split the butt, sending the slant portion of the medium into the test tube cap.

In addition to Triple Sugar Iron Agar, microbiologists employ other multitest media. Representatives, and the reactions that can be detected in them, are described in table 26.2.

Multitest Systems

Microbiologists also use sets of assorted media contained in miniaturized compartments. Each set is designed to help identify a certain group of bacteria—for example, the family **Enterobacteriaceae,** the gram-positive aerobic cocci, or the gram-positive anaerobic rods. In this exercise you inoculate, incubate, and interpret the **API 20E®** test, a set of

Figure 26.1 An abbreviated key for the tentative identification of *Salmonella* and *Shigella*

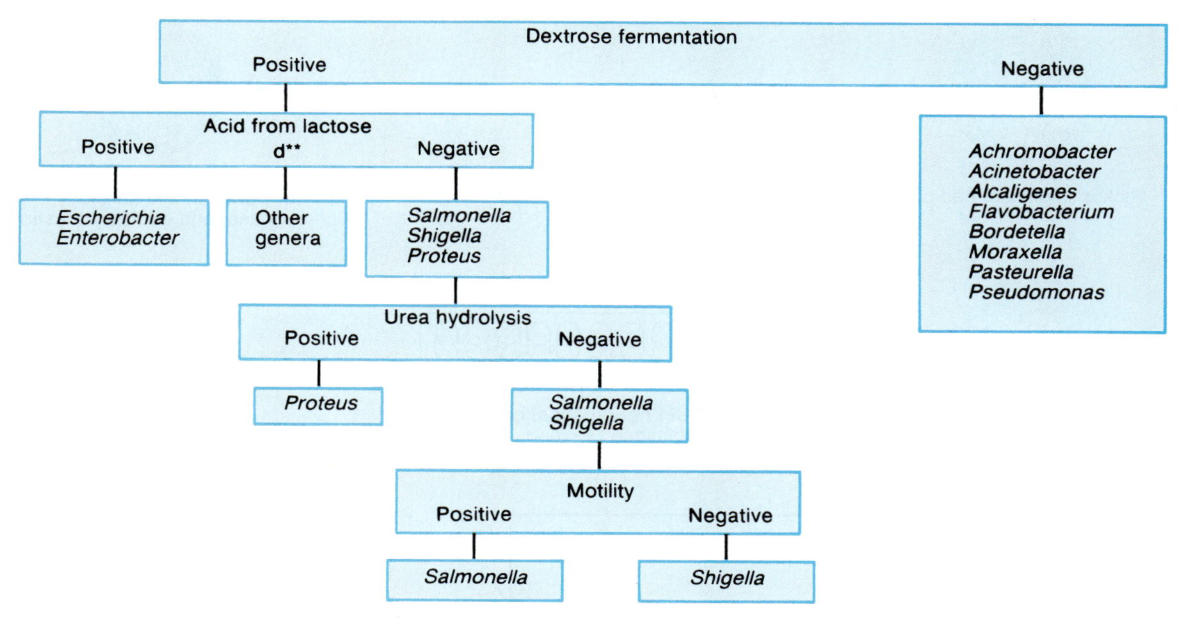

*For more complete information see Appendixes 2 and 3.

**The symbol "d" indicates that some, but not all, species are positive, or that not all strains of a species are positive.

Figure 26.2 The effects of time on representative Triple Sugar Iron Agar cultures

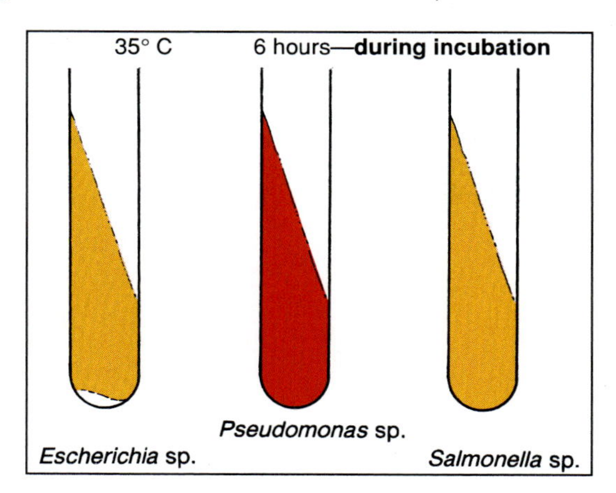

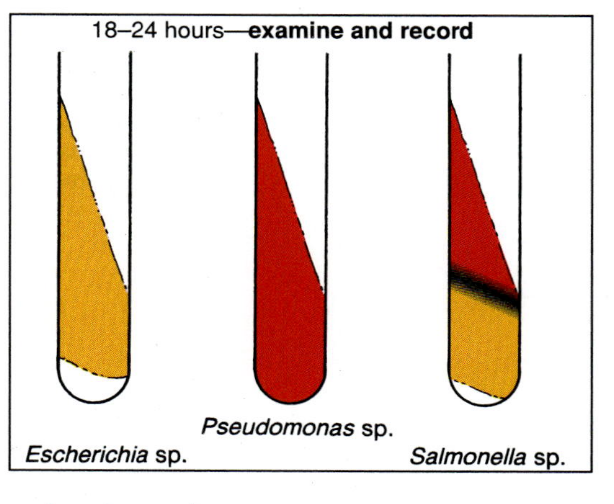

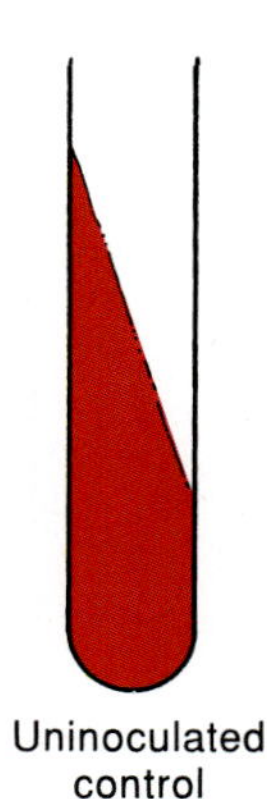

Reactions in a Triple Sugar Iron Agar culture

Yellow: Acid from carbohydrate fermentation
Butt: Dextrose fermentation
Slant: Lactose and/or sucrose fermentation

Red: Alkaline from peptone deamination
Black: H_2S production
Bubbles: Primarily CO_2 production

20 media designed to help identify gram-negative bacteria from clinical specimens. See the flow chart in figure 26.3 and typical positive and negative results in figure 26.4.

The API 20E® strip holds 20 microtubes, each containing a dehydrated medium that you rehydrate with a saline suspension of your unknown. After incubation in a humidifying tray, the strip yields information on 26 different biochemical reactions, as shown in table 26.3. The basis for each test is explained in table 26.4.

Use the information in these tables to begin or add to your media file. If you need help starting a media file, see Exercise 21, Selected Microbial Enzymes.

Figure 26.3 The API 20E® recommended procedure for the identification of Enterobacteriaceae and other gram-negative bacteria (18–24 hours)

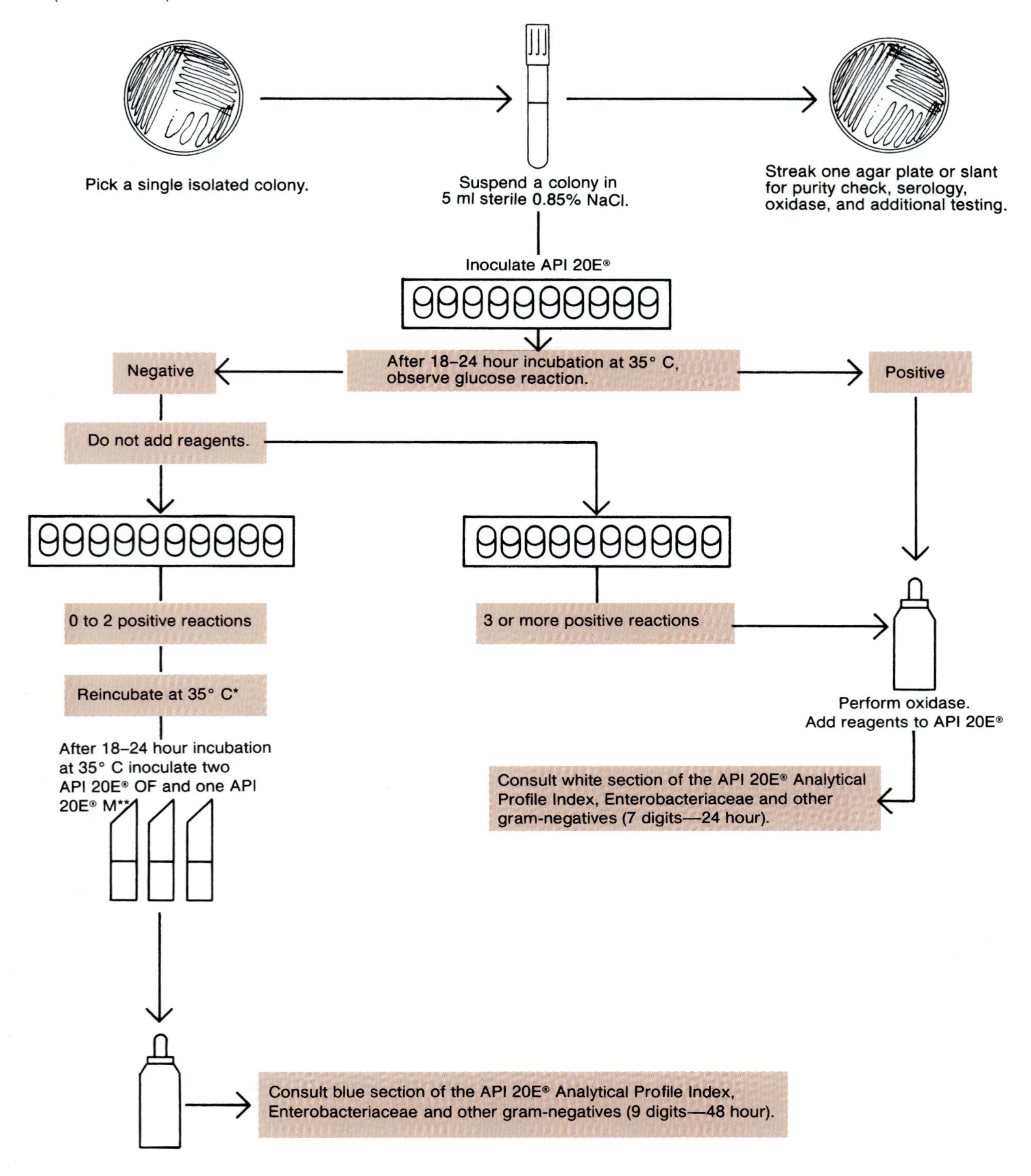

*If no reactions are positive, confirm viability by streaking a loopful of suspension from the ARA microtube onto a Blood Agar plate. Incubate for 24 hours and note growth.

**If MacConkey Agar is not used as the primary isolation medium, inoculate MacConkey medium.

Figure 26.4 Typical positive and negative results in the API 20E® test strip

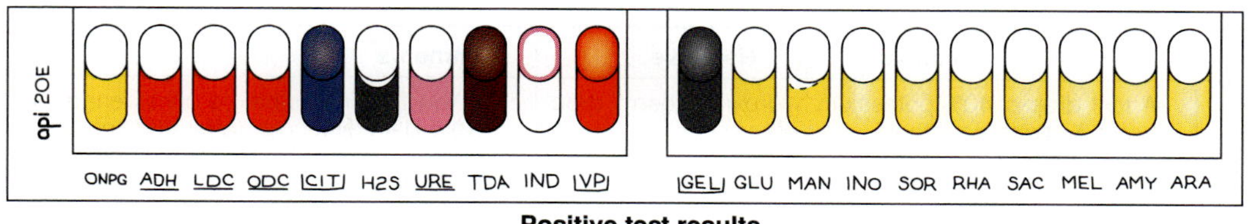

Positive test results

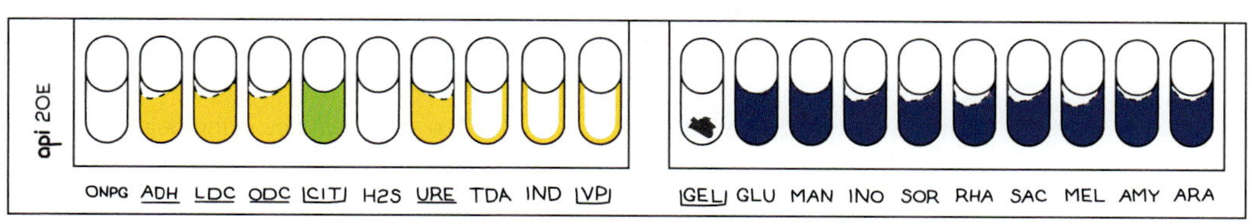

Negative test results

Table 26.3 Summary of Results: 18–24 Hour Procedure API 20E® System[1]

	Interpretation of Reactions			
Tube		**Positive**	**Negative**	**Comments**
ONPG		Yellow	Colorless	(1) Any shade of yellow is a positive reaction. (2) VP tube, before the addition of reagents, can be used as a negative control.
ADH	Incubation 18–24 hours	Red to orange	Yellow	Orange reactions occurring at 36–48 hours should be interpreted as negative.
	36–48 hours	Red	Yellow to orange	
LDC	18–24 hours	Red to orange	Yellow	Any shade of orange within 18–24 hours is a positive reaction. At 36–48 hours, orange decarboxylase reactions should be interpreted as negative.
	36–48 hours	Red	Yellow to orange	
ODC	18–24 hours	Red to orange	Yellow	Orange reactions occurring at 36–48 hours should be interpreted as negative.
	36–48 hours	Red	Yellow to orange	
CIT		Turquoise to dark blue	Light green to yellow	(1) Both the tube and cupule should be filled. (2) Reaction is read in the aerobic (cupule) area.
H_2S		Black deposit	No black deposit	(1) H_2S production may range from a heavy black deposit to a very thin black line around the tube bottom. Carefully examine the bottom of the tube before considering the reaction negative. (2) A "browning" of the medium is a negative reaction unless a black deposit is present. "Browning" occurs with TDA-positive organisms.
URE	18–24 hours 36–48 hours	Red to orange Red	Yellow Yellow to orange	A method of lower sensitivity has been chosen. *Klebsiella, Proteus,* and *Yersinia* routinely give positive reactions.
TDA		Add 1 drop of 10% ferric chloride. Brown-red	Yellow	(1) Immediate reaction. (2) Indole-positive organisms may produce a golden-orange color due to indole production. This is a negative reaction.
IND		Add 1 drop of Kovacs' reagent. Red ring	Yellow	(1) The reaction should be read within 2 minutes after the addition of the Kovacs' reagent and the results recorded. (2) After several minutes, the HCl present in Kovacs' reagent may react with the plastic of the cupule, resulting in a change from a negative (yellow) color to a brownish red. This is a negative reaction.

Table 26.3 Continued

	Interpretation of Reactions		
Tube	**Positive**	**Negative**	**Comments**
VP	Add 1 drop of 40% potassium hydroxide, then 1 drop of 6% alpha-naphthol.		(1) Wait 10 minutes before considering the reaction negative. (2) A pale pink color (after 10 minutes) should be interpreted as negative. A pale pink color that appears immediately after the addition of reagents but turns dark pink or red after 10 minutes should be interpreted as positive.
	Red	Colorless	
GEL	Diffusion of the pigment	No diffusion	Motility may be observed by hanging-drop or wet mount preparation. (1) The solid particles in the gelatin may spread throughout the tube after inoculation. Unless diffusion occurs, the reaction is negative. (2) Any degree of diffusion is a positive reaction.
GLU	Yellow or gray	Blue to blue-green	COMMENTS FOR ALL CARBOHYDRATES: **Fermentation** (Enterobacteriaceae, *Aeromonas, Vibrio*) (1) Fermentation of the carbohydrates begins in the most anaerobic portion (bottom) of the tube. Therefore these reactions should be read from the bottom of the tube to the top. (2) A yellow color located only at the bottom of the tube indicates a weak or delayed positive reaction.
MAN INO SOR RHA SAC MEL AMY ARA	Yellow	Blue to blue-green	**Oxidation** (other gram-negatives) (1) Oxidative utilization of the carbohydrates begins in the most aerobic portion (top) of the tube. Therefore, these reactions should be read from the top to the bottom of the tube. (2) A yellow color in the upper portion of the tube and a blue in the bottom of the tube indicates oxidative utilization of the sugar. This reaction should be considered positive **only** for non-Enterobacteriaceae gram-negative rods. This is a negative reaction for fermentative organisms such as Enterobacteriaceae.
GLU	After reading GLU reaction, add 2 drops of 0.8% sulfanilic acid and 2 drops of 0.5% N, N-dimethyl-alpha-naphthylamine.		(1) Before addition of reagents, observe GLU tube (positive or negative) for bubbles. Bubbles are indicative of reduction of nitrate to the nitrogen gas (N_2) state. (2) A positive reaction may take 2–3 minutes for the red color to appear. (3) Confirm a negative test by adding zinc dust of 20 mesh granular zinc. A red to orange color after 10 minutes confirms a negative reaction. A yellow color indicates bacterial reduction of nitrates to the nitrogen gas (N_2) state.
Nitrate reduction	NO_2: Red N_2 gas: Bubbles; yellow after reagents and zinc	Yellow Not red until after both reagents and zinc	
MAN, INO, or SOR for Catalase	After reading carbohydrate reaction, add 1 drop of 1.5% H_2O_2.		(1) Bubbles may take 1–2 minutes to appear. (2) Best results will be obtained if the test is run in tubes that have no gas from fermentation.
	Bubbles	No bubbles	

[1]Charts are subject to change.

Modified from the Unisept API 20E® system package insert. Courtesy of Analytab Products, Division of Sherwood Medical, 200 Express St., Plainview, NY 11803.

Table 26.4 Summary of Chemical and Physical Principles of the Tests on the API 20E® System[1]

Tube	Chemical/Physical Principles	Reactive Ingredients
ONPG	Hydrolysis of ONPG by beta-galactosidase releases yellow orthonitrophenol from the colorless ONPG; IPTG (isopropylthio-galactopyranoside) is used as the inducer.	ONPG IPTG
ADH	Arginine dihydrolase transforms arginine into ornithine, ammonia, and carbon dioxide. This causes a pH rise in the acid-buffered system and a change in the indicator from yellow to red.	Arginine
LDC	Lysine decarboxylase transforms lysine into a basic primary amine, cadaverine. This amine causes a pH rise in the acid-buffered system and a change in the indicator from yellow to red.	Lysine
ODC	Ornithine decarboxylase transforms ornithine into a basic primary amine, putrescine. This amine causes a pH rise in the acid-buffered system and a change in the indicator from yellow to red.	Ornithine
CIT	Citrate is the sole carbon source. Citrate utilization results in a pH rise and a change in the indicator from green to blue.	Sodium citrate
H_2S	Hydrogen sulfide is produced from thiosulfate. The hydrogen sulfide reacts with iron salts to produce a black precipitate.	Sodium thiosulfate
URE	Urease releases ammonia from urea; ammonia causes the pH to rise and changes the indicator from yellow to red.	Urea
TDA	Tryptophan deaminase forms indolepyruvic acid from tryptophan. Indolepyruvic acid produces a brownish-red color in the presence of ferric chloride.	Tryptophan
IND	Metabolism of tryptophan results in the formation of indole. Kovacs' reagent forms a colored complex (pink to red) with indole.	Tryptophan
VP	Acetoin, an intermediary glucose metabolite, is produced from sodium pyruvate and indicated by the formation of a colored complex. Conventional VP tests may take up to 4 days, but by using sodium pyruvate, API has shortened the required test time. Creatine intensifies the color when tests are positive.	Sodium pyruvate, Creatine
GEL	Liquefaction of gelatin by proteolytic enzymes releases a black pigment that diffuses throughout the tube.	Kohn charcoal, gelatin
GLU MAN INO SOR RHA SAC MEL AMY ARA	Utilization of the carbohydrate results in acid formation and a consequent pH drop. The indicator changes from blue to yellow.	Glucose Mannitol Inositol Sorbitol Rhamnose Sucrose Melibiose Amygdalin Arabinose
GLU Nitrate reduction	Nitrites form a red complex with sulfanilic acid and N, N-dimethyl-alpha-naphthylamine. In case of a negative reaction, addition of zinc confirms the presence of unreduced nitrates by reducing them to nitrites (red to orange color). If there is no color change after the addition of zinc, this is indicative of the complete reduction of nitrates through nitrites to nitrogen gas or to an anaerogenic amine.	Potassium nitrate
MAN, INO, or SOR for Catalase	Catalase releases oxygen gas from hydrogen peroxide.	

[1]Charts are subject to change.

Modified from the Unisept API 20E® system package insert. Courtesy of Analytab Products, Division of Sherwood Medical, 200 Express St., Plainview, NY 11803.

Table 26.5 Calculating the API 20E® Profile Index Number

The tests on the API 20E® strip are presented in 7 triplets of 3 tests each, plus nitrate reduction and catalase.
Assign a number value to each ***positive*** test in each triplet:

First position	1 point
Second position	2 points
Third position	4 points

Any ***negative*** test receives a value of 0 points.
Add the values in each column. Record the profile number as in this example:

ONPG ___	ODC ___	URE ___	VP ___	MAN ___	RHA ___	AMY ___
ADH ___	CIT ___	TDA ___	GEL ___	INO ___	SAC ___	ARA ___
LDC ___	H_2S ___	IND ___	GLU ___	SOR ___	MEL ___	OXI ___
_____	_____	_____	_____	_____	_____	_____

7-Digit Profile Number

Also record the following reactions:
Nitrate reduction _____ Catalase production _____

Assigning each positive test result a certain number of points, as outlined in the API 20E® profile recognition system, simplifies identification of an unknown; see table 26.5. These points form a 7- or 9-digit **profile index number.** By comparing the profile index number with those in the ***API 20E® Analytical Profile Index*** or the ***API 20E® Analytical Quick Index*** you rapidly identify your unknown.

Another advantage of assigning a profile index number is that the results are computerized for almost instantaneous identification of unknowns. In numerous trials, the accuracy of the computerized results from this miniaturized system has equaled accuracy of standard biochemical testing.

If the *Index* is unavailable, use **differential charts** to identify your unknown. These charts list the biochemical characteristics of selected bacteria. Reliable differential charts are found in *Bergey's Manual,* the *DIFCO Manual,* most microbiology texts, and this laboratory manual; **see Appendixes 2 and 3.**

Rapid identification systems that test for **constitutive enzymes** (those that are always present), not **inducible** ones (those that an organism produces in response to the presence of the substrate), are also available. For example, the **Micro-ID®*** system helps identify oxidase-negative, gram-negative Enterobacteriaceae within 4 hours. It consists of a strip containing 15 disks impregnated with substrates and indicators; see figure 26.5.

Multitest Plates

Other companies provide clinical microbiologists with **triplates** or **"quads,"** Petri dishes compartmentalized by plastic ridges. Each section contains a different medium. The media are selected to help differentiate a particular group of related microorganisms or a group of species that are often isolated from the same septic source.

*Produced by Organon Teknica Corp., Durham, NC 27704.

Figure 26.5 The Micro-ID® system (Courtesy of Remel Microbiology Products)

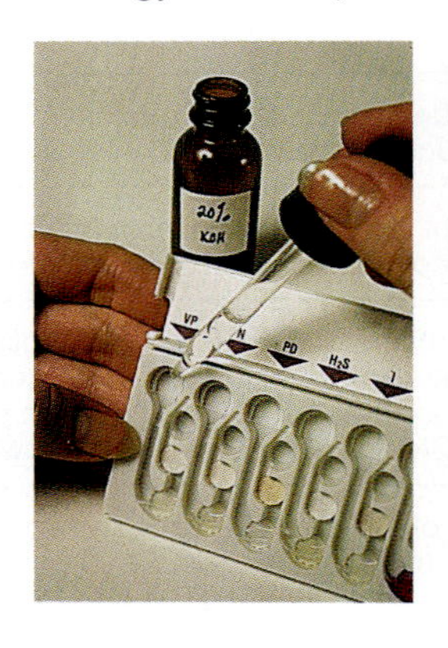

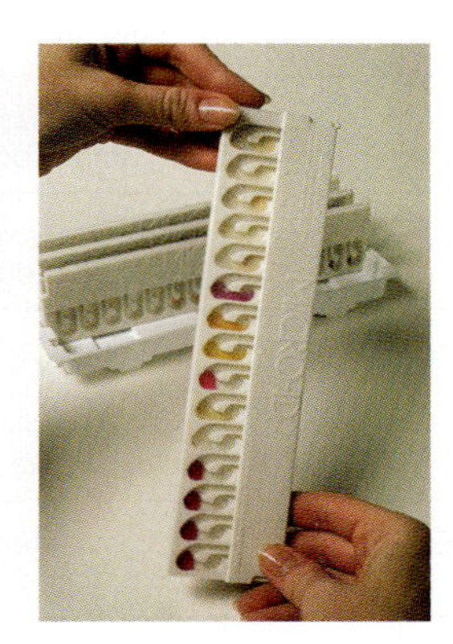

PROCEDURES EXERCISE 26

Procedure 1

A Multitest Medium, Triple Sugar Iron Agar

First Laboratory Session

> **Warning:** Your unknown may be a pathogen. Handle with care.

a Obtain a streak-plate culture; record the identifying number of this gram-negative unknown in your laboratory report.

b Label a **Triple Sugar Iron Agar** with your group identification, the date, and the number identifying your unknown.

c Using sterile technique, touch the center of a well-isolated colony with your flamed and cooled inoculating needle. (Leave enough of the colony to inoculate tests in the next procedure.) Stab the inoculum on your needle to the bottom of the Triple Sugar Iron Agar. Do not allow contact between the handle and the sterile inner wall of the test tube. Withdraw the needle tip to the agar surface. Cover the surface of the slant with a zigzag streak of the bacteria. Resterilize your needle.

d Incubate your Triple Sugar Iron Agar—with a loosened lid—at 35° C for no longer than 18 to 24 hours, then read your results.

Second Laboratory Session

a Observe the reactions in your **Triple Sugar Iron Agar** culture. Draw and interpret the results of this test in your laboratory report. After all members of the group have finished examining the slant, discard it in an appropriate fashion.

Procedure 2

Preparing for the API 20E®: Gram Stain and Oxidase Test (Optional)

> **Warning:** Your unknown may be a pathogen. Handle with care.

This procedure is optional because your unknown is already identified as a member of the Enterobacteriaceae. Therefore it is gram-negative and oxidase-negative. However, before you inoculate the API 20E® strip, **your instructor** may ask you to confirm these findings by performing Procedure 2.

a Using sterile technique, transfer a very small amount of the growth from your chosen colony to a drop of water on a microscope slide.

b After the smear is dried and heat fixed, prepare and examine a **Gram stain** of your unknown. Record the results in your laboratory report. In your report explain why this step is performed before inoculating an API 20E® test strip.

c Use alcohol-flame sterilized forceps to place an **oxidase differentiation disk** onto a drop of water on a clean microscope slide.

d With a sterile wooden applicator stick, spread some growth from the colony onto the dampened disk. (The Nichrome wire commonly used to make inoculating loops and needles sometimes causes false-positive reactions.) Discard the contaminated stick in an appropriate manner.

e A change in the color of the smear on the disk from pink to maroon to nearly black within 20 minutes constitutes a positive reaction. In your laboratory report, record the results of the oxidase test. Discard your slide appropriately.

Procedure 3

A Multitest System, the API 20E®

> **Warning:** Your unknown may be a pathogen. Handle with care.

First Laboratory Session

a Use sterile technique and a small amount of the remaining growth in your colony to prepare a suspension of your unknown in 5 ml of sterile saline.

b Label the elongated flap of your incubation tray. Add about 5 ml of water to the dimples in the bottom of the tray to provide a moist chamber for your **API 20E® test strip.**

c Remove your API 20E® strip from its sealed pouch. Reseal the pouch and label the strip. Place the strip in the incubation tray.

d Read the following steps before inoculating your strip with your saline suspension of bacteria, then carry out the procedure precisely. Filling tubes is easier if you tilt the strip and touch the tip of the pipet to the inside of each cupule. See the illustration.

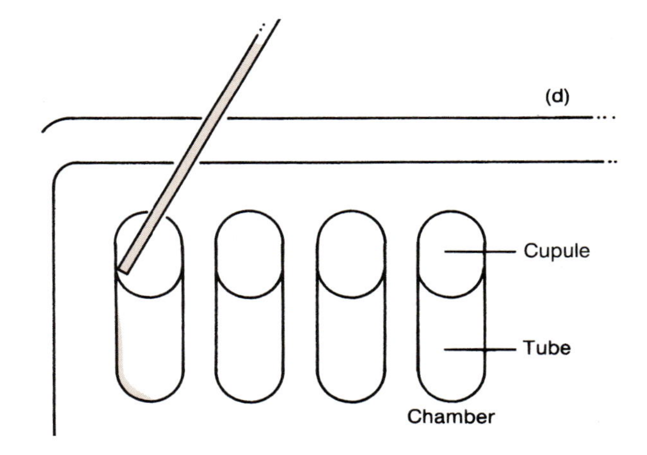

e With a Pasteur pipet, **fill the tube and cupule** of chambers marked "⌊___⌋"(⌊CIT⌋, ⌊VP⌋, ⌊GEL⌋).

f Slightly **underfill chambers** marked "___" (ADH, LDC, ODC, H_2S, and URE).

g Fill all of the remaining **tubes.**

h With a sterile pipet, completely fill the cupule of the chambers marked "___" (ADH, LDC, ODC, H_2S, and URE) with **sterile mineral oil.** In your laboratory report, explain why microbiologists sometimes layer sterile mineral oil onto biochemical test media. (Review Exercise 16, Anaerobic Culture: Oxygen Requirements, if you need help with this explanation.)

i Place the lid on the incubation tray. Incubate the tray and test strip at 35° C for no more than 18 to 24 hours, then read your results. Alternatively, you may refrigerate the system after incubation and read it as soon as all members of the group are available.

Subsequent Laboratory Sessions

a When all in your group are available, examine your API 20E® strip. Utilize the flowchart in figure 26.3 to help plan your examination of results. Examine the pictures of typical positive and negative results shown in figure 26.4, and the summary of tests and results in table 26.3, to help you perform and interpret each biochemical test.

b Record the results of each test in your laboratory report. Perform further tests if required.

c As shown in table 26.5, assign a 7- or 9-digit profile number to your unknown. If one is available, utilize an **API 20E® Analytical Profile Index** to identify your unknown. If not, consult differential charts (Appendixes 2 and 3) to identify your unknown. When all in your group have completed their identification of the unknown, discard your API 20E® strip appropriately.

POSTTEST EXERCISE 26

Part 1

True or False (Circle one, then correct every false statement.)

1. T F Triple Sugar Iron Agar contains more dextrose than sucrose or lactose.
2. T F Triple Sugar Iron Agar is designed to help differentiate the gram-negative bacteria.
3. T F In a Triple Sugar Iron Agar, *Salmonella* and *Shigella* both generally produce a yellow butt (with or without a black precipitate) and a pink slant after 24 hours of incubation at 35° C.
4. T F Glucose medium is layered into the butt of a Triple Sugar Iron Agar tube; lactose medium is layered on top.
5. T F Species that release large amounts of $CO_2\uparrow$ in a Triple Sugar Iron Agar may split the agar, sending the slant portion of the medium into the test tube cap.
6. T F An API 20E® strip is designed to help identify gram-negative, oxidase negative bacteria from clinical specimens.
7. T F To provide anaerobic test conditions, sterile mineral oil is added to some of the tubes of the API 20E® strip before a saline suspension of bacteria is pipetted into the chambers.
8. T F The API 20E® strip is inoculated with a saline suspension of a pure culture.
9. T F After assigning points for positive results on an API 20E® strip, the clinician can compute a profile index number for the culture, then utilize this number to find the identity of the unknown inoculum.
10. T F A quad is a set of four very stable bacteria that give repeatable results in a multitest system. Quads are used to standardize each API 20E® test strip before it is used.

Part 2

Short Answer

1. Explain why correct identification of microbial species is crucial for research and in medical microbiology.

2. List three multitest media, in addition to Triple Sugar Iron Agar, and the biochemical tests performed in each.

a.

b.

c.

Laboratory Report **26**

Name ______________________________

Section ______________

Multiple-Test Media and Systems

Procedure 1

A Multitest Medium, Triple Sugar Iron Agar

First Laboratory Session

(*a*) Record the identifying number of your unknown here. _______

Second Laboratory Session

(*a*) Draw and indicate the colors in your Triple Sugar Iron Agar tube.

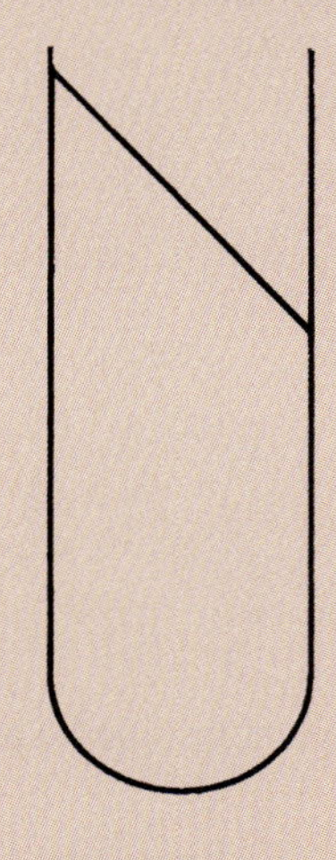

Triple Sugar Iron Agar

Fill in the following chart, showing your interpretation of the results of your Triple Sugar Iron Agar test.

Triple Sugar Iron Agar Results		
Test	**Results Positive or Negative**	**Evidence**
Glucose fermentation		
Lactose and/or sucrose fermentation		
Gas (primarily CO_2) production		
H_2S production		

Procedure 2

Preparing for the API 20E®: Gram Stain and Oxidase Test (Optional)

(*b*) Record the shape and the Gram stain reaction of your unknown here. ______________________

Why is it necessary to know the Gram stain reaction of an unknown before inoculating an API 20E® test strip?

(*e*) Record the results of your oxidase test here. ______________________

Procedure 3

A Multitest System, the API 20E®

First Laboratory Session

(*h*) Why do microbiologists sometimes layer sterile mineral oil onto biochemical test media after inoculation?

Subsequent Laboratory Sessions

(*b*) Record the results of each test in your API 20E® strip on the following chart. Fill in the 7-digit profile number of your unknown by **adding** the indicated number of points for each ***positive*** **reaction.** Fill in the additional 2 numbers if calculated. Identify your unknown.

API 20E® Unknown No. ________ Results

ONPG	ADH	LDC	ODC	CIT	H_2S	URE	TDA	IND	VP	GEL	GLU	MAN	INO	SOR	RHA	SAC	MEL	AMY	ARA	OXI
1	2	4	1	2	4	1	2	4	1	2	4	1	2	4	1	2	4	1	2	4
___	___	___	___	___	___	___	___	___	___	___	___	___	___	___	___	___	___	___	___	___
	______			______			______			______			______			______			______	

Profile Number

NO_3	N_2 Gas	MOT	MAC	OF-O	OF-F
1	2	4	1	2	4
___	___	___	___	___	___
	______			______	

Additional Digits

The 7-digit profile number of your unknown is __________________ .
The 2 additional numbers (if calculated) are __________ .
(*c*) The identity of your unknown is ________________________ ______________________ .

Medically Significant Gram-Positive Cocci

Necessary Skills

Aseptic transfer techniques, Gram staining, Culture characteristics, Streak-plate techniques, Knowledge of factors influencing microbial growth, Microbial metabolic reactions, Selective and differential media, Anaerobic culture: oxygen requirements

Materials

> **Warning:** Handle your cultures with care. Some are pathogens.

First Laboratory Session

Procedure 1

Cultures (24-hour [not over 30-hour], 35° C, Blood Agar plate streaked for isolation of colonies, 1 per group):

Staphylococcus aureus (labeled "Unknown ____")

Staphylococcus epidermidis (labeled "Unknown ____")

Other staphylococci may be utilized (with a key to the unknowns given to the **instructor**)

Procedure 2

All available streptococcal and enterococcal cultures

Procedure 3

Cultures (24-hour, 35° C, Tryptic Soy Broth or Todd Hewitt Broth, 1 per set):

Streptococcus pneumoniae

Streptococcus mitis

Procedure 4

Cultures (24-hour, 35° C, Tryptic Soy Broth, 1 per set):

Streptococcus pyogenes

Enterococcus faecium

Procedure 5

Cultures (Tryptic Soy Broth, 24-hour, 35° C, 1 per set):

Streptococcus pyogenes

Streptococcus agalactiae

Enterococcus faecium

Staphylococcus aureus (ATCC e 25923)

Supplies:

Procedure 1 (per group)

Gram stain reagents (at **instructor's** request)

Staining equipment (at **instructor's** request)

0.5 ml of Coagulase Plasma EDTA in a serology test tube

2 Microscope slides

Access to 3% hydrogen peroxide in dropper bottles

35° C Water bath

Serology test tube racks in water bath, to hold 2 tubes per group

Access to a Mannitol Salt Agar plate

Access to a *DIFCO Manual* during second laboratory session

Procedure 2 (per group)

Gram stain reagents

Staining equipment

5 Microscope slides

Procedure 3 (per group)

Blood Agar plate

2 Sterile optochin differentiation disks (ethylhydro-cupreine hydrochloride)

Forceps

Alcohol for flaming

Procedure 4 (per group)

2 Blood Agar plates

2 Bacitracin disks

Forceps

Alcohol for flaming

Quebec colony counter (for class use during second laboratory session, optional)

Procedure 5 (per group)

Blood Agar plate

Primary Objective

Identify and differentiate medically significant gram-positive cocci.

Other Objectives

1. Compare and contrast biochemical tests that are commonly used to differentiate staphylococci from micrococci and from streptococci and enterococci.
2. Correlate sites on the human body where staphylococci are commonly located and clinical specimens that often contain staphylococci.
3. Compare and contrast the pathogenic potentials of *Staphylococcus epidermidis* and of *Staphylococcus aureus*.
4. Explain how cooked food that contains no living *Staphylococcus aureus* may still cause an epidemic of staphylococcal food poisoning.
5. Perform and evaluate critical microscopic and macroscopic observations, as well as biochemical tests, that identify *Staphylococcus aureus*.
6. Identify one morphological and three chemical characteristics of the streptococci and enterococci that separate the genera from other gram-positive cocci.
7. Observe and distinguish among alpha, beta, and gamma hemolysis.
8. Identify Rebecca Lancefield's contribution to microbiology.
9. Appraise two distinguishing characteristics of pneumococcal colonies.
10. Observe and describe *Streptococcus pneumoniae* cells.
11. Compare and contrast the clinical significance of *Streptococcus pneumoniae* and of alpha-hemolytic members of the *Bergey's Manual* group, Oral Streptococci.
12. Explain why rapid diagnosis of *Streptococcus pyogenes* is important, then list five GABHS infections and three serious nonsuppurative sequelae.
13. Describe how bacitracin sensitivity, the CAMP reaction, and SXT sensitivity are used to differentiate the Lancefield groups of beta-hemolytic streptococci.
14. Perform and interpret a bacitracin sensitivity test, a test for streptolysin O, and the CAMP test.
15. Define: GABHS, nonsuppurative, sequelae, hemolytic reaction, rapid techniques, bacitracin, streptolysin O, streptolysin S, CAMP test, SXT.

INTRODUCTION

In this exercise you learn to differentiate the pathogens ***Staphylococcus aureus, Streptococcus pneumoniae,*** and ***Streptococcus pyogenes*** from other gram-positive cocci. *Bergey's Manual* places these common harbingers of disease in the **phylogenetically** (pertaining to the development, the genetics, of the group) diverse section entitled **Gram-positive Cocci;** see table 27.1.

Table 27.1 Gram-Positive Cocci

Family I. Micrococcaceae
- Genus I. *Micrococcus*
- Genus II. *Stomatococcus*
- Genus III. *Planococcus*
- Genus IV. *Staphylococcus*

Family II. Deinococcaceae
- Genus I. *Deinococcus*

Other genera
- *Streptococcus, Leuconostoc, Pediococcus, Aerococcus, Gemella, Peptococcus, Peptostreptococcus, Ruminococcus, Coprococcus, and Sarcina*

Data taken from Holt, J. G., et al., eds., *Bergey's Manual of Determinative Bacteriology,* 9th ed., Williams & Wilkins Co., Baltimore, MD, 1994.

Since **micrococci, streptococci,** and **enterococci** (a genus that was formerly included in the streptococci) are commonly isolated from many of the same clinical samples as **staphylococci,** the investigator relies upon rapid serological and/or biochemical tests to separate these genera; see table 27.2.

Medical microbiologists generally employ the **catalase** test to separate staphylococci from both streptococci and enterococci. For a review of this test, see Exercise 21, Selected Microbial Enzymes. Staphylococci are catalase producers; streptococci and enterococci are not. A less sensitive test is **colony size.** Generally, on Blood Agar after 24 hours of incubation at 35° C, staphylococcal colonies are larger than those of streptococci.

Lysis of staphylococcal cells by the enzyme **lysostaphin** occurs within 15 minutes and helps differentiate staphylococci from micrococci. The latter are not as readily lysed by this enzyme.

The **oxidase** test (see Exercise 21, Selected Microbial Enzymes) also separates staphylococci and micrococci. While most species of the genus *Micrococcus* are oxidase-positive, the common species of staphylococci are oxidase-negative.

Characterizing Selected Species of *Staphylococcus*

Medical microbiologists frequently encounter staphylococci; clinical specimens from nearly any human source may contain them. They colonize skin, skin glands, and mucous membranes of warm-blooded animals. Staphylococci are recovered from specimens including pus, purulent fluids, sputum, urine, feces, and blood.

The predominant *Staphylococcus* species on human skin is ***Staphylococcus epidermidis.*** Normally a member of the resident flora, it is an opportunistic pathogen that may cause disease in compromised tissue or in immunosuppressed patients.

In contrast to its less formidable relative, the well-armed pathogen ***Staphylococcus aureus*** produces many enzymes that help it invade, poison, and parasitize healthy

Table 27.2 Differentiation of *Staphylococcus* from *Micrococcus* and from *Streptococcus* and *Enterococcus* (Medically significant Gram-positive cocci)

Characteristic	Genera		
	Staphylococcus	***Micrococcus***	***Streptococcus* and *Enterococcus***
Catalase production	Positive	Positive	Negative
Colony, Blood Agar, 24–48 hours, 35° C	>1.0 mm Diameter, carotenoid pigments[1]	>1.0 mm Diameter, carotenoid pigments[2]	0.5–1.0 mm Diameter, not pigmented[2]
Predominant arrangements of cells (other than single cells)	Clusters, pairs	Clusters, tetrads	Chains, pairs
Oxidase test	Negative[2]	Positive[2]	Negative[2]
Lysis by lysostaphin	Positive	Negative	Negative
Sensitivity to bacitracin, 0.04 units	Resistant	Inhibited	Variable

[1]Some species.

[2]Most species.

Modified from Holt, J. G., et al., eds., *Bergey's Manual of Determinative Bacteriology*, 9th ed. Williams & Wilkins Co., Baltimore, MD, 1994.

tissue. For example, the enzyme **coagulase** converts fibrinogen in blood serum into fibrin, an insoluble, threadlike protein that forms a protective cocoon around infected tissue, guarding staphylococcal cells from phagocytosis and other body defenses.

Staphylococcus aureus is the most common etiologic agent of boils, furuncles, carbuncles, and wound infections, both surgical and nonsurgical. Among the other pathological conditions frequently caused by *Staphylococcus aureus* are impetigo, pneumonia, osteomyelitis, meningitis, endocarditis, enterocolitis, and urogenital infections.

Staphylococcus aureus is characterized as follows (other staphylococci generally lack these characteristics):

1. **Production of coagulase** (This characteristic is the primary indicator of virulence among staphylococci.)
2. **Fermentation of mannitol yielding acid**
3. **Production of heat-stable deoxyribonuclease**
4. When first isolated from clinical specimens, **production of beta hemolysis** by most strains

To differentiate *Staphylococcus aureus* from normal flora and to initiate appropriate treatment while controlling the spread of infection, prompt identification of this opportunistic pathogen is essential.

Staphylococcal Food Poisoning

Staphylococcus aureus also is the most common bacterial cause of food poisoning. Many strains produce a **heat-stable exotoxin** while growing in contaminated, room temperature food. Thorough cooking destroys the bacteria, but the toxin remains potent.

This colorless, odorless poison causes **gastroenteritis** (nausea, vomiting, abdominal cramps, diarrhea, headache, and occasionally fever) within 2 to 4 hours after ingestion. An epidemiologist tracing an outbreak of staphylococcal food poisoning may culture cooked food for *Staphylococcus aureus*. Frequently there is no growth on these cultures even though the food is toxic. Explain why.

Classifying Streptococci, Enterococci, and Lactococci

The streptococci, enterococci, and lactococci are traditionally separated from other gram-positive cocci by the following features of their chemistry and morphology:

1. They are **homofermentative** (able to ferment carbohydrates yielding primarily one end product). Their homofermentation produces **lactic acid.**
2. **Strictly fermenters,** these cocci lack a cytochrome system.
3. Under special conditions a few of these cocci produce peroxidases, but they **do not produce catalase.**
4. The cells of these cocci are arranged in **singles, pairs,** and **chains.**

To a clinical microbiologist, the ability of a colony to lyse red blood cells is often the first hint revealing its identity. When grown for 24 hours at 35° C on Blood Agar, streptococci generally produce **alpha, beta,** or **gamma hemolysis** (destruction of erythrocytes). Hemolysis is described in Exercise 25, Selective and Differential Media.

While grouping bacteria by their hemolytic reactions is often convenient, researchers generally utilize other, more chemically reliable and measurable features to separate the streptococci. For example, in the early 1930s five distinct, antigenic groups of streptococci were identified by **Rebecca Lancefield,** who named the groups A, B, C, D, and E. Since then an additional 13 antigenically unique groups have been recognized.

Table 27.3 Differential Reactions of Selected Alpha-Hemolytic Streptococcal and Enterococcal Species

Characteristics	Streptococci					Enterococci	
	Pyogenic	Oral					
	S. pneumoniae	*S. salivarius*	*S. sanguis*	*S. mitis*	*S. mutans*	*E. faecium*	*E. gallinarum*
Growth at 10° C	—	—	—	—	—	+	+
Growth at 45° C	—	d	d	d	d	+	+
Growth in 6.5% NaCl	—	—	—	—	—	+	+
Growth with pH 9.6	—	—	—	—	—	+	+
Arginine hydrolysis	+	—	+	+	—	+	d
Lancefield group	—[1]	K or —	H	—	Some E	D	D

[1]Not all alpha-hemolytic cocci possess Lancefield's group specific antigens.

The letter "d" indicates that 11% to 89% of the strains are positive.

Data modified from Holt, J. G., et al., eds., *Bergey's Manual of Determinative Bacteriology,* 9th ed. Williams & Wilkins Co., Baltimore, MD, 1994.

Classifying Alpha-Hemolytic Streptococci and Enterococci

Alpha-hemolytic streptococci and enterococci are those that partially hemolyze red blood cells. This incomplete lysis produces a greenish "halo" of color that surrounds each alpha-hemolytic colony growing on Blood Agar. Alpha-hemolytic cocci are found in many of the Lancefield groupings; see table 27.3.

Characterizing *Streptococcus pneumoniae*

In this exercise you learn to differentiate ***Streptococcus pneumoniae*** from other alpha-hemolytic streptococci, including ***Streptococcus mitis.*** You observe characteristic inhibition of **pneumococci** by **optochin** (ethylhydro-cupreine hydrochloride). Most other alpha-hemolytic streptococci are resistant to optochin.

Two characteristics of pneumococcal colonies require careful observation on your part. First, the small, convex, circular colonies formed on Blood Agar incubated at 35° C collapse as they age. Typically, a dimpled, checker-shaped colony results. Second, the alpha hemolysis produced by *Streptococcus pneumoniae* colonies is generally more intensely green than the drab alpha hemolysis produced by other streptococci, although *Streptococcus mitis* also produces a fairly intense greening of sheep erythrocytes.

Certain microscopic characteristics of *Streptococcus pneumoniae* merit your attention. Observe the characteristic "lancet" shape (somewhat pointed on one end and flattened on the other) of *Streptococcus pneumoniae* **diplococci.** You may also note the tendency of laboratory pneumococcal cultures to form chains in broths.

Clinical Significance of Alpha-Hemolytic Streptococci

Pneumococcal pneumonia, caused by *Streptococcus pneumoniae,* is currently listed among the top 10 causes of death in the United States. The disease is spread primarily by healthy **carriers,** people who are infected but not ill. The usual site of infection is the **nasopharynx,** a cavity behind the nasal passages and above the soft palate. While performing the procedures section of Exercise 29, Mouth, Throat, and Lower Respiratory System Flora, you may discover *Streptococcus pneumoniae* among your own resident microbiota.

The pneumococcus is an etiologic agent of two other important diseases. It is the most common cause of bacterial meningitis in adults and a frequent cause of otitis media, a disease that generally occurs in children.

The other species that you utilize in this exercise is *Streptococcus mitis.* While usually a harmless member of the human upper respiratory tract flora, this species and other alpha-hemolytic, oral streptococci are the most common cause of subacute bacterial endocarditis.

Use caution when handling your cultures in this exercise.

Characterizing Beta-Hemolytic Streptococci and Enterococci

The **beta-hemolytic streptococci and enterococci** are gram-positive, spherical bacteria arranged in singles, pairs, or chains. These cocci produce **beta hemolysins,** molecules that destroy red blood cells. ***Streptococcus pyogenes*** is the most frequently encountered pathogen among the beta-hemolytic streptococci.

It is estimated that 30 to 40 million throat swabbings are performed in the United States every year to find the etiologic agents of **septic sore throat** ("strep throat," **pharyngitis**). About one out of four cases of this disease is caused by *Streptococcus pyogenes.*

When this bacterium is identified, effective antibiotic treatment can be initiated. Unlike viruses, which cause over half the cases of pharyngitis, the streptococci are generally killed by antibiotics. Rapid and accurate identification of *Streptococcus pyogenes* allows the physician to initiate antibiotic therapy when it is needed while avoiding overprescription of unnecessary medication.

In the past, before antibiotic treatment of *Streptococcus pyogenes* infections was possible, streptococcal septic sore throat, impetigo, septicemia, erysipelas, and wound infections frequently caused pain and suffering. Serious **nonsuppurative** (not pus-forming) **sequelae** (postinfection conditions) including rheumatic fever, glomerulonephritis, and scarlet fever claimed many lives.

The beta-hemolytic streptococci, which are described in this exercise, are distributed among four **Lancefield groups.** *Streptococcus pyogenes* is the only member of Lancefield's group A. This pathogen is called the **group A, beta-hemolytic *Streptococcus* (GABHS).**

With modern **rapid techniques** it is possible to identify GABHS on a throat swab within 10 minutes. Some of the other groups can also be identified quickly.

Identifying GABHS

The accuracy of rapid techniques is always compared with the standard culture and staining method of identifying GABHS that you perform in this exercise. In the standard technique of identification, a specimen is streaked for isolation of colonies on Blood Agar. The primary area of the streak is **stabbed** and exposed to the antimicrobics **bacitracin** and **sulfamethoxazole-trimethoprim (SXT).** After incubation, growth pattern and hemolytic abilities of the bacteria are evaluated. Typical colonies are Gram stained and examined. An explanation of this culture and stain procedure follows.

GABHS produces two beta **hemolysins: streptolysin O,** which is sensitive to oxygen and is antigenic, and **streptolysin S,** which is stable in the presence of oxygen but is not antigenic. Some strains of *Streptococcus pyogenes* produce only one of the streptolysins, as illustrated in figure 27.1. Therefore, to identify a GABHS reliably, the microbiologist must provide both an aerobic environment and a reduced environment, one with less oxygen than is found in air.

Bacteria in the primary area of the streak are **stabbed** down into the medium to provide them with an area of reduced oxygen potential. Atmospheric levels of oxygen are available to streptococci growing on the surface of the medium; reduced conditions surround bacteria that have been pushed into a Blood Agar pocket. Now both streptolysins are able to destroy red blood cells. The resulting beta hemolysis is readily observed by the microbiologist.

The antibiotic **bacitracin** also helps the microbiologist identify GABHS since it inhibits *Streptococcus pyogenes* but does not affect most other beta-hemolytic streptococci. To observe the bacterial response to bacitracin, the microbiologist places a sterile disk impregnated with the antibiotic onto the primary area of a Blood Agar streak-plate culture.

A disk containing the antimicrobic **SXT** may also be placed in the primary area of a Blood Agar streak-plate culture. It inhibits the growth of group C, but not group A, streptococci. The differential responses are shown in figure 27.2.

After incubating the culture for 24 hours at 35° C, the microbiologist examines the plate and selects typical colonies that are 0.5 to 1.0 mm in diameter, colorless, beta-hemolytic, bacitracin-sensitive, and SXT-resistant. These colonies are Gram stained. *Streptococcus pyogenes* stains from solid media yield gram-positive cocci, approximately 1 µm in diameter, that are arranged in singles, pairs, and short chains. GABHS grown in broth culture often forms long chains.

Identifying Other Beta-Hemolytic Streptococci

Table 27.4 lists common, beta-hemolytic streptococcal groups, their pathogenic potential, and selected tests that help identify them.

Figure 27.1 *Streptococcus pyogenes* producing streptolysin O

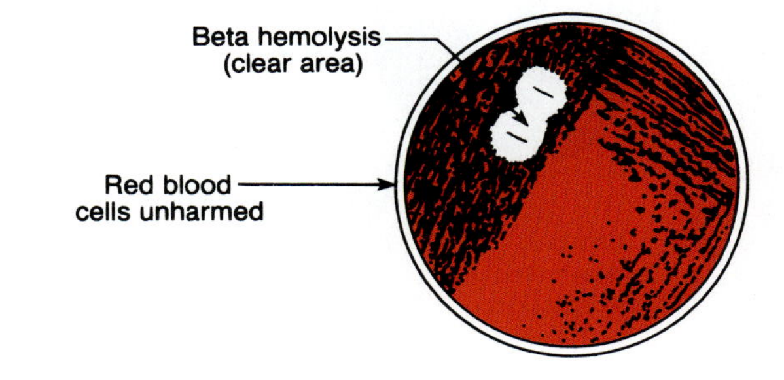

Figure 27.2 The group A and group C streptococcal responses to bacitracin and SXT

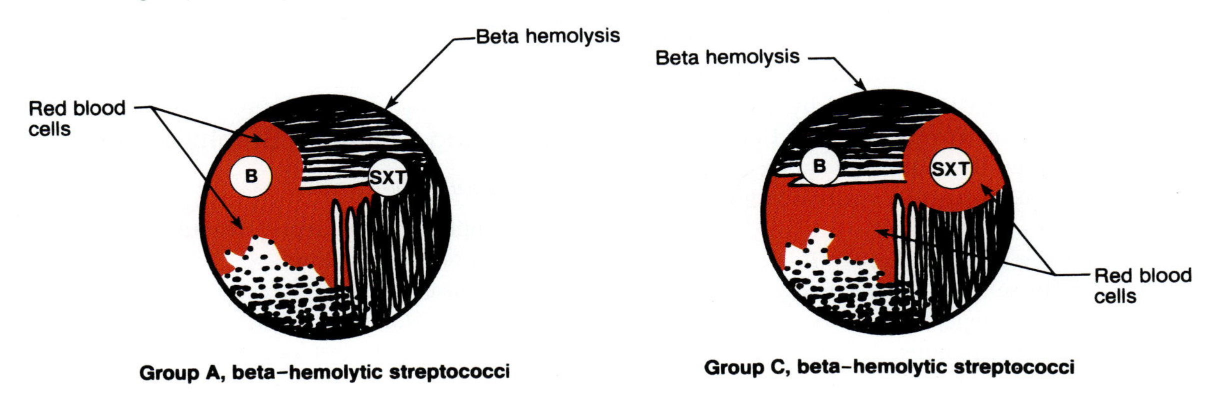

Table 27.4 Beta-Hemolytic Streptococci and Enterococci

Group Species	Hemolysis	Bacitracin Sensitivity	CAMP Reaction	SXT Sensitivity	Pathogenic Potential
A *Streptococcus pyogenes*	Beta	Sensitive[1]	Negative	Resistant	*Primary* Pharyngitis, impetigo, erysipelas, septicemia *Sequelae* Rheumatic fever, glomerulonephritis, scarlet fever
B *Streptococcus agalactiae*	Generally beta	Resistant[2]	Positive	Resistant	Puerperal sepsis, pneumonia, endocarditis, neonatal infections
C *Streptococcus equi* *Streptococcus equisimilis* *Streptococcus zooepidemicus*	Beta	Resistant[2]	Negative	Sensitive	Wound infections, puerperal sepsis, endocarditis
D *Enterococcus faecalis* *Enterococcus faecium*	Alpha, beta, or gamma	Resistant	Negative	Resistant	Urinary tract infections, peritonitis, pelvic abscesses, endocarditis

[1]Over 90% of strains are sensitive.
[2]Rare strains are sensitive.

Figure 27.3 The CAMP test. Group B, beta-hemolytic streptococci produce an area of increased hemolysis near beta-hemolytic *Staphylococcus aureus.* Group A streptococci do not.

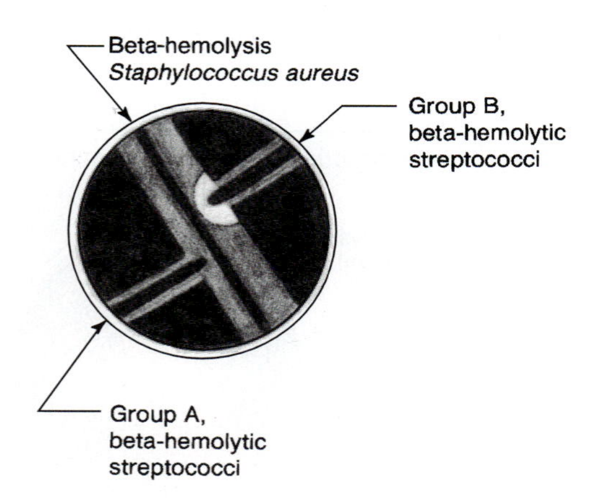

The **CAMP** test—which is named for Christie, Atkins, and Munch-Peterson, who first described the phenomenon in 1944—helps identify **group B,** beta-hemolytic streptococci. In the CAMP test, the beta hemolysis of ***Staphylococcus aureus*** is intensified by the **CAMP factor,** an extracellular substance that is produced by Lancefield's group B streptococci. An arrowhead-shaped area of increased lysis where group B streptococci approach *Staphylococcus aureus* indicates a positive reaction in the CAMP test; see figure 27.3.

Identification of the Lancefield group B streptococci is often indicated since these cocci are able to cause puerperal sepsis, endocarditis, and neonatal infections including septicemia, pneumonia, and meningitis.

Lancefield's group A and B streptococci can be differentiated from **group C,** beta-hemolytic streptococci by placing an **SXT** disk across from the bacitracin disk, as shown in figure 27.2. Groups A and B are resistant to SXT while group C, beta-hemolytic streptococci are sensitive to it.

The techniques described here separate the four antigenically distinct Lancefield groups containing beta-hemolytic streptococci and enterococci efficiently but not perfectly. To reach 100% accuracy, a microbiologist could use the original, immunological techniques. However, the original protocol employs difficult, time-consuming methods and utilizes equipment that is not generally found in a medical or teaching laboratory. The techniques you perform in this exercise are those that are most often employed by medical microbiologists.

PROCEDURES EXERCISE 27

Warning: Handle your cultures with care. Some are pathogens.

Procedure 1

Identifying Unknown Staphylococci

First Laboratory Session

a In your laboratory report, record the number that identifies your unknown culture.

Because the coagulase test can take up to 4 hours (or, alternatively, 24 hours) start it first, then continue with your other work while the test incubates.

b Perform a **coagulase** test by heavily inoculating a serology tube of Coagulase Plasma EDTA with your unknown. Incubate the slurry in a 35° C water bath for up to 4 hours. (If this is inconvenient, incubate the coagulase 24 hours in a 35° C incubator.)

c Every hour, gently examine the coagulase test in the water bath. As illustrated, look for evidence of coagulation. Do not disrupt a fragile clot with rough handling.

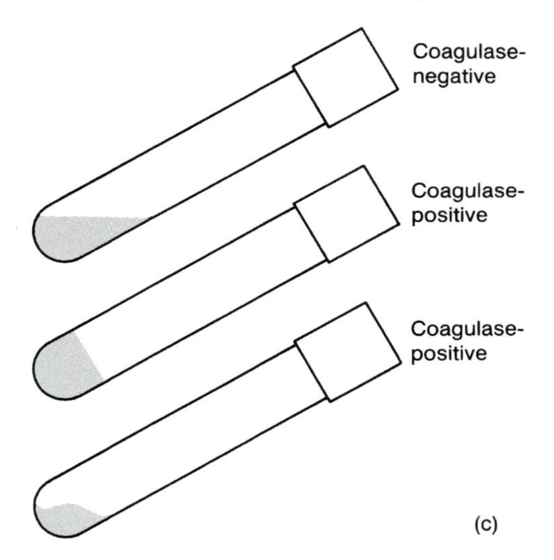

d Record your results in your laboratory report. If no clot has formed, reincubate the liquid for up to a total of 4 hours.

e Enter a complete **description of a well-isolated colony** of your unknown in your laboratory report. Note hemolysis, if any.

f **Your instructor** may ask you to perform a **Gram stain** to confirm that your unknown culture contains gram-positive cocci. If so, enter a complete description of cellular morphology and Gram stain reaction in your laboratory report.

g Perform a slide **catalase test** to verify the production of this enzyme by your unknown. For a review of this procedure, see Exercise 21, Selected Microbial Enzymes. Record your results in your laboratory report. Use caution in discarding your contaminated slide into an appropriate receptacle.

h Inoculate the appropriately labeled section of a **Mannitol Salt Agar** plate with a straight line streak of your unknown. **Your instructor** may ask you to share a plate of Mannitol Salt Agar with other groups.

i Incubate your inverted Mannitol Salt Agar plate at 35° C for 24 to 48 hours.

Second Laboratory Session

a If you incubated your **coagulase** test for 24 hours, examine it for evidence of coagulation. Record your observations in your laboratory report.

b Examine the growth on your **Mannitol Salt Agar** plate. Growth indicates tolerance of 7.5% NaCl, a characteristic of staphylococci that is not shared by most microbial genera. A yellow zone indicates that the bacteria ferment mannitol and release enough acid to alter the phenol red indicator from red to yellow.

c In your laboratory report, discuss the selective and differential properties of Mannitol Salt Agar. Also record the reactions of your unknown on this medium.

d In your laboratory report, state whether your unknown could be *Staphylococcus aureus*.

Procedure 2

Microscopic Observations

a Prepare **Gram stains** of the streptococci and enterococci.

b In your laboratory report, describe the morphology and Gram reaction of each culture.

Procedure 3

Optochin Sensitivity and Colony Description

First Laboratory Session

a As in the illustration, label a **biplate of Blood Agar.**

b After thoroughly resuspending your *Streptococcus pneumoniae* and *Streptococcus mitis* cultures, streak each half of the biplate for isolation of colonies of the appropriate species.

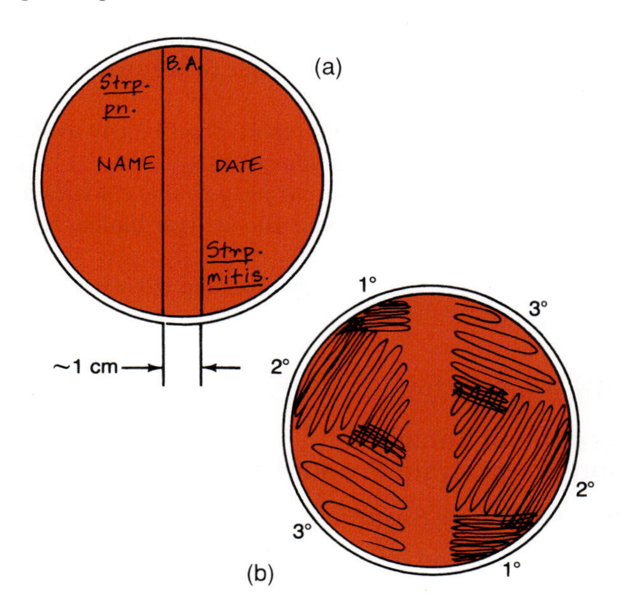

c Alcohol-flame sterilize your forceps. Using sterile technique, deliver 1 sterile optochin disk to the primary area of streak of *Streptococcus pneumoniae* on your Blood Agar biplate. Alcohol-flame sterilize your forceps.

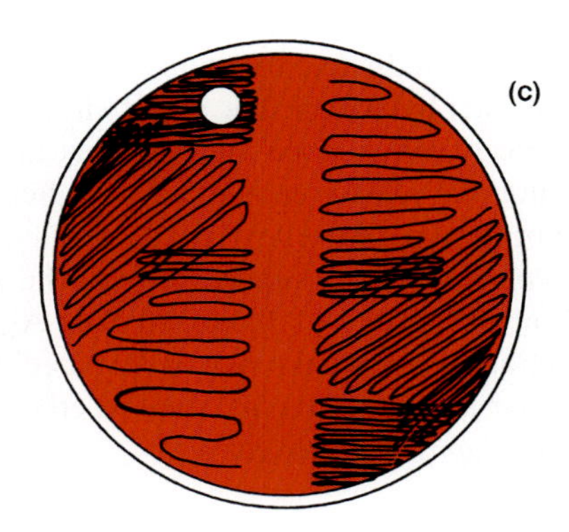
(c)

d Using sterile technique, deliver an optochin disk to the primary area of streak of *Streptococcus mitis* on your Blood Agar biplate. Touch the disks gently with the tips of your forceps. Why? Sterilize the forceps.

e Incubate the inverted plate at 35° C for 18 to 24 hours.

Subsequent Laboratory Sessions

a Carefully examine your **Blood Agar biplate.** In your laboratory report, note any visible evidence of **alpha hemolysis** surrounding either colony type. Hold the plate up toward a light to see the tiny halos of color surrounding punctiform, streptococcal colonies. If a Quebec colony counter is available, use it to help you evaluate hemolysis.

b Examine a zone of hemolysis through the low-power lens of your microscope. In your laboratory report, describe microscopic evidence of partial hemolysis.

c Note in your laboratory report whether close inspection reveals collapsed, **checker-shaped colonies.** Explain why these may be characteristic of one species but not the other.

d In your laboratory report, describe any evidence of **optochin sensitivity** that you observe.

e If your *Streptococcus pneumoniae* colonies have not collapsed, reincubate your Blood Agar biplate for another 24 hours at 35° C.

f Report your findings after the extended incubation interval.

Procedure 4

Bacitracin Sensitivity and Streptolysin Production

First Laboratory Session

a Label 2 **Blood Agar** plates with your group identification and the date. Label 1 plate to receive *Streptococcus pyogenes,* the other for *Enterococcus faecium.*

b With the appropriate species, streak each plate for isolation of colonies.

c Obtain a vial of **bacitracin** disks. Remember that the disks are sterile inside the vial and handle them with care.

d Loosen the lid of the vial, then alcohol-flame your forceps. Use the sterile forceps to place 1 bacitracin disk onto a corner of your primary streak of *Streptococcus pyogenes.* Touch the disk with the tips of your forceps to improve contact and adhesion.

e Resterilize your forceps. Place a bacitracin disk onto a corner of the primary streak of *Enterococcus faecium.* Touch the disk with the tips of your forceps. Alcohol-flame sterilize your forceps.

f Sterilize your inoculation loop; thoroughly cool the loop. Hold the instrument at a 45° angle to 1 of the Blood Agar plates, then gently **stab the loop 2 times** into the center of the primary streak, as illustrated. The 2 stabs should be about 1 cm apart. You want smooth stabs that push cocci down into a reduced environment and close completely over the bacteria as you remove the loop.

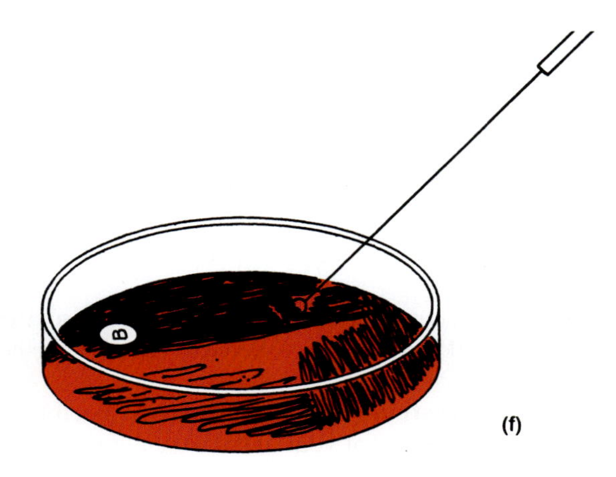

(f)

g Resterilize the loop, thoroughly cool it, and repeat the 2-stab process on the other plate.

h Incubate the inverted plates at 35° C for 24 hours.

Second Laboratory Session

a Examine your streak plates of *Streptococcus pyogenes* and *Enterococcus faecium.* In your laboratory report, describe the appearance of **isolated colonies** of each species.

b Examine the area surrounding the **bacitracin disk** on each plate. In your laboratory report, indicate where bacteria have hemolyzed the blood and where there are still red blood cells.

Bacteria that are sensitive to bacitracin are inhibited by this antibiotic. You will observe a blood-red area surrounding their disk. These sensitive microbes do not grow next to bacitracin, consequently they cannot hemolyze the red blood cells there. In contrast, resistant bacteria grow around the disk and lyse the red blood cells in the medium.

c In your laboratory report, label which species is resistant to bacitracin and which is sensitive to this antibiotic.

d Hold your *Streptococcus pyogenes* Blood Agar culture up to a light and inspect the area immediately surrounding your 2 stabs. **Streptolysin S** produces beta hemolysis on the surface of the plate and **streptolysin O,** which is oxygen-sensitive, generates hemolysis within the stabs. If a Quebec colony counter is available, use it to help you evaluate hemolysis.

e In your laboratory report, indicate the activity of streptolysins S and O (if possible) produced by *Streptococcus pyogenes.* If the strain of *Streptococcus pyogenes* that you are examining produces streptolysin S, the activity of streptolysin O may be difficult or impossible to assess. If only streptolysin O is produced, the 2 stabs will be surrounded by clear agar medium while the rest of the plate remains red.

f In your laboratory report, describe hemolysis produced by *Enterococcus faecium.*

g If possible, observe beta hemolysis around an isolated colony through the low-power lens of your microscope. In your laboratory report, describe your microscopic observations.

Procedure 5

The CAMP Test

First Laboratory Session

a Take 1 **Blood Agar** plate; turn it upside down. With your marking pen draw the largest possible capital letter "E" on the plastic bottom of the plate, but do not allow the horizontal lines to touch the vertical line. Label the plate as illustrated. Now turn the plate right side up.

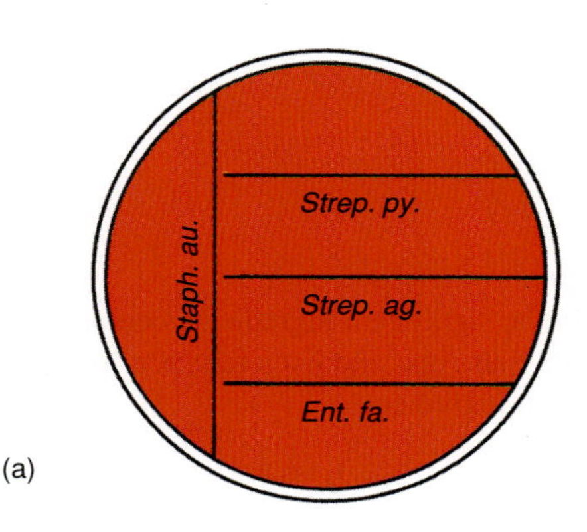

(a)

b With your inoculation loop, streak a single line of *Staphylococcus aureus* (ATCC e 25923) down the back of the "E." Sterilize your loop.

c As described in the following steps, inoculate each of the 3 cross-lines of the "E" with a different 1 of your 3 beta-hemolytic cultures.

d At the top of the "E," streak *Streptococcus pyogenes* toward the *Staphylococcus.* Bring the line very close but be careful not to touch the *Staphylococcus.* Sterilize your loop.

e Streak *Streptococcus agalactiae* toward the *Staphylococcus* in the middle of the "E." Again, the lines should almost touch. Sterilize your loop.

f For the bottom cross-line of the "E," streak with *Enterococcus faecium.* Remember to sterilize your loop before returning it to the laboratory bench.

g Invert the plate and incubate it at 35° C for 24 to 48 hours.

Second Laboratory Session

a Examine the **CAMP test.** Note any arrowhead-shaped area of intensified hemolysis where a line of streptococci or enterococci approaches the streak of *Staphylococcus aureus.* This heightening of lysis constitutes a positive CAMP reaction.

b In your laboratory report, draw and label the CAMP test streaks. Illustrate a positive CAMP reaction.

POSTTEST EXERCISE 27

Part 1

Matching

Each response may be used once, more than once, or not at all. Some statements may require more than one response. Use the immediately preceding set of responses for each set of statements.

a. *Staphylococcus epidermidis*
b. *Staphylococcus aureus*
c. Neither

1. _____ Catalase-positive
2. _____ Lysed by lysostaphin within 15 minutes
3. _____ Oxidase-positive
4. _____ The most common etiologic agent of boils, furuncles, carbuncles, wound infections, and bacterial food poisoning
5. _____ May produce beta hemolysis on Blood Agar
6. _____ Produces coagulase
7. _____ Ferments mannitol, yielding acid
8. _____ Tolerates over 5% NaCl content in media

a. All streptococci
b. No streptococci
c. *Streptococcus pneumoniae*
d. *Streptococcus mitis*

1. _____ Etiologic agent of pneumococcal pneumonia
2. _____ Alpha-hemolytic
3. _____ Strictly fermenters, lacking a cytochrome system
4. _____ Inhibited by optochin
5. _____ Frequently lancet-shaped diplococci
6. _____ Frequent cause of subacute bacterial endocarditis
7. _____ Most common cause of adult meningitis
8. _____ Homofermentative
9. _____ Catalase-positive

a. *Streptococcus pyogenes*
b. *Streptococcus agalactiae*
c. *Enterococcus faecium*
d. *Staphylococcus aureus*
e. None of these

1. _____ Group A, beta-hemolytic streptococci
2. _____ Group B, beta-hemolytic streptococci
3. _____ Bacitracin-sensitive streptococci
4. _____ SXT-resistant streptococci or enterococci
5. _____ Group C, beta-hemolytic streptococci
6. _____ Produces CAMP factor
7. _____ Shows increased hemolysis in CAMP test

Part 2

True or False (Circle one, then correct every false statement.)

1. T F *Bergey's Manual* places the staphylococci in the phylogenetically diverse section entitled Gram-positive Cocci.
2. T F Biochemical studies of DNA, cell wall composition, and RNA sequences underline the close relation of the staphylococci and micrococci to each other.
3. T F Most streptococci are catalase-negative; staphylococci are catalase-positive.
4. T F Impetigo, pneumonia, osteomyelitis, meningitis, enterocolitis, and urogenital infections are often caused by *Staphylococcus epidermidis* in previously healthy people.
5. T F *Staphylococcus aureus* enterotoxin remains poisonous even after thorough cooking of contaminated food kills the bacterial cells that produce it.
6. T F The coagulase test is considered "positive" only if a culture is able to congeal the entire tube of Coagulase Plasma EDTA.
7. T F Alpha hemolysis entails complete clearing of erythrocytes in Blood Agar.
8. T F Lancefield utilized group-specific streptococcal antigens to characterize groups of species.
9. T F New editions of *Bergey's Manual* never change established genera's names.
10. T F Streptococci ferment carbohydrates, yielding chiefly acetic acid.
11. T F *Streptococcus pneumoniae* is optochin-sensitive.
12. T F A lancet is a small surgical knife, sharp on both sides and pointed at the tip.
13. T F Pneumococcal pneumonia is spread primarily by acutely ill patients.
14. T F The nasopharynx is a cavity under the tongue where *Streptococcus pneumoniae* most often colonizes healthy carriers.
15. T F On a Blood Agar plate, optochin sensitivity of *Streptococcus pneumoniae* appears as a greenish ring around the optochin disk.
16. T F *Streptococcus mitis* is a virulent pathogen, especially when it is found among the oral flora of a human.
17. T F Streptococci often form chains in broth cultures.
18. T F After collapse due to autolysis, a *Streptococcus pneumoniae* colony appears flattened with a slightly raised outer ring.
19. T F The most frequently encountered pathogen among the beta-hemolytic streptococci is *Streptococcus pyogenes*.
20. T F GABHS infections usually respond well to antibiotic treatment.
21. T F Rapid techniques can be used to identify group A streptococci in a throat swab within minutes.
22. T F Bacitracin inhibits GABHS but generally does not inhibit other beta-hemolytic streptococci.
23. T F Streptolysin O is sensitive to oxygen and is antigenic.
24. T F Streptolysin S is sensitive to oxygen and is antigenic.
25. T F Christie, Atkins, and Munch-Peterson are credited with first describing a positive CAMP test reaction. The test is named in honor of them.

Part 3

Completion and Short Answer

1. List three general sites on the human body that are normally colonized by staphylococci.
   a.
   b.
   c.
2. List six types of clinical specimens from which pathogenic staphylococci are recovered.
   a.
   b.
   c.
   d.
   e.
   f.
3. Identify the significant ingredients contained in Mannitol Salt Agar and list a function for each significant ingredient.

4. Why is a rapid, accurate diagnosis of *Streptococcus pyogenes* important?

5. Fill in the following chart, showing the pathogenic potential of each of the four Lancefield groups that include beta-hemolytic streptococci and enterococci.

	Beta-Hemolytic Streptococci and Enterococci
Group	**Pathogenic Potential**
A	
B	
C	
D	

6. Define each of these terms.
   a. pharyngitis

   b. GABHS

   c. nonsuppurative

   d. sequelae

   e. reduced environment

   f. CAMP test

7. To test your complete mastery of the techniques that differentiate beta-hemolytic streptococci and enterococci, make a chart showing four Lancefield groups of beta-hemolytic cocci in the left column. Name one species for each group. Then, across the top of the chart, list three tests that are commonly used to separate the groups. Fill in the chart by noting the reaction of each group to each of the three tests.

Laboratory Report **27**

Name ________________________________

Section ______________

Medically Significant Gram-Positive Cocci

Procedure 1

Identifying Unknown Staphylococci

First Laboratory Session

(*a*) Record the identifying number of your unknown, staphylococcal culture. __________

(*d, e, f, g*) Fill in the following chart to characterize your unknown.

	Unknown Staphylococcus
Cellular morphology*	
Colony description	
Gram reaction* ______________	
Catalase reaction ______________	
Coagulase reaction ______________	

Second Laboratory Session

(*a*) If you incubated your coagulase test for 24 hours, record the results here. ____________________

(*c*) Discuss the selective and differential properties of Mannitol Salt Agar.

Record the reactions of your unknown on Mannitol Salt Agar.

Growth ____________________

Mannitol fermentation ____________________

(*d*) Could your unknown be *Staphylococcus aureus?* __________ Why or why not?

*Provide this information at your instructor's request.

Procedure 2

Microscopic Observations

(*b*) List the size, shape, and typical arrangements (singles, pairs, short chains, long chains, irregular clusters, tetrads, packets of 8 cells, etc.) for each species. Draw a representative oil-immersion lens field of each culture.

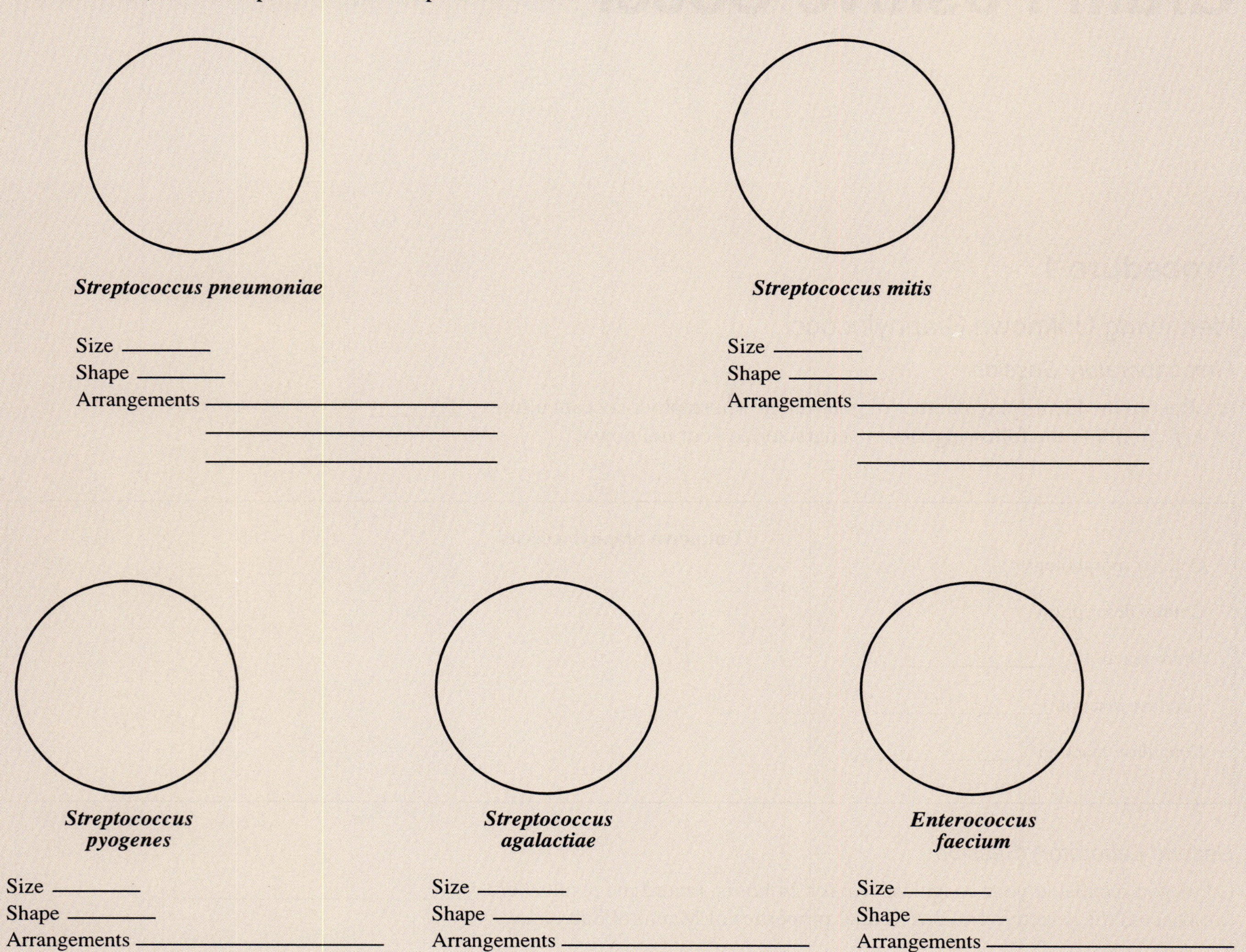

Streptococcus pneumoniae

Size ________
Shape ________
Arrangements ____________________________

Streptococcus mitis

Size ________
Shape ________
Arrangements ____________________________

Streptococcus pyogenes

Size ________
Shape ________
Arrangements ____________________________

Streptococcus agalactiae

Size ________
Shape ________
Arrangements ____________________________

Enterococcus faecium

Size ________
Shape ________
Arrangements ____________________________

Procedure 3

Optochin Sensitivity and Colony Description

Subsequent Laboratory Sessions

(*a*) Describe hemolysis as it appears around isolated colonies of your class cultures growing on Blood Agar. Note the duration and temperature of incubation. ______________________ ______________________

Streptococcus pneumoniae:

Streptococcus mitis:

(*b*) If possible, observe a zone of alpha hemolysis through your microscope. Describe the microscopic evidence of partial disruption of erythrocytes as it appears to you. Keep the Petri dish closed to avoid contaminating your microscope.

(*c*) Does close inspection of *Streptococcus pneumoniae* reveal collapsed colonies? ________
Explain why checker-shaped colonies may be characteristic of one species but not the other.

(*d*) Which species displays optochin sensitivity? _________________
Describe the visible evidence of optochin sensitivity.

(*f*) How many hours did you incubate *Streptococcus pneumoniae* on Blood Agar before collapsed colonies appeared?
______________________ .

Procedure 4

Bacitracin Sensitivity and Streptolysin Production

Second Laboratory Session

(*a*) Describe the appearance of isolated colonies of each species.
Streptococcus pyogenes:

Enterococcus faecium:

(*b*) Indicate where bacteria have hemolyzed the blood and where there are still red blood cells.

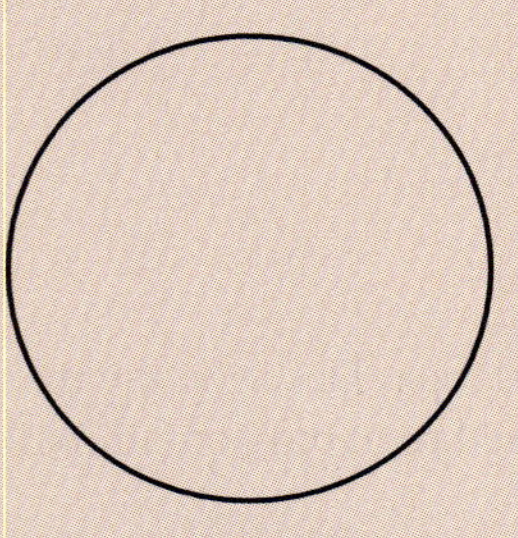

Streptococcus pyogenes

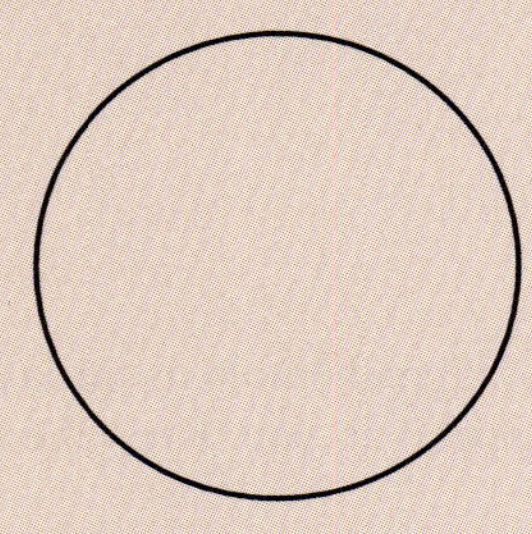

Enterococcus faecium

Evidence of Streptolysin Activity

(*c*) Which species is resistant to bacitracin? ____________________
Which species is sensitive to this antibiotic? ____________________
(*e*) Describe any evidence of streptolysin S activity that you observe on your plate of *Streptococcus pyogenes.*

Describe any evidence of streptolysin O activity that you observe on your plate of *Streptococcus pyogenes.*

(*f*) Describe any hemolysis produced by *Enterococcus faecium.*

(*g*) If possible, describe the microscopic appearance of beta hemolysis surrounding an isolated colony.

Procedure 5

The CAMP Test

Second Laboratory Session

(*a*) Is there an arrowhead-shaped area of enhanced hemolysis where one of the *Streptococcus* or *Enterococcus* species approaches *Staphylococcus aureus?* _________

If so, which species is producing this positive reaction in the CAMP test? ___________________________

Is this species producing the CAMP factor? _________

(*b*) Illustrate the reaction here; label your drawing.

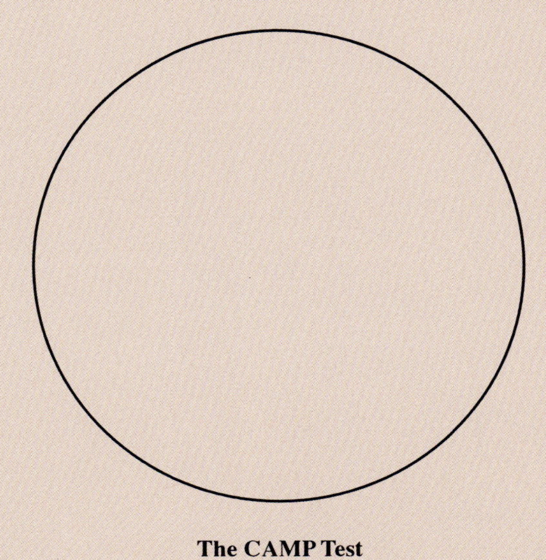

The CAMP Test

Exercise 28

Intestinal Pathogens: *Campylobacter* and Selected Enterobacteriaceae

Necessary Skills

Aseptic transfer techniques; Gram staining; Culture characteristics; Streak-plate techniques; Knowledge of: Selective and differential media, Selected microbial enzymatic reactions including cytochrome oxidase and nitrate reduction; Knowledge of other enzymatic reactions (recommended)

Materials

> **Warning:** Handle cultures with care. *Salmonella, Shigella,* and *Campylobacter* are pathogens.

> **Warning:** Hydrogen gas is flammable. Observe appropriate precautions when handling gas-generating envelopes.

First Laboratory Session

Cultures:

Procedures 1, 2, and 3 (24-hour, 35° C, Tryptic Soy Broth, 1 per set)

- *Escherichia* species
- *Salmonella* species
- *Shigella* species

Procedure 4

- *Campylobacter jejuni* (48-hour, 35° C, Campylobacter Agar plus 10% defibrinated sheep blood, with a CampyPak® in a GasPak® jar with palladium catalyst)

Supplies:

- Access to *Bergey's Manual*
- Access to the *DIFCO Manual*

Procedures 1, 2, and 3 (per group)

- Selenite Broth, at least 5 cm deep
- 4 Hektoen Enteric Agar plates
- Capped, sterile test tube
- 3 Sterile Pasteur pipets
- Bulb

Procedure 4

- 2 Campylobacter Agar (plus 10% defibrinated sheep blood) plates per group

Additional supplies (per class):

- GasPak® jar with palladium catalyst
- CampyPak® (BBL)
- 10 ml Pipet
- Pipettor
- Container of tap water (at least 10 ml per CampyPak®)
- Scissors
- Access to a 42° C incubator

Second Laboratory Session

Supplies (per group):

- 3 Enterotube II®* tubes
- Access to a package insert from Enterotube II® system
- 1 Hektoen Enteric Agar plate
- Access to the *DIFCO Manual*
- Access to refrigeration
- 3 Motility Medium S tubes (on ice)

Third Laboratory Session

Materials (per class):

- Dropper bottles of the following reagents:
  - Kovacs' reagent
  - Barritt's reagents for Voges-Proskauer testing
    - Solution A: alpha-naphthol, 5%, in 95% ethanol
    - Solution B: potassium hydroxide, 40% aq.

*Available from Roche Diagnostic Systems, 11 Franklin Ave., Belleville, NJ 07109.

Access to sharp probes
Access to a *Computer Coding and Identification System for Enterotube II®* (to be referred to in this exercise as the *Coding Manual*)
Access to a package insert from Enterotube II® system
Additional supplies (per group):
3 Computer coding pad sheets (recommended)
Gram stain reagents

Primary Objective

Grow and identify *Campylobacter jejuni, Salmonella,* and *Shigella.*

Other Objectives

1. Describe cells, colonies, and metabolic characteristics of *Campylobacter jejuni.*
2. Define: serovar, gastroenteritis, shigellosis, bacillary dysentery, salmonellosis, plasmid.
3. List characteristics shared by members of the family Enterobacteriaceae.
4. Evaluate the relatedness of *Escherichia* and *Shigella.*
5. Estimate the medical significance of ETEC and EIEC.
6. Describe nomenclature used to identify serovars of *Salmonella.*
7. Observe and evaluate the appearance of *Salmonella, Shigella,* and *Escherichia coli* on Hektoen Enteric Agar.
8. Explain and evaluate the effects of Selenite Broth on a mixture of *Salmonella, Shigella,* and *Escherichia coli.*
9. Inoculate and evaluate Enterotube II® systems.

INTRODUCTION

In this exercise you identify common intestinal pathogens including ***Campylobacter jejuni,*** a leading etiologic agent of **gastroenteritis** (inflammation of the gastrointestinal tract).

You also work with ***Salmonella*** and ***Shigella,*** pathogenic genera belonging to the family **Enterobacteriaceae.** Many common **serovars** (antigenically distinct groups) of *Salmonella* cause **salmonellosis,** a gastroenteritis. *Shigella* is the etiologic agent of **shigellosis,** also known as **bacillary dysentery,** a disease that sometimes includes ulceration of the lower intestine. Shigellosis can be fatal, especially in children. An increasing occurrence of multiple drug resistance among the Enterobacteriaceae complicates treatment.

All three genera are frequently encountered by clinical microbiologists. Transmitted by the **fecal-oral route** in food, milk, or water, *Campylobacter jejuni, Salmonella,* and *Shigella* cause pain and suffering in every country on earth.

Campylobacter

Campylobacters are slender, vibrioid cells with polar flagella and corkscrewlike motility. They are members of Bergey's tongue-twisting group **Aerobic/Microaerophilic, Motile, Helical/Vibrioid Gram-Negative Bacteria.** Watch for typical, gull wing-shaped cells of ***Campylobacter jejuni*** in your Gram stain.

Interestingly, while campylobacters obtain energy from amino acids, they neither oxidize nor ferment carbohydrates. Furthermore, they are oxidase-positive but microaerophilic. *Campylobacter jejuni* demands 3% to 10% CO_2 with 3% to 10% oxygen (atmospheric oxygen is 21%). This species flourishes at 42° C, a temperature that inhibits the growth of many other fecal bacteria.

In this exercise you cultivate *Campylobacter jejuni* on **Campylobacter Agar,** a selective medium that contains antibiotics to prevent overgrowth by competing microbes. This nutrient-rich medium is supplemented with sheep blood to improve growth and identification of the nonhemolytic *Campylobacter jejuni* colonies, which are usually grey to pinkish and mucoid. Watch for their typical tailing along your streak line.

Enterobacteriaceae

Accurate recognition of the **Enterobacteriaceae** is essential for the clinical or veterinary microbiologist. Members of the family are gram-negative, non-endospore-forming, and non-acid-fast rods; if motile, they are generally peritrichously flagellated. **Chemoorganotrophs,** they utilize organic compounds as the sole source of both energy and carbon. Most are catalase-positive and able to reduce nitrate. All are oxidase-negative.

Table 28.1 is a helpful flow chart showing one way to differentiate selected, medically significant members of the Enterobacteriaceae.

Escherichia and *Shigella*

You also investigate, in this exercise, the biochemical capabilities of ***Escherichia.*** This genus of Enterobacteriaceae is common among normal intestinal flora in warm-blooded animals and is frequently implicated in gastrointestinal disease.

Escherichia shares medically significant **plasmids** (transmissible, extrachromosomal pieces of DNA) with many other genera in the family. Some of the plasmids carry DNA that codes for enzymes and structures that increase bacterial resistance to antibiotics. Other plasmids enable the bacteria to produce **enterotoxins,** poisons that alter host intestinal mucosa causing the nausea, vomiting, and severe diarrhea of gastroenteritis.

Traditionally, microbiologists called *Salmonella* and *Shigella* the **"enteric pathogens."** Other Enterobacteriaceae, including **enterotoxigenic *Escherichia coli* (ETEC),** were categorized as **opportunistic pathogens.** An **opportunist** is a microorganism that causes disease in a **compromised** host (weakened and at high risk of infection due to some underlying condition or preexisting disease).

This categorization of *Escherichia coli* is now disputed. Epidemiologists have discovered that throughout the world, ETEC is a frequent cause of traveler's diarrhea, a severe gastroenteritis that develops in previously healthy people.

Table 28.1 Differentiation of Selected Enterobacteriaceae[1]

Enterobacteriaceae are gram-negative, non-endospore-forming, non-acid-fast rods; if motile, peritrichously flagellated; chemoorganotrophs; oxidase-negative; most are catalase-positive and able to reduce nitrate.

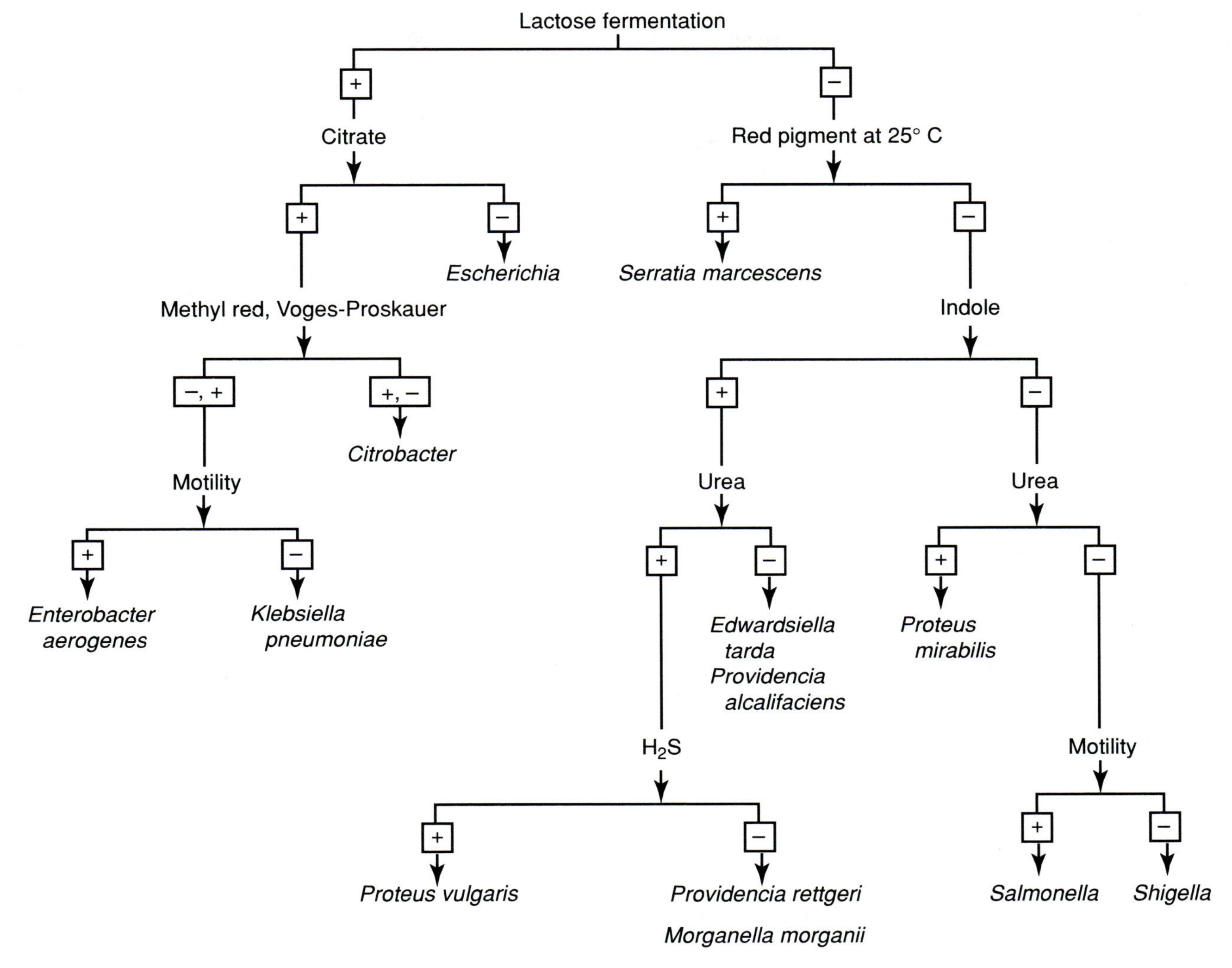

[1]Only the most common reactions are listed in this table. For an extensive review of the many exceptions, see Holt, J. G., et al., eds., *Bergey's Manual of Determinative Bacteriology,* 9th ed., Williams & Wilkins Co., Baltimore, MD, 1994.

Like the shigellae, some serovars of *Escherichia coli* are **enteroinvasive (EIEC);** they ulcerate the intestine causing dysentery. The disease produced by EIEC is identical to shigellosis.

Bergey's Manual of Systematic Bacteriology notes that, in addition to pathogenic potential, *Escherichia* and *Shigella* share similar biochemical capabilities and exhibit a high degree of DNA relatedness. So alike are the two genera that *Bergey's Manual* suggests grouping all four species of *Shigella* together with *Escherichia* as one species. The current separation of species is retained primarily for ease of communication within the medical community.

At this time, microbiologists have no convenient biochemical test to distinguish all EIEC from noninvasive representatives of the species; therefore, EIEC serovars are still categorized with the "opportunistic" genus rather than with the "pathogenic" shigellae.

In cases of gastroenteritis and even dysentery, *"Escherichia coli"* is as suspect as the "enteric pathogens." When investigating the etiology of intestinal disease, it cannot be assumed that *Escherichia coli* is a harmless member of the normal flora.

Salmonella

Currently over 2,000 closely related serovars of *Salmonella* have been identified. In Exercise 33, Serology, representative immunological techniques that help identify groups of serovars are introduced. In this exercise you examine biochemical techniques for identifying the genus and some of its serovars.

Serovars of *Salmonella* were first given specific epithets that indicated the disease and/or the animal from which they were isolated. Examples include *Salmonella typhi,* the etiologic agent of typhoid fever, and the ubiquitous *Salmonella typhimurium,* a common cause of food-borne infection. The specific epithet *"typhimurium"* indicates that this serovar was first isolated from which animals?*

As more serovars were identified, microbiologists named them for regions of origin. Look in *Bergey's Manual* or the *DIFCO Manual* to see if your area is represented among the names—which include *Salmonella chicago, Salmonella canada,* and *Salmonella texas.* The most recent isolates are simply identified by a formula indicating their antigenic makeup.

Identifying *Salmonella, Shigella,* and *Escherichia*

In this exercise you examine biochemical features of *Salmonella, Shigella,* and *Escherichia.* Some of their characteristic reactions are specified in Appendix 2.

A patient's feces usually contain many genera of bacteria, not just the etiologic agent of disease. Therefore a clinical microbiologist, looking for enteric pathogens of the family Enterobacteriaceae, generally inoculates fecal material into both an enrichment medium and an agar selective for gram-negative bacteria; see figure 28.1. Biochemical tests are performed and results confirmed with serological procedures.

In the procedures section of this exercise, you test the ability of the selective enrichment medium **Selenite Broth** to favor the growth of *Salmonella* while inhibiting the other two genera. This medium includes sodium selenite as a selective agent capable of retarding the growth of **enterococci** (spherical bacteria found in human intestines) and **coliforms** (gram-negative, facultative or aerobic rods that ferment lactose, producing acid and gas within 48 hours at 35° C). At this time, there is no effective enrichment or selective medium intended exclusively for isolation of *Shigella.*

You also attempt to isolate and examine colonies of the three bacterial genera on a medium that is both selective and differential. This medium, **Hektoen Enteric Agar,** retards the growth of gram-positive bacteria.

Certain biochemical capabilities of surviving bacteria are displayed on Hektoen Enteric Agar. Colonies that yield hydrogen sulfide have black centers. Lactose-, sucrose-, and salicin-negative (nonfermenting) colonies are greenish-blue. Bacteria that ferment any or all of these sugars form orange colonies.

* Hint: Check your dictionary for the word *murine.*

Figure 28.1 Laboratory culture of fecal specimens for retrieval of pathogenic Enterobacteriaceae

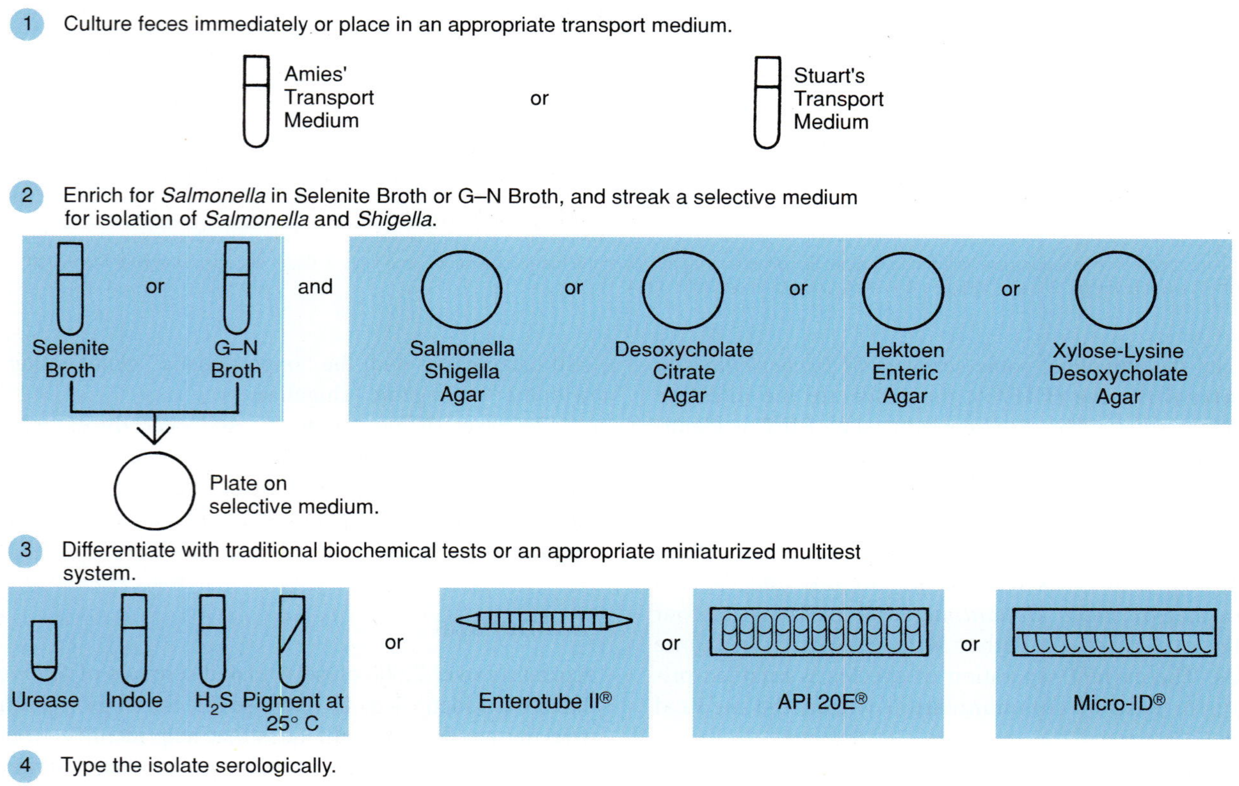

1 Culture feces immediately or place in an appropriate transport medium.

Amies' Transport Medium or Stuart's Transport Medium

2 Enrich for *Salmonella* in Selenite Broth or G–N Broth, and streak a selective medium for isolation of *Salmonella* and *Shigella.*

Selenite Broth or G–N Broth and Salmonella Shigella Agar or Desoxycholate Citrate Agar or Hektoen Enteric Agar or Xylose-Lysine Desoxycholate Agar

Plate on selective medium.

3 Differentiate with traditional biochemical tests or an appropriate miniaturized multitest system.

Urease Indole H_2S Pigment at 25° C or Enterotube II® or API 20E® or Micro-ID®

4 Type the isolate serologically.

Figure 28.2 An Enteric Quad® (Courtesy of Remel, Lenexa, KA 66215)

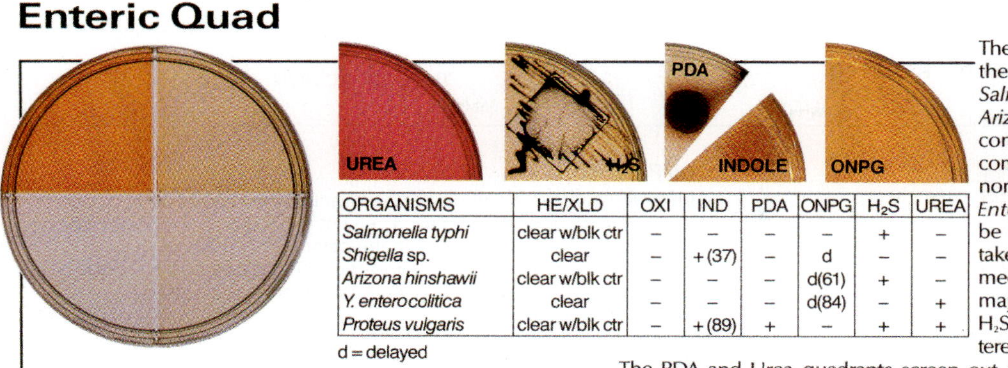

ORGANISMS	HE/XLD	OXI	IND	PDA	ONPG	H_2S	UREA
Salmonella typhi	clear w/blk ctr	–	–	–	–	+	–
Shigella sp.	clear	–	+(37)	–	d	–	–
Arizona hinshawii	clear w/blk ctr	–	–	–	d(61)	+	–
Y. enterocolitica	clear	–	–	–	d(84)	–	+
Proteus vulgaris	clear w/blk ctr	–	+(89)	+	–	+	+

d = delayed

The ENTERIC QUAD is intended for the presumptive identification of *Salmonella, Shigella, Yersinia,* and *Arizona sp.* from fecal specimens. Six conventional biochemical tests are combined into one convenient, economical unit for the identification of *Enterobacteriacea*. The plate should be inoculated with a pure culture taken from a differential or selective medium, such as HE or XLD Agar. The majority of non-lactose fermenting H_2S producing organisms encountered in fecal specimens are *Proteus* sp. The PDA and Urea quadrants screen out these non-pathogenic organisms, eliminating the need for additional biochemical and seriological testing, thus providing an economical means of screening specimens for enteric pathogens.

Figure 28.3 The Enterotube II® (Courtesy of Roche Diagnostic Systems, Division of Hoffman-Laroche, Inc., Nutley, NJ 07110)

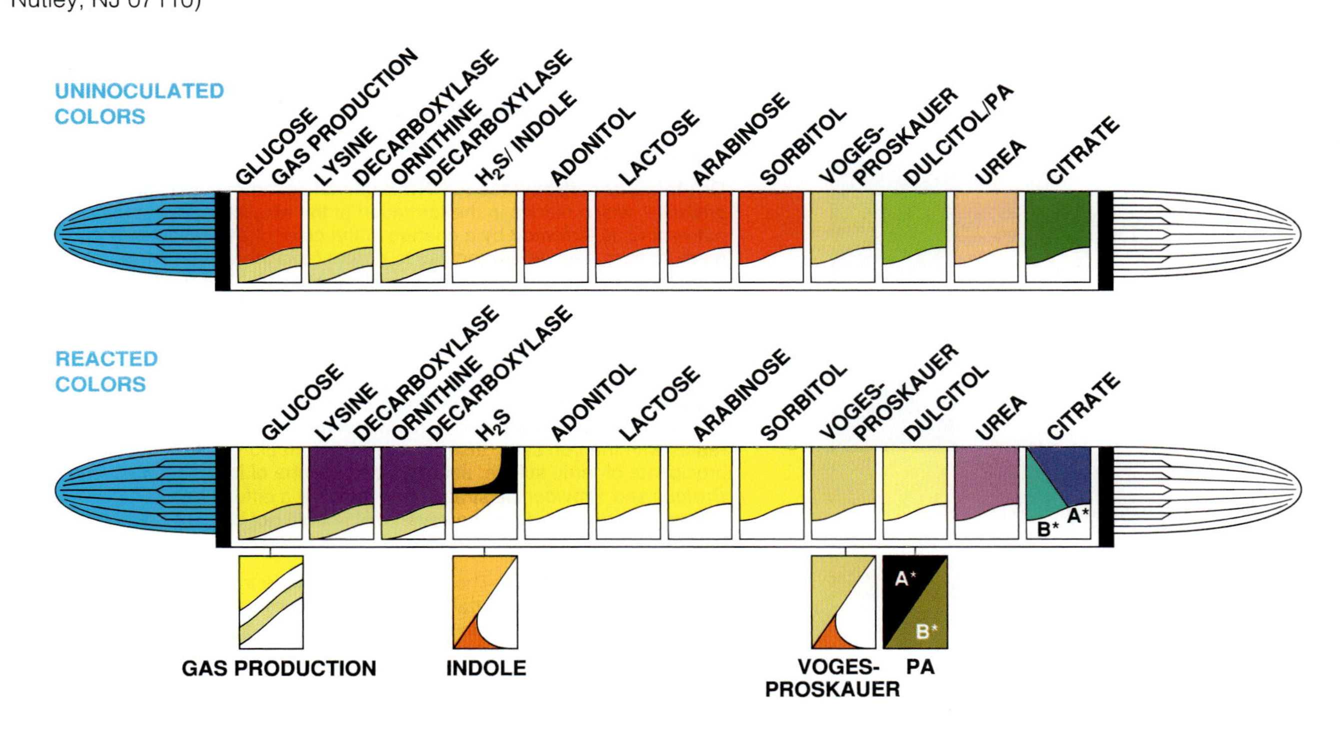

For presumptive identification of genera, many laboratories inoculate **Enteric Quad®** plates (see figure 28.2) or **Enterotube II®** systems (see figure 28.3). In this exercise you examine the biochemical capabilities of the three genera in some detail by inoculating each genus into an **Enterotube II®**.

The **Enterotube II®**, a miniaturized biochemical testing system, is inoculated with bacteria from an isolated colony on your Hektoen Enteric Agar plate. After incubation, it yields accurate results of 15 separate biochemical reactions chosen for their ability to help in the presumptive identification of the Enterobacteriaceae; see figure 28.4.

Utilizing a coding manual to help interpret the results of the Enterotube II® system, you identify serovars and species of your class cultures of *Salmonella, Shigella,* and *Escherichia.* A coding manual is also used to identify other Enterobacteriaceae including those that have been implicated in intestinal disease, such as *Yersinia enterocolitica.*

In laboratories that are not employing Enterotube II systems, microbiologists often use **Motility Medium S** to

Figure 28.4 Enterotube II® biochemical color reactions (Courtesy of Roche Diagnostic Systems, Division of Hoffman-Laroche, Inc., Nutley, NJ 07110)

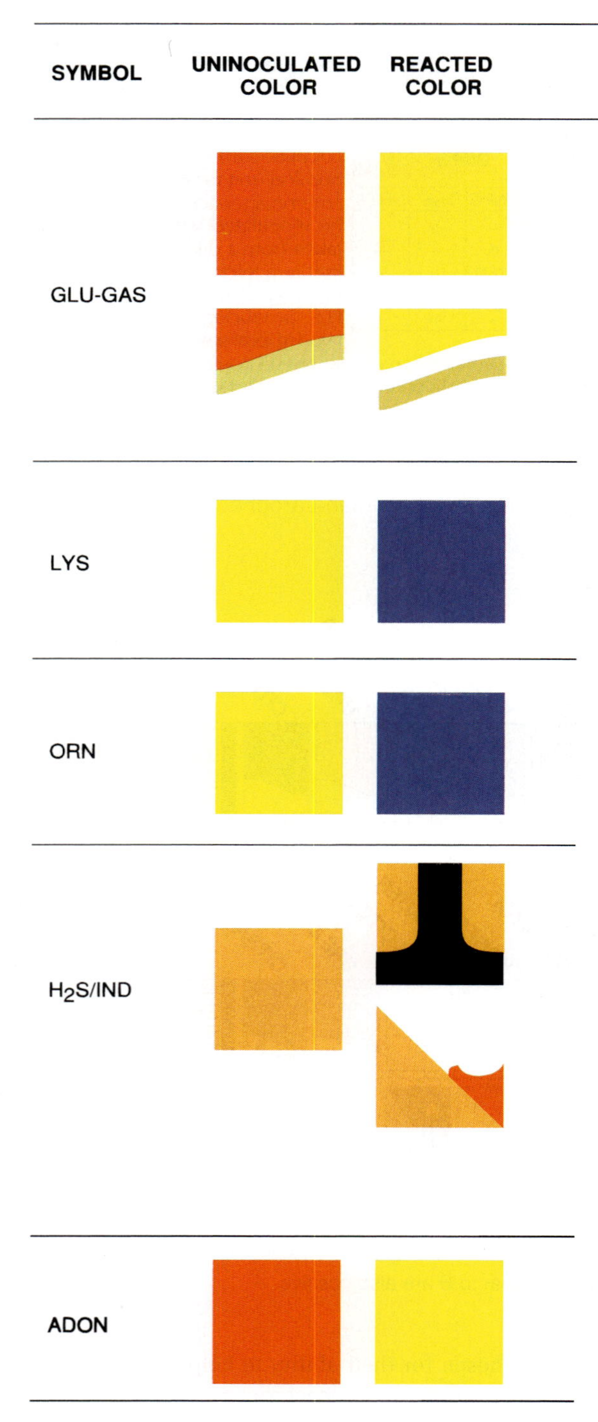

TYPE OF ACTION

COMPARTMENT 1

Glucose (GLU): The end products of bacterial fermentation of glucose are either acid, or acid and gas. The shift in pH due to the production of acid is indicated by a change in the color of the indicator in the medium from red (alkaline) to yellow (acidic). Any degree of yellow should be interpreted as a positive reaction; orange should be considered negative.

Gas production (GAS): This is evidenced by a definite and complete separation of the wax overlay from the surface of the glucose medium but not by bubbles in the medium. Since the amount of gas produced by different bacteria varies, the amount of separation between medium and overlay will also vary with the strain being tested.

COMPARTMENT 2

Lysine decarboxylase (LYS): Bacterial decarboxylation of lysine, which results in the formation of the alkaline end product cadaverine, is indicated by a change in the color of the indicator in the medium from pale yellow (acidic) to purple (alkaline). Any degree of purple should be interpreted as a positive reaction. The medium remains yellow if decarboxylation of lysine does not occur.

COMPARTMENT 3

Ornithine decarboxylase (ORN): Bacterial decarboxylation of ornithine, which results in the formation of the alkaline end product putrescine, is indicated by a change in the color of the indicator in the medium from pale yellow (acidic) to purple (alkaline). Any degree of purple should be interpreted as a positive reaction. The medium remains yellow if decarboxylation of ornithine does not occur.

COMPARTMENT 4

H_2S production (H_2S): Hydrogen sulfide is produced by bacteria capable of reducing sulfur-containing compounds such as peptones and sodium thiosulfate present in the medium. The hydrogen sulfide reacts with the iron salts also present in the medium to form a black precipitate of ferric sulfide, usually along the line of inoculation. Some *Proteus* and *Providencia* strains may produce a diffuse brown coloration in this medium; however, this should not be confused with true H_2S production (i.e., presence of black color).

Indole formation (IND): The production of indole from the metabolism of tryptophan by the bacterial enzyme tryptophanase is detected by the development of a pink to red color after the addition of Kovacs' indole reagent, which is injected into the compartment after 18 to 24 hours incubation of the tube.

COMPARTMENT 5

Adonitol (ADON): Bacterial fermentation of adonitol, which results in the formation of acidic end products, is indicated by a change in color of the indicator present in the medium from red (alkaline) to yellow (acidic). Any sign of yellow should be interpreted as a positive reaction; orange should be considered negative.

Figure 28.4 Continued

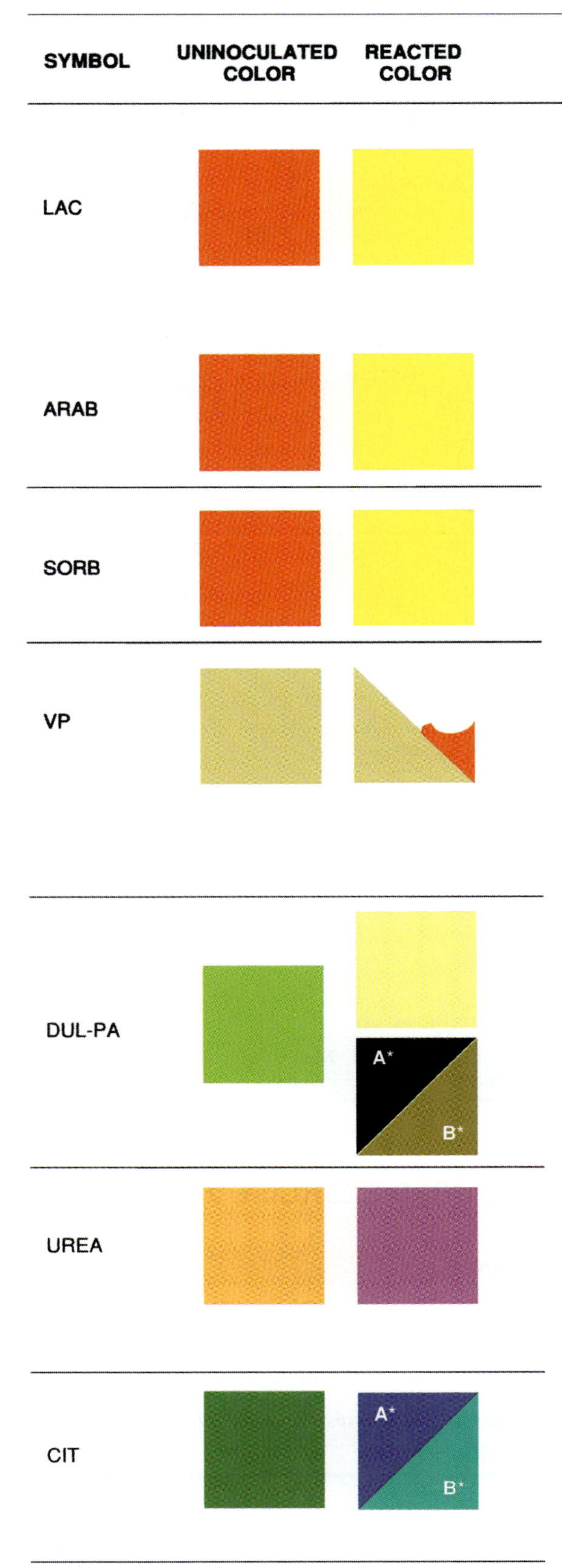

TYPE OF ACTION

COMPARTMENT 6

Lactose (LAC): Bacterial fermentation of lactose, which results in the formation of acidic end products, is indicated by a change in color of the indicator present in the medium from red (alkaline) to yellow (acidic). Any sign of yellow should be interpreted as a positive reaction; orange should be considered negative. This test is useful to confirm the lactose reaction of colonies taken from various enteric differential media (e.g. Salmonella Shigella [SS], MacConkey [MAC], Eosin Methylene Blue [EMB]).

COMPARTMENT 7

Arabinose (ARAB): Bacterial fermentation of arabinose, which results in the formation of acidic end products, is indicated by a change in color of the indicator present in the medium from red (alkaline) to yellow (acidic). Any sign of yellow should be interpreted as a positive reaction; orange should be considered negative.

COMPARTMENT 8

Sorbitol (SORB): Bacterial fermentation of sorbitol, which results in the formation of acidic end products, is indicated by a change in color of the indicator present in the medium from red (alkaline) to yellow (acidic). Any sign of yellow should be interpreted as a positive reaction; orange should be considered negative.

COMPARTMENT 9

Voges-Proskauer (VP): Acetyl methyl carbinol (acetoin) is an intermediate in the production of 2,3-butanediol from glucose fermentation. The production of acetoin is detected by the addition of two drops of a 40% w/v aqueous solution of potassium hydroxide containing 0.3% w/v of creatine and three drops of a 5% w/v solution of alpha-naphthol in absolute ethyl alcohol. The presence of acetoin is indicated by the development of a red color within 20 minutes. However, most positive reactions are evident within 10 minutes.

COMPARTMENT 10

Dulcitol (DUL): Bacterial fermentation of dulcitol, which results in the formation of acidic end products, is indicated by a change in color of the indicator present in the medium from green (alkaline) to yellow or pale yellow (acidic).

Phenylalanine deaminase (PA): This test detects the formation of pyruvic acid from the deamination of phenylalanine. The pyruvic acid formed reacts with a ferric salt present in the medium to produce a characteristic black to smoky gray color. Ferric chloride need not be added since the medium already contains an iron salt.

COMPARTMENT 11

Urea (UREA): Urease, an enzyme possessed by various microorganisms, hydrolyzes urea to ammonia causing the color of the indicator in the medium to shift from yellow (acidic) to red-purple (alkaline). The urease test is strongly positive for *Proteus* species and may be evident as early as 4 to 6 hours after incubation; it is weakly positive (light pink color) after 18 to 24 hours incubation for *Klebsiella* and some *Enterobacter* species.

COMPARTMENT 12

Citrate (CIT): This test detects those organisms which are capable of utilizing citrate, in the form of its sodium salt, as the sole source of carbon. Organisms capable of utilizing citrate produce alkaline metabolites which change the color of the indicator from green (acidic) to deep blue (alkaline). Any degree of blue should be considered positive.

*Certain bacteria do not produce the strong, positive A response. Reactions similar to B are also positive.

Figure 28.5 A flowchart of Procedures 1, 2, and 3

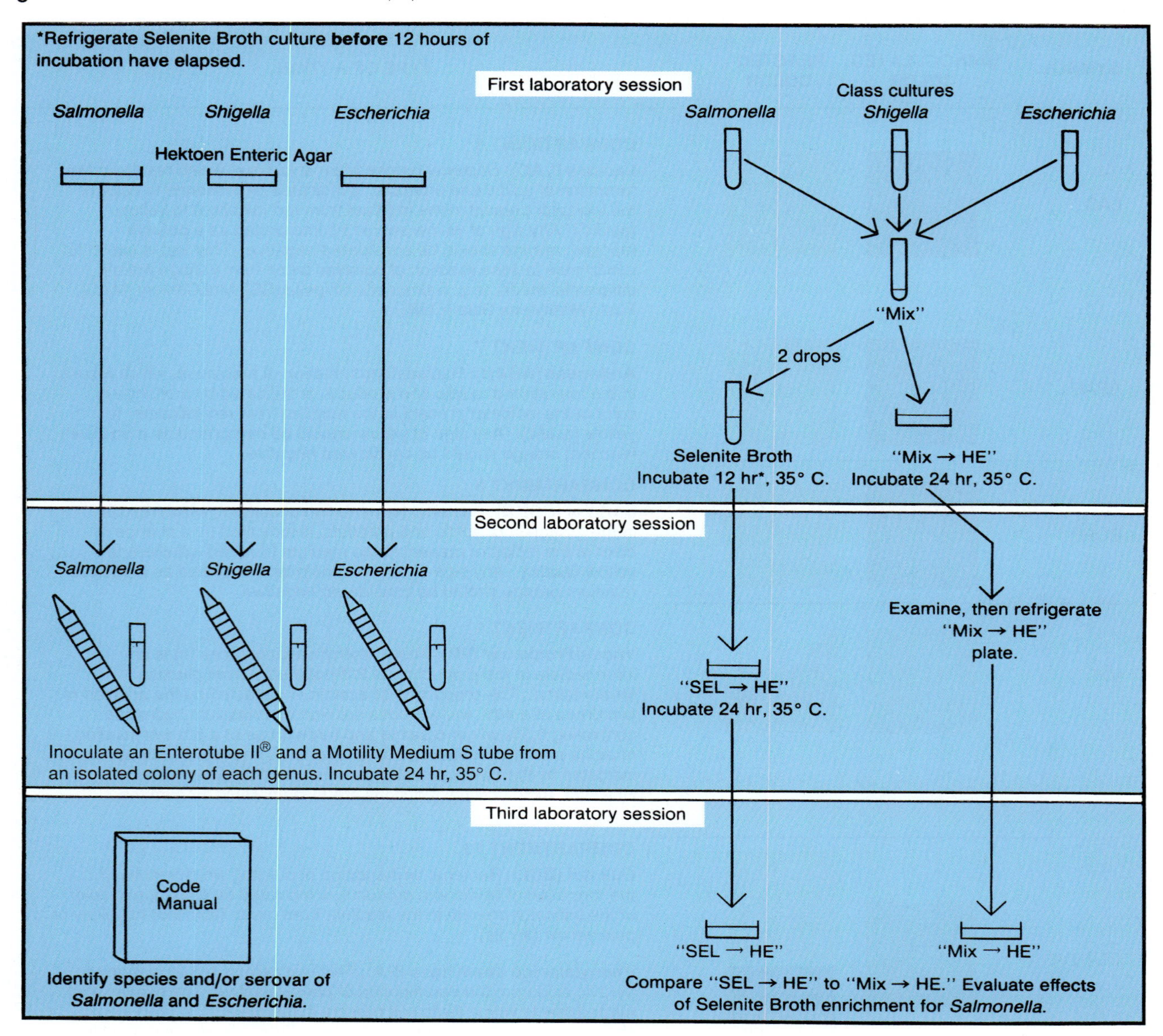

help distinguish between motile *Escherichia* and *Salmonella* and their non-motile relative, *Shigella*. Motile bacteria swim out from a short stab inoculation into this semisolid medium, **reducing** their environment as they go. The medium contains **2,3,5-triphenyltetrazolium chloride (TTC),** a **redox potential indicator** that turns red in reduced environments. In this exercise you learn to evaluate microbial motility by using Motility Medium S.

Before beginning the procedures section of this exercise, study figure 28.5, a flowchart of your laboratory work.

PROCEDURES EXERCISE 28

Procedure 1

Isolating *Salmonella, Shigella,* and *Escherichia*

Warning: Handle cultures with care. *Salmonella* and *Shigella* are pathogens.

First Laboratory Session

a Label 3 plates of **Hektoen Enteric Agar** as in the illustration. Streak the plates for isolation of the appropriate bacteria.

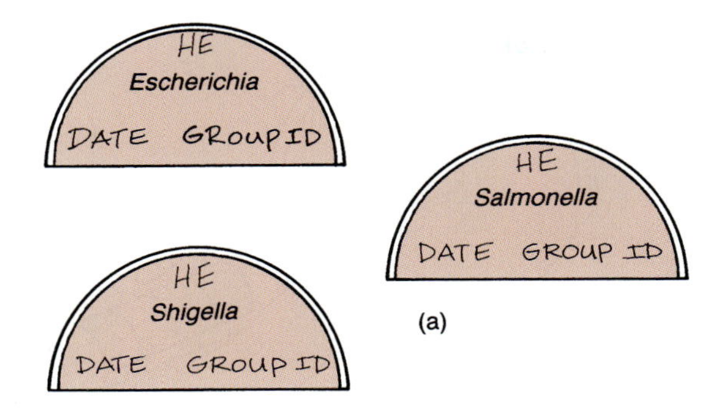

(a)

b Invert the plates and incubate them 18 to 24 hours at 35° C.

c In your laboratory report, discuss why saccharose (sucrose) and salicin are included in Hektoen Enteric Agar when neither salmonellae nor shigellae ferment either sugar. For help, see the *DIFCO Manual* and/or Appendix 2 of this laboratory manual.

Second Laboratory Session

a Examine well-isolated colonies on each **Hektoen Enteric Agar** plate. In your laboratory report, write a complete description of well-isolated *Salmonella, Shigella,* and *Escherichia* colonies as they appear on Hektoen Enteric Agar.

b Interpret the biochemical reactions evidenced by the colonies. Compare your observations with the expected results listed in the *DIFCO Manual* and with the tabulated responses listed in Appendix 2.

c Is there evidence that Hektoen Enteric Agar inhibits the growth of any of the 3 genera?

d Do not discard your plates until you have completed Procedures 2 and 3. If possible, refrigerate your inverted plates so you can refer to them during the third laboratory session.

Procedure 2

Enrichment for *Salmonella*

First Laboratory Session

a Gather 3 sterile Pasteur pipets, a bulb, and a sterile test tube.

b Using sterile technique, **prepare a mixed culture** of *Salmonella, Escherichia,* and *Shigella.* As in the illustration, try to deliver an approximately equal number of bacterial **cells** of each genus into the test tube. The number of drops you deliver from each culture will vary depending on the turbidity of the cultures.

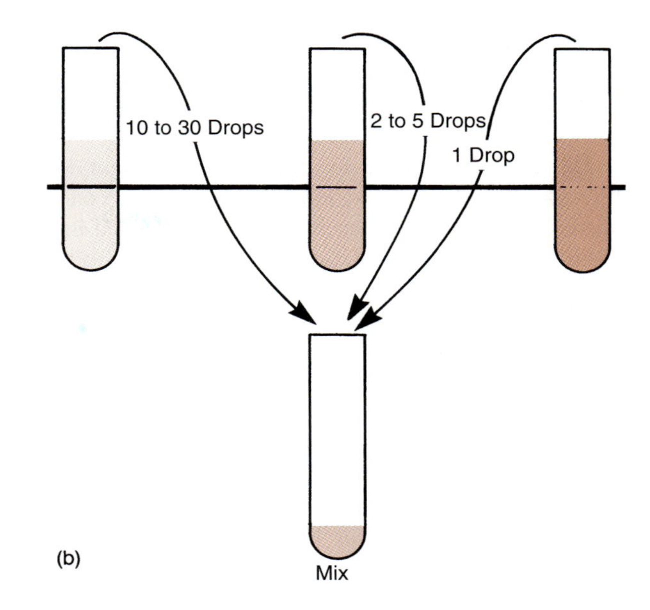

(b)

c Discard the contaminated pipets in an appropriate manner.

d Thoroughly mix the contents of the test tube.

e Label 1 **Hektoen Enteric Agar** plate and 1 **Selenite Broth** as in the illustration. With your flamed and cooled inoculating loop, prepare a 3-part streak of the mixed culture for isolation of colonies on your **"Mix→HE"** plate.

(e)

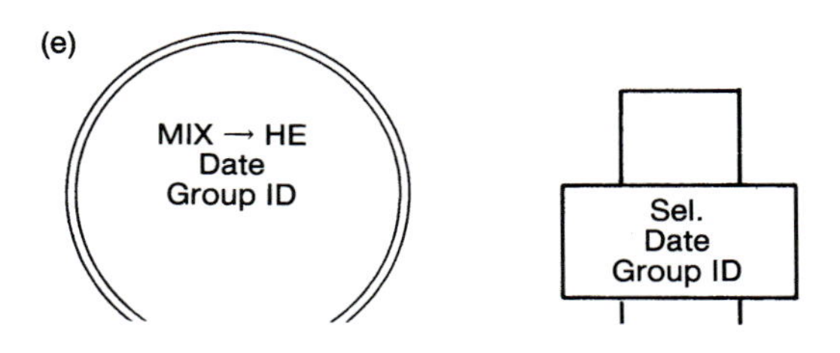

f Inoculate 2 loopfuls of the mixed culture into an appropriately labeled tube of **Selenite Broth.**

g Invert and incubate the Petri plate 18 to 24 hours at 35° C.

h Selenite Broth generally favors the growth of *Salmonella* during the first 12 hours of incubation. If possible, incubate your Selenite Broth culture for **no more than** 12 hours.

Second Laboratory Session

a Examine the growth on your **"Mix→HE"** plate prepared from your mixed culture. Carefully compare the colonies with those of your pure culture plates prepared during Procedure 1.

b Estimate the percentage of each type of colony—*Salmonella, Shigella,* and *Escherichia*—on the "Mix→HE" plate. Record your findings in your laboratory report.

c Refrigerate your inverted pure culture and mixed culture plates until the next laboratory session. You will compare the proportion of each genus in the mixed inoculum with the proportion that survives the Selenite Broth.

d In your laboratory report, note the number of hours your Selenite Broth culture was incubated at 35° C. Thoroughly resuspend the bacteria in your Selenite Broth.

e Label 1 **Hektoen Enteric Agar** plate. Include the tag **"Sel→HE."**

f With your flamed and cooled inoculating loop, prepare a 3-part streak of the **Selenite Broth** culture for isolation of colonies on your **"Sel→HE"** plate.

g Invert and incubate the plate at 35° C for 18 to 24 hours.

Third Laboratory Session

a Examine the bacterial growth on your **"Sel→HE"** plate, which was inoculated with a mixed culture from your Selenite Broth. Carefully compare these colonies with those on your pure culture plates prepared during Procedure 1. Did you recover all 3 genera of bacteria from the Selenite Broth?

b In your laboratory report, estimate the percentage of each type of colony—*Salmonella, Shigella,* and *Escherichia*—on your "Sel→HE" plate.

c Does Selenite Broth encourage the growth of *Salmonella* while discouraging coliforms and/or *Shigella?* To answer this question, compare proportions of each type of bacteria in the mix before and after incubation in Selenite Broth. In other words, compare your data from the "Mix→HE" plate with those from the "Sel→HE" plate.

Procedure 3

The Enterotube II® and Motility

Second Laboratory Session

Perform this procedure during the second laboratory session, utilizing isolated colonies on pure culture plates prepared in Procedure 1. If possible, examine the manufacturer's Enterotube II® package insert before continuing with Procedure 3.

a Label your 3 **Enterotube II®** systems with your group identification and the date. Label 1 to receive *Salmonella,* another for *Shigella,* and the third for *Escherichia.*

b Remove the caps from 1 of your labeled Enterotubes. Touch the pointed end of the inoculating wire to a well-isolated colony of the appropriate genus on a Hektoen Enteric Agar pure culture plate.

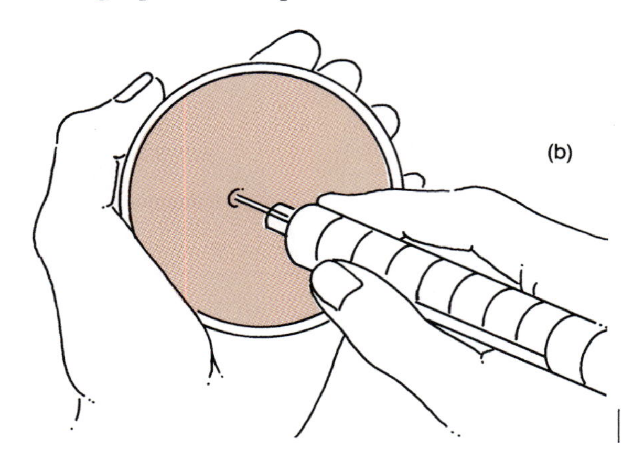

c Twist the wire and draw it back through all 12 compartments. Without flaming, reinsert the wire all the way back into the Enterotube.

d Draw the wire out until you see its tip just beginning to leave the citrate compartment. Stop. At the handle end, right where the wire protrudes from the tube, you can now see a groove in the wire.

e Bend and snap the wire off at the groove. Do not lay the wire down.

f Use the broken tip of the wire to punch a hole in each of the 8 tiny, tombstone-shaped slots on the side of your Enterotube. The holes provide aerobic conditions for enzymatic reactions in the exposed compartments.

g Discard the broken, bacteria-laden piece of inoculating wire in an appropriate manner. Replace caps on the tube.

h For incubation, lay the tube on its flat surface. Incubate at 35° C for 18 to 24 hours.

i Repeat steps (*b*) through (*h*) for each of the other 2 cultures.

j Label 3 tubes of chilled **Motility Medium S** with your group identification, the date, the name of the medium, and the name of the bacterium—*Salmonella* on 1 tube, *Shigella* on the second, and *Escherichia* on the third.

k Using aseptic technique and an inoculating needle, inoculate the appropriately labeled Motility Medium S tube with *Salmonella* by stabbing the medium about 1 cm deep. Do not stab deeper. Withdraw the needle without stirring. Sterilize the needle.

l Similarly, inoculate your other tubes of Motility Medium S, one with *Shigella* and the last with *Escherichia,* again performing short, straight stab inoculations.

m Incubate the 3 Motility Medium S cultures at 35° C for 18 to 24 hours.

Third Laboratory Session

a Interpret and record the reactions in all **Enterotube II®** compartments that do not require added reagents. Using 1 sheet per tube, mark the results of the biochemical tests on a computer coding pad. Also enter your results in your laboratory report. For help in interpreting the test results, utilize figure 28.4 and, if possible, a package insert from the Enterotube II® system.

b When all members of your group are ready, perform the **indole tests.** To do this, pierce the thin, flat plastic film covering an indole compartment with a hot, flamed, sharp probe. Sterilize the probe before replacing it on the bench top. Why sterilize the probe?

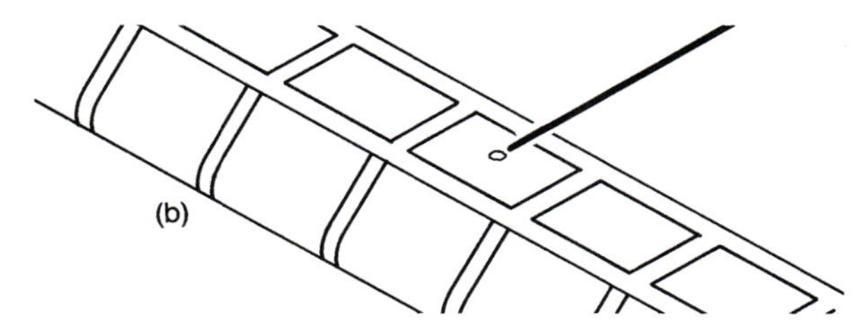

c Add a few drops of **Kovacs' reagent** to the exposed compartment. If the test is positive, a red color develops within 10 minutes.

d Record the results on the computer coding pad and in your laboratory report.

e Repeat the indole test with your 2 other Enterotube II® cultures. Record your results.

f On the computer coding pad and in your report, circle numbers that correspond to **positive** test reactions.

g Add the numbers in each bracketed section. Place each sum in the appropriate space.

h The 5 numbers form a 5-digit identification code number. Find the ID code number for each of your 3 class cultures in the *Coding Manual.* The *Coding Manual* helps you identify your *Salmonella, Shigella,* and *Escherichia* serovars or species. The *Coding Manual* may direct you to perform a Voges-Proskauer test.

i If necessary, perform the **Voges-Proskauer procedure** as a confirmatory test. To do this, pierce the thin plastic film covering the VP compartment with a hot, flamed, sharp probe. Sterilize the probe.

j To the opened compartment, add 2 drops of **Barritt's solution B** (potassium hydroxide), then 3 drops of **Barritt's solution A** (alpha-naphthol). If the test is positive, a red color develops within 20 minutes.

k Record your results on the computer coding pad and in your laboratory report.

l Repeat Voges-Proskauer testing with your 2 other Enterotube II® cultures if necessary for identification.

m In your laboratory report, record the identity of each serovar or species.

n Read about **Motility Medium S** in the *DIFCO Manual.* Without jostling them, examine your 3 Motility Medium S cultures and record your findings in your laboratory report.

Procedure 4

Cultivating *Campylobacter jejuni*

First Laboratory Session

a Label 2 **Campylobacter Agar** plates as in the illustration, 1 for aerobic incubation, the other for a somewhat reduced environment.

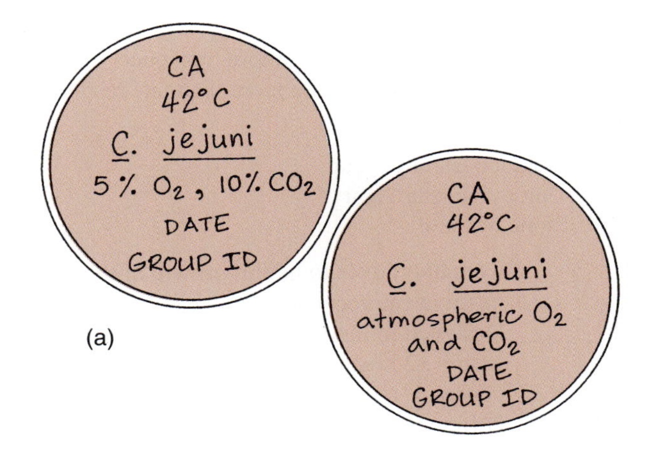

b **Your instructor,** or an assigned group, will open a GasPak® jar and remove cultures of ***Campylobacter jejuni*** for your use.

c Immediately streak your group's 2 Campylobacter Agar plates with *Campylobacter jejuni* for isolation of colonies.

d Incubate the inverted plates at 42° C for 48 hours, 1 aerobically and the other in a microaerophilic jar with a *Campylobacter* gas mixture.

e When all appropriately labeled plates are in the GasPak® jar, **your instructor** or an assigned group will pipet 10 ml of water into the open corner of a **CampyPak® envelope,** place the envelope into the jar, seal the jar immediately, and incubate it at 42° C.

f In your laboratory report, describe the conditions of incubation that are required by *Campylobacter jejuni* as compared and contrasted with those that clinical laboratories typically supply for intestinal pathogens in the family Enterobacteriaceae.

Second Laboratory Session

a Examine and evaluate the growth on your streak plates of *Campylobacter jejuni.* In your laboratory report, compare growth of the species in atmospheric oxygen and in a microaerophilic environment.

b In your laboratory report, describe typical, well-isolated, Campylobacter Agar colonies of *Campylobacter jejuni.*

c Gram stain the culture. Record the results of your Gram stain in your laboratory report.

POSTTEST EXERCISE 28

Part 1

Matching

Each response may be used once, more than once, or not at all. Some statements may require more than one response.

a. Selenite Broth
b. Hektoen Enteric Agar
c. Campylobacter Agar
d. None of these

1. ______ Contain(s) dyes to inhibit growth of the gram-positive enterococci
2. ______ Designed to favor the growth of *Salmonella* within 12 hours while inhibiting other fecal genera
3. ______ Differentiate(s) lactose-, sucrose-, and/or salicin-fermenting colonies from nonfermenters of these sugars
4. ______ Differentiate(s) H_2S producers from nonproducers
5. ______ Miniaturized multitest system designed to differentiate members of the Enterobacteriaceae
6. ______ Contain(s) antibiotics

Part 2

True or False (Circle one, then correct every false statement.)

1. T F Salmonellosis, a gastroenteritis, is caused by many serovars of *Shigella.*
2. T F *Salmonella* serovars infect a wide variety of both warm- and cold-blooded animals.
3. T F Some plasmids found in *Escherichia* carry a genetic code for resistance to antibiotics.
4. T F Enterotoxins are plasmids.
5. T F Enterotoxigenic *Escherichia coli* is a frequent cause of intestinal ulceration; this sometimes fatal condition resembles shigellosis.
6. T F Traveler's diarrhea is often caused by enterotoxigenic *Escherichia coli.*
7. T F In pathogenic potential, biochemical capabilities, and DNA composition, *Escherichia* often resembles *Shigella.*
8. T F Over 2,000 serovars of *Salmonella* have been identified.
9. T F *Salmonella typhimurium* and *Campylobacter jejuni* are common causes of food-borne infection.
10. T F An enterococcus is a gram-negative, facultative or aerobic rod that ferments lactose, producing acid and gas within 48 hours at 35° C.
11. T F In general, inoculating an Enteric Quad® plate and an Enterotube II® with the same unknown would be a waste of time and media.
12. T F The Enterotube II® is designed for use with all gram-negative bacteria including *Campylobacter jejuni.*
13. T F To obtain approximately equal numbers of bacteria in a mixture prepared from broth cultures, always add an equal number of drops from each culture to the mix.
14. T F To interpret the lactose, sucrose, and/or salicin fermentation capabilities of a fecal isolate on Hektoen Enteric Agar, examine well-isolated colonies, not areas of confluent growth.
15. T F Campylobacters are motile, oxidase-negative, microaerophilic, vibrioid bacteria.

Part 3

Completion and Short Answer

1. Fill in the following chart describing the color of well-isolated *Salmonella, Shigella,* and *Escherichia* colonies on Hektoen Enteric Agar.

Appearance of Representative Enterobacteriaceae on Hektoen Enteric Agar

Genus	Color
Salmonella	
Shigella	
Escherichia	

2. List nine characteristics shared by members of the family Enterobacteriaceae.
   a.
   b.
   c.
   d.
   e.
   f.
   g.
   h.
   i.

Laboratory Report **28**

Name ____________________

Section ____________

Intestinal Pathogens: *Campylobacter* and Selected Enterobacteriaceae

Procedure 1

Isolating *Salmonella, Shigella,* and *Escherichia*

First Laboratory Session

(*c*) Neither salmonellae nor shigellae ferment sucrose or salicin. Why are these sugars included in Hektoen Enteric Agar?

Second Laboratory Session

(*a*) Fill in the following chart with a complete description of well-isolated colonies on Hektoen Enteric Agar. If possible, include colored-pencil illustrations of the colony and the surrounding agar.

Enterobacteriaceae on Hektoen Enteric Agar

Bacteria	Appearance
Salmonella	
Shigella	
Escherichia	

(*b*) In the following chart, interpret the biochemical reactions exhibited by isolated colonies of *Salmonella, Shigella,* and *Escherichia* on Hektoen Enteric Agar.

Bacteria	Biochemical Reactions: Lactose, Sucrose, and/or Salicin Fermentation		H_2S Production	
	Positive	Negative	Positive	Negative
Salmonella				
Shigella				
Escherichia				

How do your tabulated observations compare with the expected results listed in the *DIFCO Manual* and/or Appendix 2?

(*c*) Did Hektoen Enteric Agar inhibit the growth of any of the 3 genera? ________________________
If so, which? ____

Procedure 2

Enrichment for *Salmonella*

Second Laboratory Session

(*b*) Estimate the percentage of each colony type on your "Mix→HE" plate.
Salmonella ________
Shigella ________
Escherichia ________
(*d*) How many hours did you incubate your Selenite Broth? ________

Third Laboratory Session

(*a*) Did you recover all 3 genera of bacteria from the Selenite Broth? ________
If not, which genus (genera) failed to grow? ____
How do these data compare with your results in the second laboratory session of Procedure 1, step (*c*)?

(*b*) Estimate the percentage of each colony type on your "Sel→HE" plate.
Salmonella ________
Shigella ________
Escherichia ________
(*c*) Comparing your data from step (*b*) in the second laboratory session of this procedure with your observations in step (*b*) of the third session, discuss whether Selenite Broth encourages the growth of *Salmonella* while discouraging coliforms and/or *Shigella*.

Procedure 3

The Enterotube II® and Motility

Third Laboratory Session

(*a*) Examine your Enterotube II® cultures of *Escherichia, Salmonella,* and *Shigella.* First, evaluate reactions in compartments to which you do **not** add reagents. Compare the color of the medium in each of these compartments with the appropriate block of figure 28.4. For further help in interpreting your results, see a package insert from the Enterotube II® carton. Record each result on a sheet from a computer coding pad and on the following charts. Use the Voges-Proskauer (VP) reaction as a confirmatory test if necessary.

Escherichia

GLU (2) ____	LYS (4) ____	IND (4) ____	ARAB (4) ____	PA (4) ____
GAS (1) ____	ORN (2) ____	ADON (2) ____	SORB (2) ____	UREA (2) ____
	H_2S (1) ____	LAC (1) ____	DUL (1) ____	CIT (1) ____
________	________	________	________	________

Escherichia ID Value

Salmonella

GLU (2) ____	LYS (4) ____	IND (4) ____	ARAB (4) ____	PA (4) ____
GAS (1) ____	ORN (2) ____	ADON (2) ____	SORB (2) ____	UREA (2) ____
	H_2S (1) ____	LAC (1) ____	DUL (1) ____	CIT (1) ____
________	________	________	________	________

Salmonella ID Value

Shigella

GLU (2) ____	LYS (4) ____	IND (4) ____	ARAB (4) ____	PA (4) ____
GAS (1) ____	ORN (2) ____	ADON (2) ____	SORB (2) ____	UREA (2) ____
	H_2S (1) ____	LAC (1) ____	DUL (1) ____	CIT (1) ____
________	________	________	________	________

Shigella ID Value

(*b*) After piercing the plastic Enterotube II® cover, why do you sterilize the sharp probe before replacing the instrument on the bench top?

(*d, e*) On the appropriate computer coding pad sheet, record the indole reaction of each of your cultures.
(*f*) Circle numbers on the sheets that correspond to positive test reactions.
(*g*) Add the circled numbers in each bracketed section. Place each sum in the appropriate space.
(*k, l*) Record the Voges-Proskauer reaction of each of your cultures if necessary.

(*m*) Record the 5-digit ID numbers and the serovar or species identifications in the following chart.

		Serovar or Species Identification
Genus	**ID Code No.**	**Serovar or Species**
Escherichia	______	__________________
Salmonella	______	__________________
Shigella	______	__________________

(*n*) In the *DIFCO Manual,* discover what percentage of agar is in Motility Medium S. ________
What color is 2,3,5-triphenyltetrazolium chloride (TTC) when it is in its reduced state? ____________________
Draw and describe the growth patterns of *Escherichia, Salmonella,* and *Shigella* in Motility Medium S.
Which is/are motile and how do you know?

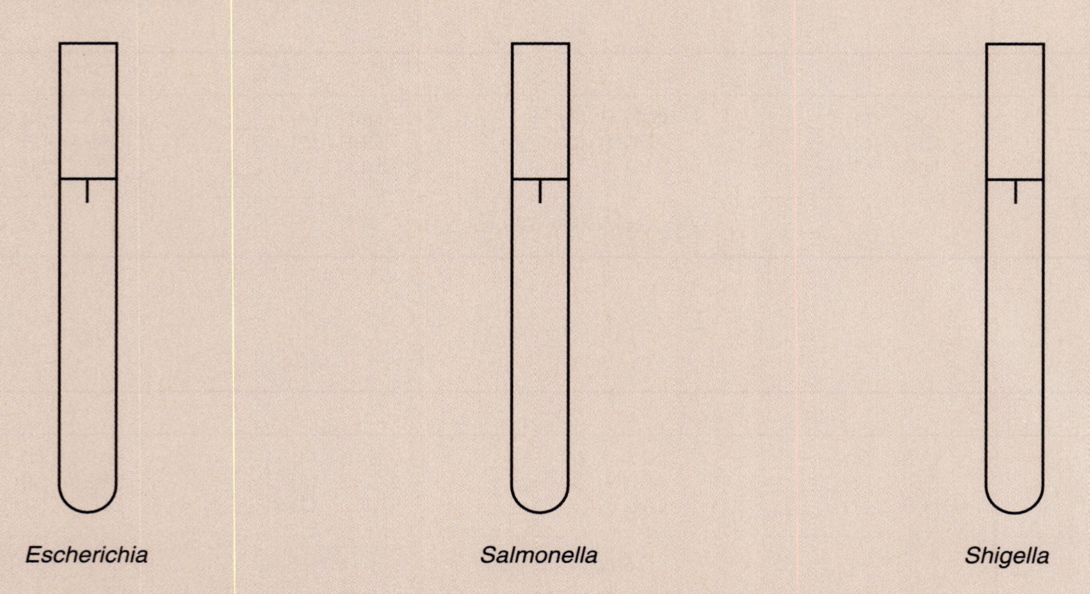

Procedure 4

Cultivating *Campylobacter jejuni*

First Laboratory Session

(*f*) Describe the conditions of incubation that are required by *Campylobacter jejuni.* Compare and contrast them with the conditions of incubation that clinicians generally provide for intestinal pathogens in the family Enterobacteriaceae.

Second Laboratory Session

(*a*) Observe and compare 48-hour, 42° C growth of *Campylobacter jejuni* in atmospheric oxygen and in a microaerophilic environment.

(*b*) Describe a typical, well-isolated *Campylobacter jejuni* colony.

Do you see a tailing effect along the streak line? ___________

(*c*) Record the results of your Gram stain. Do you observe sea gull-winged cells? _________ List the size, shape, predominant arrangement, and Gram reaction of *Campylobacter jejuni* cells.

Exercise 29

Mouth, Throat, and Lower Respiratory System Flora

Necessary Skills

Microscopy, Aseptic transfer techniques, Gram staining, Streak-plate techniques, Knowledge of: Microbial oxygen and pH requirements, Staphylococci, Streptococci, Hemolytic reactions, Antibiotic disk sensitivity testing, Selective and differential media, Selected microbial enzymes: catalase and cytochrome oxidase

Materials

Procedure 1

Supplies (per group):
- Snyder Test Agar deep
- Container to hold boiling water and Snyder Test Agar deep
- Sterile container, at least 25 ml, to hold saliva
- Sterile 1 ml pipet
- Clean block of paraffin, about 1 cm^3
- Thermometer
- Pipettor
- Access to the *DIFCO Manual*

Procedure 2

First Laboratory Session

Supplies (per group):
- Sterile calcium alginate swab
- Tongue depressor
- 2 Blood Agar plates
- Optochin differentiation disk
- Bacitracin differentiation disk
- Forceps
- Alcohol for flaming
- Mirror (optional)

Additional supplies:
- Candle jars to hold 1 plate per group
- Matches
- Petroleum jelly

Second Laboratory Session

Supplies (per class):
- Hydrogen peroxide, 3% aq., in dropper bottles
- Oxidase reagent or disks
- Access to *Schneierson's Atlas of Diagnostic Microbiology**
- Gram stain reagents and equipment

Procedure 3

Access to *Bergey's Manual*

Prepared slides:
- *Bordetella pertussis* Gram stain
- *Corynebacterium diphtheriae* Gram or methylene blue stain
- *Haemophilus influenzae* Gram stain
- *Klebsiella pneumoniae* Gram stain
- *Mycobacterium tuberculosis* in sputum, acid-fast stain

Primary Objective

Investigate microbial flora of the mouth and throat, then examine pathogenic microorganisms from the lower respiratory system.

Other Objectives

1. Compare the quantity of bacterial flora in various microenvironments within the human mouth.
2. List common members of the mouth and upper respiratory system normal flora.
3. Analyze how *Streptococcus mutans* utilizes sucrose in initiating dental caries.
4. Define: dental plaque, dental caries, dental calculus, gingivitis, periodontitis, microbial antagonism, morbidity, mortality.

---

*Bottone, E. J. et al., eds., *Schneierson's Atlas of Diagnostic Microbiology,* 9th ed. Abbott Laboratories, North Chicago, IL, 1984.

5. Propose procedures that an individual may perform to help reduce the risk of developing gingivitis or periodontitis.
6. Utilize Snyder Test Agar to appraise the cariogenic potential of your own oral flora.
7. Discuss environmental factors that encourage microbiota to inhabit the upper respiratory system; list reasons why the lower respiratory system is not generally colonized.
8. Indicate etiologic agents of pertussis, tuberculosis, laryngitis, and acute epiglottitis.
9. List and describe selected microorganisms that cause pharyngitis; that cause pneumonia.
10. Explain and demonstrate acceptable throat-swabbing technique.
11. Distinguish among colonies that may consist of *Streptococcus pyogenes, Streptococcus pneumoniae, Staphylococcus aureus, Neisseria,* or *Haemophilus* from throat cultures.
12. Prepare and then investigate a culture of your own nasopharyngeal flora for the presence of the pathogens listed in Objective 11.

Table 29.1 Selected Bacterial Genera That Are Commonly Found in the Human Mouth

Genus	Description
Actinomyces	Preferentially anaerobic, acidogenic, gram-positive rods
Bacteroides	Anaerobic, acidogenic, proteolytic, gram-negative rods
Borrelia	Anaerobic, acidogenic, gram-negative spirochetes
Corynebacterium	Generally facultatively anaerobic, some species acidogenic, gram-positive, pleomorphic rods (diphtheroids)
Fusobacterium	Anaerobic, acidogenic, proteolytic, gram-negative rods
Lactobacillus	Aerotolerant, acidogenic, gram-positive rods
Nocardia	Aerobic, acidogenic, proteolytic, gram-positive rods, some species acid-fast
Staphylococcus	Facultatively anaerobic, acidogenic, proteolytic, gram-positive cocci
Streptococcus	Aerotolerant, acidogenic, proteolytic, gram-positive cocci
Treponema	Anaerobic, acidogenic, proteolytic, gram-negative spirochetes

INTRODUCTION

In a healthy person, the **upper respiratory system,** generally defined as including the nose, sinuses, mouth, and throat, contains a large number and wide variety of microorganisms. **Streptococci, staphylococci, neisseriae, haemophili, diphtheroids** (gram-positive, pleomorphic rods resembling *Corynebacterium diphtheriae*), **yeasts, protozoans, viruses,** and many species of **anaerobic bacteria** abound in the upper respiratory system. Here, protected from desiccation and starvation by their human host, they inhabit a moist, warm environment, frequently washed with food, for the mouth is also an entryway into the **digestive system.**

The **lower respiratory system,** including the larynx, trachea, bronchi, and the alveoli of the lungs, has few if any resident microbes. These areas are protected from microbial colonization by phagocytic cells in the lungs and by the **ciliary escalator** (cilia—lining the bronchial tubes, trachea, and larynx—which move a sticky layer of mucus up and out of the lower respiratory system).

Microorganisms in the Human Mouth

Each microenvironment within your mouth contains tremendous numbers of organisms. Saliva has over a million bacteria per milliliter. **Dental plaque** (a deposit of organic material on teeth) contains billions of bacteria per gram, as does material from the **gingival sulci** (spaces between the gums and the teeth).

Selected bacterial genera that are commonly found in the human mouth are listed in table 29.1. It is most often normal flora, not invasive pathogens, that cause tooth decay and **periodontal disease** (inflammation and/or degeneration of tissues around the teeth).

Dental Caries

Acid producers, especially **streptococci** and **lactobacilli,** damage teeth. At a pH of 5.5 or lower, demineralization of the tough, protective, tooth enamel begins. As a result of this **decalcification,** the soft, inner layers of the teeth, the **dentin** and nerve-containing **pulp** are exposed; see figure 29.1.

The highly **saccharoclastic** (able to split carbohydrates) species ***Streptococcus mutans*** initiates **dental caries** (cavities; lesions in the enamel and dentin in a tooth). First, the *Streptococcus* splits **sucrose** (table sugar) into fructose and glucose; see figure 29.2. Utilizing the enzyme **glucosyltransferase,** *Streptococcus mutans* then joins the monosaccharide glucose units forming **dextrans** and other sticky polymers.

Streptococcal dextrans attach bacteria to the teeth. Here streptococci, lactobacilli, and other **cariogenic** (producing caries) microbes degrade sugars, including the fructose that *Streptococcus mutans* releases from sucrose. While gaining energy from this catabolic activity, the bacteria release acids that destroy tooth enamel.

Periodontal Disease

When dental plaque is allowed to accumulate on teeth, the deeper portions of this tenacious accretion become anaerobic. The spirochetes ***Treponema*** and ***Borrelia*** accumulate in this reduced environment, as does ***Fusobacterium,*** a gram-negative rod-shaped genus that generally has pointed ends.

High numbers of these anaerobic members of the normal oral flora are found in the **dental calculus** (concretions of calcium phosphate and calcium carbonate that form in organic material on the teeth) of people who suffer from **acute, necrotizing, ulcerative gingivitis.** The signs of

Figure 29.1 Dental caries

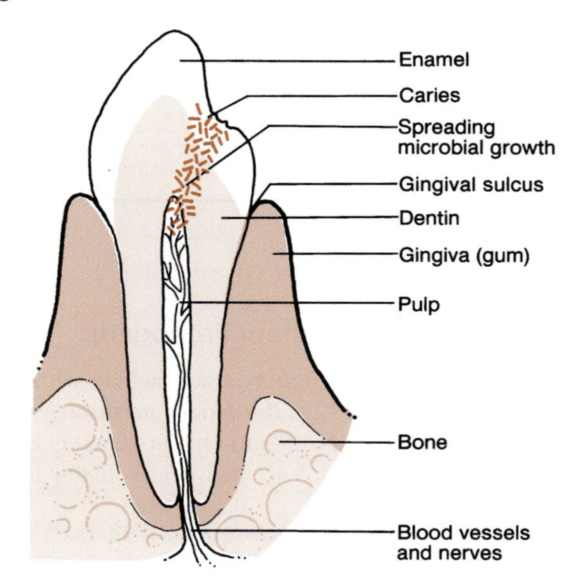

acute, necrotizing, ulcerative gingivitis, a disease that is also known as **trench mouth** or **Vincent's angina,** include rapidly developing inflammation and death of gum tissue. This disease occasionally develops in young adults during times of high anxiety and stress.

Preventing Dental Caries and Periodontal Disease

Heredity plays a role in providing people with strong teeth, but the individual can help nature. Those who brush and floss often, avoid frequent intake of sucrose, and have regular, professional cleaning greatly reduce their risk of developing caries, gingivitis, or **periodontitis** (inflammation of tissues surrounding and supporting the teeth).

In this exercise you estimate the cariogenic potential of your own oral flora. You utilize **Snyder Test Agar** to indicate the rate at which your **oral lactobacilli** generate acid.

The low pH and reduced environment provided in Snyder Test Agar inhibit most oral flora. Lactobacilli thrive, producing acid from the medium's dextrose. If your oral lactobacilli drop the pH of the medium below about 4.6, the color of the brom cresol green indicator changes from green to yellow. This color change shows that your flora produce enough acid to decalcify teeth.

The best time to collect saliva for this examination of your oral flora is before breakfast and before you brush

Figure 29.2 *Streptococcus mutans* splits sucrose then produces dextran and lactic acid.

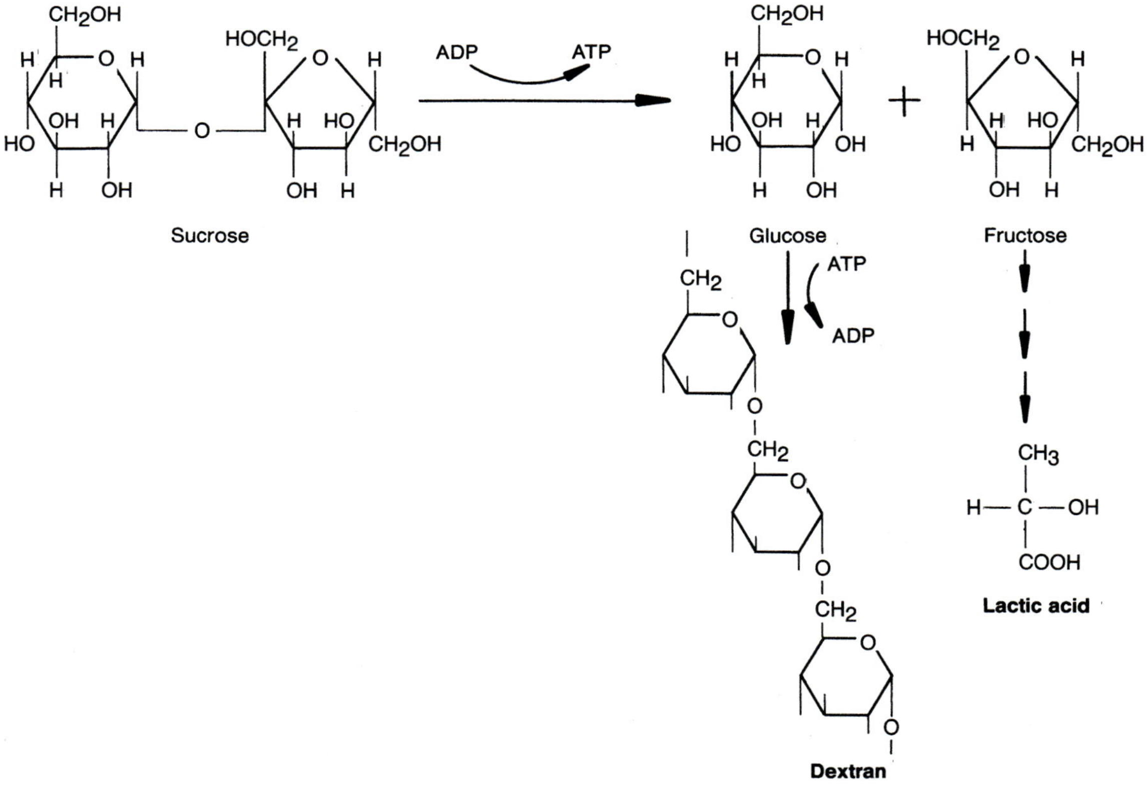

Sucrose → (ADP → ATP) → Glucose + Fructose

Glucose → (ATP → ADP) → Dextran

Fructose → → → Lactic acid

Table 29.2 Utilizing Snyder Test Agar to Estimate the Cariogenic Potential of Oral Flora

Caries Potential	Duration of Incubation		
	24 Hours	48 Hours	72 Hours
Marked	Positive[1]	—	—
Intermediate	Negative[2]	Positive	—
Slight	Negative	Negative	Positive
Insignificant	Negative	Negative	Negative

[1]Positive: Yellow is the dominant color.

[2]Negative: Green is the dominant color.

your teeth. The test should be repeated several times and your data pooled. If Snyder Test Agar is employed only once, the results simply reflect what is happening in your mouth at the time you collect the specimen. With this limitation in mind, use table 29.2 to estimate the cariogenic potential of your oral lactobacilli.

Upper Respiratory System Pathogens

The throat, nose, and sinuses, with their lifelong, repeated exposure to air, provide a natural portal for entry of disease-producing organisms. **Microbial antagonism** (production of antimicrobial substances, competition for space and food) among the normal flora of the upper respiratory system helps to protect the system from invasion by pathogens, but there is a never-ending battle. The normal flora of the upper respiratory system is constantly challenged by viruses, bacteria, and other microbiota from outside the human body.

Streptococcus pyogenes, which causes strep throat, a **pharyngitis** (inflammation of the pharynx, sore throat), is discussed in Exercise 27, Medically Significant Gram-Positive Cocci.

Corynebacterium diphtheriae also causes acute pharyngitis. This deadly, gram-positive, toxigenic, nonsporing rod is the etiologic agent of **diphtheria,** a disease that is rare now because vaccination is common.

Many cases of pharyngitis result from a spread of **upper respiratory infection** from the tonsils or sinuses. Which microorganisms infect the tonsils and/or sinuses? **Viruses** or **group A, beta-hemolytic streptococci** give rise to most cases of acute **tonsillitis.** The origin of **sinusitis** is generally **viral.** Sinusitis can also be caused by a variety of bacteria including ***Streptococcus pyogenes, Streptococcus pneumoniae,*** and ***Staphylococcus aureus*** (discussed in Exercise 27, Medically Significant Gram-Positive Cocci).

Laryngitis, an inflammation of the larynx (voice box) is also generally initiated by an acute upper respiratory infection such as the common cold, pharyngitis, tonsillitis, or sinusitis. **Acute epiglottitis,** a life-threatening disease of the upper respiratory system, is generally due to ***Haemophilus influenzae*** type b. This pleomorphic, bipolar-staining, gram-negative rod causes inflammation of the epiglottis, a cartilaginous flap that covers the larynx. A vaccine offering protection from *Haemophilus influenzae* type b is available.

Lower Respiratory System Pathogens

Diseases of the lower respiratory tract include **whooping cough** (pertussis) caused by ***Bordetella pertussis,*** a very small (approximately $0.3 \times 1.0\ \mu m$) bipolar-staining, encapsulated, gram-negative rod. Vaccination is available to prevent this disease.

Mycobacterium tuberculosis, an acid-fast, slow-growing, gram-positive rod is the etiologic agent of **tuberculosis,** a disease that usually attacks the lower respiratory system but may involve any organ of the body. Bovine and avian species of mycobacteria also infect humans.

While only one genus causes tuberculosis, many genera induce pneumonias. **Pneumonias** (acute infections of the alveoli in the lung) are most often due to ***Streptococcus pneumoniae, Staphylococcus aureus, Streptococcus pyogenes, Klebsiella pneumoniae*** (an encapsulated, gram-negative rod), or ***Haemophilus influenzae.*** Other etiologic agents include **viruses, rickettsiae** and other bacteria, **fungi,** and **protozoans;** see table 29.3.

In this exercise you swab your throat and investigate the normal flora that inhabit your own upper respiratory system. By examining prepared slides of pathogens that infect the upper and lower respiratory systems, you learn to recognize these virulent agents of **morbidity** (disease) and **mortality** (death on a large scale, as from disease).

PROCEDURES EXERCISE 29

Procedure 1

Estimating the Cariogenic Potential of Oral Flora

First Laboratory Session

a Liquefy a tube of **Snyder Test Agar** in a container of boiling water, remembering to loosen screw caps before heating. Leave the deep in the boiling water for 10 minutes to drive dissolved oxygen out of the fluid.

b Cool the medium to 45° C.

c Wash your hands, then obtain a block of paraffin. Soften the block by holding it in your mouth.

d Chew the block of paraffin for 3 minutes without swallowing. Collect your saliva in a sterile container. Chewing removes microbiota from your teeth.

Table 29.3 Selected Etiologic Agents of Pneumonia

Agent	Microbial Group	Comments on Cultivation and/or Identification
Chlamydia psittaci	Chlamydias	Intracellular, isolation in embryonated eggs or cell culture
Coccidioides immitis	Fungi	Mycological media, stained smear, intradermal skin test
Coxiella burnetii	Rickettsias	Intracellular, isolation in embryonated eggs or cell culture
Haemophilus influenzae	Gram-negative rods	Satellitism, require X (hemin) and V (NAD), high CO_2
Histoplasma capsulatum	Fungi	Mycological media, stained smear, intradermal skin test
Klebsiella pneumoniae	Gram-negative rods	Large capsule, mucoid colonies
Legionella pneumophila	Gram-negative rods	Selective yeast extract medium
Mycoplasma pneumoniae	Small bacteria lacking cell walls	Minute colonies on enriched media, fluorescent-antibody identification
Pneumocystis carinii	Protozoans (Sporozoa)	Stained slide of sputum or lung tissue
Pseudomonas aeruginosa	Gram-negative rods	Oxidase-positive, nonfermentative, produce pyocyanin
Staphylococcus aureus	Gram-positive cocci	Catalase-, coagulase-, and mannitol-positive
Streptococcus pneumoniae	Gram-positive cocci	Alpha-hemolytic, optochin-sensitive, catalase-negative diplococci
Viral	Viruses	Adenoviruses, coxsackieviruses, echoviruses, herpes viruses, influenza viruses, respiratory syncytial viruses, rhinoviruses

e Shake your saliva specimen to resuspend microorganisms.

f Pipet 0.2 ml of your saliva into your liquefied, tempered tube of Snyder Test Agar.

g Rotate the tube between your palms to distribute the saliva without introducing excess oxygen.

h Incubate your deep at 35° C for 72 hours, recording your observation of color development every 24 hours.

i In your laboratory report, explain why Snyder Test Agar employs the pH indicator **brom cresol green** instead of brom cresol purple or phenol red. For help, read about dyes and indicators in the *DIFCO Manual.*

Subsequent Laboratory Sessions

a After your **Snyder Test Agar** saliva culture has incubated 24 hours, examine it for color change. Record your observations in your laboratory report.

b Repeat step (*a*) the next 2 days.

c To evaluate the cariogenic potential of your oral flora at the time you collected saliva, compare your results with table 29.2.

Procedure 2

Flora of the Pharynx: Throat Culture

Wash your hands immediately before swabbing a throat.

First Laboratory Session

a As in the illustration, use a sterile, calcium alginate swab to collect a specimen from your laboratory partner's **nasopharyngeal area** *gently and quickly.* Rotating the swab, touch only the posterior wall of the pharynx, behind the uvula and between the tonsillar pillars. If necessary, utilize a tongue depressor and have your "patient" phonate an "ahhh" sound to lower the tongue, raise the uvula, and reduce the gag reflex. Avoid contaminating the swab with your patient's saliva.

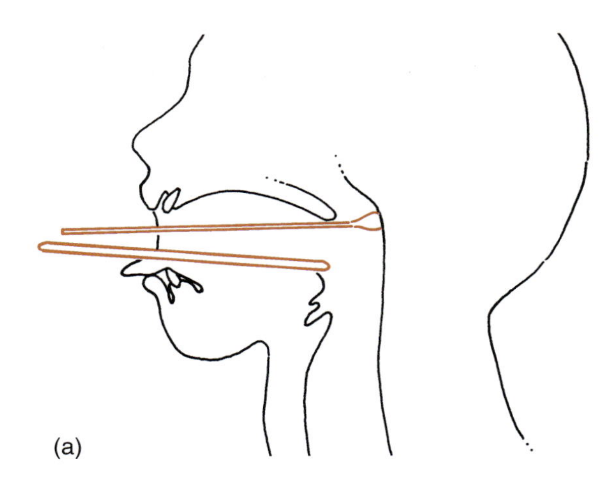
(a)

b Give the specimen to your partner; cultivate your own oral flora. Swab the primary areas of 2 **Blood Agar** plates with your own nasopharyngeal specimen. Dispose of the swab as directed by **your instructor.**

c With your sterilized inoculating loop, streak the secondary and tertiary areas of your Blood Agar plates for the isolation of colonies.

d Label the plates with your name, the date, and the word "Throat." Label 1 plate "Aerobic," the other "Candle Jar."

e Alcohol-flame sterilize your forceps. Use them to deliver 1 **optochin** differentiation disk aseptically to one end of the primary area of your **"Aerobic"** incubation plate.

f Alcohol-flame sterilize your forceps again. Use them to deliver 1 **bacitracin** differentiation disk aseptically to the other end of the primary area of your **"Aerobic"** incubation plate. Resterilize the forceps.*

*In a clinical laboratory, optochin and bacitracin disks should be used on pure cultures of suspected pathogens, not on the primary inoculation plate. How does the correct practice avoid false reactions and inaccurate interpretations of results?

g Use your flamed and cooled inoculating loop. Hold it at a 45° angle to the surface of the **"Aerobic"** incubation plate. Stab the loop twice into the center of the primary streak.

h Incubate your plates at 35° C for 24 hours as indicated on their labels—1 aerobically and the other in a candle jar.

i In your laboratory report, explain why you perform steps (*e*), (*f*), (*g*), and (*h*) on Blood Agar streak plates prepared from a throat specimen. For help, refer to the introduction section of this exercise and to Exercise 27, Medically Significant Gram-Positive Cocci.

Second Laboratory Session

a Examine and compare your 2 **Blood Agar throat culture plates.** In your laboratory report, note the approximate number of different kinds of colonies on each plate.

b If you had a sore throat or were taking oral antibiotics at the time of the throat swabbing, **your instructor** may ask you to examine your Blood Agar throat cultures, then donate them to serve as demonstration plates for the class. If so, clearly label your plates "Pharyngitis" or "Antibiotics." In your laboratory report, compare the microbial growth that develops on "Pharyngitis," "Antibiotics," and normal throat plates.

c Utilize *Schneierson's Atlas of Diagnostic Microbiology* to help you identify a possible *Neisseria* colony on your "Candle Jar" plate. Gram stain this colony.

d If your colony contains gram-negative diplococci, test it for oxidase production. You may wish to refer to the discussion of cytochrome oxidase in Exercise 21, Selected Microbial Enzymes. In your laboratory report, record and analyze your findings.

e Utilizing *Schneierson's Atlas of Diagnostic Microbiology* as a guide, examine your "Candle Jar" plate for typically tiny, satellite colonies of *Haemophilus* surrounding larger colonies of other microorganisms.

f If you see possible *Haemophilus* colonies, Gram stain one. Look for thin, gram-negative rods in singles, pairs, or chains.

g Examine your "Aerobic," Blood Agar plate for possible *Streptococcus pyogenes, Streptococcus pneumoniae,* and *Staphylococcus aureus* colonies. Refer to appropriate exercises in this laboratory manual and *Schneierson's Atlas of Diagnostic Microbiology* for help. After performing the required stains and the catalase test, record your findings in your laboratory report.

h Note what further testing is required for the identification of *Staphylococcus aureus.*

i To characterize the normal flora of your own throat, examine 5 unique colonies on your Blood Agar plates. These may include the possible pathogens you have already investigated in steps (*c*), (*d*), (*e*), (*f*), and (*g*). Number each colony on the back of its plate. In your laboratory report, describe each of the chosen colonies and the bacteria or fungi that it contains. Approximate what percentage of the total number of colonies on the plate are similar to it.

j Gram stain each of the 5 colonies. Record the shape and Gram reaction of each species in your laboratory report next to the corresponding colony description. What is the predominant type of microbial flora you find in your own nasopharyngeal area?

Procedure 3

Examining Prepared Slides of Respiratory Tract Pathogens

a With the oil-immersion lens of your microscope, examine the respiratory tract pathogens listed here.

b In your laboratory report, draw and describe each, noting morphological characteristics that help you differentiate these organisms from each other. Check your findings in *Bergey's Manual.* Examine the following bacteria:

Bordetella pertussis Gram stain
Corynebacterium diphtheriae Gram or methylene blue stain
Haemophilus influenzae Gram stain
Klebsiella pneumoniae Gram stain
Mycobacterium tuberculosis in sputum, acid-fast stain

POSTTEST EXERCISE 29

Part 1

Matching

Each response may be used once, more than once, or not at all. Utilize the preceding set of responses for each group of statements. Some statements may require more than one response.

a. Dental plaque
b. Dental caries
c. Dental calculus
d. Gingival sulci
e. None of these

1. _____ Accumulations of calcium carbonate and calcium phosphate attached to the teeth
2. _____ Inflammation and destruction of bone and other supporting tissue around the teeth
3. _____ The spaces between the gums and the teeth
4. _____ Lesions in the enamel and dentin of teeth

a. Gram-positive rods
b. Acid-fast rods
c. Gram-negative rods
d. Gram-positive cocci
e. Fungi
f. Protozoans
g. Other, none of these

1. _____ *Neisseria*
2. _____ *Streptococcus*
3. _____ *Corynebacterium*
4. _____ *Haemophilus*
5. _____ *Bordetella*
6. _____ *Klebsiella*
7. _____ *Mycobacterium*
8. _____ *Staphylococcus*
9. _____ *Pneumocystis*
10. _____ *Histoplasma*

Part 2

True or False (Circle one, then correct every false statement.)

1. T F Only pathogenic fungi, protozoans, and viruses are found in the human oral region. Their presence indicates disease.
2. T F Dental plaque contains billions of bacteria per gram.
3. T F Streptococci, lactobacilli, spirochetes, and diphtheroids are among the normal bacterial flora of the mouth.
4. T F *Streptococcus mutans,* in the **absence** of sucrose, does **not** initiate tooth decay.
5. T F Glucosyltransferase splits fructose into glucose and sucrose.
6. T F *Borrelia, Treponema,* and *Fusobacterium* are anaerobes. These bacterial genera are found only in diseased mouths.
7. T F Acute, necrotizing, ulcerative gingivitis involves inflammation and even death of gum tissue.
8. T F Spirochetes and gram-negative rods with pointed ends abound on the gums of patients who suffer from Vincent's angina.
9. T F Brushing and flossing frequently; having regular, professional cleaning; and avoiding frequent intake of sucrose help protect against trench mouth but are useless in preventing dental caries.
10. T F To gain reliable information from Snyder Test Agar, it is best to collect several saliva specimens on consecutive evenings after eating dinner.
11. T F Diphtheroids—gram-positive, pleomorphic rods—are among the normal flora of the lower respiratory tract.
12. T F Yeasts, protozoans, and viruses are among the normal flora of the upper respiratory tract.
13. T F Microbial antagonism is defined as the attack of virulent pathogens that precipitates disease in the host.
14. T F Vaccination has made diphtheria a rare disease in the United States.
15. T F Viruses and *Streptococcus pyogenes* are the most common etiologic agents of tonsillitis.
16. T F Sinusitis is a strictly viral disease.
17. T F Acute epiglottitis, a life-threatening disease, is most often caused by *Staphylococcus aureus.*
18. T F *Haemophilus influenzae* is the tiny, bipolar-staining, encapsulated, gram-negative, toxigenic rod that causes pertussis.
19. T F *Streptococcus pneumoniae* is the sole etiologic agent of pneumonia.
20. T F *Mycobacterium tuberculosis* causes tuberculosis.
21. T F *Streptococcus, Haemophilus,* and *Neisseria* from the human throat all grow well in the high carbon dioxide, low oxygen concentration atmosphere of a candle jar.

Laboratory Report **29**

Name ______________________________

Section ______________

Mouth, Throat, and Lower Respiratory System Flora

Procedure 1

Estimating the Cariogenic Potential of Oral Flora

First Laboratory Session

(*i*) Explain why Snyder Test Agar includes the pH indicator brom cresol green instead of brom cresol purple or phenol red.

Subsequent Laboratory Sessions

(*a,b*) In the following chart, record the reactions you observe in your Snyder Test Agar saliva culture. Use table 29.2 and your data to estimate the cariogenic potential of your oral flora at the time of sampling.

Snyder Test Agar Saliva Culture

Duration of Incubation	Reaction*	
	Positive	Negative
24 Hours		
48 Hours		
72 Hours		

*Positive indicates that the medium is yellow; green is no longer the dominant color.
Negative indicates little or no change in color; the medium is still green.

Procedure 2

Flora of the Pharynx: Throat Culture

First Laboratory Session

(*i*) Why did you place an optochin differentiation disk on the primary area of your Blood Agar throat culture?

Why did you utilize a bacitracin differentiation disk?

Why did you stab your loop into the center of the primary streak of your Blood Agar throat culture?

Which genera, common in the human nasopharynx, are likely to grow well on a Blood Agar plate during candle jar incubation?

Second Laboratory Session

(*a*) Examine and compare the growth on your "Aerobic" and "Candle Jar" Blood Agar throat culture plates.
Which plate displays heavier growth? ______________________
Which plate displays a greater variety of colonies? ______________________
(*b*) If possible, compare the quantity and variety of colonies on plates prepared from healthy throats with those on plates streaked with samples from throats of students with pharyngitis.

Examine the quantity and variety of colonies on plates made from throats of students who are **not** taking oral chemotherapeutics. If possible, compare your observations with data you collect from the throat specimen streak plates of students who were taking oral antibiotics at the time of the throat swabbing.

(*c, d*) Utilizing *Schneierson's Atlas of Diagnostic Microbiology,* describe a typical *Neisseria meningitidis* colony.

Describe the results of your search for *Neisseria* colonies among your own pharyngeal flora.

(*e, f*) Describe the results of your search for *Haemophilus* colonies among your own pharyngeal flora.

(*g*) Was there inhibition of small, beta-hemolytic colonies around your bacitracin differentiation disk? _______________
Explain the medical significance of your answer.

Was there inhibition of small, alpha-hemolytic colonies around your optochin differentiation disk? _______________
Explain the medical significance of your answer.

Describe the results of your search for *Staphylococcus aureus* among the flora of your throat.

(*h*) What further testing is required for the identification of *Staphylococcus aureus?*

(*i, j*) To characterize the normal flora of your own throat, fill in the following chart.

Nasopharyngeal Flora					
Colony No.	**Colony Description***	**Gram Reaction**	**Cell Shape**	**Cell Arrangement**	**Frequency of Occurrence**
1					
2					
3					
4					
5					

What is the predominant type of microbial flora you find in your own nasopharyngeal area?

Procedure 3

Examining Prepared Slides of Respiratory Tract Pathogens

(*b*) To develop skill in differentiating respiratory system pathogens microscopically, fill in the following chart. Check your answers in *Bergey's Manual.* Reexamine your slides if you discover discrepancies.

Respiratory Tract Pathogens				
Pathogen	**Cell Shape**	**Cell Arrangement**	**Cell Size**	**Illustration (Note Distinguishing Features)**
Bordetella pertussis				Gram reaction __________
Corynebacterium diphtheriae				Gram reaction __________
Haemophilus influenzae				Gram reaction __________
Klebsiella pneumoniae				Gram reaction __________
Mycobacterium tuberculosis				Acid-fast reaction __________

*Along with the description, note whether the colony is on your "Aerobic" or "Candle Jar" Blood Agar plate.

Exercise **30**

Urinary Tract Normal Flora

Necessary Skills

Microscopy, Aseptic transfer techniques, Gram staining, Culture characteristics, Selective and differential media, Quantifying bacteria

Materials

First Laboratory Session

Supplies (per group):
- Sterile, tightly capped container for collecting urine
- 4 Antiseptic towelette packages
- 2 Sterile, capped test tubes
- Calibrated 0.001 ml (1 μL) loop or disposable, calibrated 1 μL loop
- Blood Agar **or** Tryptic Soy Agar plate
- Bacturcult®
- Sterile 10 ml pipet
- Pipettor
- Access to a tray to hold Bacturcults® in the incubator
- Access to sanitary containment for contaminated, disposable loops and urine sample containers

Second Laboratory Session

Supplies (per group):
- Calibrated 1 μL loop (or 2 disposable, calibrated 1 μL loops)
- 2 Blood Agar **or** Tryptic Soy Agar plates
- Access to sanitary containment for contaminated, disposable loops
- Access to a package insert from Bacturcult® system
- Access to a Bacturcult® counting strip
- Access to an uninoculated control Bacturcult®

Primary Objective

Perform and evaluate a quantitative analysis and presumptively qualitative bacterial analysis of urine.

Other Objectives

1. Define: bacteriuria, uropathogens.
2. List normal flora of the distal urethra.
3. Evaluate the significance of finding over 100,000 or under 10,000 bacteria per milliliter in a clean-catch, midstream urine sample.
4. Propose instances when the numbers of bacteria per milliliter of urine need not exceed 100,000 per milliliter to indicate that a urinary tract infection may be present.
5. Explain how to produce a clean-catch, midstream urine sample.
6. Demonstrate and appraise the importance of immediately processing or refrigerating urine samples in a clinical microbiology laboratory.

INTRODUCTION

Urine is the second most common medical specimen received by clinical microbiology laboratories. Only throat cultures are requested more often by doctors. The numbers of microorganisms that are present in a properly collected urine sample indicate whether there is a urinary tract infection.

In this exercise you cultivate microorganisms from your own (presumably) healthy urinary tract. You perform quantitative and presumptively qualitative bacterial analyses. Then you evaluate the effects of storage on the microbial profile of your specimen. Experimentally storing your urine sample at room temperature will demonstrate the importance of either immediately cultivating a urine specimen or storing it under refrigeration.

Flora of the Urethra

Normal flora of the distal urethra include **coagulase-negative staphylococci** and **diphtheroids** from the skin, **alpha-** and **nonhemolytic streptococci, enterococci,**

yeasts, coliforms, and other **fecal microbiota.** Nonpathogenic ***Neisseria*** and ***Mycobacterium*** species are also present. The upper urethra, urinary bladder, ureters, and kidneys of a healthy person are generally sterile.

Urine, as it exits the healthy bladder, is almost always sterile. But, as it passes through the urethra and out of the body, the liquid flushes out microorganisms. Clinical microbiologists differentiate normal levels of **bacteriuria** (the presence of bacteria in urine) from pathological conditions. The clinician must also recognize **uropathogens** (organisms that cause disease of the urinary tract).

The Urine Specimen

To ensure an accurate estimate of the numbers of bacteria per milliliter of urine, the patient needs to avoid contaminating the urine specimen. A **clean-catch, midstream urine collection** helps prevent contamination of the sample.

To produce the clean-catch, midstream urine sample, the patient washes skin surrounding the external urethral opening with soapy water or a disinfectant solution before urinating. After urinating into the toilet to flush out **endogenous** (arising from within) urethral microbiota, the patient urinates into a sterile collection vessel until it is no more than two-thirds full. To prevent soiling the container, care is taken to avoid touching its sterile surfaces with clothing or any part of the body.

After the collection vessel is securely capped, its outer surfaces are disinfected. Then the urine specimen is delivered to the clinical microbiology laboratory, where it is examined within 30 minutes or refrigerated.

Bacteriuria

Bacteriuria may indicate infection of the urinary tract, contamination by **exogenous** (originating outside) bacteria, or the removal of normal urethral flora with the first flow of urine. Properly obtaining a clean-catch, midstream urine sample greatly reduces the likelihood of contamination. But, to allow for slight contamination, a finding of **fewer than 10,000 bacteria per milliliter** of voided urine is generally considered to be normal. Discovering **over 100,000 bacteria per milliliter** of voided urine indicates **significant bacteriuria,** a presumably pathological condition.

If between 10,000 and 100,000 bacteria per milliliter of voided urine are present in the sample, the numbers may indicate that the specimen was contaminated or inadequately refrigerated.

The microbial count need not exceed 100,000 bacteria per milliliter of urine to be significant. Since the urinary bladder of a healthy person is normally sterile, any bacteria found in a specimen collected by catheterization may indicate infection. While catheterization provides a sterile means of obtaining a urine sample, the process is avoided, when possible, since it carries the risk of introducing infection.

Certain microorganisms are never considered normal flora of the urinary tract, even if they are isolated in extremely low numbers. These uropathogens include **beta-hemolytic streptococci, coagulase-positive staphylococci,** and ***Mycobacterium tuberculosis.***

Quantitation

Traditionally, quantitation of microorganisms in urine is accomplished by spreading a measured amount of the fluid onto an all-purpose medium, incubating the plate at 35° C for 18 to 24 hours, then counting the number of colonies that develop. Each colony is assumed to represent the descendants of one cell.

In this exercise you employ a **calibrated loop** that holds 1 μL (10^{-3} ml) of liquid. The amount of urine that you apply to the plate is called the plate dilution.

To determine the number of bacteria per milliliter of urine, multiply the number of colonies that develop on the plate times the inverse of the plate dilution. For further information, see Exercise 15, Quantifying Bacteria.

If you deliver 10^{-3} ml of voided urine to a plate of Blood Agar, and 37 colonies of fecal microbiota appear the next day, is there significant bacteriuria?*

A Diagnostic Culture System

The **Bacturcult®** is one of many systems for conveniently cultivating and evaluating urine samples.

Urine is poured into the Bacturcult®, then quickly poured back out. A residue of fluid is left on the surface of the medium that lines the inside of the Bacturcult®.

After incubation, the numbers and kinds of colonies that develop in the Bacturcult® are evaluated. A Bacturcult® counting strip with an inscribed circle is placed around the tube. Colonies within the circle are counted. Fewer than 25 colonies in the circle indicates no infection. Over 50 colonies in the circled area indicates significant bacteriuria and represents the presence of over 100,000 bacteria per milliliter of urine.

Since the Bacturcult® medium contains **lactose** and **phenol red** pH indicator, the color of the incubated culture aids in the presumptive identification of the recovered organisms. If the medium becomes yellow, indicating lactose fermentation, ***Escherichia coli, Enterobacter,*** and/or ***Citrobacter*** are suspected. If the medium remains rose to orange, ***Klebsiella pneumoniae,*** **staphylococci, streptococci,** and/or **enterococci** are indicated. If the medium becomes magenta, purplish red, indicating an alkaline reaction, ***Proteus*** and/or ***Pseudomonas*** are the most likely dominant flora.

*No. Significant bacteriuria is generally indicated by over 100,000 bacteria per milliliter of voided urine, and there are 37×10^3 in 1 ml of this urine sample; that is, $37 \times 1{,}000 = 37{,}000$ bacteria per milliliter.

PROCEDURES EXERCISE 30

Procedure 1

Analyzing Normal Urine

First Laboratory Session

Your instructor may provide more detailed instructions for obtaining a clean-catch, midstream urine sample.

a Take a sterile container and 4 antiseptic towelette packages to a restroom. Wash your hands; then, with 1 of the towelettes, thoroughly cleanse 1 side of your external genitalia surrounding the outer opening of your urethra. Wipe from front to back. With the second towelette, cleanse the other side of your external genitalia from front to back. With the third towelette, cleanse the urethral opening. Discard each towelette after use.

b Women should keep the labia separated throughout the process. Men, if uncircumcised, should retract the foreskin. Begin urinating into the toilet, then half fill your sterile container with a midstream, clean-catch urine specimen. Use caution to avoid touching the sterile surfaces of the container to your body or clothing.

c Wipe the outer surfaces of your closed container with the fourth towelette. After washing your hands, take your urine sample to the microbiology laboratory.

d Agitate the **urine sample** in its sealed container to mix the contents. Using aseptic technique, a sterile 10 ml pipet, and a pipettor, *transfer 5 ml of urine into each of* **2 sterile test tubes.** Label the tubes, 1 for refrigeration, the other for room temperature storage. Include your group identification and the date on the labels.

e *Refrigerate* the appropriately labeled tube of urine; deposit the other tube in a safe place for *room temperature storage* until your next laboratory session.

f Again, agitate your original urine sample in its sealed container. Dip the end of a sterile, calibrated 1 μL (0.001 ml) loop into your urine sample. Submerge only the loop, not the handle. Deliver the loopful of urine to your **Blood Agar *or* Tryptic Soy Agar** plate, drawing a line of urine across the surface.

(g)

g Spread the urine over the surface of the medium, zigzagging your loop back and forth through the line of urine as illustrated. If using a disposable loop, discard the loop as directed by **your instructor.** Label the plate; invert and incubate it at 35° C for 18 to 24 hours.

h Agitate your original urine sample in its sealed container. Pour some of the remaining urine into a **Bacturcult® tube,** almost filling the tube. Immediately pour the urine back out of the Bacturcult® into the original container. Drain the Bacturcult® well.

i Close your Bacturcult®, turn the cap until it is tight, then loosen it 1 full turn. Label the Bacturcult® lid. Incubate your Bacturcult®, cap down, at 35° C for 18 to 24 hours.

j **Your instructor** will discuss sanitary disposal of urine samples and containers.

Second Laboratory Session

a Using a sterile, calibrated, 0.001 ml loop, inoculate a **Blood Agar *or* Tryptic Soy Agar** plate with a straight line of urine from your thoroughly resuspended, *refrigerated tube of urine.*

b Carefully spread the urine over the surface of the medium with your inoculating loop. If the loop is disposable, discard it in a sanitary manner. Label the plate with the date and your group identification. Add the words "Stored Under Refrigeration" to the label.

c With your second **Blood Agar *or* Tryptic Soy Agar** plate, repeat steps (*a*) and (*b*), this time with your well-agitated, *room temperature urine sample.* Use a new sterile loop if you are employing disposable loops. Label the plate with the date and your group identification. Add the words "Stored at Room Temperature" to the label.

d Incubate the 2 plates at 35° C for 18 to 24 hours.

e Examine the **urine culture** that you prepared during the last laboratory session. If possible, count the number of colonies. Examine the plate closely for evidence of spreading, confluent growth. Enter the number of colonies in your laboratory report and calculate the number of bacteria per milliliter in the urine specimen.

f Examine your **Bacturcult®.** Place the **counting strip** around the tube and center the circle over an area that contains a typical number of well-distributed colonies. Count and record the number of colonies. Repeat the count in another area; record your results.

g If no discrete colonies are present, place the tube in front of a light. See if the medium has become cloudy with a confluent growth of bacteria.

h Compare your results with those in the Bacturcult® package insert and with the uninoculated control. Evaluate Bacturcult® results in your laboratory report.

Third Laboratory Session

a Examine your 24-hour **urine cultures,** 1 prepared from urine that was refrigerated, the other from urine that was stored at room temperature. If possible, count the number of colonies on each plate. Enter these numbers in your laboratory report and calculate the number of bacteria per milliliter in each urine specimen.

POSTTEST EXERCISE 30

Part 1

True or False (Circle one, then correct every false statement.)

1. T F Bacteriuria is the presence of bacteria in urine.
2. T F A uropathogen is defined as any organism that survives in urine.
3. T F In a healthy person, there are usually about 10,000 bacteria per milliliter of urine within the urinary bladder.
4. T F Bacteriuria always indicates infection of the urinary bladder.
5. T F Catheterization of the urinary bladder carries the risk of introducing infection.
6. T F Over 1,000 bacteria per milliliter of voided urine usually indicates significant bacteriuria.
7. T F After 24 hours of incubation, the number of colonies on a Blood Agar plate streaked with 10^{-3} ml of urine is called the plate dilution.
8. T F The number of colonies on an all-purpose medium, after 24 hours of incubation at 35° C, times the inverse of the plate dilution approximately equals the number of bacteria in the sample of urine.
9. T F If there are 284 colonies on a plate that has a plate dilution of 10^{-3}, and the plate was prepared from a clean-catch, midstream, voided urine sample, the patient has significant bacteriuria.
10. T F If there are 227 colonies on a plate prepared with 10^{-2} ml of urine from a clean-catch, midstream, voided urine sample, the urine specimen had over 100,000 bacteria per milliliter.
11. T F A patient's urine that is accidentally stored at room temperature for 24 hours should be discarded and a fresh sample of urine obtained for microbiological testing.
12. T F Zigzag a calibrated loopful of urine back and forth across a plate of a selective medium to obtain well-distributed colonies for a quantitative analysis of urine.
13. T F Because the Bacturcult® contains phenol red and lactose, *Proteus* and/or *Pseudomonas* change the medium's color to magenta while *Klebsiella pneumoniae,* staphylococci, streptococci, or enterococci change it to orange or rose.
14. T F An inoculated, incubated Bacturcult® that turns yellow indicates negative bacteriuria.

Part 2

Completion and Short Answer

1. List normal flora of the opening and distal end of the urethra.
   a.
   b.
   c.
   d.
   e.
   f.
   g.
   h.
   i.
   j.

2. ____________________ , ____________________ , and ____________________ are microorganisms that are never considered to be normal flora of the urinary tract.

Laboratory Report **30**

Name ______________________________

Section ______________

Urinary Tract Normal Flora

Procedure 1

Analyzing Normal Urine

Second Laboratory Session

(*e*) Did you observe isolated colonies on your urine culture plate? __________
If so, how many colonies grew? __________
How many bacteria per milliliter in the urine specimen does this number of colonies represent? ________________
Does the number of bacteria per milliliter of your urine represent significant bacteriuria? __________
Did you observe spreading confluent growth? __________
(*f*) How many colonies are inside the circle on the counting strip?
First count __________
Second count __________
Average __________
(*g*) Is the medium cloudy with confluent growth? __________
(*h*) Does the average number of colonies in the circles, or the evidence of confluent growth, indicate that you have significant bacteriuria? __________
What color is the Bacturcult®? __________
If there is evidence of growth, what organisms are most likely to be responsible for the color of your Bacturcult®?

Third Laboratory Session

(*a*) How many colonies grew on your 24-hour culture prepared from refrigerated urine? _________ (If there are over 300 colonies record TNTC, which means "Too Numerous to Count.")

How many colonies grew on your 24-hour culture prepared from urine stored at room temperature? _________ (Again, utilize the term TNTC if necessary.)

Compare these results with each other and with your observations of the culture you prepared from your **freshly** voided urine specimen. Appraise the effects of refrigerated and room temperature storage on the microbial profile of your specimen.

Bacterial Unknowns: Isolation, Stock Culture, and Morphology

Necessary Skills

Microscopy; Microscopic measurement; Aseptic transfer techniques; Staining techniques; Streak-plate techniques; Knowledge of: Oxygen, Temperature, pH requirements, Selective and differential media

Materials

Warning: Handle cultures with care. Unknowns may be pathogens.

First Laboratory Session

Cultures: Mixtures are numbered

The instructor receives a key to the identity of the unknown bacteria.

1 Mixed unknown per group

The mixture is **prepared immediately before distribution** to students. It contains broth cultures of 2 different bacterial species that do **not** have any of the following attributes: spiral or curved, anaerobic, strictly microaerophilic, sulfate or sulfur reducing, rickettsial, chlamydial, mycoplasmal, endosymbiontic, nocardioform, gliding, sheathed, budding, appendaged, chemoautotrophic, archaeobacterial, cyanobacterial, slow growing, fastidious, swarming, gram-variable.

Supplies (per group):
- 2 Tryptic Soy Agar plates

Additional supplies:
- Staining reagents for Gram and acid-fast procedures
- Microscope slides
- Staining supplies
- Ocular micrometer disks
- Access to a stage micrometer
- Access to *Bergey's Manual*
- Access to the *DIFCO Manual*

Second Laboratory Session

Supplies:
- 4 Tryptic Soy Agar slants per group
- Tryptic Soy Agar plate at student's request
- Levine EMB Agar **or** MacConkey Agar plate at student's request
- Columbia CNA Agar plate at student's request
- Endospore Agar slant (Nutrient Agar plus $MnCl \cdot 4H_2O$ [0.002%]) at student's request
- Staining reagents for Gram, acid-fast, and endospore procedures
- Microscope slides
- Staining supplies
- Ocular micrometer disks
- Access to a stage micrometer
- Access to *Bergey's Manual*
- Access to the *DIFCO Manual*

Subsequent Laboratory Sessions

Supplies:
- Tryptic Soy Agar, Levine EMB Agar **or** MacConkey Agar, and Columbia CNA Agar plates at student's request
- Endospore Agar slant at student's request
- Access to *Bergey's Manual*
- Access to the *DIFCO Manual*

Primary Objective

Distinguish between and isolate two assigned, unknown bacterial species from a mixed broth culture.

Other Objectives

1. Prepare and evaluate a working stock culture slant and a reserve stock culture slant of each isolate.
2. Describe the colony and cell morphology of each isolate.

INTRODUCTION

In this exercise you isolate two bacterial species from a mixed broth culture. You then prepare stock culture slants and describe the morphology of each isolate. **Your instructor** may ask you to follow the protocol outlined in Exercise 32, Identifying Bacterial Unknowns, and identify both of your unknown isolates.

The experience of isolating and identifying bacteria from mixed samples simulates part of the day-to-day work of clinical microbiologists. Medical laboratories receiving throat, sputum, urine, stool, and other specimens must quickly and accurately isolate and identify pathogenic microorganisms from among a milieu of normal flora and contaminating microbes.

A look at the materials section of this exercise, with its list of attributes that your unknown species does **not** have, may lead you to wonder which bacteria are left. Examine table 31.1. It tabulates selected **sections** (large groups of similar bacteria) into which the procaryotes are divided by *Bergey's Manual of Determinative Bacteriology.* Your unknowns are selected from sections listed in this table. Many of the world's most deadly, pathogenic bacteria are found among bacteria in these sections; handle your unknown with care.

Preparing Stock Cultures

Figure 31.1 is a flow chart designed to help you isolate, cultivate, and describe the morphology of your unknowns. As in figure 31.1, stain your mixed broth culture to obtain your first clues leading to the identity of your unknown bacteria. Then streak the mixed culture for isolation of colonies. Incubate the streak plates, one at 30° C, the other at 35° C. Grow all subsequent cultures of each unknown at whichever temperature you discover to be closest to the species' optimum growth temperature.

After you have isolated colonies of both your unknowns, you prepare a **working stock** and a **reserve stock** slant for each species. The **working stock** is your source of bacteria for stains that help you identify the morphological characteristics of your unknowns. If you proceed to Exercise 32 and identify your bacterial unknowns, all your inoculations will be made from your working stocks. In contrast, your **reserve stock** slants are incubated, then refrigerated, unopened. These slants are your source of pure, uncontaminated bacteria.

If one of your working stocks is compromised, prepare a new one from the corresponding reserve stock slant. Reserve stocks of laboratory reference cultures are routinely **passed** (restreaked) on a schedule determined by the nutritional requirements of the individual cultures. If you are allowed more than 2 weeks to identify your unknown cultures, you may need to prepare a new reserve stock. To do this, restreak your reserve stock onto a plate of Tryptic Soy Agar, thus verifying its purity, then prepare a new reserve stock slant from an isolated, typical colony.

Table 31.1 Unknowns Are Selected from the Following Sections in *Bergey's Manual of Determinative Bacteriology*

Section No.	Section Title
4	Gram-Negative Aerobic/Microaerophilic Rods and Cocci
5	Facultatively Anaerobic Gram-Negative Rods
17	Gram-Positive Cocci
18	Endospore-Forming Gram-Positive Rods and Cocci
19	Regular, Nonsporing, Gram-Positive Rods
20	Irregular, Nonsporing, Gram-Positive Rods
21	Mycobacteria

Data taken from *Bergey's Manual of Determinative Bacteriology,* 9th ed., Holt, J. G., et al, eds., Williams & Wilkins Co., Baltimore, MD, 1994.

Staining

Differential stains, the Gram stain and acid-fast procedure, are best performed on **young** (showing first visible growth) cultures while bacterial cell walls are still intact. To discover when your unknown is "young," check your working stocks frequently. The first growth may be visible to your naked eye and ready for staining after 18, 24, or 48 hours of incubation.

To save time, effort, and expense, perform an endospore stain only on gram-positive bacteria and an acid-fast stain only on gram-positive rods. Why? (If you need help answering this question, see Exercises 9 and 10, Acid-Fast Staining and Endospore Staining.)

Troubleshooting

If you have difficulty obtaining well-isolated colonies of two different unknown species, utilizing selective, or selective and differential, media may help. Plates of Levine EMB Agar (or MacConkey Agar) and Columbia CNA Agar are available at your request. The properties of these agars are described in the *DIFCO Manual.*

When colonies on your initial streak plates are not well isolated or if they all look alike, Gram stain bacteria from several of the colonies. Use these stains plus a Gram stain of your original mixed culture to help you decide what to do next. You may wish to restreak the mixture on Tryptic Soy Agar, or you may choose to employ one of the selective and differential media.

For example, what should you do if your unknowns happen to be two species of gram-negative rods that form similar colonies on Tryptic Soy Agar and exhibit nearly identical microscopic morphology?*

If your mixed broth culture contains both gram-positive and gram-negative bacteria, living cells of the inhibited species may be present in what appears to be an

*Streak the mixture for isolation of colonies on Levine EMB Agar or MacConkey Agar. Colonies of the two species may be differentiated on these media.

Figure 31.1 Isolation and morphological description of bacterial unknowns

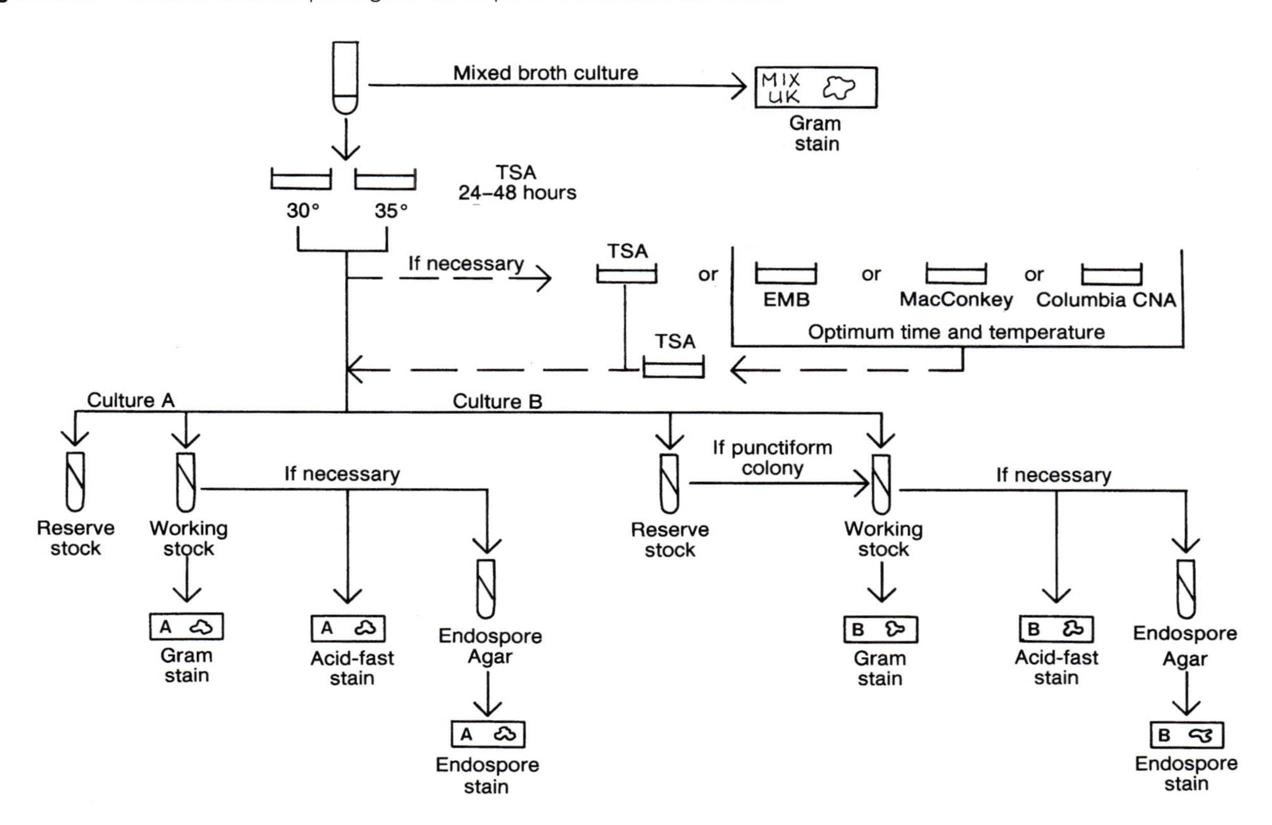

isolated colony on a **selective** medium. Restreak a well-isolated colony onto a **nonselective** medium to check purity before inoculating your stock slants.

In this exercise you must make many decisions on your own—when to stain, which stains to perform, what media to employ, how long to incubate bacteria and at what temperature, when to prepare stock cultures, and when to subculture. These decisions prepare you to embark upon the identification of your unknowns, a challenging educational experience that is introduced in Exercise 32.

PROCEDURES EXERCISE 31

Warning: Handle cultures with care. Unknowns may be pathogens.

Procedure 1

Isolating Unknowns

First Laboratory Session

Unknowns are processed according to many protocols. **Your instructor** may suggest variations on the methods that are presented in this procedure.

a Using sterile technique, *streak your mixed culture* of unknown bacterial species onto 2 plates of **Tryptic Soy Agar.**

b Incubate the cultures, 1 at 30° C, the other at 35° C. Check your plates for adequate growth after 18 to 24 hours. One or both species may require up to 48 hours to develop visible colonies.

c *Gram stain your original mixed broth culture* and examine the stained bacteria.

d In your laboratory report, record the identifying number of your mixture. Record shape, arrangement, and Gram stain reaction of the bacteria in your mixed culture.

Note that although your unknowns were mixed together immediately before class, the broths may not have been "young." Differential stains of your own young cultures during subsequent laboratory sessions may give you more reliable results.

e Label and refrigerate your mixed broth culture.

Second and Subsequent Laboratory Sessions

a To prepare stock cultures, you need an isolated colony of each bacterial species growing on **Tryptic Soy Agar** or some other nonselective medium. If you have *well-isolated colonies of 2 distinct types* growing on your line of streak, skip step (*b*), and continue with step (*c*) of this procedure.

b If you have not isolated colonies of both types on your first attempt, try again after reading this step. **Gram stain** bacteria from your streak plates. Compare the results of these stains with your observations of stains prepared from your mixed broth culture. Use the information you gain from your stains to help you decide whether to restreak on Tryptic Soy Agar, Levine EMB Agar, MacConkey Agar, and/or Columbia CNA Agar. Also, decide whether to streak several plates, each from a nearly isolated colony, or to utilize your original mixed culture as the inoculum. Incubate your plates at an appropriate temperature for an adequate length of time. Record your work and your logic in your laboratory report.

c As soon as you have isolated a colony of 1 species on Tryptic Soy Agar, assign it a working title, "Culture A" or "Culture B."

d In your laboratory report, describe the colonial morphology on Tryptic Soy Agar of your unknown species.

e As in figure 31.1, *prepare a working stock culture slant and a reserve stock culture slant* from 1 well-isolated, typical colony of your unknown. Incubate the slants at an appropriate temperature for an adequate length of time.

f If portions of **the same colony** remain, you may wish to prepare a Gram stain and, if relevant, an acid-fast stain from this colony. To obtain sufficient numbers of bacteria, it is often necessary to wait until your working stock has grown, then prepare stains from it.

g In your laboratory report, *describe the microscopic morphology* of each unknown species. Remember that while it is best to prepare Gram and acid-fast stains from a young culture, endospores are most likely to appear among older cells or on Endospore Agar (available at your request).

h **Your instructor** may require you to turn in your colony and microscopic morphology reports as well as your reserve stock slant cultures. **Your instructor** may also ask you to continue on to Exercise 32 and identify your unknown cultures.

POSTTEST EXERCISE 31

There is no written posttest for this exercise. You demonstrate your mastery of the subject by isolating unknowns from a mixture, preparing stock cultures, and accurately describing the morphology of the bacteria.

Laboratory Report **31**

Name ______________________________

Section ______________

Bacterial Unknowns: Isolation, Stock Culture, and Morphology

Procedure 1

Isolating Unknowns

First Laboratory Session

(*d*) Record the identifying number of your mixed unknown. ______________________ Draw and color a typical oil-immersion field of the Gram stain you prepared from your mixed broth culture. Record the shape, arrangement, and Gram stain reaction of the bacteria. Try to differentiate your 2 assigned species.

Description: ______________________________

____

____

____

____

Gram Stain of Mixed Culture

Subsequent Laboratory Sessions

(*b, c*) Record all the steps you take to isolate 2 distinct types of colonies from your mixed culture. Explain why you utilized each medium.

(*d*) Describe the conditions of incubation and the colonial morphology on Tryptic Soy Agar of both unknown species.
Culture A: Incubation duration _________ , temperature _________
Complete colony description:

Culture B: Incubation duration _________ , temperature _________
Complete colony description:

(*g*) In the following chart, describe the microscopic morphology of both unknown species.

Microscopic Morphology

	Culture A	**Culture B**
Shape	____________________	____________________
Size	____________________	____________________
Arrangement	____________________	____________________
Gram stain reaction	____________________	____________________
Acid-fast stain reaction*	____________________	____________________
Endospores present?**	____________________	____________________

*Record acid-fast stain reaction for gram-positive rods only.
**Record endospore presence for gram-positive bacteria.

Exercise **32**

Identifying Bacterial Unknowns

Necessary Skills

Microscopy; Microscopic measurement; Aseptic techniques; Gram, acid-fast, and endospore staining; Streak-plate techniques; Understanding of factors that influence and control microbial growth; Understanding of microbial metabolic reactions; Introduction to medical microbiology

Materials

Warning: Handle cultures with care. Unknowns may be pathogens.

Cultures (2 cultures per group):

Each group should have reserve stock culture and working stock culture Tryptic Soy Agar slants of two unknowns. The group prepared the 4 slant cultures while completing Exercise 31, Bacterial Unknowns: Isolation, Stock Cultures, and Morphology. Each unknown mixture was numbered and the **the instructor** has a key for identifying the unknowns.

Supplies:

Nearly all stains, media, reagents, equipment, antibiotic differentiation disks, etc. used in previous laboratory exercises will be available at the student's request. However, it may take several days to prepare and distribute some of these items. No multiple-test systems are employed. **The instructor** may limit student use of media that are both selective and differential.

Primary Objective

Through independent judgment and microbiological skill, identify unknown bacterial cultures.

INTRODUCTION

In this exercise you identify the genus and specific epithet of each of two unknown bacterial cultures. These are probably the same bacteria that you isolated in the previous exercise. It is assumed, therefore, that you already know the size, shape, arrangement, Gram and acid-fast reactions, and endospore-producing capabilities of each isolate. You already have determined the optimum temperature and duration of incubation for each unknown, and you have prepared a complete colony description for each culture grown on Tryptic Soy Agar.

Selecting Tests

You work independently while completing this exercise. Much of your guidance will come from *Bergey's Manual.* You will also benefit from the information contained in several of the tables and appendixes in this laboratory manual. Table 31.1 lists the sections of *Bergey's Manual of Determinative Bacteriology* from which your unknowns are selected. Appendixes 2 and 3 in this laboratory manual differentiate many of the gram-negative rods. Following the flow chart in figure 32.1 will help to start your journey of discovery down appropriate avenues. Be sure to check other flow charts that are diagramed in previous exercises.

Keep efficiency in mind; identify your unknown species without using excessive amounts of media. Accomplish this feat by cleverly choosing tests that split all possible species into two large groups. In general, avoid tests that single out only one species until you near the end of your identification project.

For example, while you may feel absolutely sure that your gram-positive coccus is ***Staphylococcus aureus,*** it is inappropriate to perform a coagulase test on the bacteria before discovering whether they produce catalase. Refer to the exercise in this manual that discusses **staphylococci,**

Figure 32.1 Flowchart for the preliminary identification of selected bacteria*

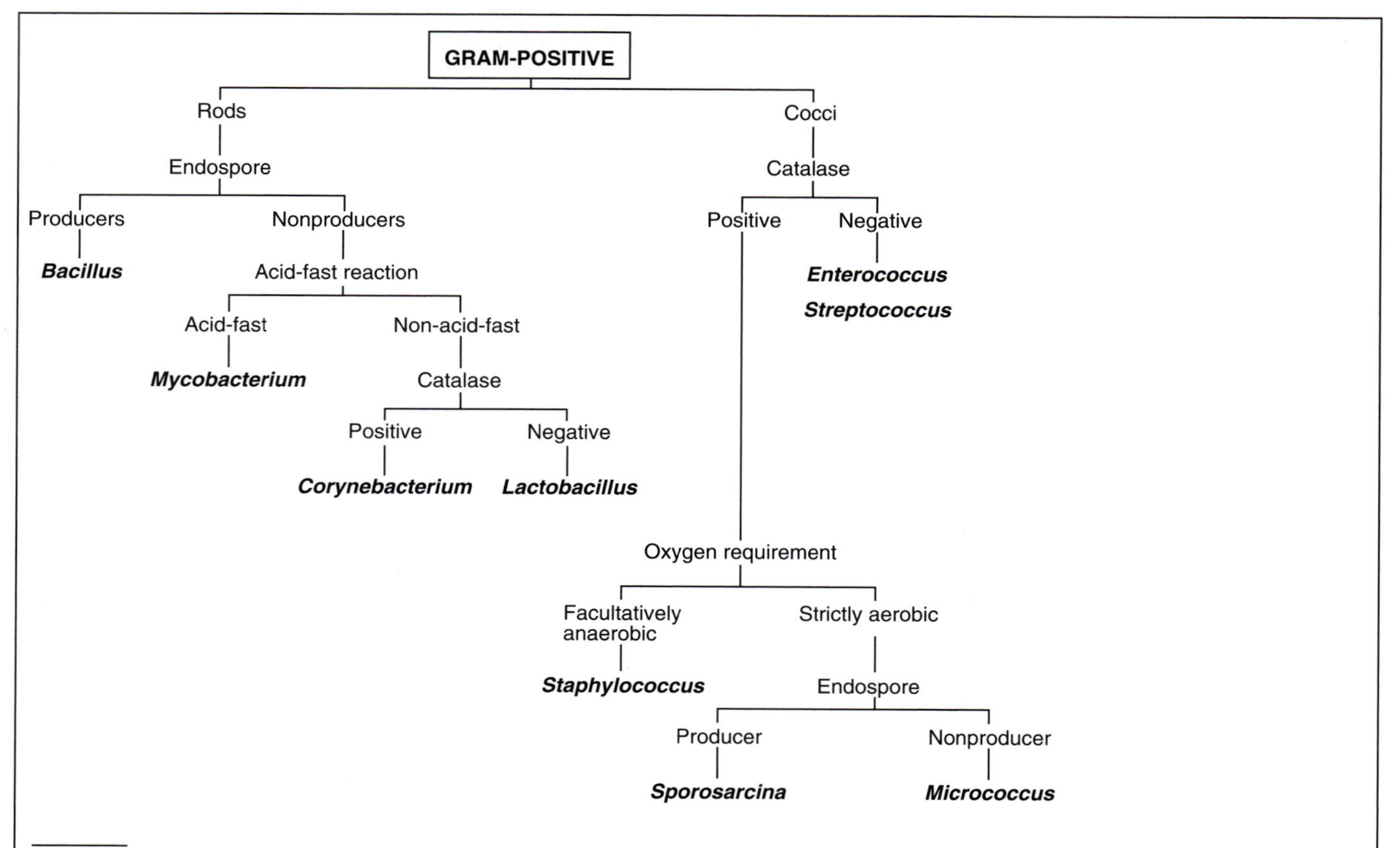

*Only genera eligible to be unknowns are included; that is, bacterial genera including species that do **not** have any of the following attributes: spiral or curved, anaerobic, strictly microaerophilic, sulfate- or sulfur-reducing, rickettsial, chlamydial, mycoplasmal, endosymbiontic, nocardioform, gliding, sheathed, budding, appendaged, chemoautotrophic, archaeobacterial, cyanobacterial, slow growing, fastidious, swarming, gram-variable.

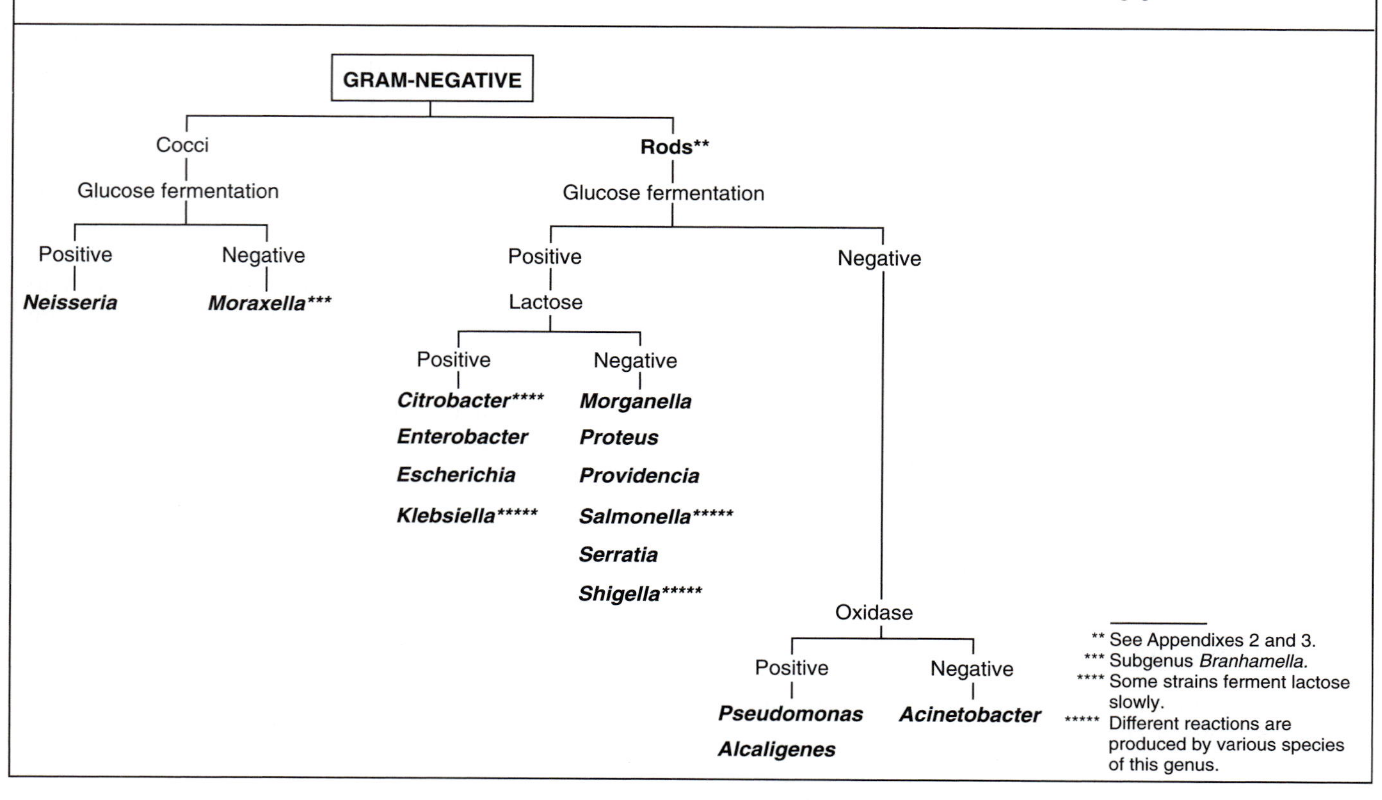

** See Appendixes 2 and 3.
*** Subgenus *Branhamella*.
**** Some strains ferment lactose slowly.
***** Different reactions are produced by various species of this genus.

enterococci, and **streptococci** if you do not understand why catalase testing precedes coagulase testing of gram-positive cocci.

Report Sheet

Your instructor may ask you to turn in a report sheet for each of your unknowns. On the sheet, list each test you perform, your logic in selecting that test, and the reaction that you observe. **Your instructor** may also require you to list your next request for media. In many teaching laboratories, a student's media request is initialed by **the instructor** or another authority before it can be filled.

Preparing a report sheet helps you avoid a "shotgun approach" to identifying unknowns. This is the wasteful and time-consuming practice of inoculating every medium in sight and hoping that you will be able to distill accurate drops of data from the flood of superfluous information that results.

In this exercise you identify bacteria by utilizing bits of information gleaned from many of the preceding exercises. You have probably noticed that not all biochemical tests are needed to identify every species. The enzymatic systems that distinguish one group of bacteria are frequently inconsequential when it comes to identifying another group. Keep this concept in mind when you select a biochemical test. Be perspicacious.

Troubleshooting

The two most common problems that students encounter while identifying unknowns are contamination and confusion. Avoid contamination by following all the rules of sterile technique. Reduce confusion by recording the results of tests as soon as you observe them. Keep an accurate log of your work in your laboratory report.

A third item that may challenge your sleuthing skill is the existence of **variants,** bacteria that resemble the other members of their species except for a few bits of enzymatic individuality.

What if one of your unknowns is a variant strain? *Bergey's Manual* generally lets you know where to expect variation in enzymatic capability. For example, *Bergey's Manual* tells you that **Shigella** species vary in indole production. Performing the indole test will not help you identify the specific epithet of a *Shigella.* To avoid performing tests that yield little or no useful information, do thorough research before constructing a logical plan for identifying your unknowns.

If, even after extensive testing, one of your unknowns does not conform to the identification of any known species, consider the possibility that it may be a variant strain. Its variation from the norm may occur in an enzyme system that you tested near the beginning of your work. When one of your first tests is inaccurate (whether it is you or the bacteria that caused the problem) all your subsequent test results may be meaningless.

When all else fails, go back, construct a different path for identifying your unknown, and try again.

PROCEDURES EXERCISE 32

Procedure 1

Identifying Unknowns

Warning: Handle cultures with care. Unknowns may be pathogens.

Unknowns are processed according to many protocols. **Your instructor** may suggest variations on the methods that are presented in this procedure.

a In your laboratory report, record the number that identifies your mixture. Also, record preliminary information concerning the morphology of each of your unknowns.

b Utilize this laboratory manual, *Bergey's Manual,* and any other reference books that are available to help you choose the next appropriate test to perform on each of your unknowns.

c If **your instructor** requires it, fill out a report sheet for each unknown stating what media you need and why you selected these tests. Obtain initialed proof of approval, gather the media, and perform the required procedures.

d Record all results as you make your observations. Continue to work independently until you have identified the genus and the specific epithet of each of your unknowns.

e **Your instructor** may require you to turn in your report sheet along with a justification for each test and a record of the results.

POSTTEST EXERCISE 32

There is no written posttest for this exercise. On your report sheets, you demonstrate your ability to identify unknown bacterial cultures logically and efficiently.

For a complete description, see Appendix 4, on page 352.

Laboratory Report **32**

Name ______________________________

Section ____________

Identifying Bacterial Unknowns

REPORT SHEET

"CULTURE A" MIXTURE No. ________

Genus and specific epithet of "Culture A" ______________________ ______________________

Gram reaction ____________

Shape ____________

Arrangement ______________________________

Endospores? ____________

Acid-fast? ____________

Colony description:

Initialed OK	**Test**	**Logic**	**Reaction**

REPORT SHEET

"CULTURE B" MIXTURE No. ______________

Genus and specific epithet of "Culture B" ________________________________ ________________________________

Gram reaction ______________

Shape ______________

Arrangement ____________________________________

Endospores? ______________

Acid-fast? ______________

Colony description:

Initialed OK	Test	Logic	Reaction

Serology

Necessary Skills

Aseptic transfer techniques, Knowledge of intestinal pathogens of the family Enterobacteriaceae, Knowledge of beta-hemolytic streptococci

Materials

Warning: Handle unknowns with care. Some are pathogens.

Procedure 1

Cultures: 1 pair of streaked Blood Agar biplates (35° C, 24 hour) per group, labeled as unknowns 1 through 4, with a key provided for **the instructor**
- *Streptococcus agalactiae* (Lancefield group B)
- *Streptococcus equisimilis* (Lancefield group C)
- *Enterococcus faecalis* (Lancefield group D)
- *Streptococcus pyogenes* (Lancefield group A)

Supplies (per class):
- BBL® Strep Grouping Kit (40842) for Lancefield groups A, B, C, F, and G streptococci: Slide Test (75 tests per kit)
- Matching strip of colored tape on each vial of reagent and its dropper. Use a different color for each pair. Do not cover labels on vials.
- Ice bath to hold reagent set
- Container of disinfectant for used glass plates

Supplies **to instructor** for demonstration of positive and negative controls:
- Glass plate or slides with at least 6 raised rings
- Lint-free cloth for dusting glass plate

Supplies (per group):
- Glass plate or slides with at least 12 raised rings; VDRL plates may be used
- Access to a lint-free cloth for dusting glass plate

Procedure 2

Cultures: 2 per group, 24-hour, 35° C, Tryptic Soy Agar slants, labeled as "Unknown X" and "Unknown Q," with key provided for **the instructor**
- *Salmonella* species
- *Escherichia coli*

Supplies:
- 1 Triple depression slide per group
- 1 Single depression slide per group
- *Salmonella* O Antigen group B (DIFCO No. 2840–56), several dropper bottles per class
- *Salmonella* O Antiserum Poly A-I (DIFCO No. 2264–47), several dropper bottles per class
- Phenolized saline (0.85% NaCl in 0.5% phenol, aq.), several dropper bottles per class
- 6 Toothpicks per group
- Container of disinfectant for depression slides

Procedure 3

1 Artificial blood typing kit* **or**

1 Monoclonal blood typing kit** per 50 students **or** the next 7 items:
- A, B, and D antisera in dropper bottles
- 1 Alcohol swab per student
- 1 Sterile microlancet per student
- 1 Blood test card or microscope slide per student
- 3 Mixing sticks or toothpicks per student
- 1 Student instruction sheet** per student (optional)
- 1 Small Band-Aid or gauze per student (optional)

---

*Available from Ward's Biology and Lab Supplies, Rochester, NY 14692.

**Available from Carolina Biological Supply Company, Burlington, NC 27215.

Procedure 4

Supplies:
- 1 Directigen Flu A 20 test kit per class
- 1 Dacron- or rayon-tipped nasopharyngeal swab per group
- 1 ml of sterile saline in a capped tube per group
- 1 Sterile 0.2 ml pipet per group
- Pipettors
- Disposable gloves

Primary Objective

Perform and interpret serological tests.

Other Objectives

1. Define: serology, antigen, antibody, *in vitro,* immunoglobulin, antiserum, serotyping, RPR, VDRL, reagin, nonspecific antibody, cardiolipin, FTA-ABS, FITC-anti-HGG, EIA, antigen sandwich.
2. Discuss the nature of antigen and antibody interactions in a homologous system.
3. Compare and contrast agglutinins and precipitins.
4. Illustrate how the condition of H substance on erythrocytes determines an individual's ABO blood type.
5. Utilize monoclonal antibodies for blood grouping; determine the ABO and Rh reactions of your own blood.
6. Diagram and explain the production of monoclonal antibodies.
7. Characterize plasma cells, myeloma cells, and hybridoma cells.
8. Characterize the O and H antigens of *Salmonella,* and employ multivalent *Salmonella* O antiserum to identify *Salmonella.*
9. Evaluate and explain three serological tests for syphilis.
10. Explain the function of latex beads in agglutination reactions, and utilize a latex agglutination test to identify Lancefield groups of streptococci and enterococci.
11. Compare and contrast the direct and the indirect enzyme immunoassays, diagraming each test.

INTRODUCTION

Serology is the study of **antigen** and **antibody** interactions ***in vitro*** (in a test tube or other artificial environment). **Antibodies** are **immunoglobulins.** These protein molecules, found in animal blood, react with **antigens,** the molecules that provoke antibody synthesis by the animal.

Antibody-antigen reactions are nearly as specific as enzyme-substrate interactions. Only the initiating antigen, or a substance very similar to it in chemical structure, reacts with a corresponding antibody. An antibody and its complementary antigen are a **homologous** (matched) system.

Immunoglobulins form one of the body's major defenses against invasion by foreign agents including bacteria, viruses, toxins, and even blood. Antibodies are classified by their mode of action. **Agglutinins** clump **particulate matter. Precipitins** precipitate **dissolved antigens,** such as toxins, out of solution.

Blood Groups

The surfaces of red blood cells are covered with antigens. One set of antigens that coats every human erythrocyte is the **ABO** group. Another set is the **Rh** array.

The basis of the ABO system is a glycolipid, **substance H.** It is attached in many places to the surface of each erythrocyte. On the red blood cell, substance H may stand naked or it may combine with a specific mucopeptide or sugar.

The condition of substance H, whether it is naked or combined, determines a person's ABO blood type; see figure 33.1. If the glycolipid is attached to ***N*-acetylgalactosamine,** the blood is type A. If **galactose** is attached to substance H, the blood is type B. When the glycolipid antigen is **unmodified,** the blood is type O. Individuals with type AB blood have *N*-acetylgalactosamine on some of their H substance, galactose on the rest.

Humans form antibody that agglutinates whichever ABO group antigen they lack; see figure 33.2. In this exercise you find your own ABO blood group by mixing your blood with **antisera** (sera containing antibody). How is it possible to distinguish four blood groups—A, B, AB, and O—using only two antisera, anti-A and anti-B? See figure 33.2 for the answer.

You determine your Rh type, positive or negative, by mixing your blood with anti-D. You are "Rh positive" if the Rh_D antigen is present in your blood. Since there are far more ABO antigenic sites on each erythrocyte than Rh sites, the Rh reaction is weaker than the ABO responses to antisera.

Monoclonal Antibody

In this exercise you employ **monoclonal antibody** for blood typing. As illustrated in figure 33.3, researchers make monoclonal antibody by injecting a mouse with a *selected antigen.* In response to this immunological attack, the mouse produces *antigen-specific antibody.* The scientists then join the mouse's *specific antibody-producing* **plasma cells** together with **myeloma** (a type of cancer) **cells** that produce large amounts of cellular products, multiply rapidly, and are immortal.

Fused myeloma and plasma cells are called **hybridoma cells.** Showing traits from both of their parent cells, the immortal hybridomas multiply rapidly, and they produce large amounts of pure, specific antibody that may be used to treat cancer, diagnose disease, or type blood, depending upon which antigen was originally injected into the mouse.

Each selected hybridoma forms a **clone** (the genetically identical descendants of a cell), and every cell in the clone produces exactly the same antibody. These exceedingly pure and specific products of genetic engineering are called monoclonal antibodies.

Figure 33.1 The ABO blood-grouping antigens

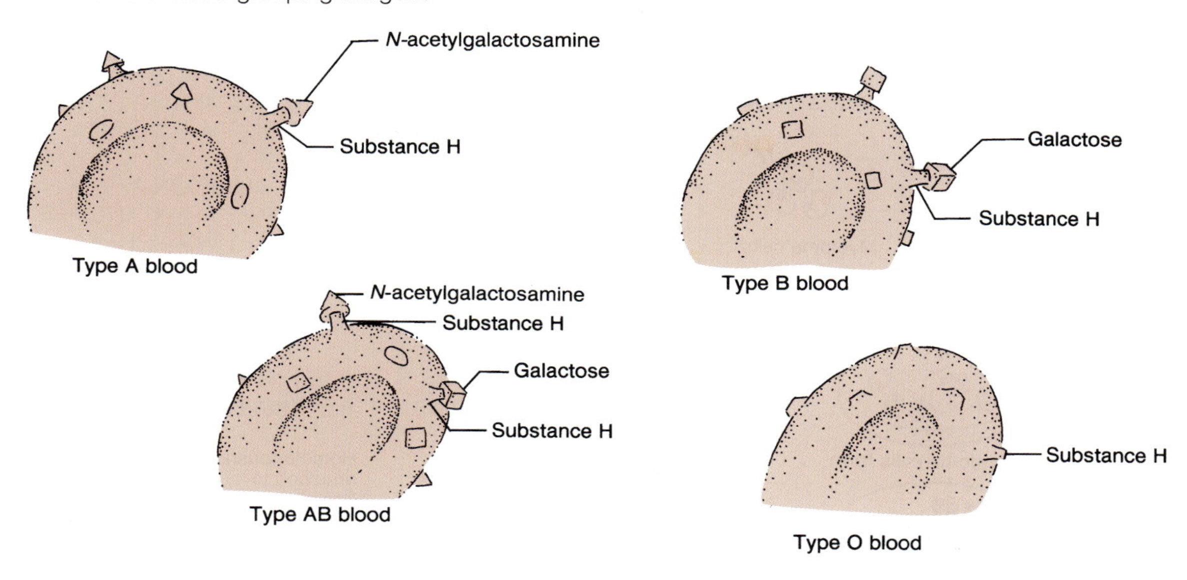

Figure 33.2 ABO blood grouping

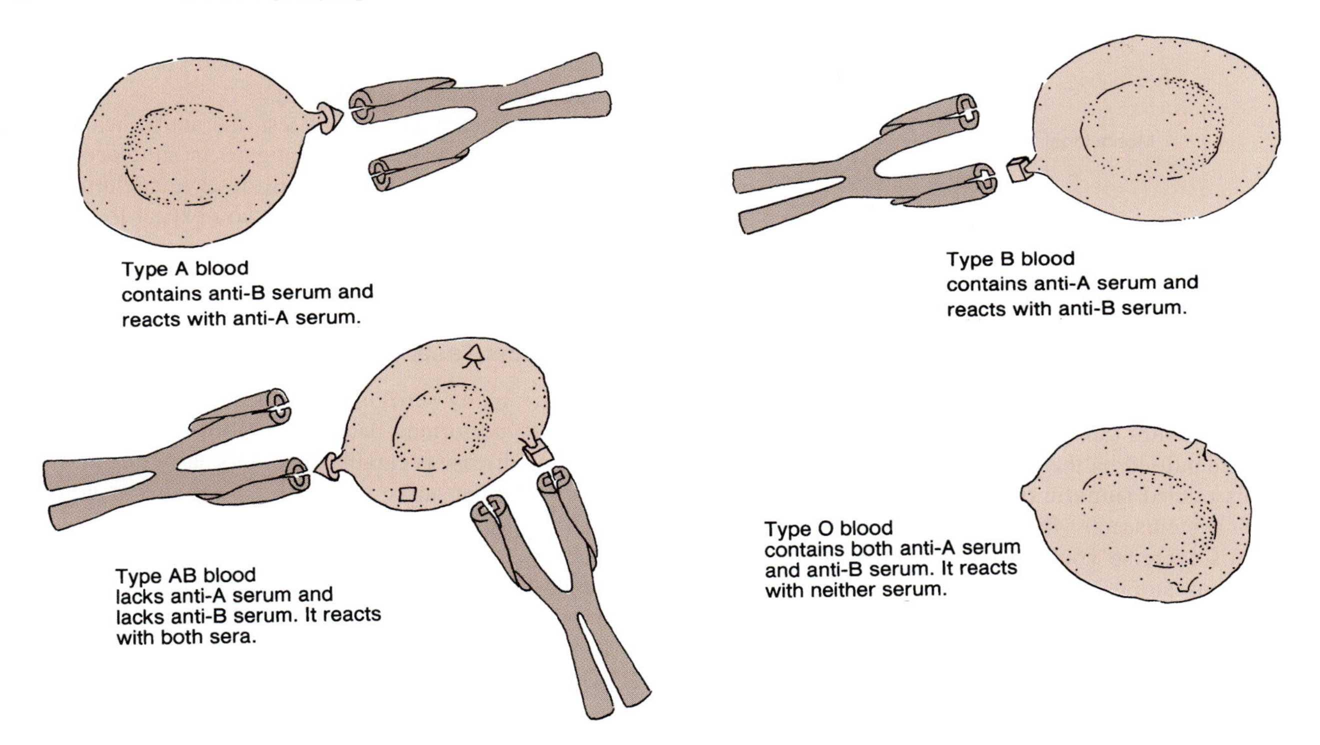

Serological Diagnostics

Traditionally, microbiologists cultivate organisms at least overnight, then identify them with biochemical and staining techniques. Rapid, accurate identification of a pathogen enables physicians to initiate appropriate treatment. It allows epidemiologists to trace an organism's route through the population so they can suggest ways to prevent further spread of the disease.

As you demonstrate in this exercise, **serotyping** (identifying the species or strain of a bacterium by its antigens) is faster and easier than the traditional methods of microbial identification.

Serology offers medical microbiologists choices. The patient's serum can be tested against commercially available antigens. Pathogens attacking the patient can be cultivated, isolated, and then identified with monoclonal antibodies and other commercial antisera. Or, as in the case of

Figure 33.3 Monoclonal antibody production

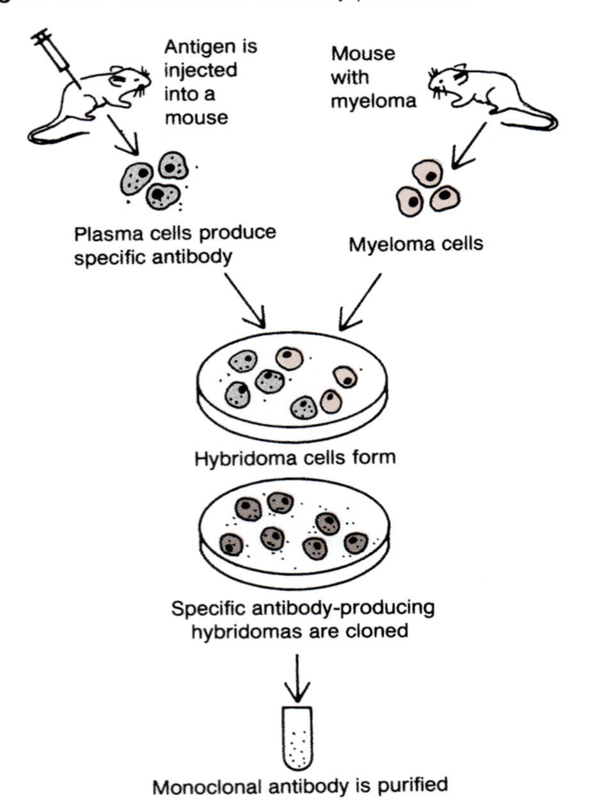

Figure 33.4 Latex bead agglutination

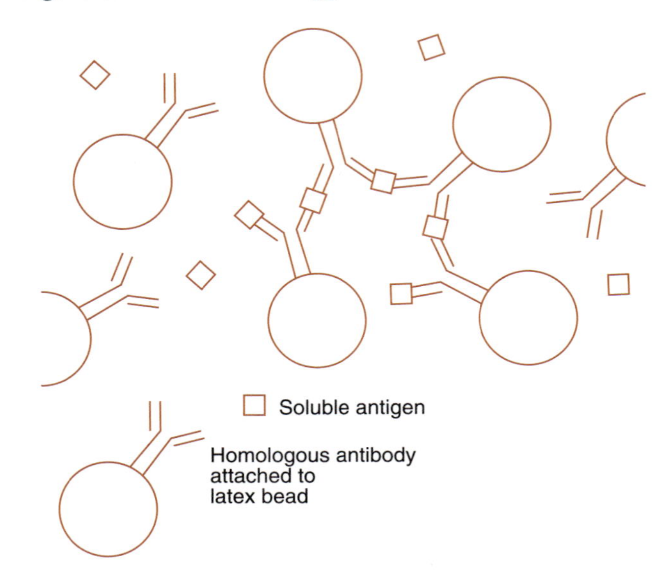

Streptococcus pyogenes, the etiologic agent of strep throat and an assortment of other pathological conditions, a throat swab specimen from the patient can be tested in less than ten minutes in the doctor's office for the presence of specific, identifying antigens.

Serotyping is much more precise than biochemical methods of bacterial identification. For example, over a thousand **serovars** (antigenically distinct types) of ***Salmonella*** are identified by the polysaccharide antigens on their cell walls (called **somatic** or **O antigens**) in combination with their proteinaceous, **flagellar (H) antigens.**

In this exercise you employ the **multivalent** (active against many strains of an organism) antibody in *Salmonella* O Antiserum to agglutinate commercially prepared *Salmonella* O Antigen. After observing this positive reaction, you employ the antiserum to distinguish between killed suspensions of ***Salmonella*** and ***Escherichia.***

Latex Bead Agglutination

Sometimes, when antigens interact with antisera, the reaction is invisible. If the antiserum is first attached to **latex beads,** small clumps of the plastic appear wherever the immune reaction occurs; see figure 33.4. These complexes of antigen, antibody, and latex beads form quickly and are easy to see.

In this exercise you employ ***Streptococcus*** grouping reagents, antibodies coated with a latex bead suspension, to identify the Lancefield groups to which your streptococcal or enterococcal unknowns belong.

In the latex agglutination test, the antibodies attach to group-specific, carbohydrate antigens in streptococcal and enterococcal cell walls. You will observe agglutination in the homologous system. What do you expect to see in the **heterologous** (nonmatching) mixtures of antigens and coated antibodies?*

Serological Tests for Syphilis

There are many serological tests for **syphilis.** They are frequently performed because this life-threatening, fetus-damaging disease is curable only in its early stages and also because the causative spirochete, ***Treponema pallidum,*** cannot be grown outside of living cells.

Easy and inexpensive **screening tests** (designed for mass testing of a population) for syphilis—including the two most common ones, the **Rapid Plasma Reagin (RPR) Card** test and the **Venereal Disease Research Laboratory (VDRL)** flocculation test—occasionally give **false "positive"** reactions but rarely miss finding antibodies that are present in the patient.

Both of these procedures check the patient's serum for the presence of **reagin,** a **nonspecific antibody** (one that combines not only with its initial antigen but also with other structurally similar substances). The syphilitic patient's body produces reagin in response to a phospholipid

*No agglutination occurs unless the latex-coated antibody matches the extracted, carbohydrate, cell wall antigen.

Figure 33.5 The positive Fluorescent Treponemal Antibody Absorption test

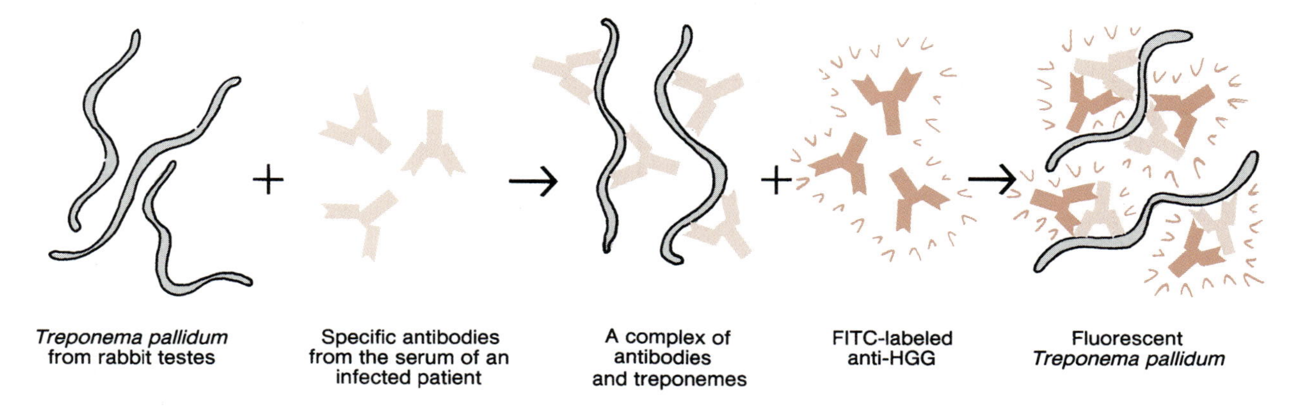

antigen. This antigenic lipid is formed within the patient's body as part of his or her reaction to an infection with *Treponema pallidum.*

Reagin is classified as a nonspecific antibody because it responds to the initiating lipid and also to an inexpensive, readily available, soluble extract of beef heart called **cardiolipin.** Both the RPR and VDRL tests contain cardiolipin. Remember, neither the RPR nor the VDRL procedure tests for the presence of *Treponema pallidum,* and neither test checks for antitreponemal antibodies.

Because the **Fluorescent Treponemal Antibody Absorption (FTA-ABS)** test generates fewer than 1% false positives, this more accurate but expensive and difficult procedure is frequently employed as a **confirmatory test** for syphilis; see figure 33.5. In FTA-ABS testing, a microscope slide is smeared with *Treponema pallidum* cells that were grown in rabbit testes. A patient's serum is then layered onto the smear, and after incubation a saline rinse removes any unbound antibody.

If the patient's serum contains **antitreponemal antibodies,** these human globulin molecules react with the rabbit-grown *Treponema pallidum* and—despite the rinse—they remain attached to the bacteria on the slide.

The captured, homologous antibody molecules are not visible through a microscope, so stained **anti-human gamma globulin (anti-HHG)** antibody is applied. Before being used in an FTA-ABS test, the anti-HGG is attached chemically to molecules of **fluorescein isothiocyanate (FITC),** a fluorescent dye that emits a yellow-green glow.

On the slide, the **FITC-anti-HGG** links to any human gamma globulin that is present. This human antibody is firmly attached to *Treponema pallidum* in the smear, or it would have washed away with the saline rinse. Thus, at the end of the staining procedure, when the smear is examined with a **fluorescence microscope,** only a positive test glows.

The light indicates that the patient's serum contains specific, antitreponemal antibodies because the patient is (or was) infected with syphilis. In a negative test, no glowing color develops because the unbound antibody was rinsed off the slide.

Enzyme Immunoassay (EIA)

An **enzyme immunoassay (EIA)** is a diagnostic, serological technique that employs **antibodies** that have been chemically attached to **enzymes;** see figure 33.6. Both the enzymes and the antibodies stay active. In other words, the **enzyme-linked antibodies** maintain their ability to attach to homologous antigens, and the **conjugated** (coupled) **enzymes** (usually **horseradish peroxidase** or **alkaline phosphatase**) are still active catalysts. These enzymes react with their **substrates** to produce color. Thus, an EIA makes an antibody-antigen reaction visible to a spectrophotometer or to the naked eyes.

In a **direct EIA,** a specific type of monoclonal antibody is adsorbed onto a tiny well, bead, or membrane. The patient's unidentified antigen is then added, and if this **test antigen** is homologous, it attaches to the antibody. It reacts, but the complex is still invisible.

To see the reaction, the alert clinician makes an **antigen sandwich** in "antibody bread" by adding enzyme-linked monoclonal antibody that is specific for the test antigen. After rinsing away any unattached material, this wily scientist adds substrate for the enzyme. If color develops, the microbiologist has identified the patient's antigen. In a negative test, no color develops because the unbound antigen was rinsed away.

Look at figure 33.6 again and note that the direct, sandwich method of EIA requires a specific, enzyme-labeled antibody for every type of test antigen. Ingenious scientists have developed an **indirect enzyme immunoassay** that analyzes a patient's antibodies. The **test antibody** may attach to an ever-growing array of commercially available antigens. The indirect EIA employs monoclonal antibody called **enzyme-linked anti-human globulin.**

One common screening test for **human immunodeficiency virus (HIV),** which causes **acquired immunodeficiency syndrome (AIDS),** is an indirect EIA. Many other viral, bacterial, and parasitic diseases are also diagnosed with this immunological technique.

Figure 33.6 Direct and Indirect Enzyme Immunoassays (EIAs)

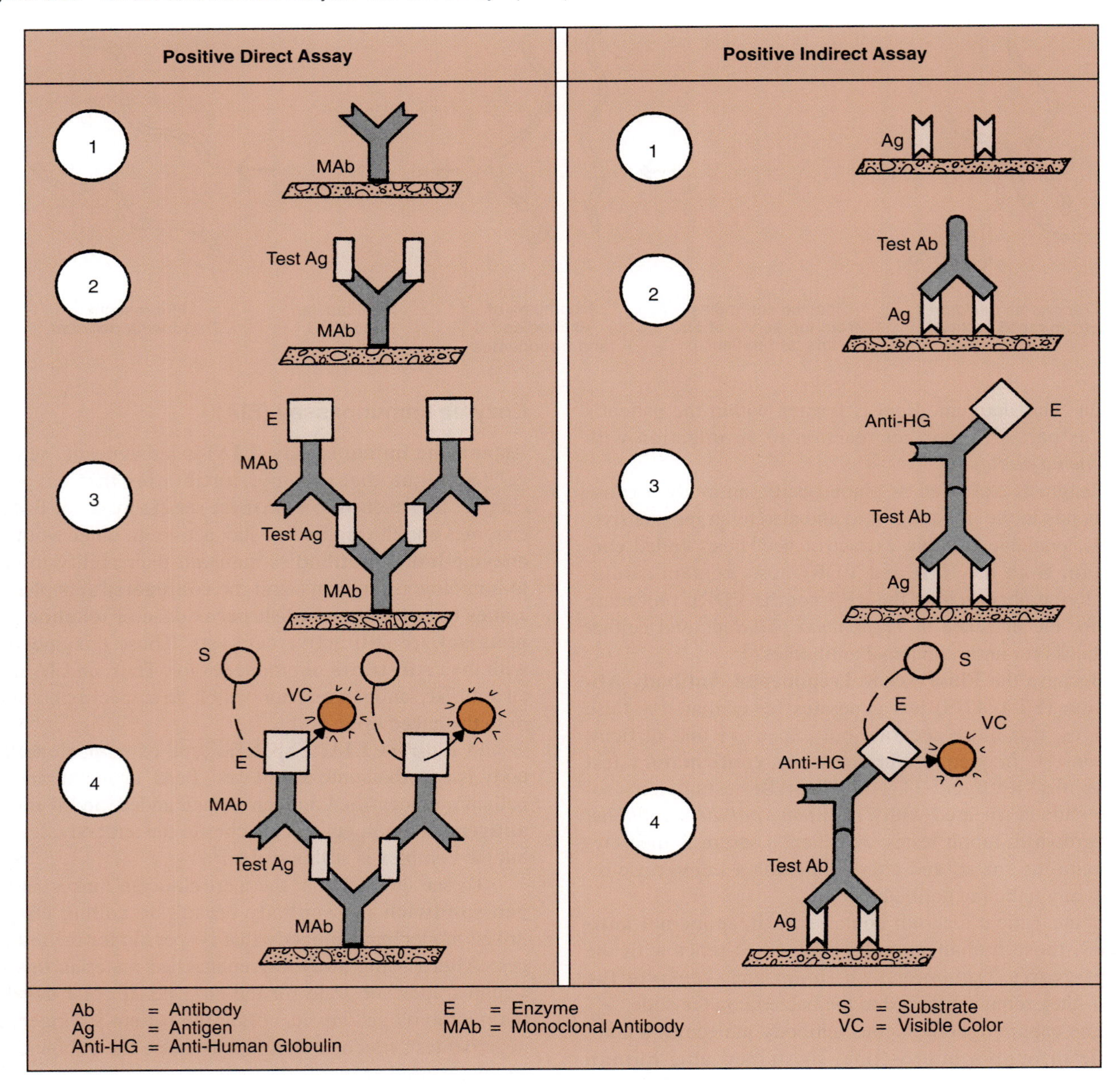

PROCEDURES EXERCISE 33

Procedure 1

Latex Bead Agglutination

Warning: Handle cultures with care. Some are pathogens.

a Your group will test all 4 streptococcal unknowns with each of the 3 antisera. **Your instructor,** or an assigned group, will prepare **positive** and **negative controls** for class observation; see steps (*j*) through (*p*). Before proceeding, wipe your glass plate or slides with rings thoroughly with a lint-free cloth. Oil from your fingers may interfere with correct interpretation of your results, so avoid touching the glass within the rings.

b Label a glass plate or slides with rings as shown in the illustration.

c Without spilling any liquid, mix the **antibody-coated latex bead suspension** in each vial just before using it.

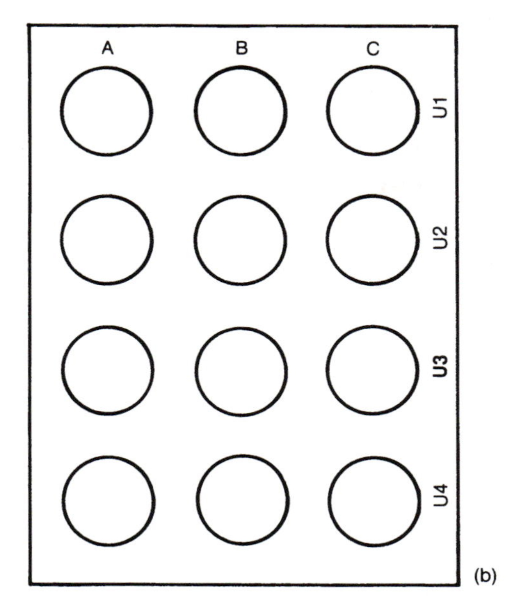

d Utilize the dropper in the lid of a vial to dispense 1 tiny drop of a latex suspension into each ring of the appropriately labeled column. Dispense all 3 latex reagents.

e Use a sterile loop to dispense a visible amount of confluent growth from 1 unknown **streptococcal or enterococcal culture** into a drop of latex reagent. Mix well, then flame and cool your loop. Dispense a total of 3 loopfuls of the bacteria into the appropriately labeled row, flaming and cooling your loop after each addition.

f Repeat step (*e*) with each of your 3 remaining streptococcal or enterococcal unknowns. There are now 12 latex reagent drops mixed with 12 loopfuls of bacteria on your plate.

g Mix antigens and antibodies by gently rocking the slide for exactly 2 minutes or rotating it at a speed of 95 to 100 rpm for 2 minutes; avoid splashing.

h With a bright light, immediately read the rings for **agglutination.** Record your results in your laboratory report.

i Disinfect your glass plate or slides.

j **Positive** and **negative controls** for the class are to be prepared by **the instructor** or an assigned group. To prepare positive and negative controls, label a glass plate or slides with rings as shown in the illustration.

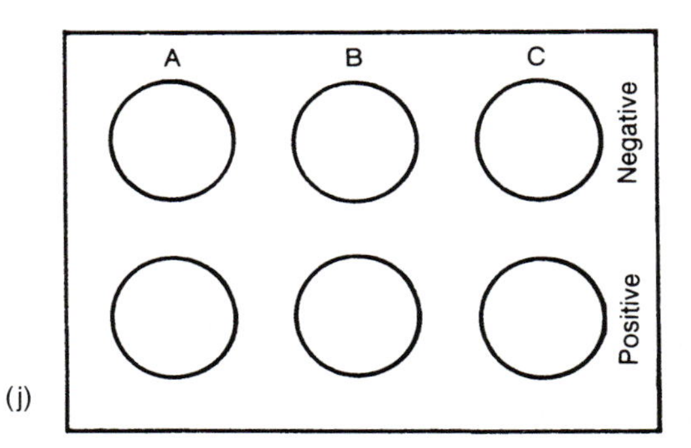

Figure 33.7 The slide agglutination test

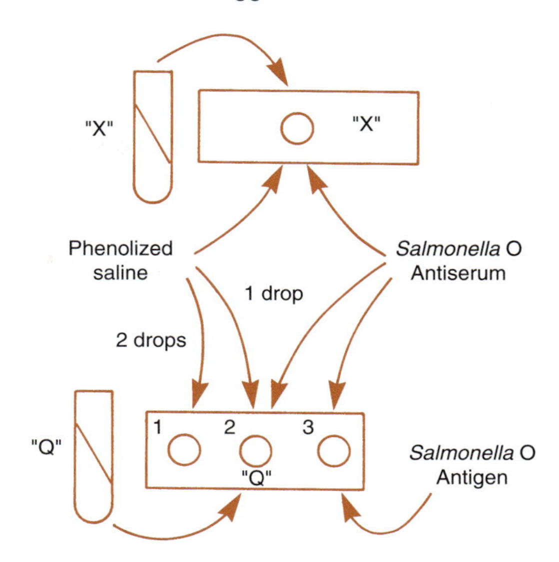

k For the **negative control,** mix the antibody-coated latex bead suspension in each vial just before using it. Utilize the dropper in the lid of each vial to dispense 1 drop of its latex suspension into the appropriately labeled ring.

l Use a sterile loop to dispense a visible amount of confluent growth from your *Enterococcus faecalis* slant into a drop of latex reagent. Mix well, then flame and cool your loop. Dispense a total of 3 loopfuls of *Enterococcus faecalis* across the row. Flame and cool your loop after each addition.

m Rock the slide for 2 minutes, or rotate it at a speed of 95 to 100 rpm for 2 minutes; avoid splashing. With a bright light, all members of the class should immediately read the rings for agglutination. Record results in your laboratory reports.

n For the **positive control,** mix 1 drop of the **antigenic Polyvalent Positive Control reagent,** which is provided with the kit, with 1 drop of each of the **antibody-coated latex suspensions** in the appropriately labeled rings.

o Rock the slide for 2 minutes, or rotate it at a speed of 95 to 100 rpm for 2 minutes; avoid splashing. With a bright light, all members of the class should immediately read the rings for agglutination.

p Record your results in your laboratory report. Disinfect your slides or glass plate in an appropriate manner.

Procedure 2

Slide Agglutination Test; see figure 33.7

a Label a triple depression slide as in figure 33.7.

b Transfer a drop of **phenolized saline** into depressions 1 and 2.

c With a flamed and thoroughly cooled loop, mix a loopful of 1 **unknown** *X* or *Q* into depression 2. Stir the suspension with your loop until no clumps of bacteria remain. Sterilize your loop.

d Add another drop of **phenolized saline** to depression 1.

e *Without touching the dropper to the slide,* place a drop of ***Salmonella* O Antiserum** into depression 2 and another drop into depression 3.

f Finally, *without touching the dropper to the slide,* add a drop of ***Salmonella* O Antigen** to depression 3.

g Mix the contents of each depression with a separate, clean toothpick. Discard the toothpicks as directed by **your instructor.** Compare the reactions in the 3 depressions.

h In your laboratory report, describe the reactions, identifying the positive and negative controls. State whether or not this unknown is a *Salmonella.*

i Label a single depression slide as in figure 33.7.

j Transfer a drop of **phenolized saline** into the depression.

k With a flamed and thoroughly cooled loop, mix a loopful of your other *X* or *Q* **unknown** in the depression. Stir the suspension with your loop until no clumps of bacteria remain. Flame your loop.

l *Without touching the dropper to the slide,* place a drop of ***Salmonella* O Antiserum** into the depression.

m Mix the contents of the depression with a clean toothpick. Discard the toothpick appropriately. Compare the reactions in this depression with the results on your triple depression slide. In your laboratory report, describe the reaction. State whether or not this unknown is a *Salmonella.* Disinfect your depression slides.

Procedure 3

Blood Typing with Monoclonal Antibodies*

> **Warning:** Wash your hands before obtaining a blood sample.

a Obtain a blood test card or draw and label 3 dime-sized circles on a microscope slide.

b Place a drop of monoclonal **anti-A serum** in the first circle, **anti-B serum** in the second circle, and **anti-D serum** in the third circle. Replace caps.

c Wash your hands thoroughly before obtaining a blood sample.

d Swab the pad of the little finger on your non-dominant hand with alcohol and allow the alcohol to dry. Then pierce your finger pad with a sterile microlancet.

e Let 1 **drop of blood** flow freely next to each drop of monoclonal antiserum in its circle.

f Mix the blood and reagent in each circle with a separate stirring stick. Use a total of 3 stirrers. **Do not cross-contaminate samples.** Start timing.

g Gently tilt the card back and forth and from side to side for 1 minute.

h Read the card by tilting it to one side and allowing the blood to pool. **Do not allow drops to run together.** In your laboratory report, record which circles show agglutination (clumping).

i Then gently rock the card for 4 minutes more to double check any negative readings.

j Do not confuse any drying around the edges with agglutination. Record your final observations of agglutination in your laboratory report.

k Discard your card or slide, your 3 stirrers, and your microlancet in a receptacle for adequate disinfection as directed by **your instructor.**

Remember, human blood may carry hepatitis viruses, AIDS viruses, and other deadly agents. Discard the blood appropriately; ***never*** throw human blood in the trash can.

Procedure 4

EIA, The Directigen Flu A Test

a Wash your hands. You may be taking a sample from a human nasopharynx; see Exercise 29, Mouth, Throat, and Lower Respiratory System Flora. Handle specimens as if they were potentially infectious samples.

b **Your instructor** may assign the positive control or the negative control as your group's specimen. If you are not assigned a control, swab your group's "volunteer nasopharynx" and then ream the swab on the inside of the saline tube. Thoroughly mix the specimen into the saline. Discard the swab as directed by **your instructor.**

c Before proceeding with your experiment, it is a good idea to read steps (*d*) through (*s*) and also read the Directigen Flu A package insert.

d Seat the **flow controller** snugly into the **ColorPAC** device.

e Pipet 125 μl of your well-mixed **nasopharynx sample** (or 4 drops of your well-mixed positive or negative **control sample**) into the **DispensTube.**

f Mix **reagent A** gently. (Reagent A **extracts influenza A antigen.**) Dispense 8 drops of reagent A into the DispensTube with your sample. Mix well.

g Insert a **dropper tip** into your DispensTube. Invert the tube over a **ColorPAC test well.** Avoiding bubbles, gently squeeze the DispensTube so that all of the extracted sample drops into the ColorPAC test well. Allow the sample to absorb completely. This should take under 5 minutes.

h **Reagents 1** and **2** are **washes.** Gently mix **reagent 1.** Rapidly fill the ColorPAC test well with approximately 10 drops of reagent 1 and allow the reagent to absorb completely.

i Remove the **flow controller** and discard it as directed by your instructor.

j Gently mix **reagent 2.** Dispense 4 drops of reagent 2 onto the ColorPAC membrane and allow it to absorb completely.

*An artificial blood typing kit and its directions can be substituted for this procedure.

k **Reagent 3** contains **anti-influenza A monoclonal murine antibodies** and **conjugated enzyme.** Gently mix reagent 3. Dispense 4 drops of reagent 3 onto the ColorPAC membrane and allow it to absorb completely.

l Wait 2 minutes before adding the next **wash (reagent 4).** After 2 minutes, gently mix reagent 4, then rapidly add approximately 12 drops until the ColorPAC well is filled. Allow reagent 4 to absorb completely.

m **Reagent 5** is a **wash.** Gently mix it, then dispense 4 drops onto the ColorPAC membrane and allow it to absorb completely.

n **Reagents 6** and **7** contain **chromogenic substrates A** and **B.** Gently mix **reagent 6.** Dispense 4 drops of reagent 6 onto the ColorPAC membrane and allow it to absorb completely. Watch as the membrane turns yellow.

o Gently mix **reagent 7.** Dispense 4 drops of reagent 7 onto the ColorPAC membrane and allow it to absorb completely.

p Wait 5 minutes and then either **read** the test **or** add 4 drops of **reagent 8 (a stop solution)** onto the ColorPAC membrane. The stop solution extends the readout time period up to 12 hours. Without the stop solution, the Directigen Flu A test must be evaluated within 30 minutes.

q Evaluate the test. In a **positive test,** antigen is present. A pale to dark purple triangle appears on the grayish white background color of the ColorPAC membrane. In a **negative test,** no antigen is detected. No purple triangle appears on the grayish white background of the ColorPAC membrane.

r A purple control dot should appear on both the positive and negative tests unless it is obscured by a dark purple positive test. The dot indicates that you performed the test procedure correctly and that the reagents functioned adequately.

s Examine the ColorPAC membranes of the positive and negative controls as well as those of other student groups. Draw and evaluate your observations in the laboratory report. Discard all your materials as directed by **your instructor.**

POSTTEST EXERCISE 33

Part 1

True or False (Circle one, then correct every false statement.)

1. T F Serology is the study of antigen and antibody reactions in the human body.
2. T F Antibodies are proteins that are abundant in blood.
3. T F Every human erythrocyte has numerous substance H projections.
4. T F Substance H is a glycolipid, *N*-acetylgalactosamine is a mucopeptide, and galactose is a sugar. Together they make up the ABO group of antigens.
5. T F Hybridoma cells consist of fused antibody-producing plasma cells and myeloma cells.
6. T F Anti-D serum agglutinates Rh positive blood.
7. T F Serotyping is generally accomplished by staining and performing biochemical tests on pure cultures of bacteria.
8. T F Multivalent *Salmonella* O Antiserum reacts with both *Escherichia* and *Salmonella.*
9. T F To kill bacteria in a slide agglutination test, phenol is added to the saline.
10. T F The microbiologist attaches latex beads to streptococci and enterococci before performing the latex bead agglutination test.
11. T F In the *Streptococcus* and *Enterococcus* grouping procedure, the Polyvalent Control serum containing cell wall extracts from strains of groups A, B, C, F, and G is utilized as the negative control.
12. T F Monoclonal antibodies are produced by mouse plasma cells *in vivo.*
13. T F The RPR and the VDRL are screening tests for syphilis.
14. T F *Treponema pallidum* produces reagin.
15. T F The FTA-ABS test checks for antitreponemal antibodies.
16. T F In the direct EIA, a patient's enzyme-linked antigen is added to substrate-linked, monoclonal antibody.
17. T F The Directigen Flu A test is an indirect EIA. It utilizes the patient's serum as a source of test antibody.

Part 2

Completion and Short Answer

1. In a ______________________ system, the antibody and antigen match; in a __________________ system, there is no reaction since the antibody does not complement the antigen.
2. ______________________________ are antibodies that clump particles; __________________________ are antibodies that precipitate dissolved antigens out of solution.
3. The complexes formed in a positive latex bead agglutination test contain ___________________ , ___________________ , and ___________________ .
4. If anti-A and anti-D (not anti-B) antisera agglutinate your blood, what is your ABO group, and what is your Rh reaction?

   _______________ _______________
5. Label and illustrate positive direct and indirect enzyme immunoassays. Use page 328.

Laboratory Report **33**

Name ____________________

Section ____________

Serology

Procedure 1

Latex Bead Agglutination

(*e*) Why should you take care to avoid touching first 1 drop of latex reagent then the next with your loopful of bacteria?

(*h, m, p*) In the following chart, record agglutination or no reaction in each of the blanks. Use these data to identify, if possible, the Lancefield group to which each of your streptococcal or enterococcal unknowns belongs.

Latex Agglutination Test

Antibody			Unknown or Control	Lancefield Group
A	B	C		
______	______	______	1	______
______	______	______	2	______
______	______	______	3	______
______	______	______	4	______
______	______	______	Negative	______
______	______	______	Positive	______

Procedure 2

Slide Agglutination Test

(*h*) Describe the reactions, identifying the positive and negative controls.

Which unknown is this, "X" or "Q"? _________
Is this unknown a *Salmonella?* _________
(*m*) Describe the reaction.

Which unknown is this, "X" or "Q"? _________
Is this unknown a *Salmonella?* _________

Procedure 3

Blood Typing with Monoclonal Antibodies

(*h*) What are your preliminary results? ____
(*g*) Record your final observations. What is the ABO blood type of your sample? _________
What is the Rh reaction of your sample? _________

Procedure 4

EIA, The Directigen Flu A Test

(*s*) Examine the ColorPAC membranes of the positive and negative class controls as well as your own test and those of another student group. Draw and label your observations here. Interpret your results. Note that if there is neither a purple dot nor a purple triangle on the ColorPAC membrane, the test is recorded as "uninterpretable."

Nasopharyngeal Specimens

Positive Control

Negative Control

Exercise **34**

Genetic Engineering and Bacterial Transformation

Necessary Skills

Aseptic transfer techniques, Knowledge of the metric system of measurement, Quantifying bacteria

Materials

Cultures (24-hour, 35° C, Tryptic Soy Agar [TSA] slant, 1 per group):
  Escherichia coli, ATCC #33625, Stratagene AGl,*
Supplies (per class of 6 groups):
E. coli AGl, the next 11 items, and instructions for their preparation are available in the Stratagene Educational Products kit, *Bacterial Transformation,* catalog number 756–022A.*
  1 Vial containing 3 ml Tryptic Soy Broth (TSB)
  6 Vials containing 2 ml TSB
  6 Vials containing 1 ml sterile, 0.1 M calcium chloride (1.11 g $CaCl_2$ per total volume of 100 ml of $_dH_2O$)
  6 Vials containing 10 μL of 0.1 ng/μL purified plasmid DNA (pUC18)
  30 Sterile 60 × 15 mm Petri dishes
  2 Bottles, each containing 100 ml TSA**
  1 Vial containing 2 ampicillin tablets
  1 Vial containing 6 sterile, 10 μL micropipets with plungers
  40 Sterile transfer pipets
  20 Sterile 5 ml culture tubes
  6 Glass spreaders for bacteria
Supplies that are not in the kit:
  Tub of ice for the class
  100 ml Beaker placed in a larger container labeled "Ice for $CaCl_2$," 1 set per group
  250 ml Beaker labeled "75% Ethanol," 1 per group
  Foil cover for ethanol beaker
  42° C Water bath, 1 per class
  At least 300 ml of 75% ethanol per class
  Rack to hold 5 ml culture tubes, 1 per group

*Available from Stratagene: 800–424–5444; 11011 N. Torrey Pines Rd., La Jolla, CA 92037; E-mail address, tech_services@stratagene. com

**Liquefy the agar. Cool it to 50° C in a water bath. Using 1 bottle, pour 5 or 6 ml of TSA into 12 or more 60 × 15 mm Petri dishes. Label the TSA plates.
  For the TSA plus ampicillin plates, mix 2 ampicillin tablets into 3 ml TSB. Transfer the suspension into the second bottle of liquefied, tempered TSA. Roll the upright bottle back and forth between your palms to mix the suspension well without introducing air bubbles. Pour 12 or more TSA plus ampicillin plates and label them. The inverted plates should be refrigerated if they must be stored longer than 48 hours before being used.

Primary Objective

Perform and evaluate a transformation of bacterial cells.

Other Objectives

1. Define: Genetics, recombinant DNA, plasmid.
2. Compare the functions of restriction endonucleases and DNA ligases.
3. Explain and evaluate the function and importance of DNA.
4. Review the functions of proteins in cells.
5. Compare and contrast three procaryotic procedures for passing DNA from one cell to another.
6. Employ genetic engineering techniques of plasmid manipulation to prepare a bacterial transformation of *Escherichia coli.* Examine and evaluate your results.
7. Calculate the transformation efficiency of your experimental protocol.

INTRODUCTION

Genetics is the study of how inheritable traits are passed from one generation to the next in a **deoxyribonucleic acid (DNA) code.** DNA carries a pattern for making **proteins** in a cell. Some of these proteins are **enzymes**—organic catalysts that make reactions go faster. Other proteins are **structural,** serving as part of each cellular membrane, for example.

As you can see, among their many functions, various proteins determine the structure of a cell and which reactions go forward within each cell. That is why the **genetic code,** by specifying which proteins a cell makes, is essential for life as we know it.

To examine genetics, scientists can spend generations observing and recording human, inheritable traits; years studying plants and their seedlings; or they can gather information overnight from the 20-minute generation times of ***Escherichia coli*** and a number of other single celled species. Needless to say, microbial genetics attracts many creative, energetic researchers; the field is mushrooming!

Bacterial DNA Transfer

Most bacteria simply divide to form the next generation. In this **asexual** form of reproduction, each daughter cell receives a nearly identical copy of the single parent cell's DNA. But occasionally bacterial cells in nature use one of three procaryotic procedures to combine DNA from several sources.

Thus a new generation is produced that can be genetically different from either parent. When any of the three processes occurs, DNA from one bacterial cell enters another. From within that second cell, the newly combined nucleic acid can pass on to the next generation and the next. This redistribution of DNA assures **genetic diversity.**

The three DNA exchange processes include **conjugation, transduction,** and **transformation.** In conjugation DNA passes from a donor bacterial cell to a recipient, and contact between the two cells is required; see figure 34.1. **Transduction** occurs when a departing viral visitor snatches a bit of bacterial host DNA and carries it, wrapped safely inside the viral coat, into the next host. In **transformation,** naked bits of foreign DNA enter **competent** (receptive) **bacterial cells.** The foreign DNA does not need to be bacterial; plants, animals, fungi, and others may offer their own genetic donations.

Bacterial Transformation

Genetic engineers move DNA from one genus into another in their laboratories creating **recombinant DNA.** Microbes carrying recombinant DNA now supply us with human insulin, human growth hormone, bovine growth hormone, and other useful products.

In the laboratory geneticists insert intact **plasmids** (small, extrachromosomal circles of DNA) into various microorganisms. Since the plasmids make copies (**clones**) of themselves and are fused to bits of foreign DNA that they carry into a new host, they are called **cloning vectors.**

Microbiologists cut open plasmids with **restriction endonucleases** (enzymes that cleave DNA). Then they glue foreign DNA into place and reseal the cloning vectors with other enzymes, the **DNA ligases,** in a process called **ligation.**

In this exercise you perform a bacterial transformation of *Escherichia coli* with a plasmid that carries a gene for resistance to the antibiotic ampicillin; see figure 34.2. First you suspend **ampicillin sensitive *Escherichia coli* (*E. coli*$^{\text{Amp sens}}$)** cells in two tubes of chilled calcium chloride ($CaCl_2$) solution. To one of the tubes, you add **pUC18, the plasmid,** with its **ampicillin resistance gene.** Then you heat shock both tubes of *E. coli*$^{\text{Amp sens}}$ to make the bacteria competent.

If transformation occurs, you will recover colonies of **ampicillin resistant *Escherichia coli* (*E. coli*$^{\text{Amp res}}$).** Remember that the original bacterial culture was sensitive to ampicillin (*E. coli*$^{\text{Amp sens}}$), and the transformed cells that you create in this experiment are resistant to the antibiotic. How will you discover whether a cell is ampicillin resistant or not?

To see if transformation took place, you inoculate, incubate, and then examine ampicillin-containing agar plates spread with bacteria from each tube. Why? You also inspect inoculated, ampicillin-free plates; will there be growth? Why do you add pUC18 to only one of the tubes, not both? To answer these questions, investigate the experimental protocol used in this exercise by studying figure 34.2. Then, if necessary, review the **scientific method of investigation** described in Exercise 11, Ubiquity and Culture Characteristics.

Genetic engineers need to compare the effectiveness of various transformation protocols, so they calculate a procedure's **transformation efficiency.** The transformation efficiency is the number of **transformants** (in this case, *E. coli*$^{\text{Amp res}}$) per nanogram of the plasmid that is used in an experiment. You will calculate the transformation efficiency of your work in this exercise.

PROCEDURES EXERCISE 34

Procedure 1*

Genetic Engineering and Bacterial Transformation

First Laboratory Session

a Examine figure 34.2 thoroughly before continuing with this procedure.

b Label 2 sterile 5 ml tubes, one is **"Transformation,"** and the other is **"Control."** Get a set of containers for ice and a beaker of alcohol. Cover the alcohol beaker with foil to slow evaporation.

c Using sterile technique, transfer 250 μL (5 drops) of ice-cold **calcium chloride** ($CaCl_2$) solution into each tube. Keep both tubes on ice.

*Available from Stratagene: 800–424–5444; 11011 N. Torrey Pines Rd., La Jolla, CA 92037; E-mail address, tech_services@stratagene. com

Figure 34.1 Selected forms of procaryotic deoxyribonucleic acid transfer

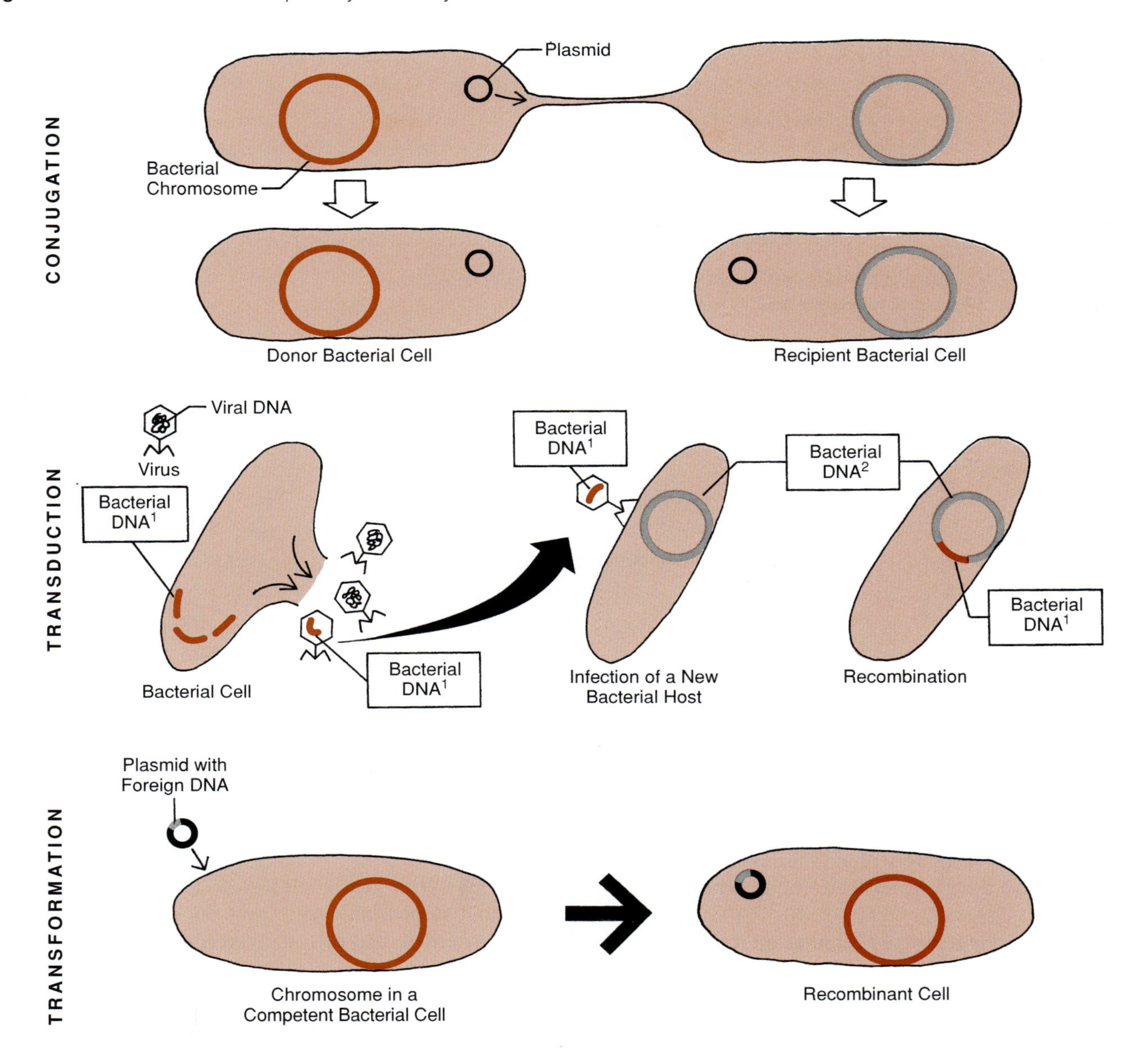

d Label a sterile 5 ml tube ***"E. coli."*** Add 250 μL (5 drops) of **TSB** to the tube. Using sterile technique, transfer 2 loopfuls of ***Escherichia coli***$^{\text{Amp sens}}$ from the slant culture into the broth, then gently vortex the closed broth to thoroughly resuspend the bacteria.

e Add 50 μL (a drop) of well-suspended cells to 1 $CaCl_2$ tube. Gently suction the mixture in and out of a sterile transfer pipet until the bacteria are well suspended. Cap the tube and repeat this step with the other $CaCl_2$ tube. Keep the closed tubes iced.

f Use a sterile 10 μL micropipet and plunger to transfer 10 μL of 0.1 ng/μL **pUC18 (plasmid DNA)** solution into your "Transformation" tube. Cap and vortex briefly to mix the DNA into your bacterial suspension.

g Keep your closed "Transformation" and "Control" tubes iced for 15 to 30 minutes.

h While you wait, label 2 **TSA** and 2 **TSA plus ampicillin plates** with your group identification and the date. Label one plate of each medium **"T."** These plates will receive bacteria from your "Transformation" tube.

Figure 34.2 Bacterial transformation experiment

1. Mix each tube, then chill the cell suspensions on ice for 15 to 30 minutes.
2. Heat shock the cells in a 42° C water bath for 45 seconds.
3. Ice the cells for 2 minutes.
4. Add Tryptic Soy Broth to each tube; return the tubes to room temperature.
5. Transfer bacteria from the transformation tube to a Tryptic Soy Agar (TSA) plus ampicillin plate and to a TSA plate.
6. Transfer bacteria from the control tube to a TSA plus ampicillin plate and to a TSA plate.
7. Spread the bacteria. Incubate them, then examine and evaluate your results.

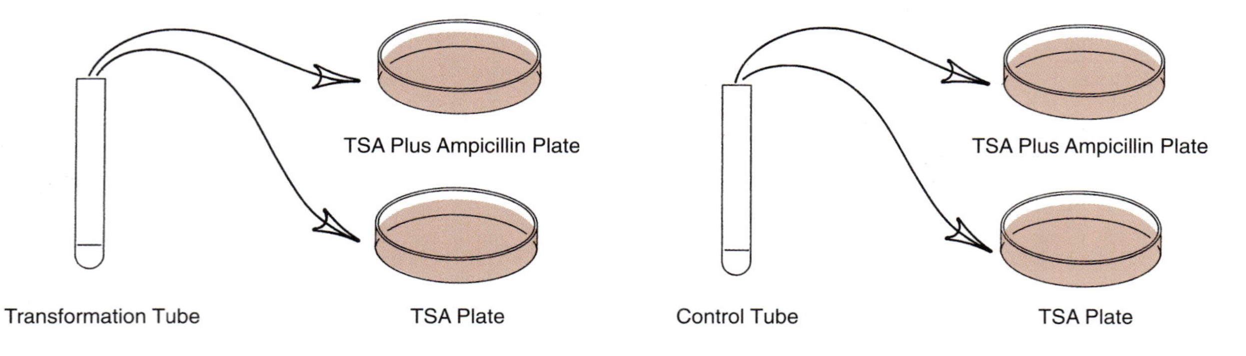

Label one plate of each medium **"C."** These plates are to receive bacteria from your "Control" tube.

i In your laboratory report, discuss the function of each plate in the design of this bacterial transformation experiment.

j After their 15 to 30 minutes on ice, **heat shock** your bacterial cultures by removing them from the ice and immediately placing them in a 42° C water bath for 45 seconds.

k Return the tubes to ice for 2 minutes.

l Using sterile technique, pipet 250 μL (5 drops) **TSB** into each tube. Vortex the tubes' contents by gently thwacking each closed tube with your finger. From this point on, keep the tubes at room temperature.

m Employing sterile technique, transfer 250 μL (5 drops) from your "Transformation" tube onto 1 appropriately labeled **TSA plus ampicillin plate.** Using the same pipet, also transfer 250 μL from the "Transformation" tube onto your appropriately labeled **TSA plate.**

n Alcohol flame sterilize your glass cell spreader by dipping it into alcohol then passing it through and out of your Bunsen burner flame. Hold the ignited spreader horizontally to avoid burning your hands or bench top with dripping, flaming alcohol. Avoid lighting the alcohol in your beaker.

o With your cooled glass spreader, distribute the bacterial suspension evenly over the surface of one plate. Flame sterilize and cool the spreader, then distribute the bacterial suspension over the surface of the other plate. Flame sterilize the spreader again.

p Employing sterile technique, pipet 250 μL (5 drops) from your "Control" tube onto your appropriately labeled **TSA plus ampicillin plate.** Use the same pipet to transfer 250 μL from your "Control" tube onto your appropriately labeled **TSA plate.**

q Repeat steps (*n*) and (*o*).

r Wait a few minutes before inverting the plates and incubating them at 35° C for 24 hours.

s In your laboratory report, record how much pUC18 you transferred in step (*f*).

Second Laboratory Session

a Examine microbial growth on each of the 4 Petri plates in your bacterial transformation experiment. Count colonies on countable plates—those that display 10 to 100 colonies. Which plates have colonies that are too numerous to count (TNTC) and which have too few colonies to count (TFTC)?

b Compare and contrast growth on the Petri dishes and explain the results that you observe.

c Calculate the **transformation efficiency** of this procedure.

d Discuss factors that might affect the transformation efficiency of a plasmid suspension.

POSTTEST EXERCISE 34

Part 1

True or False (Circle one, then correct every false statement.)

1. T F Genetics is the study of how inheritable traits are passed from one generation to the next in an enzyme code.
2. T F Enzymes can be structural or protein.
3. T F Bacterial cells generally carry DNA from two parent cells. Occasionally, however, a single cell simply divides to form two daughter (or son?) cells.
4. T F Conjugation involves the transfer of bits of naked DNA into a bacterial cell. So does transduction, but transformation requires a sexual exchange and generally occurs at the beach or on a ski slope.
5. T F The foreign DNA involved in transformation does not have to be bacterial.
6. T F Genetic engineers have created recombinant DNA and inserted it into microbial cells that now produce human and bovine proteins.
7. T F Plasmids are small, circular bits of extrachromosomal DNA that copy themselves and can carry pieces of foreign DNA.
8. T F Restriction endonucleases cut DNA, and DNA ligases glue it back together.
9. T F The plasmid pUC18 can transform ampicillin-sensitive *Escherichia coli,* rendering the bacterium resistant to ampicillin.
10. T F By using an experimental protocol that yields a high transformation efficiency, you can expect to see many colonies of transformed cells.

Part 2

Short Essay

Study the laboratory report section of this exercise. Then calculate the transformation efficiency of the experimental protocol described in this exercise. Assume that 38 colonies grew on a student's TSA plus ampicillin "T" plate and none grew on the TSA plus ampicillin "C" plate. Show all of your calculations.

Laboratory Report **34**

Name ______________________

Section ____________

Genetic Engineering and Bacterial Transformation

Procedure 1

Bacterial Transformation

First Laboratory Session

(*i*) Discuss the function of each plate in the design of this bacterial transformation experiment. Why did you need to include each control, and which plate was the "experimental plate"?

TSA plus ampicillin "T"

TSA "T"

TSA plus ampicillin "C"

TSA "C"

(*s*) Use the following formula to calculate the mass in nanograms (ng) of pUC18 that you transferred in step (*f*):

concentration × volume = mass

_______ ng/μL × _______ μL = _______ ng of pUC18 transferred

Second Laboratory Session

(*a*) Gather data from your transformation experiment by examining any *Escherichia coli* growth on your 4 Petri plates. Count colonies on countable plates. Record TNTC or TFTC for plates with too many or too few colonies to count. Record 0 for plates that show no *Escherichia coli* growth.

Bacterial Transformation

Plate	**Amount of Growth**
TSA plus ampicillin "T"	
TSA "T"	
TSA plus ampicillin "C"	
TSA "C"	

(*b*) Compare and contrast growth on each of the plates and explain the results that you observe.

(*c*) Calculate the **transformation efficiency**—the number of transformants per nanogram of plasmid—of this experimental protocol.

(*i*) What was the total mass in nanograms of pUC18 that you added to your bacterial cell suspension? ________

(*ii*) Use the following equation to calculate approximately what fraction of the volume of your cell suspension (plus pUC18) you spread onto a TSA plus ampicillin plate.

volume of suspension spread / total volume of suspension = fraction spread

________ μL (step *p*) / ________ μL (steps *c* and *l*) = fraction spread

(*iii*) Use the following equation to calculate the mass of the pUC18 in the bacterial suspension that you spread onto a TSA plus ampicillin plate.

total mass of pUC18 (*i*) × fraction spread (*ii*) = mass of pUC18 spread

________ ng × ________ = ________ ng of pUC18 spread

(*iv*) If possible, calculate the **transformation efficiency** of your experiment using the following equation to determine the number of transformant colonies that developed per nanogram of pUC18. Then explain how growth on your TSA plus ampicillin "C" plate alters your results.

number of colonies on TSA plus ampicillin "T" plate / mass of pUC18 spread (*iii*) = transformation efficiency

________ colonies / ________ ng = transformation efficiency

(*d*) Discuss experimental factors that might influence the transformation efficiency of a plasmid suspension.

Appendix **1**

Answers to the Posttests

Appendix **2**

Differentiation of Enterobacteriaceae by API 20E® Test Results

Appendix **3**

Differentiation of Representative Non-Enterobacteriaceae, Gram-Negative Bacteria by API 20E® Test Results

Appendix **4**

Computer Simulation of Bacterial Identification with *Identibacter interactus* CD-ROM

Appendix **1**

Answers to the Posttests

Exercise 1: The Compound Microscope

Part 1

1. a
2. b

Part 2

1. F	6. F	11. T
2. T	7. T	12. T
3. T	8. T	13. F
4. F	9. F	14. F
5. F	10. T	15. T

Part 3

1. microscope
2. numerical aperture
3. working
4. field
5. Empty magnification does not reveal any further detail; it simply increases the size of an image.
6. Total magnification = objective magnification × ocular magnification
7. Resolving power is better when the number RP in the equation $RP = \lambda/(2 \times NA)$ is smaller. In the equation a shorter wavelength of light (λ) yields a smaller number for RP. This change indicates that a shorter wavelength improves resolving power. Thus a blue filter enhances resolution, since blue light has a short wavelength.
8. See figure 1.3 and table 1.1.

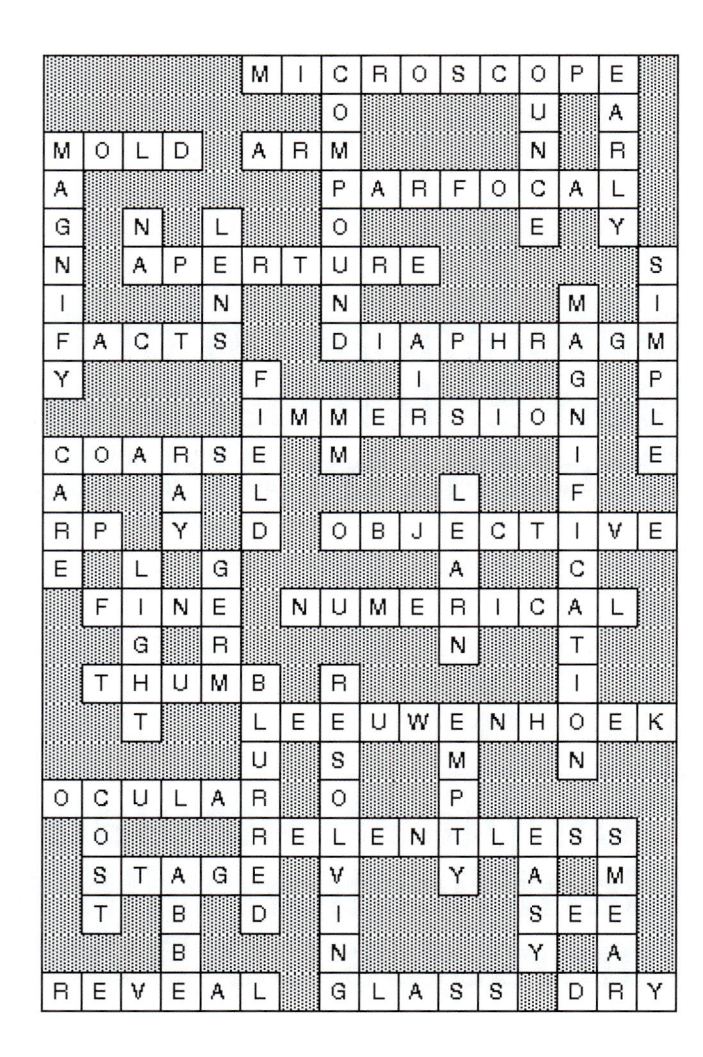

Exercise 2: Microscopic Measurement

1. T
2. F
3. F
4. T
5. T

Exercise 3: Aseptic Transfer Techniques

Part 1

1. c
2. e
3. a
4. d
5. b

Part 2

1. F
2. F
3. F
4. T
5. T
6. T
7. F
8. F
9. F
10. T
11. T
12. F
13. T
14. T
15. F
16. T

Part 3

1. culture, you, environment
2. a. Wash your hands.
   b. Disinfect your area.
   c. Sterilize your transfer instruments.
3. a. Do not talk.
   b. Work near the flame.
   c. Slant test tubes toward the flame.
   d. Hold caps, do not lay them on the bench top.
   e. Protect Petri dish cultures with the lid.
   f. Work quickly.
4. a. Close the can when you are not using it.
   b. Do not reach over an open can.
   c. Touch only one pipet at a time.
   d. Lift out the pipet. Do not drag it out.
   e. Use a sterile pipet immediately.
   f. Never return a pipet to a can of sterile pipets.
5. fresh nutrients, their own waste products

Exercise 4: Preparing a Wet Mount and Observing Microorganisms

Part 1

1. a, b, c
2. c
3. c
4. c
5. a, b, d

Part 2

1. T
2. F
3. F
4. F
5. T
6. T
7. T
8. F
9. T
10. T
11. T
12. F
13. T
14. F
15. T
16. F

Part 3

1. a. It is generally quicker and easier to prepare a wet mount than a stain.
   b. Heat fixation associated with staining shrinks microbes. To measure them accurately, use an unstained preparation.
   c. The microscopist needs living microbes to observe motility. Staining procedures often kill cells.
   Extra. The arrangement of cells in relation to each other is least disturbed in unstained preparations.
2. Brownian movement is random motion caused by molecular bombardment. True motility is directed motion to or away from a stimulus.
3. Biodetritus is composed of disintegrating, decomposing fragments of once living material.

Exercise 5: Algae

Part 1

1. F
2. F
3. T
4. F
5. T

Part 2

1. intertidal, neritic, phytoplankton
2. blooms
3. red algae (Rhodophyta)

Exercise 6: Fungi, Protozoa, and Microscopic Animalia

Part 1

1. F
2. T
3. T
4. F
5. F
6. T
7. T
8. F
9. T
10. T
11. F
12. F
13. T
14. T
15. T
16. F
17. T
18. T
19. T

Part 2

1. mushrooms, bracket fungi, puffballs, stinkhorns, truffles
2. hyphal
3. A mycologist is a scientist who studies fungi.
4. a. unicellular
   b. eucaryotic cell
   c. wall-less
   d. colorless
   e. chemoheterotrophic
5. a. Sarcodina, pseudopodia (amoeboid motion)
   b. Mastigophora, flagella
   c. Ciliata, cilia
   d. Sporozoa, mature stages not motile

Exercise 7: Preparing a Smear and Simple Stain

Part 1

a. 10
b. 5
c. 3, 6
d. 8
e. 1
f. 2
g. 4, 7
h. 13
i. 9
j. 12
k. 11
l. —
m. —

List: e, f, c, g, b, c, g, d, i, a, k, j, h

Part 2

1. T
2. F
3. T
4. F
5. F
6. T
7. T
8. F
9. F
10. F

Part 3

1. Evaluate bacterial morphology by considering cell size, shape, arrangement, and structural detail.
2. Common bacterial arrangements include singles, pairs (diplococci and diplobacilli), clusters, short chains, long chains, tetrads, palisades, X's, and Y's.
3. The chromophore of a basic stain carries a positive charge. Bacteria usually carry a net negative charge. The opposite charges are attracted to each other and, as a result, color coats the bacteria.

   "Basic" stains do not necessarily have a high pH. Actually, most have a neutral pH of around 7. They are called basic because negatively charged (basic) cell components attract the chromophore.
4. a. A vortex is a whirlpool.
   b. If you can read the label on a yellow pencil through a well-suspended broth culture, the bacteria in the culture are exhibiting light growth.
   c. Tetrads are groups of four cocci that form when a coccus divides in two planes yielding a pair of cocci stacked on another pair.
   d. Bipolar staining is evident when dark spots of color appear on both ends of a rod.
   e. Acid dyes are those stains that carry a negative charge on the chromophore.
   f. Xylol is a flammable, organic solvent that removes oil from a smear but does not alter the stain.

Exercise 8: Gram Staining

Part 1

1. b
2. b
3. b
4. a
5. b

Part 2

1. T
2. F
3. F
4. T
5. T
6. T
7. F
8. F
9. T
10. T

Part 3

1. differential stain
2. purple, pink
3. faster
4. boils, wound infections, diphtheria, septic sore throat, scarlet fever, gas gangrene, some pneumonia
5. dysentery, typhoid, bubonic plague
6. *Neisseria, Moraxella*
7. *Mycoplasma*
8. cell wall structure
9. old culture, overheated during heat fixation, decolorized too long, rinsed too long with water
10. crowding, decolorization too brief

Part 4

I. Obtain a thin, dried, heat-fixed smear of bacteria on a microscope slide.
II. Gram stain the slide as follows:
   A. Apply the primary stain, crystal violet, for 1 minute. Rinse the slide with water.
   B. Apply the mordant, Gram's iodine, for 1 minute. Rinse the slide with water.
   C. Decolorize.
      1. Apply the decolorizer, 95% ethanol, for 1 second. Rinse the slide with water.
      2. Reapply the decolorizer for up to 5 seconds, until the purple color no longer rises freely from the smear. Rinse the slide with water.
   D. Apply the counterstain, safranin, for 1 minute. Rinse the slide with water.
III. Blot or air dry the slide. Examine the smear with the oil-immersion lens of your microscope.

Exercise 9: Acid-Fast Staining

Part 1

1. c
2. a
3. f
4. d
5. b

Part 2

1. T
2. F
3. F
4. F
5. T
6. T
7. T
8. F
9. T
10. T

Part 3

1. Paul Ehrlich, *Mycobacterium tuberculosis*
2. tuberculosis, nocardiosis, leprosy
3. sputum
4. body fluids

Exercise 10: Endospore Staining

Part 1

1. d
2. c
3. a
4. b

Part 2

1. F
2. F
3. F
4. T
5. T
6. T
7. T
8. F

Part 3

1. cryptobiotic
2. desiccation, ultraviolet radiation, strong acids, strong bases, disinfectants
3. bacitracin
4. altering cell membranes

Exercise 11: Ubiquity and Culture Characteristics

Part 1

1. c
2. a
3. f
4. d
5. b
6. e

Part 2

1. e
2. b
3. a
4. d
5. c
6. f

Part 3

1. m
2. a, l
3. b, d, h
4. i, k
5. —
6. c, e
7. f, j
8. g

Part 4

1. F
2. F
3. T
4. F
5. T
6. T
7. F
8. F
9. F
10. T

Exercise 12: Pour-Plate Techniques

Part 1

1. F
2. T
3. T
4. T
5. T
6. F
7. F
8. F
9. F
10. T
11. F
12. T
13. T
14. T
15. F

Part 2

1. prepare media, dilute bacteria
2. pure culture
3. use, production
4. deep or pour
5. Quebec colony counter
6. A loop dilution series is a number of containers holding liquid into which bacteria have been transferred from the first receptacle into the second, and then from the second into the third, and so on, by an inoculating loop.
7. TFTC means "too few to count."
   TNTC means "too numerous to count."
   NG means "no growth."
8. The extinction point is the place or time where a species ceases to exist.

Part 3

1. The same microorganism must be present in every case of a disease.
2. The organism must be grown in pure culture.
3. The organism must be used to produce the disease in a healthy, susceptible experimental animal.
4. The same pathogen must be isolated from the experimental animal that has developed the disease.

Exercise 13: Streak-Plate and Spread-Plate Techniques

Part 1

1. F
2. T
3. F
4. T
5. T
6. T
7. F
8. F
9. T
10. T
11. F
12. F

Part 2

1. micromanipulation
2. inoculum
3. sputum, pus, feces
4. binary fission
5. bacterial growth

Exercise 14: Hand Washing

Part 1

1. T
2. F
3. F
4. T
5. F
6. T
7. F
8. F
9. F
10. F
11. T
12. T

Part 2

1. $$\begin{array}{lcl} CH_2OOC\mathbf{R} & & CH_2OH \\ | & & | \\ CHOOC\mathbf{R} + 3\,NaOH_2 & \rightarrow & CHOH + 3\,Na^+\mathbf{R}COO^- \\ | & & | \\ CH_2OOC\mathbf{R} & & CH_2OH \end{array}$$

   Fat or oil + Alkali (sodium hydroxide) → Glycerol + Soap
2. transient, resident
3. calcium, magnesium
4. hydrocarbon, carboxylate ion

Exercise 15: Quantifying Bacteria

Part 1

1. a
2. a
3. a
4. b
5. b

Part 2

1. F
2. T
3. F
4. T
5. T
6. T
7. F
8. F
9. T

Part 3

1. See figure 15.1.
2. more
3. transmittance, transmittance, uninoculated control

Exercise 16: Anaerobic Culture: Oxygen Requirements

Part 1

1. a, d
2. d
3. c
4. b
5. e
6. a, b, c, d
7. d
8. b
9. e
10. c

Part 2

1. T
2. F
3. T
4. T
5. F
6. F
7. F
8. T
9. F
10. F
11. F
12. T
13. F
14. T
15. F
16. F

Part 3

1. a. *botulinum,* botulism; *tetani,* tetanus; *perfringens,* gas gangrene and food poisoning
2. a. agar shake culture
   b. stab culture
3. See figure 16.2.
4. methylene blue, resazurin, litmus
5. a. 21
   b. 16
   c. 0.3
   d. 4

Exercise 17: Bacteriophage

1. F
2. F
3. F
4. T
5. F
6. F
7. F
8. T
9. F
10. T
11. F
12. T

Exercise 18: Temperature and pH

Part 1

1. b
2. c
3. a
4. a
5. b

Part 2

1. T
2. T
3. F
4. F
5. T
6. F
7. F
8. T
9. T
10. T
11. F
12. T
13. F
14. T
15. F

Part 3

1. 63, 30, 72, 15
2. optimum temperature range
3. thermal death time
4. thermal death point

Exercise 19: Disinfectants and Antiseptics

Part 1

1. c
2. b
3. d
4. a
5. e

Part 2

1. T
2. T
3. F
4. T
5. F
6. F
7. F

Part 3

1. See table 19.1.
2. See table 19.2.

Exercise 20: Antibiotic Disk Sensitivity Testing

Part 1

1. T	4. T	7. F
2. F	5. T	8. T
3. F	6. F	9. F

Part 2

1. *Bacillus, Streptomyces, Penicillium*
2. sulfonamides
3. Paul Ehrlich, selective toxicity
4. See table 20.1.
5. sensitivity of the pathogen, density of microbial growth, concentration of the agent, rate of diffusion of the agent, thickness of medium, kind of medium, viscosity of medium, temperature of incubation, duration of incubation
6. resistant, sensitive, intermediate or inconclusive
7. See figure 20.2.

Exercise 21: Selected Microbial Enzymes

Part 1

1. c	3. b	5. e
2. b	4. a	

Part 2

1. F	8. F	14. F
2. T	9. F	15. F
3. T	10. T	16. T
4. F	11. F	17. T
5. F	12. T	18. F
6. T	13. T	19. F
7. F		

Part 3

1. See figure 21.2.
2. $$2H_2O_2 \xrightarrow{\text{catalase}} 2H_2O + O_2\uparrow$$
   $$\begin{pmatrix}\text{hydrogen}\\ \text{peroxide}\end{pmatrix} \rightarrow (\text{water}) + \begin{pmatrix}\text{oxygen}\\ \text{gas}\end{pmatrix}$$
3. See figure 21.4.
4. a. starch
   b. egg yolk lipid including lecithin
   c. lactose, dextrose, peptone (and sodium thiosulfate, an additional source of sulfide ions), ferrous sulfate

Exercise 22: Enzymatic Attacks on Nitrogen-Containing Substrates

Part 1

1. a	3. d	5. a, b
2. d	4. c	

Part 2

1. T	5. T	8. F
2. F	6. T	9. T
3. F	7. T	10. T
4. F		

Part 3

1. See figure 22.1 or table 22.1.

Exercise 23: Carbohydrate Fermentation

Part 1

1. F	5. F	8. T
2. F	6. T	9. F
3. F	7. F	10. F
4. F		

Part 2

1. See figure 23.1.
2. organic, organic, inorganic
3. acids: acetic, butyric, formic, lactic, propionic, succinic, etc.; neutral pH substances: acetone, butanediol, butanol, ethanol, isopropyl alcohol, etc.; gases: carbon dioxide, hydrogen, methane
4. gases

Exercise 24: The IMViC Tests

Part 1

IMViC Chart A

- Indole: 1% Tryptone aq.; tryptophane
- Methyl red: MR-VP Medium; dextrose
- Voges-Proskauer: MR-VP Medium; dextrose
- Citrate: Simmons Citrate Agar; citrate, ammonium salts, brom thymol blue

IMViC Chart B

- Indole: Kovacs' (para-dimethylaminobenzaldehyde); positive, red surface layer; negative, no red color in layer
- Methyl red: methyl red; positive red, negative yellow to orange
- Voges-Proskauer: Barritt's solution A (alpha-naphthol), then Barritt's solution B (KOH); positive, pink to red; negative, no pink develops
- Citrate: no added reagents; positive, growth and generally blue color development; negative, no growth and no blue color

Part 2

1. F	5. F	8. T
2. T	6. T	9. T
3. T	7. F	10. F
4. F		

Part 3

1. enterics
2. water, milk, sewage effluent
3. mixed acid
4. acetyl methyl carbinol
5. citrate, ammonium ions

Exercise 25: Selective and Differential Media

Part 1

1. a, b	3. a, b	5. c
2. b, c	4. b	

Part 2

1. F	4. T	7. F
2. T	5. F	8. F
3. T	6. T	9. T

Part 3

1. all-purpose
2. selective
3. temperature, duration, humidity, medium, oxygen pressure, osmotic pressure, atmospheric pressure
4. Blood Agar, Brain Heart Infusion Agar, Nutrient Agar and Broth, Sabouraud Dextrose Agar (for fungi), Tryptic Soy Agar and Broth, Tryptose Phosphate Broth
5. To isolate a pure culture from growth on a **selective solid medium,** touch your sterilized and cooled needle to the center of an isolated colony on the selective medium. Transfer the bacteria onto an **all-purpose solid medium.** Use your sterilized and cooled loop to streak the bacteria on the all-purpose plate for isolation of cells. Each well-isolated colony that develops is considered to be a pure culture.

   To isolate a pure culture from growth on a **selective liquid medium,** transfer a loopful onto a **solid all-purpose medium.** Streak for isolation of cells.
6. the medium itself
7. One solution is to stroke a sterile loop across a 48-hour, 35° C, Levine EMB Agar culture of *Staphylococcus epidermidis.* Streak a plate of an all-purpose medium with the loop, then incubate the plate appropriately and examine it for staphylococcal growth.

Exercise 26: Multiple-Test Media and Systems

Part 1

1. F	5. T	8. T
2. T	6. T	9. T
3. T	7. F	10. F
4. F		

Part 2

1. Correct identification of microbial species is crucial for repeatability of experiments in research. It is essential for the initiation of appropriate treatment as well as control measures in the clinical setting.
2. See table 26.2.

Exercise 27: Medically Significant Gram-Positive Cocci

Part 1

1. a, b	4. b	7. b
2. a, b	5. b	8. a, b
3. c	6. b	

1. c	4. c	7. c
2. c, d	5. c	8. a, c, d
3. a, c, d	6. d	9. b

1. a	4. a, b, c	6. b
2. b	5. e	7. d
3. a, rarely b		

Part 2

1. T	10. F	18. T
2. F	11. T	19. T
3. T	12. T	20. T
4. F	13. F	21. T
5. T	14. F	22. T
6. F	15. F	23. T
7. F	16. F	24. F
8. T	17. T	25. T
9. F		

Part 3

1. skin, skin glands, mucous membranes
2. pus, purulent fluids, sputum, urine, feces, blood
3. Peptone and beef extract support growth of mannitol nonfermenters.

   Mannitol is a fermentable carbohydrate.

   Sodium chloride (7.5%) selects for salt-tolerant microbes.

   Phenol red dye indicates if acid is produced from mannitol fermentation.

   Agar provides a solid surface for colony isolation.
4. Accurate diagnosis of *Streptococcus pyogenes* is important medically because it is a common pathogen that can be treated effectively with antibiotics. Rapid treatment relieves pain and helps prevent the development of serious, nonsuppurative sequelae.
5. See table 27.4.

6. a. inflammation of the pharynx
   b. group A, beta-hemolytic *Streptococcus*
   c. not pus-forming
   d. postinfection conditions
   e. a place where electrons have been gained, and there is a lower concentration of oxygen than is found in air
   f. a demonstration of the production of the CAMP factor by Lancefield's group B streptococci
7. See table 27.4.

Exercise 28: Intestinal Pathogens

Part 1

1. b
2. a
3. b
4. b
5. e
6. c

Part 2

1. F
2. T
3. T
4. F
5. F
6. T
7. T
8. T
9. T
10. F
11. T
12. F
13. F
14. T
15. T

Part 3

1. **Genus** — **Appearance on Hektoen Enteric Agar**

Genus	Appearance on Hektoen Enteric Agar
Salmonella	Greenish-blue*
Shigella	Greenish-blue
Escherichia	Orange

2. gram-negative, non-endospore-forming, non-acid-fast, rods, peritrichous flagellation if motile, chemoorganotrophic, catalase-positive, nitrate reducers, oxidase-negative

Exercise 29: Mouth, Throat, and Lower Respiratory System Flora

Part 1

1. c
2. e
3. d
4. b

1. g
2. d
3. a
4. c
5. c
6. c
7. a, b
8. d
9. f
10. e

Part 2

1. F
2. T
3. T
4. T
5. F
6. F
7. T
8. T
9. F
10. F
11. F
12. T
13. F
14. T
15. T
16. F
17. F
18. F
19. F
20. T
21. T

Exercise 30: Urinary Tract Normal Flora

Part 1

1. T
2. F
3. F
4. F
5. T
6. F
7. F
8. F
9. T
10. F
11. T
12. F
13. T
14. F

Part 2

1. coagulase-negative staphylococci, diphtheroids, alpha-hemolytic streptococci, nonhemolytic streptococci, enterococci, yeasts, coliforms, other fecal microorganisms, nonpathogenic *Neisseria* species, nonpathogenic *Mycobacterium* species
2. beta-hemolytic streptococci, coagulase-positive staphylococci, *Mycobacterium tuberculosis*

Exercise 33: Serology

Part 1

1. F
2. T
3. T
4. T
5. T
6. T
7. F
8. F
9. T
10. F
11. F
12. F
13. T
14. F
15. T
16. F
17. F

Part 2

1. homologous, heterologous
2. agglutinins, precipitins
3. antibodies, antigens, latex beads
4. A, positive
5. See figure 33.6.

Exercise 34: Genetic Engineering and Bacterial Transformation

Part 1

1. F
2. F
3. F
4. F
5. T
6. T
7. T
8. T
9. T
10. T

Part 2

Complete section (c) of the laboratory report and find a transformation efficiency of roughly 76 colonies per nanogram of pUCl8. Then reexamine the protocol to find the more accurate answer.

*Most serovars yield black centers.

Appendix **2**

Differentiation of Enterobacteriaceae by API 20E® Test Results

Differentiation of Enterobacteriaceae

	ORGANISM	ONPG	ADH	LDC	ODC	CIT	H_2S	URE	TDA	IND	VP
Escherichieae	*Escherichia coli*	98.7	3.0	82.8	75.7	0.0	0.0	0.0	0.0	92.6	0.0
	E. fergusonii	93.3	0.0	100.0	100.0	0.0	0.0	0.0	0.0	100.0	0.0
	E. hermannii	99.9	0.0	0.1	99.9	3.2	0.0	0.0	0.0	99.9	0.0
	E. vulneris	100.0	22.2	45.0	0.0	0.0	0.0	0.0	0.0	0.0	0.0
	Shigella dysenteriae	33.3	0.0	0.0	0.0	0.0	0.0	0.0	0.0	33.0	0.0
	Sh. flexneri	0.9	0.0	0.0	0.9	0.0	0.0	0.0	0.0	36.2	0.0
	Sh. boydii	10.4	0.0	0.0	9.1	0.0	0.0	0.0	0.0	50.0	0.0
	Sh. sonnei	87.6	0.0	0.0	92.9	0.0	0.0	0.0	0.0	0.0	0.0
	Edwardsiella tarda	0.0	0.0	100.0	100.0	0.0	94.4	0.0	0.0	99.0	0.0
Salmonelleae	*Citrobacter freundii*	92.3	21.3	0.0	52.4	65.1	60.7	1.0	0.0	10.0	0.0
	C. amalonaticus	95.0	40.0	0.0	98.0	72.0	0.0	0.0	0.0	99.9	0.0
	C. diversus	97.5	63.6	0.0	98.8	91.0	0.0	0.0	0.0	98.8	0.0
	Salmonella enteritidis	1.9	69.2	96.2	95.7	75.4	85.7	0.0	0.0	3.0	0.0
	Sal. typhi	0.0	5.8	99.0	0.0	0.0	8.3	0.0	0.0	0.0	0.0
	Sal. cholerae suis	0.0	18.4	98.0	98.0	4.1	65.3	0.0	0.0	0.0	0.0
	Sal. paratyphi A	0.0	0.0	0.0	100.0	0.0	0.6	0.0	0.0	0.0	0.0
	Sal. subgroup 3	98.7	48.1	96.1	97.4	50.7	98.1	0.0	0.0	0.0	0.0
Klebsielleae	*Klebsiella pneumoniae*	99.0	0.2	74.8	0.9	74.9	0.0	58.6	0.0	0.0	87.4
	Kl. oxytoca	100.0	0.0	86.7	0.0	86.7	0.0	60.0	0.0	100.0	92.4
	Kl. ozaenae	90.0	23.3	32.2	1.1	40.0	0.0	6.7	0.0	0.0	0.0
	Kl. rhinoscleromatis	0.0	0.0	0.0	0.0	0.0	0.0	0.0	0.0	0.0	0.0
	Enterobacter aerogenes	99.5	0.0	98.8	99.0	83.7	0.0	0.3	0.0	0.0	93.6
	Ent. cloacae	99.0	93.5	0.1	97.3	85.5	0.0	0.4	0.0	0.0	96.6
	Ent. agglomerans	98.3	1.0	0.0	0.0	55.2	0.0	3.4	3.7	50.0	33.2
	Ent. gergoviae	94.0	0.0	28.8	99.9	82.0	0.0	100.0	0.0	0.0	95.2
	Ent. sakazakii	100.0	99.0	0.0	85.7	85.7	0.0	0.0	0.0	5.0	73.7
	Ent. taylorae	98.1	86.8	0.0	92.5	88.7	0.0	0.0	0.0	0.0	92.1
	Ent. amnigenus 1	78.6	0.0	0.0	78.6	5.0	0.0	0.0	0.0	0.0	28.6
	Ent. amnigenus 2	100.0	95.0	0.0	100.0	33.3	0.0	0.0	0.0	0.0	10.0
	Ent. intermedius	100.0	0.0	10.0	90.0	0.0	0.0	0.0	0.0	0.0	10.0
	Serratia liquefaciens	98.1	0.6	82.4	99.0	88.7	0.0	5.0	0.0	0.0	57.8
	Ser. marcescens	94.2	0.0	98.5	95.7	74.9	0.0	29.0	0.0	0.0	65.9
	Ser. rubidaea	100.0	0.0	68.5	0.0	81.5	0.0	3.3	0.0	0.0	63.5
	Ser. odorifera 1	95.0	0.0	99.0	100.0	95.0	0.0	0.0	0.0	99.9	50.0
	Ser. odorifera 2	95.0	0.0	99.0	0.0	90.9	0.0	0.0	0.0	99.9	54.5
	Ser. fonticola	99.0	0.0	66.7	97.6	21.4	0.0	0.0	0.0	0.0	0.0
	Ser. plymuthica	99.0	0.0	0.0	0.0	16.7	0.0	0.0	2.0	0.0	75.0
	Hafnia alvei	71.2	1.8	100.0	99.0	10.6	0.0	1.0	0.0	0.0	15.4
Proteeae	*Proteus penneri*	0.5	0.0	0.0	0.0	41.2	73.2	98.9	99.6	0.0	0.0
	Prot. vulgaris	0.5	0.0	0.0	0.0	41.2	83.2	98.9	99.6	100.0	0.0
	Prot mirabilis	0.2	0.6	2.0	98.4	57.8	83.4	99.0	98.7	1.9	2.4
	Morganella morganii	0.5	0.0	0.5	99.9	2.2	0.0	99.0	91.8	97.2	0.0
	Providencia rettgeri	1.0	0.0	0.0	0.0	70.7	0.0	99.0	99.0	97.4	0.0
	Prov. alcalifaciens	0.0	0.0	0.0	0.0	97.5	0.0	0.0	100.0	99.0	0.0
	Prov. stuartii Urea-	1.0	0.0	0.0	0.0	85.1	0.0	0.0	91.7	97.1	0.0
	Prov. stuartii Urea+	1.0	0.0	0.0	0.0	68.6	0.0	100.0	75.7	88.6	0.0
Yersiniae	*Yersinia enterocolitica*	81.1	0.0	0.0	85.1	0.0	0.0	93.7	0.0	69.4	1.0
	Y. intermedia	95.1	0.0	0.0	100.0	0.0	0.0	97.6	0.0	97.6	2.4
	Y. frederiksenii	95.6	0.0	0.0	100.0	0.0	0.0	100.0	0.0	99.0	0.0
	Y. kristensenii	87.5	0.0	0.0	87.5	0.0	0.0	100.0	0.0	99.0	0.0
	Y. pseudotuberculosis	77.1	0.0	0.0	0.0	13.3	0.0	96.2	0.0	0.0	0.0
	Presumptive *Y. pestis*	68.6	0.0	0.0	0.0	0.0	0.0	0.0	0.0	0.0	8.6
	Y. ruckeri (AN)	73.5	0.0	93.9	95.9	0.0	0.0	0.0	0.0	0.0	0.0
	Cedecea davisae	50.0	86.4	0.0	95.5	36.4	0.0	0.0	0.0	0.0	36.4
	Cedecea lapagei	83.3	66.7	0.0	0.0	66.7	0.0	0.0	0.0	0.0	25.0
	Cedecea sp. 3	90.0	90.0	0.0	0.0	75.0	0.0	0.0	0.0	0.0	0.0
	Cedecea neteri	99.0	99.0	0.0	0.0	80.0	0.0	0.0	0.0	0.0	66.7
	Cedecea sp. 5	90.0	90.0	0.0	90.0	25.0	0.0	0.0	0.0	0.0	75.0
	Tatumella ptyseos	0.0	0.0	0.1	0.0	1.0	0.0	0.0	1.0	0.0	50.0
	Kluyvera sp.	91.7	0.0	56.7	90.0	91.7	0.0	0.0	0.0	99.0	0.0
	Moellerella wisconsensis	75.0	0.0	0.0	0.0	37.5	0.0	0.0	0.0	0.0	12.5
	Ewingella americana	90.9	0.0	0.0	0.0	27.7	0.0	0.0	0.0	0.0	95.5
	CDC enteric grp 17	99.9	5.3	0.0	99.9	26.1	0.0	0.0	0.0	0.1	21.1
	CDC enteric grp 41	100.0	0.0	0.0	0.0	0.0	0.0	0.0	0.0	88.9	0.0

*These tests may be required for differentiation of *Enterobacteriaceae* from other Gram-negative bacteria.

By Biochemical Tests Provided on API 20E®

Figures indicate the percentage of positive reactions after 18–24 hours of incubation at 35–37° C.

API 20E													Supplementary Tests*			
GEL	GLU	MAN	INO	SOR	RHA	SAC	MEL	AMY	ARA	OXI	NO_3	N_2 GAS	MOT	MAC	OF-O	OF-F
0.0	99.0	99.0	0.5	89.2	87.9	41.8	67.1	9.2	90.8	0.0	99.7	0.0	62.1	100.0	100.0	100.0
0.0	100.0	100.0	0.0	0.0	80.0	0.0	0.0	100.0	100.0	0.0	100.0	0.0	93.3	100.0	100.0	100.0
0.0	99.9	99.0	0.0	0.0	96.8	58.1	0.0	96.8	99.9	0.0	99.0	0.0	99.0	99.0	99.0	99.0
0.0	100.0	100.0	0.0	0.0	83.3	5.6	94.4	94.4	100.0	0.0	100.0	0.0	100.0	100.0	100.0	100.0
0.0	95.9	2.5	0.0	18.6	22.2	0.0	0.0	0.0	16.7	0.0	99.7	0.0	0.0	100.0	99.9	99.9
0.0	99.0	91.8	0.0	24.0	2.6	0.0	24.9	0.0	61.6	0.0	99.8	0.0	0.0	100.0	100.0	100.0
0.0	99.4	94.9	0.0	60.0	0.6	1.3	21.0	0.0	70.8	0.0	99.0	0.0	0.0	100.0	100.0	100.0
0.0	99.9	99.0	0.0	2.2	80.0	0.0	0.0	0.0	93.8	0.0	100.0	0.0	0.0	100.0	100.0	100.0
0.0	99.9	0.0	0.0	0.0	0.0	0.0	0.0	0.0	1.1	0.0	100.0	0.0	98.2	100.0	100.0	100.0
0.0	99.0	99.9	10.5	96.0	91.3	53.5	65.7	59.1	96.5	0.0	98.6	0.0	95.7	100.0	100.0	100.0
0.0	99.0	99.0	0.0	99.0	99.0	24.4	28.0	95.0	97.0	0.0	99.0	0.0	99.0	100.0	100.0	100.0
0.0	99.0	99.0	4.5	92.6	95.1	24.7	1.2	96.3	95.1	0.0	100.0	0.0	92.9	100.0	100.0	100.0
0.0	99.9	98.7	33.6	93.2	93.0	2.3	78.2	0.0	94.6	0.0	100.0	0.0	94.6	100.0	100.0	100.0
0.0	99.9	99.0	0.0	100.0	0.0	0.0	94.4	0.0	0.0	0.0	100.0	0.0	100.0	100.0	100.0	100.0
0.0	99.9	99.0	0.0	89.8	95.9	0.0	20.4	0.0	0.0	0.0	100.0	0.0	100.0	100.0	100.0	100.0
0.0	99.9	99.0	0.0	99.0	99.0	0.0	97.0	0.0	99.0	0.0	100.0	0.0	94.6	100.0	100.0	100.0
0.0	99.9	99.0	0.0	99.0	96.1	0.0	64.9	0.0	99.0	0.0	100.0	0.0	100.0	100.0	100.0	100.0
0.2	99.0	99.0	96.5	99.0	97.9	99.0	99.0	99.0	99.0	0.0	99.9	0.0	0.0	100.0	100.0	100.0
1.0	99.0	99.0	96.7	99.0	96.7	99.0	96.7	99.0	96.7	0.0	99.9	0.0	0.0	100.0	100.0	100.0
0.0	97.8	92.2	61.1	44.4	66.7	21.1	83.3	90.0	67.8	0.0	92.0	0.0	0.0	100.0	100.0	100.0
0.0	96.2	99.0	83.0	64.2	39.6	39.6	18.9	92.5	1.9	0.0	100.0	0.0	0.0	100.0	100.0	100.0
0.4	99.0	99.0	92.8	97.0	99.0	98.1	99.0	99.0	99.0	0.0	100.0	0.0	97.3	100.0	100.0	100.0
0.6	99.0	99.0	13.4	92.6	80.3	99.0	91.3	99.0	99.0	0.0	100.0	0.0	94.5	100.0	100.0	100.0
2.1	98.7	98.4	16.4	41.5	82.3	73.1	68.1	85.7	96.2	0.0	85.9	0.0	89.4	100.0	100.0	100.0
0.0	100.0	98.2	19.3	0.0	100.0	98.2	100.0	100.0	100.0	0.0	100.0	0.0	100.0	100.0	100.0	100.0
0.0	100.0	100.0	72.0	0.0	99.0	99.0	99.0	99.0	99.0	0.0	100.0	0.0	94.0	100.0	100.0	100.0
0.0	99.9	99.9	0.0	0.0	99.9	1.9	0.0	99.9	99.9	0.0	99.0	0.0	83.0	99.0	99.0	99.0
0.0	100.0	100.0	0.0	28.6	92.9	78.6	71.4	85.7	85.7	0.0	100.0	0.0	100.0	100.0	100.0	100.0
0.0	100.0	100.0	0.0	100.0	100.0	0.0	100.0	100.0	100.0	0.0	100.0	0.0	100.0	100.0	100.0	100.0
0.0	100.0	90.0	0.0	80.0	90.0	40.0	100.0	100.0	100.0	0.0	100.0	0.0	90.0	100.0	100.0	100.0
60.4	99.0	99.0	71.7	98.7	22.4	99.0	77.5	99.0	93.8	0.0	100.0	0.0	93.3	100.0	100.0	100.0
85.5	99.0	99.0	75.0	91.3	0.0	98.5	63.1	97.1	15.9	0.0	95.8	0.0	98.6	100.0	100.0	100.0
62.9	96.7	98.9	42.4	4.3	2.3	84.8	82.6	94.6	85.8	0.0	100.0	0.0	88.0	100.0	100.0	100.0
99.0	99.0	99.0	99.0	99.0	99.0	100.0	99.0	99.0	99.0	0.0	99.0	0.0	99.0	100.0	100.0	100.0
99.0	99.0	99.0	99.0	99.0	99.0	0.0	99.0	99.0	99.0	0.0	99.0	0.0	87.5	100.0	100.0	100.0
0.0	99.0	99.0	88.1	97.6	95.2	0.0	99.0	97.6	92.9	0.0	99.0	0.0	99.0	100.0	100.0	100.0
66.7	99.0	99.0	50.0	95.0	0.5	100.0	66.7	99.0	99.0	0.0	99.0	0.0	95.0	99.0	100.0	100.0
0.0	99.0	95.2	0.0	1.0	71.5	0.0	2.9	11.5	90.2	0.0	100.0	0.0	93.0	100.0	100.0	100.0
52.8	97.3	0.4	1.3	0.0	2.7	89.6	0.0	65.2	0.5	0.0	100.0	0.0	94.7	100.0	100.0	100.0
52.8	97.3	0.4	1.3	0.0	2.7	89.6	0.0	65.2	0.4	0.0	100.0	0.0	94.7	100.0	100.0	100.0
76.6	96.3	0.4	0.0	0.4	0.0	0.7	0.1	1.0	0.2	0.0	93.8	0.0	95.9	100.0	100.0	100.0
0.0	97.0	0.2	0.0	0.0	0.0	0.3	0.0	0.0	1.2	0.0	88.5	0.0	87.7	100.0	100.0	100.0
0.4	99.0	84.1	78.8	0.0	41.2	34.4	0.0	33.4	1.5	0.0	98.8	0.0	94.4	100.0	100.0	100.0
0.0	99.0	2.5	2.5	0.0	0.0	2.5	0.0	0.0	2.9	0.0	100.0	0.0	96.5	100.0	100.0	100.0
0.4	99.0	0.0	99.0	0.0	0.0	3.7	0.0	0.0	3.3	0.0	100.0	0.0	87.0	100.0	100.0	100.0
0.0	99.9	14.0	85.7	0.0	0.0	62.9	0.0	0.0	0.0	0.0	100.0	0.0	87.0	100.0	100.0	100.0
0.0	99.0	99.1	25.2	98.2	0.0	100.0	0.9	87.8	76.6	0.0	98.7	0.0	0.0	100.0	100.0	100.0
0.0	99.0	99.0	63.4	95.1	100.0	100.0	100.0	99.0	46.3	0.0	98.7	0.0	0.0	100.0	100.0	100.0
0.0	99.0	99.0	11.1	95.6	100.0	100.0	0.0	97.8	57.8	0.0	98.7	0.0	0.0	100.0	100.0	100.0
0.0	99.0	99.0	62.5	99.0	0.0	0.0	0.0	99.0	87.5	0.0	98.7	0.0	0.0	100.0	100.0	100.0
0.0	98.1	97.1	0.0	0.0	77.1	0.0	9.5	0.0	29.5	0.0	95.0	0.0	0.0	100.0	100.0	100.0
0.0	99.0	97.1	0.0	71.4	0.0	0.0	0.0	11.4	0.0	0.0	47.9	0.0	0.0	100.0	100.0	100.0
0.0	83.7	95.9	0.0	0.0	0.0	0.0	0.0	0.0	0.0	0.0	50.0	0.0	0.0	100.0	100.0	100.0
0.0	99.0	99.0	27.3	0.0	0.0	100.0	0.0	99.0	0.9	0.0	99.0	0.0	99.0	99.0	99.0	99.0
0.0	99.0	99.0	0.0	0.0	0.0	0.0	0.0	99.0	0.0	0.0	99.0	0.0	80.0	99.0	99.0	99.0
0.0	99.0	99.0	0.1	0.1	0.1	99.0	90.0	99.0	0.1	0.0	90.0	0.0	90.0	90.0	99.0	99.0
0.0	99.0	99.0	0.0	99.0	0.0	99.0	0.0	99.0	0.0	0.0	99.0	0.0	90.0	99.0	99.0	99.0
0.0	100.0	99.0	0.0	95.0	0.0	99.0	50.0	99.0	0.0	0.0	90.0	0.0	90.0	90.0	99.0	99.0
0.0	99.0	0.0	0.0	0.0	0.0	80.0	90.0	80.0	0.0	0.0	98.0	0.0	0.1	99.0	99.0	99.0
0.0	99.0	91.7	0.0	5.0	85.0	66.7	87.0	99.0	90.0	0.0	95.0	0.0	97.0	100.0	100.0	100.0
0.0	100.0	5.0	0.0	0.0	0.0	100.0	100.0	0.0	0.0	0.0	100.0	0.0	0.0	100.0	100.0	100.0
31.8	100.0	81.8	0.0	0.0	0.0	0.0	0.0	13.6	4.5	0.0	90.9	0.0	40.9	100.0	100.0	100.0
0.0	99.9	99.9	21.1	84.2	0.1	99.9	0.0	99.9	99.9	0.0	99.0	0.0	5.0	99.0	99.0	99.0
0.0	100.0	100.0	0.0	11.1	100.0	66.7	100.0	100.0	100.0	0.0	100.0	0.0	100.0	100.0	100.0	100.0

Modified from the API 20E® system package insert. Courtesy of Analytab Products, Division of Sherwood Medical, 200 Express St., Plainview, NY 11803.

Appendix **3**

Differentiation of Representative Non-Enterobacteriaceae, Gram-Negative Bacteria by API 20E® Test Results

Differentiation of Some Non-Enterobacteriaceae Gram-Negative Bacteria

ORGANISM	TIME	ONPG	ADH	LDC	ODC	CIT	H_2S	URE	TDA	IND	VP
Pseudomonas aeruginosa	24th	0.0	75.8	0.0	0.0	78.8	0.0	24.0	0.0	0.0	2.0
	48th	0.0	99.2	0.0	0.0	98.8	0.0	49.3	0.0	0.0	6.4
Ps. fluorescens	24th	0.0	51.0	0.0	0.0	48.0	0.0	1.9	0.0	0.0	9.4
	48th	0.0	91.3	0.0	0.0	93.5	0.0	2.2	0.0	0.0	26.1
Ps. putida	24th	0.0	59.0	0.0	0.0	63.6	0.0	2.9	0.0	0.0	20.6
	48th	0.0	88.0	0.0	0.0	92.8	0.0	6.3	0.0	0.0	28.1
Burkholderia cepacia	24th	61.1	0.0	16.3	8.1	75.0	0.0	0.0	0.0	0.0	4.6
	48th	76.2	0.0	63.1	30.4	96.4	0.0	1.2	0.0	0.0	16.0
Ps. maltophilia	24th	73.8	0.0	76.3	0.0	76.2	0.0	0.0	0.0	0.0	0.0
	48th	71.9	0.0	73.0	0.0	93.3	0.0	1.1	0.0	0.0	2.3
Ps. putrefaciens	24th	0.0	0.0	0.0	87.5	90.6	90.6	0.0	0.0	0.0	6.3
	48th	0.0	0.0	0.0	90.6	96.9	93.8	0.0	0.0	0.0	6.3
Ps. stutzeri	24th	0.0	0.0	0.0	0.0	18.0	0.0	0.0	0.0	0.0	9.4
	48th	0.0	1.9	0.0	0.0	59.3	0.0	0.0	0.0	0.0	11.1
Burkholderia pseudomallei	24th	0.0	0.0	0.0	0.0	0.0	0.0	0.0	0.0	0.0	0.0
	48th	0.0	70.8	0.0	0.0	29.2	0.0	0.0	0.0	0.0	4.2
Ps. paucimobilis	24th	80.0	0.0	0.0	0.0	10.0	0.0	5.0	0.0	0.0	15.0
	48th	86.4	0.0	0.0	0.0	54.6	0.0	5.0	0.0	0.0	31.8
Other *Pseudomonas* spp.	24th	0.4	0.7	0.0	0.0	18.0	0.0	0.7	0.0	0.0	9.7
	48th	1.9	1.9	0.0	0.0	60.9	0.0	1.9	0.0	0.0	15.2
Acinetobacter calcoaceticus var. *anitratus*	24th	0.0	0.0	0.0	0.0	28.2	0.0	0.0	0.0	0.0	14.9
	48th	0.0	0.0	0.0	0.0	54.0	0.0	0.0	0.0	0.0	22.0
Acinetobacter calcoaceticus var. *lwoffii*	24th	0.0	0.0	0.0	0.0	7.8	0.0	2.2	0.0	0.0	11.1
	48th	0.0	0.0	0.0	0.0	19.1	0.0	2.4	0.0	0.0	11.9
Flavobacterium spp. (II B)	24th	25.0	0.0	0.0	0.0	10.0	0.0	78.3	0.0	77.5	0.0
	48th	33.9	0.0	0.0	0.0	85.4	0.0	92.7	0.0	81.9	0.0
Flav. meningosepticum	24th	64.0	0.0	0.0	0.0	20.9	0.0	0.0	0.0	81.1	0.0
	48th	86.0	0.0	0.0	0.0	84.9	0.0	2.7	0.0	84.9	0.0
Flav. odoratum	24th	0.0	0.0	0.0	0.0	23.6	0.0	45.6	0.0	0.0	0.0
	48th	0.0	0.0	0.0	0.0	82.0	0.0	53.5	0.0	0.0	0.0
Flav. breve	24th	0.0	0.0	0.0	0.0	0.0	0.0	0.0	0.0	100.0	0.0
	48th	0.0	0.0	0.0	0.0	90.0	0.0	0.0	0.0	100.0	0.0
Flav. multivorum	24th	87.0	0.0	0.0	0.0	0.0	0.0	79.2	0.0	0.0	50.0
	48th	96.2	0.0	0.0	0.0	65.4	0.0	96.2	0.0	0.0	75.0
Flav. spiritivorum	24th	75.0	0.0	0.0	0.0	0.0	0.0	0.0	0.0	0.0	0.0
	48th	100.0	0.0	0.0	0.0	0.0	0.0	0.0	0.0	0.0	6.2
Bordetella bronchiseptica	24th	0.0	0.0	0.0	0.0	10.6	0.0	63.0	0.0	0.0	17.0
	48th	0.0	0.0	0.0	0.0	69.4	0.0	86.3	0.0	0.0	38.1
Alcaligenes spp.	24th	0.0	0.0	0.0	0.0	37.0	0.0	3.0	0.0	0.0	8.9
	48th	0.0	0.0	0.0	0.0	74.8	0.0	5.8	10.0	0.0	40.3
Moraxella spp.	24th	0.0	0.0	0.0	0.0	5.2	0.0	7.0	0.0	0.0	2.9
	48th	0.0	0.0	0.0	0.0	11.5	0.0	11.5	0.0	0.0	2.6
Pasteurella multocida	24th	3.7	0.0	0.0	13.0	0.0	0.0	0.0	0.0	88.9	0.0
	48th	3.9	0.0	0.0	13.0	0.0	0.0	0.0	0.0	96.2	0.0
Past. aerogenes	24th	70.0	0.0	0.0	95.0	0.0	0.0	95.0	0.0	0.0	0.0
	48th	70.0	0.0	0.0	95.0	0.0	0.0	95.0	0.0	0.0	0.0
Pasteurella-Actinobaccillus spp.	24th	39.3	0.0	0.0	0.0	0.0	0.0	19.7	0.0	6.5	6.6
	48th	44.4	0.0	0.0	0.0	0.0	0.0	36.1	0.0	5.6	6.9
Chromobacterium	24th	0.0	96.2	0.0	0.0	57.7	0.0	0.0	0.0	19.2	0.0
	48th	0.0	96.2	0.0	0.0	95.8	0.0	0.0	0.0	19.2	0.0
Achromobacter xylosoxidans	24th	0.0	0.0	0.0	0.0	32.7	0.0	0.0	0.0	0.0	0.0
	48th	0.0	0.0	0.0	0.0	89.1	0.0	0.0	0.0	0.0	9.1
Achromobacter spp. (Vd)	24th	0.0	0.0	0.0	0.0	24.3	0.0	62.2	0.0	0.0	0.0
	48th	0.0	0.0	0.0	0.0	75.9	0.0	91.3	4.4	0.0	3.1
Agrobacterium radiobacter	24th	69.7	0.0	0.0	0.0	18.6	0.0	11.6	0.1	0.0	0.0
	48th	95.4	0.0	0.0	0.0	69.8	0.0	46.5	2.3	0.0	0.0
Brucella spp.	24th	0.0	0.0	0.0	0.0	0.0	0.0	88.9	0.0	0.0	0.0
	48th	0.0	0.0	0.0	0.0	0.0	0.0	94.4	0.0	0.0	0.0
Eikenella corrodens	24th	0.0	0.0	33.3	76.3	0.0	0.0	0.0	0.0	0.0	0.0
	48th	0.0	0.0	37.0	94.0	0.0	0.0	0.0	0.0	0.0	0.0
CDC Group II F	24th	0.0	0.0	0.0	0.0	8.6	0.0	0.0	0.0	71.4	0.0
	48th	0.0	0.0	0.0	0.0	71.4	0.0	0.0	0.0	71.4	0.0
CDC Group II J	24th	0.0	0.0	0.0	0.0	0.0	0.0	86.0	0.0	50.0	0.0
	48th	0.0	0.0	0.0	0.0	4.4	0.0	98.0	0.0	50.0	0.0
CDC Group IV C-2	24th	0.0	0.0	0.0	0.0	32.4	0.0	17.7	0.0	0.0	2.9
	48th	0.0	0.0	0.0	0.0	84.9	0.0	84.9	0.0	0.0	9.1
CDC Group IV E	24th	0.0	0.0	0.0	0.0	10.0	0.0	75.0	0.0	0.0	0.0
	48th	0.0	0.0	0.0	0.0	38.0	0.0	87.0	0.0	0.0	0.0
CDC Group V E-1	24th	86.5	59.5	0.0	0.0	77.0	0.0	0.0	0.0	0.0	43.2
	48th	86.5	80.6	0.0	0.0	98.0	0.0	2.8	0.0	0.0	58.3
CDC Group V E-2	24th	0.0	0.0	0.0	0.0	79.0	0.0	0.0	0.0	0.0	60.6
	48th	0.0	0.0	0.0	0.0	96.0	0.0	0.0	0.0	0.0	78.6
Aeromonas hydrophila group	24th	97.8	90.6	50.0	0.0	45.0	0.0	0.0	0.0	85.6	61.7
Aeromonas salmonicida (25c)	24th	11.1	88.9	100.0	0.0	0.0	0.0	0.0	0.0	0.0	0.0
Plesiomonas shigelloides	24th	95.5	95.5	100.0	100.0	0.0	0.0	0.0	0.0	100.0	0.0
Vibrio cholerae	24th	94.4	1.9	94.4	96.3	63.0	0.0	0.0	0.0	100.0	40.7
V. alginolyticus	24th	0.0	0.0	97.5	62.5	41.0	0.0	2.5	0.0	100.0	15.0
V. parahemolyticus	24th	0.0	0.0	100.0	89.7	58.8	0.0	8.6	0.0	100.0	5.2
V. vulnificus	24th	100.0	0.0	75.0	90.0	71.4	0.0	0.0	0.0	100.0	18.2
V. fluvialis	24th	99.0	92.8	0.0	0.0	21.4	0.0	0.1	0.0	78.4	0.0
V. mimicus	24th	99.0	0.2	90.4	90.3	63.0	0.0	0.0	0.0	90.0	0.0
V. damsela	24th	7.1	100.0	71.4	0.0	0.0	0.0	85.7	0.0	0.0	0.0
V. hollisae	24th	0.0	0.0	0.0	0.0	0.0	0.0	0.0	0.0	85.7	0.0

By Biochemical Reactions Provided on API 20E®

Figures indicate the percentage of positive reactions after incubation at 35–37° C*.

API 20E													Supplementary Tests			
GEL	GLU	MAN	INO	SOR	RHA	SAC	MEL	AMY	ARA	OXI	NO_3	N_2 GAS	MOT	MAC	OF-O	OF-F
75.5	57.8	0.0	0.0	0.0	0.0	0.0	2.7	0.0	10.9	98.6	12.8	56.8	86.4	99.2	98.4	0.0
87.2	61.6	0.0	0.0	0.0	0.0	0.8	12.0	0.0	29.6	99.2	12.8	87.2	86.4	99.2	98.4	0.0
37.7	28.3	0.0	0.0	0.0	0.0	0.0	7.6	0.0	13.2	100.0	50.0	0.0	78.3	99.0	97.8	0.0
60.9	43.5	0.0	0.0	0.0	2.2	0.0	19.6	0.0	26.1	100.0	50.0	0.0	78.3	99.0	97.8	0.0
0.0	41.2	0.0	0.0	0.0	2.9	0.0	11.8	0.0	8.8	97.9	9.4	0.0	93.8	96.9	93.8	0.0
0.0	46.9	0.0	0.0	0.0	3.1	0.0	31.3	0.0	18.8	100.0	9.4	0.0	93.8	96.9	93.8	0.0
46.5	70.5	2.3	0.0	1.2	0.0	13.2	0.0	24.4	13.8	90.7	40.5	0.0	67.9	88.1	97.6	0.0
64.3	97.6	3.6	1.2	1.2	0.0	26.2	2.4	36.9	39.2	90.7	40.5	0.0	67.9	88.1	97.6	0.0
79.0	1.9	0.0	0.0	0.9	0.0	0.0	0.0	0.9	0.0	4.8	27.0	0.0	88.8	91.0	49.4	0.0
92.1	1.9	0.0	0.0	1.1	0.0	0.0	0.0	1.1	0.0	4.8	27.0	0.0	88.8	91.0	49.4	0.0
93.8	3.1	0.0	0.0	0.0	0.0	3.1	0.0	0.0	0.0	100.0	96.9	0.0	93.8	96.9	9.4	0.0
96.9	3.1	0.0	0.0	0.0	0.0	9.4	0.0	0.0	0.0	100.0	96.9	0.0	93.8	96.9	9.4	0.0
22.6	7.6	0.0	0.0	0.0	0.0	0.0	3.8	0.0	1.9	98.1	44.4	51.9	68.5	98.1	81.5	0.0
25.9	13.0	0.0	0.0	0.0	0.0	0.0	5.6	0.0	1.9	100.0	8.2	91.9	68.5	98.1	81.5	0.0
91.7	37.5	79.2	75.0	79.2	0.0	54.2	0.0	12.5	0.0	100.0	0.0	100.0	95.8	100.0	100.0	0.0
100.0	100.0	100.0	91.7	100.0	0.0	100.0	0.0	79.2	41.7	100.0	0.0	100.0	95.8	100.0	100.0	0.0
0.0	5.0	0.0	0.0	0.0	0.0	5.0	0.0	0.0	5.0	50.0	0.0	0.0	40.9	0.0	95.5	0.0
9.1	9.1	0.0	0.0	0.0	4.6	18.2	0.0	4.6	5.0	59.1	0.0	0.0	40.9	0.0	95.5	0.0
9.0	0.4	0.0	0.0	0.0	0.4	0.0	0.7	0.4	0.7	99.3	68.4	9.4	62.1	85.9	25.1	0.0
28.9	1.6	0.0	0.0	0.0	0.4	0.0	0.7	0.8	0.7	99.6	68.4	9.4	62.1	85.9	25.1	0.0
12.6	87.4	0.0	0.0	0.0	0.0	1.2	88.5	0.0	79.3	0.0	3.2	0.0	0.0	90.5	98.4	0.0
12.6	87.4	0.0	0.0	0.0	0.0	1.2	89.1	0.0	79.7	0.0	3.2	0.0	0.0	90.5	98.4	0.0
3.3	0.0	0.0	0.0	0.0	0.0	0.0	0.0	0.0	0.0	0.0	7.1	0.0	0.0	70.2	0.0	0.0
11.9	0.0	0.0	0.0	0.0	0.0	0.0	0.0	0.0	0.0	0.0	7.1	0.0	0.0	70.2	0.0	0.0
85.0	0.0	0.0	0.0	0.0	0.0	0.0	0.0	0.0	0.0	100.0	0.0	0.0	0.0	57.5	90.0	10.0
100.0	0.0	0.0	0.0	0.0	0.0	0.0	0.0	0.0	0.0	100.0	4.0	22.0	0.0	57.5	90.0	10.0
87.9	0.0	0.0	0.0	0.0	0.0	0.0	0.0	0.0	0.0	100.0	6.1	0.0	0.0	48.5	93.9	6.1
98.8	0.0	0.0	0.0	0.0	0.0	0.0	0.0	0.0	0.0	100.0	6.1	0.0	0.0	48.5	93.9	6.1
58.6	0.0	0.0	0.0	0.0	0.0	0.0	0.0	0.0	0.0	100.0	0.0	0.0	0.0	84.4	0.0	0.0
96.0	0.0	0.0	0.0	0.0	0.0	0.0	0.0	0.0	0.0	100.0	0.0	0.0	0.0	84.4	0.0	0.0
20.0	0.0	0.0	0.0	0.0	0.0	0.0	0.0	0.0	0.0	100.0	0.0	0.0	0.0	99.0	20.0	0.0
99.0	0.0	0.0	0.0	0.0	0.0	0.0	0.0	0.0	0.0	100.0	0.0	0.0	0.0	100.0	20.0	0.0
10.7	46.4	0.0	0.0	0.0	0.0	25.0	0.0	7.1	17.9	96.4	0.0	0.0	15.4	84.6	96.2	0.0
10.7	61.5	0.0	0.0	0.0	15.4	84.6	0.0	38.5	23.1	96.4	0.0	0.0	15.4	84.6	96.2	0.0
0.0	0.0	0.0	0.0	0.0	0.0	0.0	0.0	0.0	0.0	100.0	0.0	0.0	0.0	0.0	99.9	31.2
0.0	0.0	0.0	0.0	0.0	0.0	0.0	0.0	0.0	0.0	100.0	0.0	0.0	0.0	0.0	99.9	31.2
0.0	0.0	0.0	0.0	0.0	0.0	0.0	0.0	0.0	0.0	95.7	77.8	0.0	88.9	97.2	0.0	0.0
2.8	0.0	0.0	0.0	0.0	0.0	0.0	0.0	0.0	0.0	100.0	77.8	0.0	88.9	97.2	0.0	0.0
1.5	0.0	0.0	0.0	0.0	0.0	0.0	0.0	0.0	0.0	99.3	43.9	8.6	84.9	91.4	0.0	0.0
10.1	0.0	0.0	0.0	0.0	0.0	0.0	0.0	0.0	0.0	100.0	43.9	8.6	84.9	91.4	0.0	0.0
2.2	0.0	0.0	0.0	0.0	0.0	0.0	0.0	0.0	0.0	100.0	9.7	0.0	0.0	23.9	0.0	0.0
11.5	0.0	0.0	0.0	0.0	0.0	0.0	0.0	0.0	0.0	100.0	9.7	0.0	0.0	23.9	0.0	0.0
0.0	29.6	74.1	0.0	68.5	0.0	77.8	0.0	0.0	0.0	81.5	52.4	0.0	0.0	2.0	19.6	19.6
0.0	30.8	74.1	0.0	65.4	0.0	78.9	0.0	0.0	0.0	82.7	52.4	0.0	0.0	2.0	19.6	19.6
0.0	99.0	0.0	90.0	0.0	10.0	99.0	0.0	0.0	80.0	85.0	100.0	0.0	0.0	100.0	100.0	100.0
0.0	100.0	0.0	90.0	0.0	10.0	100.0	0.0	0.0	80.0	85.0	100.0	0.0	0.0	100.0	100.0	100.0
1.6	13.1	0.0	0.0	0.0	0.0	3.3	0.0	0.0	1.6	91.8	59.7	0.0	0.0	9.7	25.0	25.0
2.8	13.1	0.0	0.0	0.0	0.0	3.3	0.0	0.0	1.6	97.2	59.7	0.0	0.0	9.7	25.0	25.0
92.3	96.2	0.0	0.0	0.0	0.0	11.5	0.0	0.0	0.0	95.8	75.0	0.0	95.8	99.0	99.0	99.0
100.0	96.2	0.0	0.0	0.0	0.0	16.7	0.0	0.0	0.0	95.8	75.0	0.0	95.8	99.0	99.0	99.0
0.0	0.0	0.0	0.0	0.0	0.0	0.0	0.0	0.0	0.0	100.0	65.5	43.6	65.5	99.0	47.3	0.0
1.8	1.8	0.0	0.0	0.0	0.0	0.0	0.0	0.0	0.0	100.0	65.5	43.6	65.5	99.0	47.3	0.0
0.0	0.0	2.7	2.7	2.7	2.7	5.4	0.0	5.4	0.0	100.0	15.6	71.9	99.0	99.0	53.1	0.0
0.0	0.0	12.5	12.5	12.5	12.5	18.8	0.0	5.4	15.6	100.0	28.1	91.3	99.0	99.0	53.1	0.0
0.0	0.0	0.0	0.0	0.0	0.0	0.0	0.0	2.3	4.6	100.0	18.2	27.2	86.0	100.0	51.0	0.0
0.0	0.0	0.0	0.0	0.0	7.0	9.3	2.3	14.0	20.9	100.0	23.3	37.2	86.0	100.0	62.8	0.0
0.0	0.0	0.0	0.0	0.0	0.0	0.0	0.0	0.0	0.0	100.0	85.0	0.0	0.0	0.0	0.0	0.0
0.0	0.0	0.0	0.0	0.0	0.0	0.0	0.0	0.0	0.0	100.0	85.0	0.0	0.0	0.0	0.0	0.0
0.0	0.0	0.0	0.0	0.0	0.0	0.0	0.0	0.0	0.0	100.0	3.7	0.0	0.0	0.0	0.0	0.0
0.0	0.0	0.0	0.0	0.0	0.0	0.0	0.0	0.0	0.0	100.0	3.7	0.0	0.0	0.0	0.0	0.0
88.6	0.0	0.0	0.0	0.0	0.0	0.0	0.0	0.0	0.0	99.0	0.0	0.0	0.0	2.0	0.0	0.0
89.8	0.0	0.0	0.0	0.0	0.0	0.0	0.0	0.0	0.0	99.0	0.0	0.0	0.0	2.0	0.0	0.0
4.1	0.0	0.0	0.0	0.0	0.0	0.0	0.0	0.0	0.0	100.0	0.0	0.0	0.0	2.4	0.0	0.0
78.3	0.0	0.0	0.0	0.0	0.0	0.0	0.0	0.0	0.0	100.0	0.0	0.0	0.0	2.4	0.0	0.0
0.0	0.0	0.0	0.0	0.0	0.0	0.0	0.0	0.0	0.0	100.0	0.0	0.0	72.7	90.9	0.0	0.0
0.0	0.0	0.0	0.0	0.0	0.0	0.0	0.0	0.0	0.0	100.0	0.0	0.0	72.7	90.9	0.0	0.0
0.0	0.0	0.0	0.0	0.0	0.0	0.0	0.0	0.0	0.0	100.0	0.0	66.0	29.0	73.1	0.0	0.0
0.0	0.0	0.0	0.0	0.0	0.0	0.0	0.0	0.0	0.0	100.0	8.0	66.0	29.0	73.1	0.0	0.0
13.5	86.5	0.0	13.5	0.0	16.2	2.7	13.5	2.7	78.4	0.0	30.6	0.0	63.9	91.7	94.4	0.0
77.8	86.5	0.0	13.9	0.0	30.6	8.3	16.1	5.6	91.7	2.7	63.0	0.0	63.9	91.7	94.4	0.0
6.1	45.5	0.0	15.2	0.0	0.0	0.0	3.0	0.0	81.8	0.0	7.1	0.0	96.4	99.0	99.0	0.0
64.3	50.0	0.0	14.3	3.6	0.0	0.0	3.0	0.0	92.9	0.0	7.1	0.0	96.4	99.0	99.0	0.0
94.4	98.9	96.7	0.0	12.7	10.0	82.2	5.6	61.1	61.1	99.0	98.7	0.0	96.0	99.0	99.0	99.0
100.0	88.9	100.0	0.0	11.1	0.0	0.0	0.0	0.0	0.0	100.0	100.0	0.0	0.0	99.0	100.0	100.0
0.0	100.0	0.0	100.0	0.0	0.0	0.0	0.0	0.0	0.0	100.0	99.0	0.0	95.5	99.0	99.0	99.0
92.6	98.2	98.2	0.0	0.0	0.0	100.0	0.0	5.6	0.0	100.0	96.2	0.0	99.0	96.2	99.0	99.0
55.0	100.0	100.0	0.0	0.0	0.0	100.0	0.0	5.0	7.5	100.0	47.4	0.0	97.4	99.0	94.7	94.7
68.8	100.0	96.6	0.0	0.0	3.5	1.7	0.0	12.1	41.2	100.0	63.8	0.0	98.3	98.3	99.0	99.0
100.0	100.0	36.4	0.0	0.0	0.0	0.0	0.0	93.0	0.0	100.0	54.6	0.0	99.0	99.0	99.0	99.0
21.4	100.0	100.0	0.0	7.1	0.0	99.0	0.0	50.0	99.0	100.0	100.0	0.0	50.0	100.0	100.0	100.0
92.6	98.1	95.1	0.0	0.0	0.0	0.0	0.0	0.6	0.0	100.0	92.0	0.0	95.0	96.2	100.0	100.0
0.0	100.0	0.0	0.0	0.0	0.0	7.1	0.0	0.0	0.0	100.0	92.9	0.0	100.0	0.0	100.0	100.0
0.0	0.0	0.0	0.0	0.0	0.0	0.0	0.0	0.0	0.0	100.0	14.3	0.0	100.0	0.0	7.1	7.1

Other *Pseudomonas* spp.

Ps. vesicularis
Ps. testosteroni
Ps. diminuta
Ps. pseudoalcaligenes

***Moraxella* spp.**

M. osloensis
M. phenylpyruvica
M. nonliquefaciens
M. lacunata
M. atlantae
CDC Group M5

***Brucella* spp.**

B. abortus
B. suis
B. melitensis
B. canis

***Flavobacterium* spp.**

Flav. spp. II B

***Pasteurella-Actinobacillus* spp.**

Past. haemolytica
Past. ureae
Past. pneumotropica
A. suis
A. equuli
A. lignieresii

Chromobacterium

C. violaceum

fluorescent *Pseudomonas* group

Ps. aeruginosa
Ps. fluorescens
Ps. putida

Ps. stutzeri

CDC Group Vb-1,3
Ps. mendocina

***Alcaligenes* spp.**

Al. faecalis
Al. odorans
Al. denitrificans

***Achromobacter* spp.**

CDC Vd-1,2

*Positive by oxidative metabolism.

Modified from the API 20E® system package insert. Courtesy of Analytab Products, Division of Sherwood Medical, 200 Express St., Plainview, NY 11803.

Appendix **4**

Computer Simulation of Bacterial Identification with *Identibacter interactus* CD-ROM

An exercise in the identification of an unknown culture is standard in most microbiology laboratory courses. This laboratory experience teaches students about the types of characteristics that are used to distinguish bacterial species and how classification schemes for bacteria are organized. This knowledge is of more than academic interest because these identification strategies are also used in medical diagnostic laboratories.

Because of time and/or money constraints, a student usually identifies only one or two unknown cultures. Although this is a valuable learning experience, it may be insufficient for the student to develop the deductive reasoning skills involved in bacterial identification or to understand how phenotypic information is organized in a reference such as *Bergey's Manual of Determinative Bacteriology.*

With the computer simulation on the ***Identibacter interactus*** CD-ROM, students can choose from pull-down menus (see the following list) more than 50 tests to run on their assigned unknown. A color image of the test result is displayed on the screen, and the student interprets it. Online help is available for consultation on test results or to aid in determining what additional test(s) are necessary to complete the identification. When sufficient information has been collected, the student can attempt to identify the unknown strain. An audit trail of the student's choices can be saved to disk or printed and evaluated by the instructor.

In the CD-ROM simulation, first-time users can identify an unknown in 15 to 30 minutes, and experienced users can solve unknowns more quickly. Thus, students can repeat the exercise a number of times with new unknowns and become familiar with the organization of bacterial identification schemes and the utility of specific tests for distinguishing species.

The organism database contains nearly 60 species of chemoheteroptrophic bacteria, including some human pathogens that would be difficult to use in an introductory microbiology laboratory; see the following list. The phenotypic characteristics for species are taken from data in *Bergey's Manual of Determinative Bacteriology.*

IDENTIBACTER INTERACTUS

A computer simulation of microbial identification by Allan Konopka, Paul Furbacher, and Clark Gedney, all of Purdue University.

1997 • CD-ROM • Wm. C. Brown Publishers •
ISBN 0–697–29387–4

SYSTEM REQUIREMENTS

Macintosh

PowerMac or [Alter] 040 Macintosh with CD-ROM player (2×; 4× is better)
Mac OS system 7.0 or better
QuickTime 2.0 installed
14′′ HiRes monitor
16-bit color
Mouse
12 MB RAM
512 K (1 MB preferred) free hard disk space

PCs

486 (586 preferred) system with CD-ROM (2×; 4× is better)
Windows 95 with QuickTime for Windows
14′′ HiRes monitor
16-bit color (thousands of colors—set color dept to 16- or 24-bit and restart)
Mouse
12 MB RAM
512 K (1 MB preferred) free hard disk space

ORGANISM DATABASE

Aeromonas hydrophila
Arthrobacter globiformis
Bacillus subtilis
Bifidobacterium bifidum
Bordetella bronchiseptica
Bordetella pertussis
Brucella abortus
Citrobacter freundii
Clostridium butyricum
Corynebacterium bovis
Corynebacterium diphtheriae
Corynebacterium renale
Edwardsiella tarda
Enterobacter aerogenes
Enterobacter cloacae
Enterococcus faecalis
Erwinia amylovora
Escherichia coli
Haemophilus influenzae
Haemophilus parahaemolyticus
Hafnia alvei
Klebsiella oxytoca
Klebsiella pneumoniae
Lactobacillus delbrueckii
Leuconostoc lactis
Listeria monocytogenes
Micrococcus luteus
Moraxella lacunata
Neisseria gonorrhoeae
Neisseria meningitidis
Nocardia asteroides
Pasteurella multocida
Propionibacterium acnes
Proteus mirabilis
Proteus vulgaris
Providencia alcalifaciens
Providencia rettgeri
Pseudomonas aeruginosa
Pseudomonas mendocina
Pseudomonas putida
Pseudomonas stutzeri
Salmonella choleraesuis subsp. *arizonae*
Salmonella choleraesuis subsp. *choleraesuis*
Serratia marcescens
Shigella dysenteriae
Shigella sonnei
Staphylococcus aureus
Staphylococcus epidermidis
Streptococcus pneumoniae
Streptococcus pyogenes
Streptococcus salivarius
Vibrio cholerae
Xanthomonas campestris
Xanthomonas fragariae
Yersinia enterocolitica
Yersinia pestis
Yersinia pseudotuberculosis

TEST OPTIONS

Microscopy
- Gram stain
- Phase contrast
- Motility
- Endospore heat test

Hydrolysis
- Esculin
- Gelatin
- Hippurate
- Lipid
- Starch

Metabolism
- Fluid thioglycollate
- H_2S production
- Indole
- KCN inhibition
- Methyl red
- Nitrate reduction
- O-F reaction
- Voges-Proskauer

Fermentation
- Adonitol
- Arabinose
- Dulcitol
- Glucose
- Glycerol
- Inositol
- Inulin
- Lactose
- Maltose
- Mannitol
- Raffinose
- Rhamnose
- Salicin
- Sorbitol
- Sucrose
- Trehalose
- Xylose

Carbon Source
- Arginine
- Citrate
- Glutamate
- Succinate
- Tartrate

Miscellaneous
- MacConkey Agar
- Blood Agar

Enzymes
- Beta-galactosidase
- Catalase
- Coagulase
- DNase
- Decarboxylases
- Arginine dihydrolase
- Oxidase
- Phenylalanine deaminase
- Urease

Index